CHALLENGES AND PARADIGMS IN APPLIED ROBUST CONTROL

Edited by **Andrzej Bartoszewicz**

Challenges and Paradigms in Applied Robust Control
Edited by Andrzej Bartoszewicz

Published by InTech

Janeza Trdine 9, 51000 Rijeka, Croatia

Publishing Process Manager Sandra Bakic

Technical Editor Teodora Smiljanic

Cover Designer InTech Design Team

Additional hard copies can be obtained from orders@intechopen.com

Challenges and Paradigms in Applied Robust Control, Edited by Andrzej Bartoszewicz
p. cm.
ISBN 978-953-307-338-5

For EU product safety concerns:
IN TECH d.o.o., Prolaz Marije Krucifikse Kozulić 3, 51000 Rijeka, Croatia,
info@intechopen.com or visit our website at intechopen.com.

Meet the editor

Andrzej Bartoszewicz received M.Sc. degree in 1987 and Ph.D. degree in 1993, both from Technical University of Łódź, Poland. Then he obtained the degree of Assistant Professor in control engineering and robotics from Academy of Mining and Metallurgy in Cracow, Poland. He was visiting scholar at Purdue University, West Lafayette, In., USA and at Strathclyde University, Glasgow, UK. Then for one year he was at the University of Leicester, UK. Currently he is Professor at Technical University of Łódź, vice dean of Faculty of Electrical, Electronic, Computer and Control Engineering, head of Electric Drive and Industrial Automation Group and vice-director of Institute of Automatic Control. He has published three monographs and over 250 papers, primarily in the fields of sliding mode control and congestion control in data transmission networks.

Contents

Preface

The main purpose of control engineering is to steer the regulated plant in such a way that it operates in a required manner. The desirable performance of the plant should be obtained despite the unpredictable influence of the environment on all parts of the control system, including the plant itself, and no matter if the system designer knows precisely all the parameters of the plant. Even though the parameters may change with time, load and external circumstances, still the system should preserve its nominal properties and ensure the required behavior of the plant. In other words, the principal objective of control engineering is to design and implement regulation systems which are robust with respect to external disturbances and modeling uncertainty. This objective may very well be obtained in a number of ways which are discussed and demonstrated in this book.

Book is divided into five sections. In section 1 selected aircraft, vehicle and automotive applications are presented. That section begins with a contribution on rotorcraft control. The first chapter presents input-output linearization based on sliding mode controller for a quadrotor. Chapter 2 gives a comparison of different advanced control architectures for transonic phase of space re-entry vehicle flight. Then chapter 3 discusses the problem of robust fault tolerant, vertical motion control of modular underwater autonomous robot for environment sampling. The last three chapters in section 1 present solutions of the most important control problems encountered in automotive industry. They describe the second order sliding mode control of spark ignition engine idle speed, new active suspension control method reducing the passenger's seat vibrations and advanced adaptive cruise control system design.

Section 2 begins with a chapter on H-infinity active controller design for minimizing mechanical vibration of structures. Then it focuses on robust control of mechanical systems, i.e. uncertain Lagrangian systems with partially unavailable state variables, and adaptive back-stepping control of electro-hydraulic actuators. The last chapter in that section is concerned with the control of active magnetic bearing suspension system for high-speed rotors.

Section 3 consists of three contributions on the control of distillation and multi-step evaporation processes. The first chapter, concerned with a generic double feed two-reactant two-product ideal reactive distillation and the methyl acetate reactive distillation systems, demonstrates the implications of the nonlinearity, and in particular input and output multiplicity, on the open and closed loop distillation system operation. The next chapter shows that the desirable closed-loop performance can be achieved for an ill-conditioned high-purity distillation column by the use of a decentralized PID controller and the structured uncertainty model describing the column dynamics within its entire operating range. Then the last chapter of section 3 analyses a complex multi-stage evaporation process and presents a new full order Kalman filter based scheme to obtain full loop transfer recovery for the process.

Section 4 comprises two chapters on the control of power plants and power systems. The first of the two chapters studies the problem of reducing environmental effects by operational control of nitrogen oxide and carbon monoxide emissions from thermal power plants. The second chapter is concerned with damping of inter-area oscillations in electric power systems. For that purpose a mixed H2/H-infinity output-feedback control with pole placement is applied.

Section 5 presents a number of other significant developments in applied robust control. It begins with a noteworthy contribution on networked control which demonstrates that robust control system design not only requires a proper selection and tuning of control algorithms, but also must involve careful analysis of the applied communication protocols and networks, to ensure that they are appropriate for real-time implementation in distributed environment. A similar issue – in the context of force bilateral tele-operation – is discussed in the next chapter of that section, where it is shown that H-infinity design offers good robustness with reference to network induced time delays. Then the section discusses selected problems in resource allocation and control. These include development of robust controllers for single unit resource allocation systems with unreliable resources and real world natural resource robust management with the special focus on fisheries. The monograph concludes with the presentation of H-infinity synchronizer design and its application to improve the robustness of chaotic communication systems with respect to delays in the transmission line.

In conclusion, the main objective of this book is to present a broad range of well worked out, recent engineering and non-engineering application studies in the field of robust control system design. We believe, that thanks to the authors, reviewers and the editorial staff of InTech Open Access Publisher this ambitious objective has been successfully accomplished. The editor and authors truly hope that the result of this joint effort will be of significant interest to the control community and that the contributions presented here will enrich the current state of the art, and encourage and stimulate new ideas and solutions in the robust control area.

Andrzej Bartoszewicz
Technical University of Łódź
Poland

Part 1

Robust Control in Aircraft, Vehicle and Automotive Applications

1

Sliding Mode Approach to Control Quadrotor Using Dynamic Inversion

Abhijit Das, Frank L. Lewis and Kamesh Subbarao
Automation and Robotics Research Institute
The University of Texas at Arlington
USA

1. Introduction

Nowadays unmanned rotorcraft are designed to operate with greater agility, rapid maneuvering, and are capable of work in degraded environments such as wind gusts etc. The control of this rotorcraft is a subject of research especially in applications such as rescue, surveillance, inspection, mapping etc. For these applications, the ability of the rotorcraft to maneuver sharply and hover precisely is important (Koo and Sastry 1998). Rotorcraft control as in these applications often requires holding a particular trimmed state; generally hover, as well as making changes of velocity and acceleration in a desired way (Gavrilets, Mettler, and Feron 2003). Similar to aircraft control, rotorcraft control too involves controlling the pitch, yaw, and roll motion. But the main difference is that, due to the unique body structure of rotorcraft (as well as the rotor dynamics and other rotating elements) the pitch, yaw and roll dynamics are strongly coupled. Therefore, it is difficult to design a decoupled control law of sound structure that stabilizes the faster and slower dynamics simultaneously. On the contrary, for a fixed wing aircraft it is relatively easy to design decoupled standard control laws with intuitively comprehensible structure and guaranteed performance (Stevens and F. L. Lewis 2003). There are many different approaches available for rotorcraft control such as (Altug, Ostrowski, and Mahony 2002; Bijnens et al. 2005; T. Madani and Benallegue 2006; Mistler, Benallegue, and M'Sirdi 2001; Mokhtari, Benallegue, and Orlov 2006) etc. Popular methods include input-output linearization and back-stepping. The 6-DOF airframe dynamics of a typical quadrotor involves the typical translational and rotational dynamical equations as in (Gavrilets, Mettler, and Feron 2003; Castillo, Lozano, and Dzul 2005; Castillo, Dzul, and Lozano 2004). The dynamics of a quadrotor is essentially a simplified form of helicopter dynamics that exhibits the basic problems including under-actuation, strong coupling, multi-input/multi-output, and unknown nonlinearities. The quadrotor is classified as a rotorcraft where lift is derived from the four rotors. Most often they are classified as helicopters as its movements are characterized by the resultant force and moments of the four rotors. Therefore the control algorithms designed for a quadrotor could be applied to a helicopter with relatively straightforward modifications. Most of the papers (B. Bijnens et al. 2005; T. Madani and Benallegue 2006; Mokhtari, Benallegue, and Orlov 2006) etc. deal with either input-output linearization for decoupling pitch yaw roll or back-stepping to deal with the under-actuation problem. The problem of coupling in the yaw-pitch-roll of a helicopter, as well as the problem of coupled dynamics-kinematic underactuated system, can be solved by back-stepping (Kanellakopoulos, Kokotovic, and Morse 1991; Khalil 2002; Slotine and Li 1991). Dynamic inversion (Stevens and F. L. Lewis 2003; Slotine and Li 1991; A. Das et al. 2004) is effective in the control of both linear and nonlinear systems and involves an inner inversion loop (similar to feedback linearization) which results in tracking if the residual or internal dynamics is stable. Typical usage requires the selection of the output control variables so that the internal dynamics is guaranteed to be stable. This implies that the tracking control cannot always be guaranteed for the original outputs of interest.

The application of dynamic inversion on UAV's and other flying vehicles such as missiles, fighter aircrafts etc. are proposed in several research works such as (Kim and Calise 1997; Prasad and Calise 1999; Calise et al. 1994) etc. It is also shown that the inclusion of dynamic neural network for estimating the dynamic inversion errors can improve the controller stability and tracking performance. Some other papers such as (Hovakimyan et al. 2001; Rysdyk and Calise 2005; Wise et al. 1999; Campos, F. L. Lewis, and Selmic 2000) etc. discuss the application of dynamic inversion on nonlinear systems to tackle the model and parametric uncertainties using neural nets. It is also shown that a reconfigurable control law can be designed for fighter aircrafts using neural net and dynamic inversion. Sometimes the inverse transformations required in dynamic inversion or feedback linearization are computed by neural network to reduce the inversion error by online learning.

In this chapter we apply dynamic inversion to tackle the coupling in quadrotor dynamics which is in fact an underactuated system. Dynamic inversion is applied to the inner loop, which yields internal dynamics that are not necessarily stable. Instead of redesigning the output control variables to guarantee stability of the internal dynamics, we use a sliding mode approach to stabilize the internal dynamics. This yields a two-loop structured tracking controller with a dynamic inversion inner loop and an internal dynamics stabilization outer loop. But it is interesting to notice that unlike normal two loop structure, we designed an inner loop which controls and stabilizes altitude and attitude of the quadrotor and an outer loop which controls and stabilizes the position (x,y) of the quadrotor. This yields a new structure of the autopilot in contrast to the conventional loop linear or nonlinear autopilot. Section 2 of this chapter discusses the basic quadrotor dynamics which is used for control law formulation. Section 3 shows dynamic inversion of a nonlinear state-space model of a quadrotor. Sections 4 discuss the robust control method using sliding mode approach to stabilize the internal dynamics. In the final section, simulation results are shown to validate the control law discussed in this chapter.

2. Quadrotor dynamics

Fig. 1 shows a basic model of an unmanned quadrotor. The quadrotor has some basic advantage over the conventional helicopter. Given that the front and the rear motors rotate counter-clockwise while the other two rotate clockwise, gyroscopic effects and aerodynamic torques tend to cancel in trimmed flight. This four-rotor rotorcraft does not have a swash-plate (P. Castillo, R. Lozano, and A. Dzul 2005). In fact it does not need any blade pitch control. The collective input (or throttle input) is the sum of the thrusts of each motor (see Fig. 1). Pitch movement is obtained by increasing (reducing) the speed of the rear motor while reducing (increasing) the speed of the front motor. The roll movement is obtained similarly using the lateral motors. The yaw movement is obtained by increasing (decreasing) the speed of the front and rear motors while decreasing (increasing) the speed of the lateral motors (Bouabdallah, Noth, and Siegwart 2004).

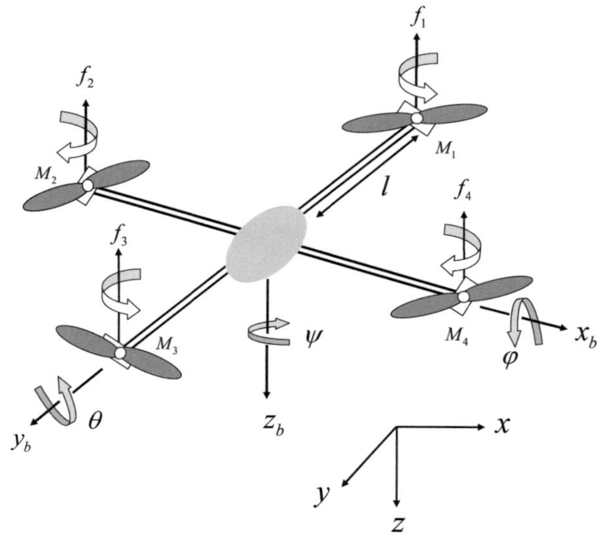

Fig. 1. A typical model of a quadrotor helicopter

In this section we will describe the basic state-space model of the quadrotor. The dynamics of the four rotors are relatively much faster than the main system and thus neglected in our case. The generalized coordinates of the rotorcraft are $q = (x, y, z, \psi, \theta, \varphi)$, where (x, y, z) represents the relative position of the center of mass of the quadrotor with respect to an inertial frame $\Im$, and (ψ, θ, φ) are the three Euler angles representing the orientation of the rotorcraft, namely yaw-pitch-roll of the vehicle.

Let us assume that the transitional and rotational coordinates are in the form $\xi = (x, y, z)^T \in R^3$ and $\eta = (\psi, \theta, \varphi) \in R^3$. Now the total transitional kinetic energy of the rotorcraft will be $T_{trans} = \frac{m}{2}\dot{\xi}^T \dot{\xi}$ where m is the mass of the quadrotor. The rotational kinetic energy is described as $T_{rot} = \frac{1}{2}\dot{\eta}^T J \dot{\eta}$, where matrix $J = J(\eta)$ is the auxiliary matrix expressed in terms of the generalized coordinates η. The potential energy in the system can be characterized by the gravitational potential, described as $U = mgz$. Defining the Lagrangian $L = T_{trans} + T_{rot} - U$, where $T_{trans} = (m/2)\dot{\xi}^T \dot{\xi}$ is the translational kinetic energy, $T_{rot} = (1/2)\omega^T I \omega$ is the rotational kinetic energy with ω as angular speed, $U = mgz$ is the potential energy, z is the quadrotor altitude, I is the body inertia matrix, and g is the acceleration due to gravity.

Then the full quadrotor dynamics is obtained as a function of the external generalized forces $F = (F_\xi, \tau)$ as

$$\frac{d}{dt}\frac{\partial L}{\partial \dot{q}} - \frac{\partial L}{\partial q} = F \tag{1}$$

The principal control inputs are defined as follows. Define

$$F_R = \begin{pmatrix} 0 \\ 0 \\ u \end{pmatrix} \tag{2}$$

where u is the main thrust and defined by

$$u = f_1 + f_2 + f_3 + f_4 \tag{3}$$

and f_i's are described as $f_i = k_i \omega_i^2$, where k_i are positive constants and ω_i are the angular speed of the motor i. Then F_ξ can be written as

$$F_\xi = \bar{R} F_R \tag{4}$$

where $\bar{R}$ is the transformation matrix representing the orientation of the rotorcraft as

$$\bar{R} = \begin{pmatrix} c_\theta c_\psi & s_\psi s_\theta & -s_\theta \\ c_\psi s_\theta s_\varphi - s_\psi c_\varphi & s_\psi s_\theta s_\varphi + c_\psi c_\varphi & c_\theta s_\varphi \\ c_\psi s_\theta c_\varphi + s_\psi s_\varphi & s_\psi s_\theta c_\varphi - c_\psi s_\varphi & c_\theta c_\varphi \end{pmatrix} \tag{5}$$

The generalized torque for the η variables are

$$\tau = \begin{pmatrix} \tau_\psi \\ \tau_\theta \\ \tau_\varphi \end{pmatrix} \tag{6}$$

where

$$\tau_\psi = \sum_{i=1}^{4} \tau_{M_i} = c(f_1 - f_2 + f_3 - f_4) \tag{7}$$

$$\tau_\theta = (f_2 - f_4) l \tag{8}$$

$$\tau_\varphi = (f_3 - f_1) l \tag{9}$$

Thus the control distribution from the four actuator motors of the quadrotor is given by

$$\begin{pmatrix} u \\ \tau_\varphi \\ \tau_\theta \\ \tau_\psi \end{pmatrix} = \underbrace{\begin{pmatrix} 1 & 1 & 1 & 1 \\ -l & 0 & l & 0 \\ 0 & l & 0 & -l \\ c & -c & c & -c \end{pmatrix}}_{C} \underbrace{\begin{pmatrix} f_1 \\ f_2 \\ f_3 \\ f_4 \end{pmatrix}}_{\underline{f}} \tag{10}$$

where l is the distance from the motors to the center of gravity, τ_{M_i} is the torque produced by motor M_i, and c is a constant known as force-to-moment scaling factor. So, if a required thrust and torque vector are given, one may solve for the rotor force using (10).

The final dynamic model of the quadrotor is described by (11)-(14),

$$m\ddot{\xi} + \begin{pmatrix} 0 \\ 0 \\ mg \end{pmatrix} = F_R \tag{11}$$

$$J(\eta)\ddot{\eta} + \frac{d}{dt}\{J(\eta)\}\dot{\eta} - \frac{1}{2}\frac{\partial}{\partial\eta}(\dot{\eta}^T J(\eta)\dot{\eta}) = \tau \tag{12}$$

$$J(\eta)\ddot{\eta} + \frac{d}{dt}\{J(\eta)\}\dot{\eta} - \bar{C}(\eta,\dot{\eta}) = \tau \tag{13}$$

$$J(\eta)\ddot{\eta} + C(\eta,\dot{\eta}) = \tau \tag{14}$$

where, $F_R = u\begin{pmatrix} -\sin\theta \\ \cos\theta\sin\varphi \\ \cos\theta\cos\varphi \end{pmatrix}$, auxiliary Matrix $\{J(\eta)\} = J = T_\eta^T I T_\eta$ with

$$T_\eta = \begin{pmatrix} -\sin\theta & 0 & 1 \\ \cos\theta\sin\psi & \cos\psi & 0 \\ \cos\theta\cos\psi & -\sin\psi & 0 \end{pmatrix}.$$

Now finally the dynamic model of the quadrotor in terms of position (x,y,z) and rotation (φ,θ,ψ) is written as,

$$\begin{pmatrix} \ddot{x} \\ \ddot{y} \\ \ddot{z} \end{pmatrix} = \begin{pmatrix} 0 \\ 0 \\ -g \end{pmatrix} + \frac{1}{m}\begin{pmatrix} -\sin\theta \\ \cos\theta\sin\varphi \\ \cos\theta\cos\varphi \end{pmatrix} u \tag{15}$$

$$\begin{pmatrix} \ddot{\varphi} \\ \ddot{\theta} \\ \ddot{\psi} \end{pmatrix} = f(\varphi,\theta,\psi) + g(\varphi,\theta,\psi)\tau \tag{16}$$

where,

$f(\varphi,\theta,\psi) = \begin{pmatrix} \dot{\theta}\dot{\psi}\left(\frac{I_y - I_z}{I_x}\right) - \frac{J_p}{I_x}\dot{\theta}\Omega \\ \dot{\varphi}\dot{\psi}\left(\frac{I_z - I_x}{I_y}\right) + \frac{J_p}{I_y}\dot{\varphi}\Omega \\ \dot{\varphi}\dot{\theta}\left(\frac{I_x - I_y}{I_z}\right) \end{pmatrix}$, $g(\varphi,\theta,\psi) = \begin{pmatrix} \frac{l}{I_x} & 0 & 0 \\ 0 & \frac{l}{I_y} & 0 \\ 0 & 0 & \frac{l}{I_z} \end{pmatrix}$, $u \in R^1$ and $\tau = \begin{bmatrix} \tau_\varphi \\ \tau_\theta \\ \tau_\psi \end{bmatrix} \in R^3$ are the control inputs, $I_{x,y,z}$ are body inertia, J_p is propeller/rotor inertia and $\Omega = \omega_2 + \omega_4 - \omega_1 - \omega_3$. Thus, the system is the form of an under-actuated system with six outputs and four inputs.

Comment 2.1: *In this chapter we considered a generalized state space model of quadrotor derived from Lagrangian dynamics. Design autopilot with actual Lagrangian model of quadrotor is discussed in* (Abhijit Das, Frank Lewis, and Kamesh Subbarao 2009).

3. Partial feedback linearization for Quadrotor model

Dynamic inversion (Stevens and F. L. Lewis 2003) is an approach where a feedback linearization loop is applied to the tracking outputs of interest. The residual dynamics, not directly controlled, is known as the internal dynamics. If the internal dynamics are stable, dynamic inversion is successful. Typical usage requires the selection of the output control variables so that the internal dynamics is guaranteed to be stable. This means that tracking cannot always be guaranteed for the original outputs of interest.

In this chapter we apply dynamic inversion to the system given by (15) and (16) to achieve station-keeping tracking control for the position outputs (x,y,z,ψ). Initially we select the convenient output vector $y_{di}=(z,\varphi,\theta,\psi)$ which makes the dynamic inverse easy to find. Dynamic inversion now yields effectively an inner control loop that feedback linearizes the system from the control $u_{di}=(u,\tau_\varphi,\tau_\theta,\tau_\psi)$ to the output $y_{di}=(z,\varphi,\theta,\psi)$. Note that the output contains attitude parameters as well as altitude of the quadrotor.

Note however that y_{di} is not the desired system output. Moreover, dynamic inversion generates a specific internal dynamics, as detailed below, which may not always be stable. Therefore, a second outer loop is designed to generate the required values for $y_{di}=(z,\varphi,\theta,\psi)$ in terms of the values of the desired tracking output (x,y,z,ψ). An overall Lyapunov proof guarantees stability and performance. The following background is required. Consider a nonlinear system of the form

$$\dot{q}=f\left(q,u_q\right) \tag{17}$$

where $u_q\in R^m$ is the control input and $q\in R^n$ is state vector. The technique of designing the control input u using dynamic inversion involves two steps. First, one finds a state transformation $z=z(q)$ and an input transformation $u_q=u_q(q,v)$ so that the nonlinear system dynamics is transformed into an equivalent linear time invariant dynamics of the form

$$\dot{z}=az+bv \tag{18}$$

where $a\in R^{n\times n}, b\in R^{n\times m}$ are constant matrices with v is known as new input to the linear system. Secondly one can design v easily from the linear control theory approach such as pole placement etc. To get the desired linear equations (18), one has to differentiate outputs until input vector u_{di} appears. The procedure is known as dynamic inversion.

3.1 Dynamic inversion for inner loop

The system, (15)→(16) is an underactuated system if we consider the states $(x,y,z,\varphi,\theta,\psi)$ as outputs and $u_{di}=\begin{bmatrix}u & \tau_\varphi & \tau_\theta & \tau_\psi\end{bmatrix}^T$ as inputs. To overcome these difficulties we consider four outputs $y_{di}=(z,\varphi,\theta,\psi)$ which are used for feedback linearization. Differentiating the output vector twice with respect to the time we get from (15) and (16) that,

$$\ddot{y}_{di} = M_{di} + E_{di}u_{di} \tag{19}$$

where,

$$M_{di} = \begin{bmatrix} -g \\ \dot{\theta}\dot{\psi}\left(\dfrac{I_y - I_z}{I_x}\right) - \dfrac{J_p}{I_x}\dot{\theta}\Omega \\ \dot{\varphi}\dot{\psi}\left(\dfrac{I_z - I_x}{I_y}\right) + \dfrac{J_p}{I_y}\dot{\varphi}\Omega \\ \dot{\varphi}\dot{\theta}\left(\dfrac{I_x - I_y}{I_z}\right) \end{bmatrix} \in \Re^4, \; E_{di} = \begin{bmatrix} -(1/m)\cos\theta\cos\varphi & 0 & 0 & 0 \\ 0 & \dfrac{l}{I_x} & 0 & 0 \\ 0 & 0 & \dfrac{l}{I_y} & 0 \\ 0 & 0 & 0 & \dfrac{l}{I_z} \end{bmatrix} \in \Re^{4\times 4}$$

The number $(r = 8)$ of differentiation required for an invertible E_{di} is known as the relative degree of the system and generally $r < n = 12$; if $r = n$ then full state feedback linearization is achieved if E_{di} is invertible. Note that for multi-input multi-output system, if number of outputs is not equal to the number of inputs (under-actuated system), then E_{di} becomes non-square and is difficult to obtain a feasible linearizing input u_{di}.

It is seen that for non-singularity of E_{di}, $0 \le \theta, \varphi < 90^\circ$. The relative degree of the system is calculated as 8 whereas the order of the system is 12. So, the remaining dynamics $(= 4)$ which does not come out in the process of feedback linearization is known as internal dynamics. To guarantee the stability of the whole system, it is mandatory to guarantee the stability of the internal dynamics. In the next section we will discuss how to control the internal dynamics using a PID with a feed-forward acceleration outer loop. Now using (19) we can write the desired input to the system

$$u_{di} = E_{di}^{-1}\left(-M_{di} + v_{di}\right) \tag{20}$$

which yields

$$\ddot{y}_{di} = v_{di} \tag{21}$$

where, $v_{di} = \begin{pmatrix} v_z & v_\varphi & v_\theta & v_\psi \end{pmatrix}^T$. This system is decoupled and linear. The auxiliary input v_{di} is designed as described below.

3.2 Design of linear controller

Assuming the desired output to the system is $y_d = \begin{pmatrix} z_d & \varphi_d & \theta_d & \psi_d \end{pmatrix}^T$, the linear controller v_{di} is designed in the following way

$$v_{di} = \begin{bmatrix} v_z \\ v_\varphi \\ v_\theta \\ v_\psi \end{bmatrix} = \begin{bmatrix} \ddot{z}_d - K_{1_z}(\dot{z} - \dot{z}_d) - K_{2_z}(z - z_d) \\ \ddot{\varphi}_d - K_{1_\varphi}(\dot{\varphi} - \dot{\varphi}_d) - K_{2_\varphi}(\varphi - \varphi_d) \\ \ddot{\theta}_d - K_{1_\theta}(\dot{\theta} - \dot{\theta}_d) - K_{2_\theta}(\theta - \theta_d) \\ \ddot{\psi}_d - K_{1_\psi}(\dot{\psi} - \dot{\psi}_d) - K_{2_\psi}(\psi - \psi_d) \end{bmatrix} \tag{22}$$

where, $K_{1_\varphi}, K_{2_\varphi}$,.... etc. are positive constants so that the poles of the error dynamics arising from (23) and (24) are in the left half of the $s-$plane. For hovering control, z_d and ψ_d are chosen depending upon the designer choice.

3.3 Defining sliding variable error

Let us define the state error $e_1 = (z_d - z \quad \varphi_d - \varphi \quad \theta_d - \theta \quad \psi_d - \psi)^T$ and a sliding mode error as

$$r_1 = \dot{e}_1 + \Lambda_1 e_1 \tag{23}$$

where, Λ_1 is a diagonal positive definite design parameter matrix. Common usage is to select Λ_1 diagonal with positive entries. Then, (23) is a stable system so that e_1 is bounded as long as the controller guarantees that the filtered error r_1 is bounded. In fact it is easy to show (F. Lewis, Jagannathan, and Yesildirek 1999) that one has

$$\|e_1\| \le \frac{\|r_1\|}{\sigma_{\min}(\Lambda_1)}, \|\dot{e}_1\| \le \|r_1\| \tag{24}$$

Note that $\dot{e}_1 + \Lambda_1 e_1 = 0$ defines a stable sliding mode surface. The function of the controller to be designed is to force the system onto this surface by making r_1 small. The parameter Λ_1 is selected for a desired sliding mode response

$$e_1(t) = e_1^{-\Lambda_1 t} e_1(0) \tag{25}$$

We now focus on designing a controller to keep $\|r_1\|$ small. From (23),

$$\dot{r}_1 = \ddot{e}_1 + \Lambda_1 \dot{e}_1 \tag{26}$$

Adding an integrator to the linear controller given in (22), and now we can rewrite (22) as

$$v_{di} = \ddot{y}_{di_d} + K_1 \dot{e}_1 + K_2 e_1 + K_3 \int_0^t r_1 dt \tag{27}$$

where, $y_{di_d} = \left[\ddot{z}_d, \ddot{\varphi}_d, \ddot{\theta}_d, \ddot{\psi}_d\right]^T$ and $K_i = diag(K_{i_z}, K_{i_\varphi}, K_{i_\theta}, K_{i_\psi}) > 0$, $i = 1,2,3,4$.

Now using equation (20) and (27) we can rewrite the equation (19) in the form of error dynamics as

$$\ddot{e}_1 + K_1 \dot{e}_1 + K_2 e_1 + K_3 \int_0^t r_1 dt = 0 \tag{28}$$

Thus equation (26) becomes

$$\dot{r}_1 = -K_1 \dot{e}_1 - K_2 e_1 - K_3 \int_0^t r_1 dt + \Lambda_1 \dot{e}_1 \tag{29}$$

If we choose $K_1 = (\Lambda_1 + R), K_2 = \Lambda_1 R$, then equation (29) will look like

$$\dot{r}_1 = -R r_1 - K_3 \int_0^t r_1 dt \tag{30}$$

Note that $R>0$ is also a diagonal matrix.

4. Sliding mode control for internal dynamics

The internal dynamics (Slotine and Li 1991) for the feedback linearizes system given by

$$\ddot{x} = -\frac{u}{m}\sin\theta \tag{31}$$

$$\ddot{y} = \frac{u}{m}\cos\theta\sin\varphi \tag{32}$$

For the stability of the whole system as well as for the tracking purposes, x,y should be bounded and controlled in a desired way. Note that the altitude z of the rotorcraft a any given time t is controlled by (20),(22).
To stabilize the zero dynamics, we select some desired θ_d and φ_d such that (x,y) is bounded. Then that (θ_d,φ_d) can be fed into (22) as a reference. Using Taylor series expansion about some nominal values θ_d^*, φ_d^* and considering up to first order terms

$$\begin{aligned}\sin\theta_d &= \sin\theta_d^* + \cos\theta_d^*(\theta_d - \theta_d^*)\\ \cos\theta_d &= \cos\theta_d^* - \sin\theta_d^*(\theta_d - \theta_d^*)\\ \sin\varphi_d &= \sin\varphi_d^* + \cos\varphi_d^*(\varphi_d - \varphi_d^*)\end{aligned} \tag{33}$$

Using (33) on (31) we get

$$\ddot{x} = -\frac{u}{m}\left\{\sin\theta_d^* + \cos\theta_d^*(\theta_d - \theta_d^*)\right\} \tag{34}$$

$$\ddot{y} = \frac{u}{m}\left\{\cos\theta_d^* - \sin\theta_d^*(\theta_d - \theta_d^*)\right\}\left\{\sin\varphi_d^* + \cos\varphi_d^*(\varphi_d - \varphi_d^*)\right\} \tag{35}$$

For hovering of a quadrotor, assuming the nominal values $\theta_d^* \approx 0$, $\varphi_d^* \approx 0$, (31) and (32) becomes

$$\ddot{x} = -\frac{u}{m}\theta_d \tag{36}$$

$$\ddot{y} = \frac{u}{m}\varphi_d \tag{37}$$

Define the state error

$$e_2 = \left(x_d - x \quad y_d - y\right)^T \tag{38}$$

and the sliding mode error for the internal dynamics as

$$r_2 = \dot{e}_2 + \Lambda_2 e_2 \tag{39}$$

where, Λ_2 is a diagonal positive definite design parameter matrix with similar characteristic of Λ_1. Also

$$\|e_2\| \le \frac{\|r_2\|}{\sigma_{\min}(\Lambda_2)}, \|\dot{e}_2\| \le \|r_2\| \tag{40}$$

Therefore according to (40), designing a controller to keep $\|r_2\|$ small will guarantee that $\|e_2\|$ and $\|\dot{e}_2\|$ are small. Differentiating r_2 we get

$$\dot{r}_2 = \ddot{e}_2 + \Lambda_2 \dot{e}_2 \tag{41}$$

Let the choice of the control law is as follows

$$\theta_d = \frac{m}{u}\left[-\ddot{x}_d - c_{11}(\dot{x}_d - \dot{x}) - c_{12}(x_d - x) - c_{13}\int_0^t r_2 dt - \beta_x \operatorname{sgn}(r_2)\right], \beta_x > 0 \tag{42}$$

$$\varphi_d = \frac{m}{u}\left[\ddot{y}_d + c_{21}(\dot{y}_d - \dot{y}) + c_{22}(y_d - y) + c_{23}\int_0^t r_2 dt + \beta_y \operatorname{sgn}(r_2)\right], \beta_y > 0 \tag{43}$$

where, $\begin{bmatrix} c_{11} & 0 \\ 0 & c_{21} \end{bmatrix} = C_1 > 0$, $\begin{bmatrix} c_{12} & 0 \\ 0 & c_{22} \end{bmatrix} = C_2 > 0$, $\begin{bmatrix} c_{13} & 0 \\ 0 & c_{23} \end{bmatrix} = C_3 > 0$ and $\begin{bmatrix} \beta_x & 0 \\ 0 & \beta_y \end{bmatrix} = \beta > 0$.

Combining the equations (36) to (43)

$$\ddot{e}_2 + C_1 \dot{e}_2 + C_2 e_2 + C_3 \int_0^t r_2 dt + \beta \operatorname{sgn}(r_2) = 0 \tag{44}$$

Therefore

$$\dot{r}_2 = -C_1 \dot{e}_2 - C_2 e_2 - C_3 \int_0^t r_2 dt - \beta \operatorname{sgn}(r_2) + \Lambda_2 \dot{e}_2 \tag{45}$$

Let

$$C_1 = \Lambda_2 + S_0 \tag{46}$$

$$C_2 = \Lambda_2 S_0 \tag{47}$$

Therefore

$$\dot{r}_2 = -(\Lambda_2 + S_0)\dot{e}_2 - \Lambda_2 S_0 e_2 - C_3 \int_0^t r_2 dt - \beta \operatorname{sgn}(r_2) + \Lambda_2 \dot{e}_2 \tag{48}$$

$$\dot{r}_2 = -S_0 r_2 - C_3 \int_0^t r_2 dt - \beta \operatorname{sgn}(r_2) \tag{49}$$

5. Controller structure and stability analysis

The overall control system has two loops and is depicted in Fig. 2. The following theorem details the performance of the controller.

Definition 5.1: *The equilibrium point* x_e *is said to be uniformly ultimately bounded (UUB) if there exist a compact set* $S \subset R^n$ *so that for all* $x_0 \in S$ *there exist a bound B and a time* $T(B,x_0)$ *such that* $\|x(t)-x_e(t)\| \le B \forall t \ge t_0 + T$.

Theorem 5.1: *Given the system as described in (15) and (16) with a control law shown in Fig. 2. and given by (20), (27), (42), (43). Then, the tracking errors* r_1 *and* r_2 *and thereby* e_1 *and* e_2 *are UUB if (53) and (54) are satisfied and can be made arbitrarily small with a suitable choice of gain parameters. According to the definition given by (23) of* r_1 *and (39) of* r_2*, this guarantees that* e_1 *and* e_2 *are UUB since*

$$\begin{aligned} \|e_1\| &\le \frac{\|r_1\|}{\sigma_{\min}(\Lambda_1)} \le \frac{b_{r_1}}{\sigma_{\min}(\Lambda_1)} \qquad b_{r_1} > 0 \\ \|e_2\| &\le \frac{\|r_2\|}{\sigma_{\min}(\Lambda_2)} \le \frac{b_{r_2}}{\sigma_{\min}(\Lambda_2)} \qquad b_{r_2} > 0 \end{aligned} \tag{50}$$

where $\sigma_{\min}(\Lambda_i)$ *is the minimum singular value of* $\Lambda_i, i = 1,2$.

Proof: Consider the Lyapunov function

$$L = \frac{1}{2} r_1^T P_1 r_1 + \frac{1}{2} r_2^T Q_1 r_2 + \frac{1}{2}\left[\int_0^t r_1^T dt\right] P_2 \left[\int_0^t r_1 dt\right] + \frac{1}{2}\left[\int_0^t r_2^T dt\right] Q_2 \left[\int_0^t r_2 dt\right] \tag{51}$$

with symmetric matrices $P_1, P_2, Q_1, Q_2 > 0$

Therefore, by differentiating L we will get the following

$$\begin{aligned} \dot{L} = &-r_1^T P_1 R r_1 - r_1^T P_1 K_3 \int_0^t r_1 dt + r_1 P_2 \int_0^t r_1 dt - r_2^T Q_1 S_0 r_2 - \\ &r_2^T Q_1 C_3 \int_0^t r_2 dt + r_2^T Q_2 \int_0^t r_2 dt - r_2^T Q_1 \beta \operatorname{sgn}(r_2) \end{aligned} \tag{52}$$

Define,

$$P_2 = P_1 K_3 \tag{53}$$

$$Q_2 = Q_1 C_3 \tag{54}$$

then integration term vanishes.

$$\dot{L} = -r_1^T P_1 R r_1 - r_2^T Q_1 S_0 r_2 - r_2^T Q_1 \beta \operatorname{sgn}(r_2) \tag{55}$$

The Equation (55) can be written as

$$\dot{L} \le -\sigma_{\max}(P_1 R) r_1^2 - \sigma_{\max}(Q_1 S_0) r_2^2 - \sigma_{\max}(Q_1 \beta)|r_2| \le 0 \tag{56}$$

where, $\sigma_{\max}(\cdot)$ denotes the maximum singular value. ■

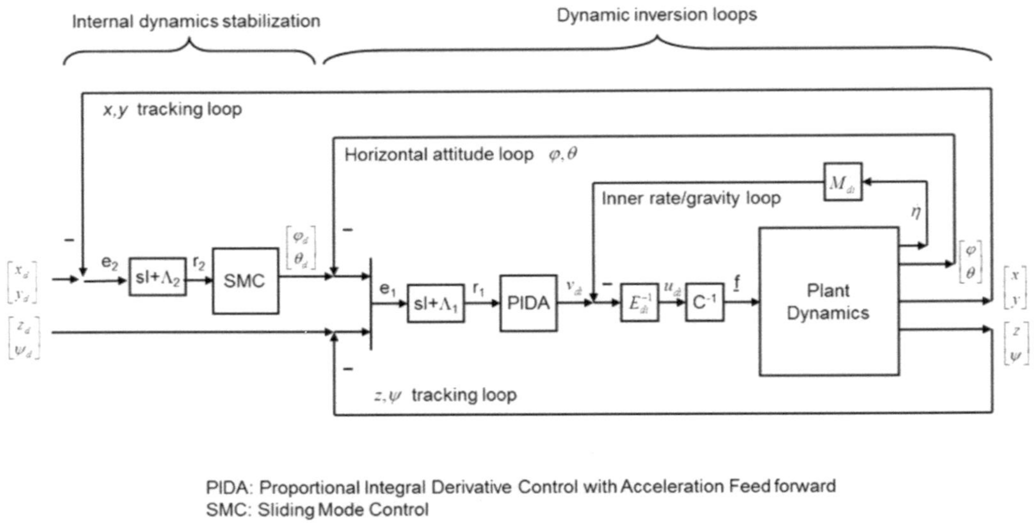

PIDA: Proportional Integral Derivative Control with Acceleration Feed forward
SMC: Sliding Mode Control

Fig. 2. Control configuration

Comment 5.1: *Equations* (31)-(32) *can also be rewritten* as

$$\ddot{x} = -\frac{u}{m}\theta_d - \frac{u}{m}(-\tilde{\theta}) \tag{57}$$

$$\ddot{y} = \frac{u}{m}\varphi_d - \frac{u}{m}\tilde{\varphi} \tag{58}$$

where $\tilde{\theta} = \theta_d - \sin\theta$ *and* $\tilde{\varphi} = \varphi_d - \cos\theta\sin\varphi$. *According to* (A. Das, K. Subbarao, and F. Lewis 2009) *there exist a robustifying term* V_r *which would modify the* v_{di} *as*

$$v_{di} = \begin{bmatrix} v_z \\ v_\varphi \\ v_\theta \\ v_\psi \end{bmatrix} = \begin{bmatrix} \ddot{z}_d - K_{1_z}(\dot{z} - \dot{z}_d) - K_{2_z}(z - z_d) \\ \ddot{\varphi}_d - K_{1_\varphi}(\dot{\varphi} - \dot{\varphi}_d) - K_{2_\varphi}(\varphi - \varphi_d) \\ \ddot{\theta}_d - K_{1_\theta}(\dot{\theta} - \dot{\theta}_d) - K_{2_\theta}(\theta - \theta_d) \\ \ddot{\psi}_d - K_{1_\psi}(\dot{\psi} - \dot{\psi}_d) - K_{2_\psi}(\psi - \psi_d) \end{bmatrix} + V_r \tag{59}$$

and thereby one can easily show that $\dot{L} \le 0$ *by suitbale choice of* V_r. *For this book chapter we considered the Eq. (36),(37), whcih in fact a simpler version of (31),(32). But we belive, for designing autopilot for quadrotor, the proposed mehtod discussed in this chapter can be used without loss of any genreality.*

6. Simulation results

6.1 Rotorcraft parameters

Simulation for a typical quadrotor is performed using the following parameters (SI unit):

$$M_1 = \begin{bmatrix} 1 & 0 & 0 \\ 0 & 1 & 0 \\ 0 & 0 & 1 \end{bmatrix}; \qquad J = \begin{bmatrix} 5 & 0 & 0 \\ 0 & 5 & 0 \\ 0 & 0 & 15 \end{bmatrix}; \qquad g = 9.81 .$$

6.2 Reference trajectory generation

As outlined in Refs (Hogan 1984; Flash and Hogan 1985), a reference trajectory is derived that minimizes the jerk (rate of change of acceleration) over the time horizon. The trajectory ensures that the velocities and accelerations at the end point are zero while meeting the position tracking objective. The following summarizes this approach:

$$\dot{x}_d(t) = a_{1_x} + 2a_{2_x} t + 3a_{3_x} t^2 + 4a_{4_x} t^3 + 5a_{5_x} t^4 \tag{60}$$

Differentiating again,

$$\ddot{x}_d(t) = 2a_{2_x} + 6a_{3_x} t + 12a_{4_x} t^2 + 20a_{5_x} t^3 \tag{61}$$

As we indicated before that initial and final velocities and accelerations are zero; so from Eqs. (60) and (61) we can conclude the following:

$$\begin{bmatrix} d_x \\ 0 \\ 0 \end{bmatrix} = \begin{bmatrix} 1 & t_f & t_f^2 \\ 3 & 4t_f & 5t_f^2 \\ 6 & 12t_f & 20t_f^2 \end{bmatrix} \begin{bmatrix} a_{3_x} \\ a_{4_x} \\ a_{5_x} \end{bmatrix} \tag{62}$$

Where, $d_x = \left(x_{d_f} - x_{d_0}\right) / t_f^3$. Now, solving for coefficients

$$\begin{bmatrix} a_{3_x} \\ a_{4_x} \\ a_{5_x} \end{bmatrix} = \begin{bmatrix} 1 & t_f & t_f^2 \\ 3 & 4t_f & 5t_f^2 \\ 6 & 12t_f & 20t_f^2 \end{bmatrix}^{-1} \begin{bmatrix} d_x \\ 0 \\ 0 \end{bmatrix} \tag{63}$$

Thus the desired trajectory for the x direction is given by

$$x_d(t) = x_{d_0} + a_{3_x} t^3 + a_{4_x} t^4 + a_{5_x} t^5 \tag{64}$$

Similarly, the reference trajectories for the y and z directions are gives by Eq. (65) and Eq. (66) respectively.

$$y_d(t) = y_{d_0} + a_{3_y} t^3 + a_{4_y} t^4 + a_{5_y} t^5 \tag{65}$$

$$z_d(t) = z_{d_0} + a_{3_z} t^3 + a_{4_z} t^4 + a_{5_z} t^5 \tag{66}$$

The beauty of this method lying in the fact that more demanding changes in position can be accommodated by varying the final time. That is acceleration/torque ratio can be controlled smoothly as per requirement. For example,
Let us assume at $t = 0,\ x_{d_0} = 0$ and at $t = 10\,\text{sec}, x_{d_f} = 10$. Therefore $d_x = 0.01$ and the trajectory is given by Eq. (67) and shown in Fig. 3 for various desired final positions.

$$x_d(t) = 0.1t^3 - 0.015t^4 + 0.0006t^5 \tag{67}$$

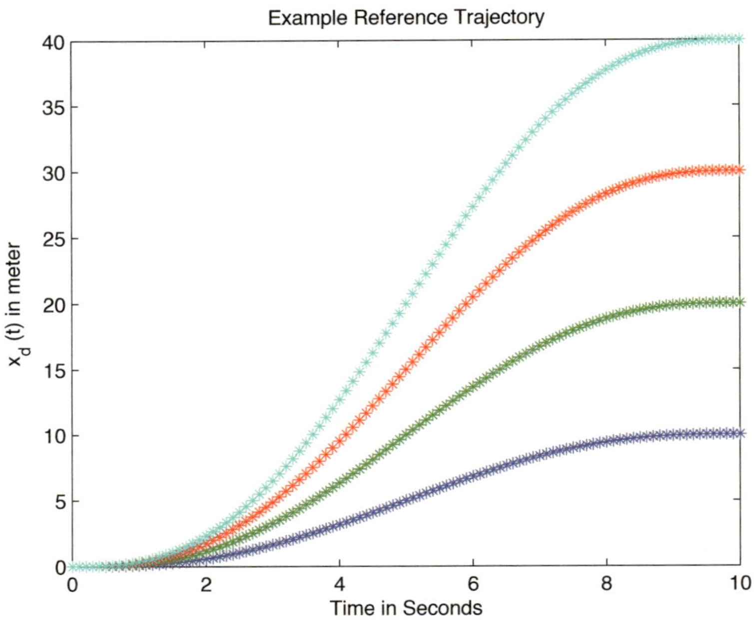

Fig. 3. Example trajectory simulation for different final positions

6.3 Case 1: From initial position at $(0,5,10)$ to final position at $(20,-5,0)$

Figure 4 describes the controlled motion of the quadrotor from its initial position $(0,5,10)$ to final position $(20,-5,0)$ for a given time (20 seconds). The actual trajectories $x(t), y(t), z(t)$ match exactly their desired values $x_d(t), y_d(t), z_d(t)$ respectively nearly exactly. The errors along the three axes are also shown in the same figure. It can be seen that the tracking is almost perfect as well as the tracking errors are significantly small. Figure 5 describes the attitude of the quadrotor φ, θ along with their demands φ_d, θ_d and attitude errors in radian. Again the angles match their command values nearly perfectly. Figure 6 describes the control input requirement which is very much realizable. Note that as described before the control requirement for yaw angle is $\tau_\psi = 0$ and it is seen from Fig. 6.

6.4 Case 2: From initial position at $(0,5,10)$ to final position at $(20,5,10)$

Figures 7-8 illustrates the decoupling phenomenon of the control law. Fig. 7 shows that $x(t)$ follows the command $x_d(t)$ nearly perfectly unlike $y(t)$ and $z(t)$ are held their initial values. Fig. 8 shows that the change in x does not make any influence on φ. The corresponding control inputs are also shown in Fig. 9 and due to the full decoupling effect it is seen that τ_φ is almost zero.

The similar type of simulations are performed for y and z directional motions separately and similar plots are obtained showing excellent tracking.

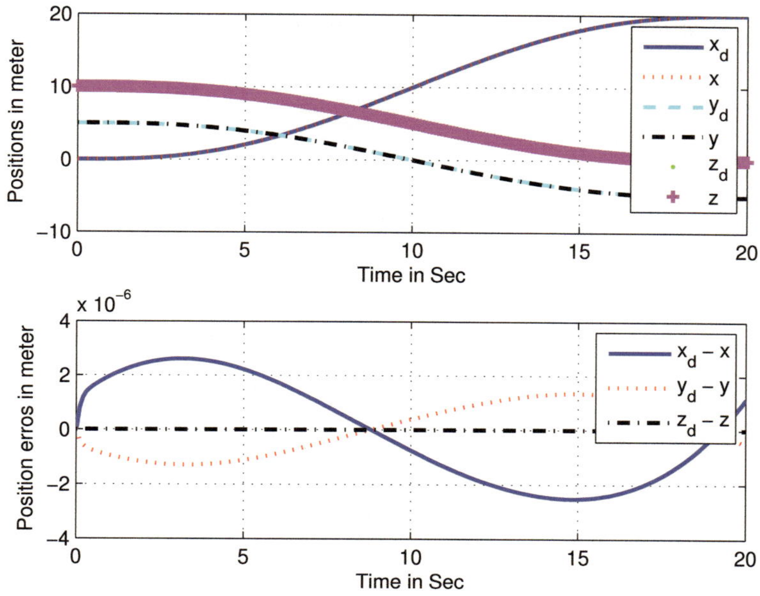

Fig. 4. Three position commands simultaneously

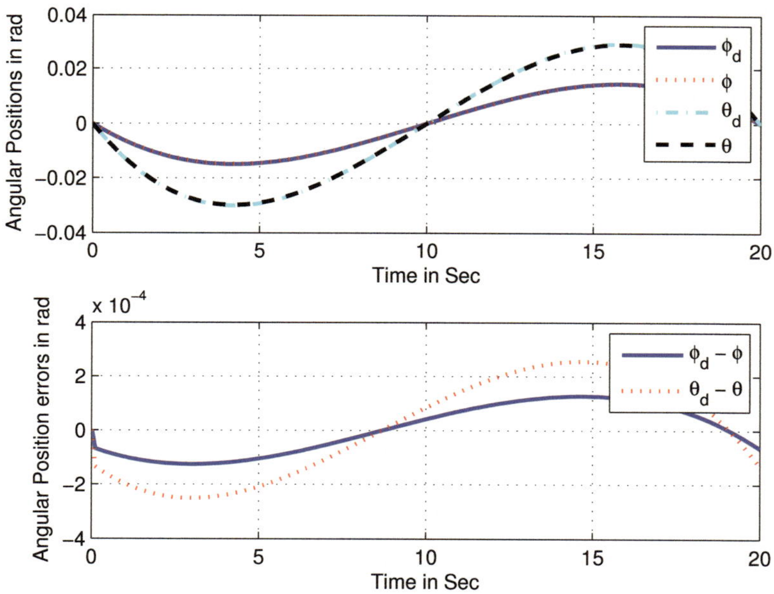

Fig. 5. Resultant angular positions and errors

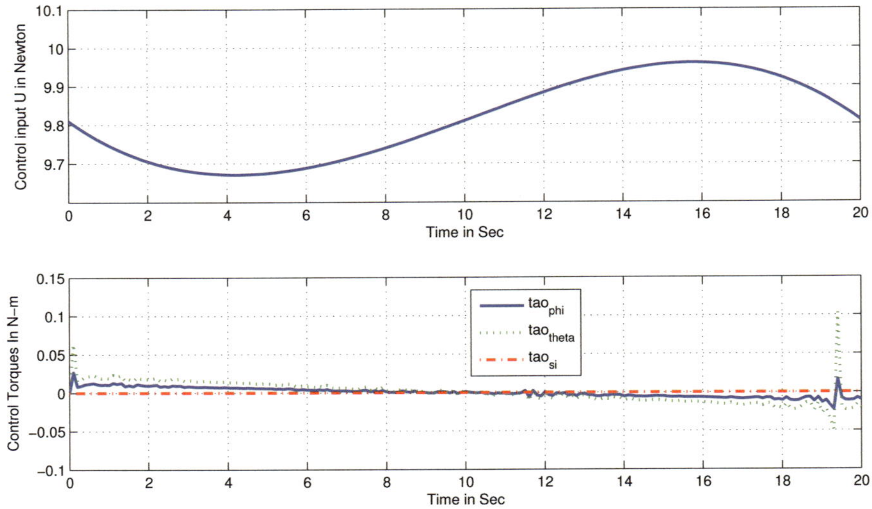

Fig. 6. Input commands for Case I

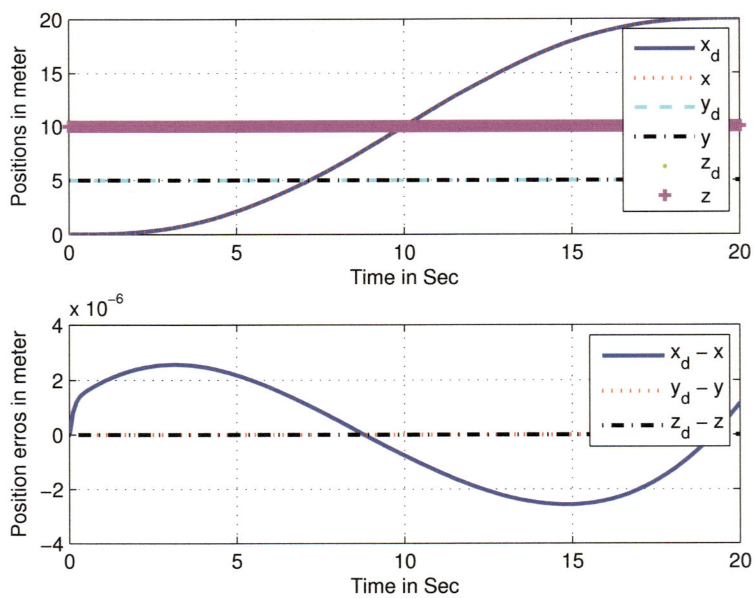

Fig. 7. Plots of position and position tracking errors for x command only

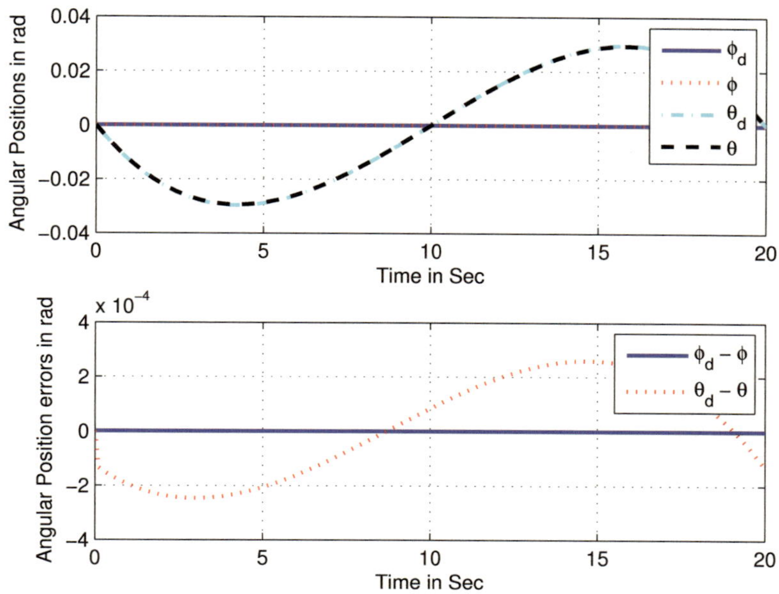

Fig. 8. Angular variations due to change in x

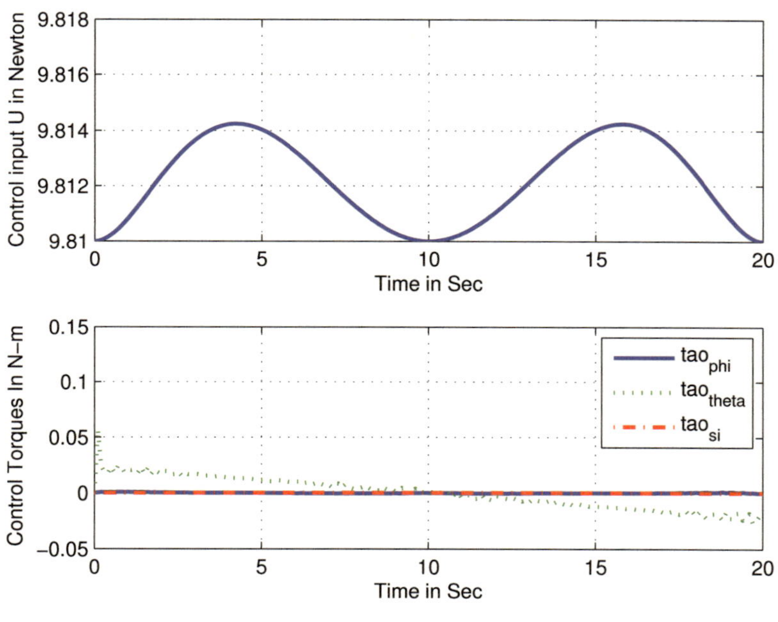

Fig. 9. Input commands for variation in x (Case II)

6.5 Simulation with unmodeled input disturbances

The simulation is performed to verify its robustness properties against unmodeled input disturbances. For this case we simulate the dynamics with high frequency disturbance 0.1* $\sin(5t)$ (1% of maximum magnitude of force) for force channel and $0.01\sin(5t)$ (~15% of maximum angular acceleration) for torque channel.

6.6 Case-3: From initial position at $(0,5,10)$ to final position at $(20,-5,0)$ with disturbance

Fig. 10-11 describes the motion of the quadrotor from its initial position $(0,5,10)$ to final position $(20,-5,0)$ for a given time (20 seconds) with input disturbances. It can be seen from Fig 10 that the quadrotor can track the desired position effectively without any effect of high input disturbances. From Fig 10 and Fig 11, it is also seen that the position errors are bounded and small. Fig. 12 shows the bounded variation of control inputs in presence of disturbance. Similar tracking performance is obtained for other commanded motion.

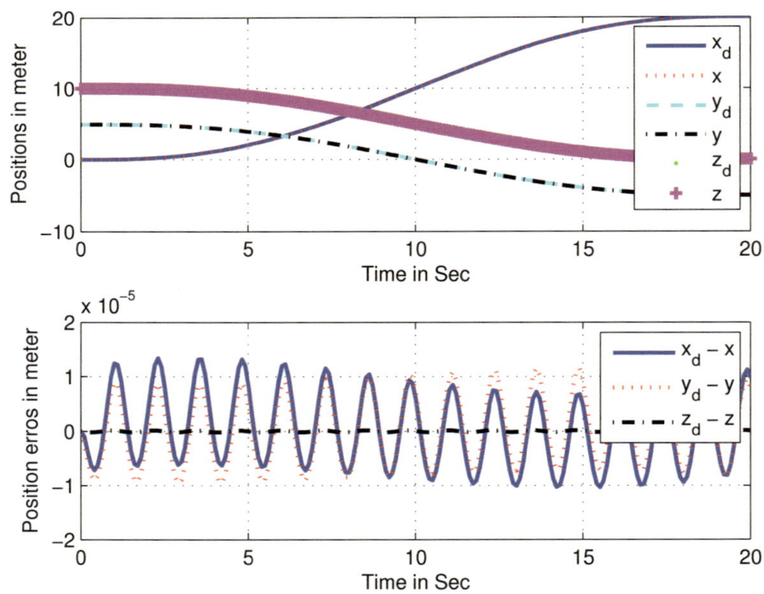

Fig. 10. Position tracking - Simultaneous command in x, y and z + Input disturbances

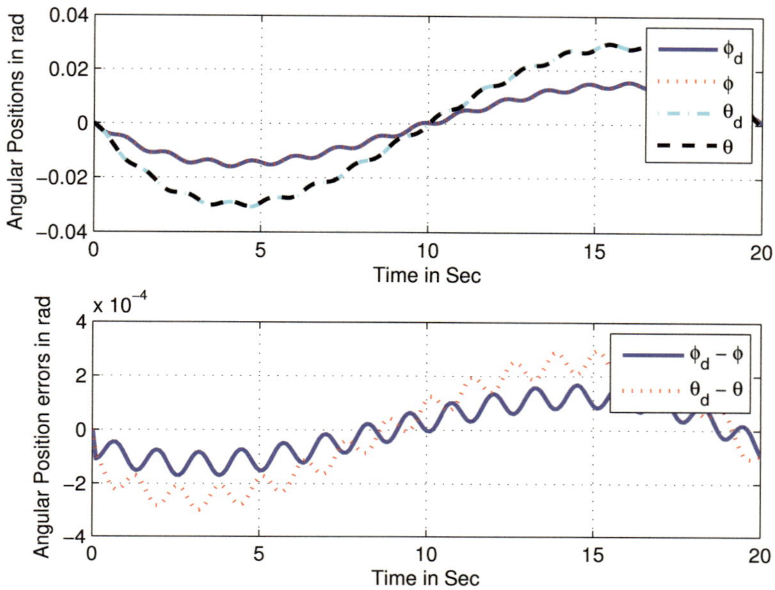

Fig. 11. Angular variations, errors and velocities (with input disturbances)

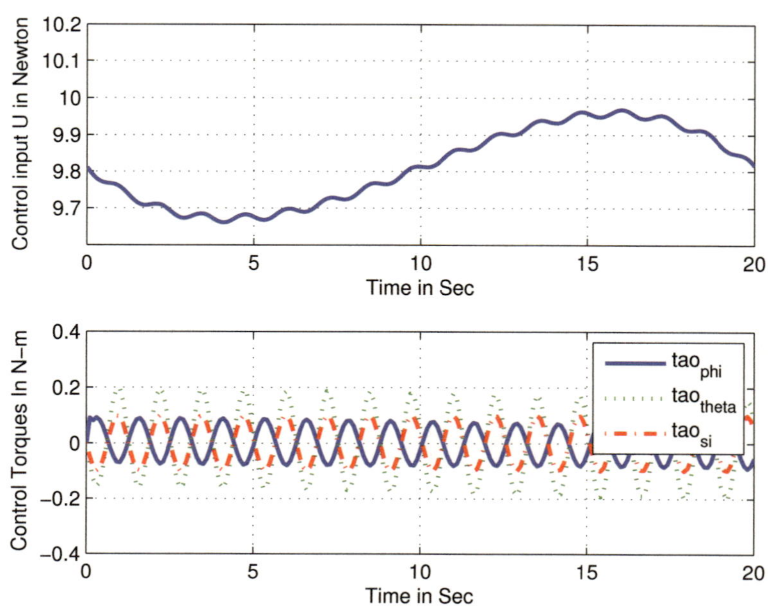

Fig. 12. Force and torque input variations (with input disturbances)

7. Conclusion

Sliding mode approach using input-output linearization to design nonlinear controller for a quadrotor dynamics is discussed in this Chapter. Using this approach, an intuitively structured controller was derived that has an outer sliding mode control loop and an inner feedback linearizing control loop. The dynamics of a quadrotor are a simplified form of helicopter dynamics that exhibits the basic problems including under-actuation, strong coupling, multi-input/multi-output. The derived controller is capable of deal with such problems simultaneously and satisfactorily. As the quadrotor model discuss in this Chapter is similar to a full scale unmanned helicopter model, the same control configuration derived for quadrotor is also applicable for a helicopter model. The simulation results are presented to demonstrate the validity of the control law discussed in the Chapter.

8. Acknowledgement

This work was supported by the National Science Foundation ECS-0801330, the Army Research Office W91NF-05-1-0314 and the Air Force Office of Scientific Research FA9550-09-1-0278.

9. References

Altug, Erdinc, James P. Ostrowski, and Robert Mahony. 2002. Control of a Quadrotor Helicopter Using Visual Feedback ID - 376. In . Washington DC, Virginia, June.

B. Bijnens, Q. P. Chu, G. M. Voorsluijs, and J. A. Mulder. 2005. AIAA Guidance, Navigation, and Control Conference and Exhibit. In . San Francisco, California.

Bijnens, B., Q. P. Chu, G. M. Voorsluijs, and J. A. Mulder. 2005. Adaptive Feedback Linearization Flight Control for a Helicopter UAVID - 199.

Bouabdallah, Samir, AndrÂ´e Noth, and Roland Siegwart. 2004. International Conference on Intelligent Robots and Systems. In , 3:2451-2456. Sendal, Japan: IEEE.

Calise, A. J., B. S. Kim, J. Leitner, and J. V. R. Prasad. 1994. Helicopter adaptive flight control using neural networks. In . Lake Buena Vista, FL.

Campos, J., F. L. Lewis, and C. R. Selmic. 2000. Backlash Compensation in Discrete Time Nonlinear Systems Using Dynamic Inversion by Neural Networks. In . San Francisco, CA.

Castillo, P., A. Dzul, and R. Lozano. 2004. Real-time Stabilization and Tracking of a Four-Rotor Mini Rotorcraft. *IEEE Transaction on Control System Technology* 12: 510-516.

Castillo, P., R. Lozano, and A. Dzul. 2005. *Modelling and Control of Mini Flying Machines.* Springer-Verlag.

Das, A., T. Garai, S. Mukhopadhyay, and A. Patra. 2004. Feedback Linearization for a Nonlinear Skid-To-Turn Missile Model. *First India annual conference, Proceedings of the IEEE INDICON 2004*: 586-589.

Das, A., K. Subbarao, and F. Lewis. 2009. Dynamic inversion with zero-dynamics stabilisation for quadrotor control. *Control Theory & Applications, IET* 3, no. 3 (March): 303 - 314.

Das, Abhijit, Frank Lewis, and Kamesh Subbarao. 2009. Backstepping Approach for Controlling a Quadrotor Using Lagrange Form Dynamics. *Journal of Intelligent and Robotic Systems* 56, no. 1-2 (4): 127-151. doi:10.1007/s10846-009-9331-0.

Flash, T., and N. Hogan. 1985. The Coordination of Arm Movements: an Experimentally Confirmed Mathematical Model. *Journal of Neuro Science* 5: 1688-1703.

Gavrilets, V., B. Mettler, and E. Feron. 2003. Dynamic Model for a Miniature Aerobatic Helicopter. *MIT-LIDS report* LIDS-P-2580.

Hogan, N. 1984. Adaptive Control of Mechanical Impedance by Coactivation of Antagonist Muscles. *IEEE Transaction of Automatic Control* 29: 681-690.

Hovakimyan, N., F. Nardi, A. J. Calise, and H. Lee. 2001. Adaptive Output Feedback Control of a Class of Nonlinear Systems Using Neural Networks. *International Journal of Control* 74: 1161-1169.

Kanellakopoulos, I., P. V. Kokotovic, and A. S. Morse. 1991. Systematic Design of Adaptive Controllers for Feedback Linearizable Systems. *IEEE Transaction of Automatic Control* 36: 1241-1253.

Khalil, Hassan K. 2002. *Nonlinear Systems*. 3rd ed. Upper Saddle River, N.J: Prentice Hall.

Kim, B. S., and A. J. Calise. 1997. Nonlinear flight control using neural networks. *Journal of Guidance Control Dynamics* 20: 26-33.

Koo, T. J., and S. Sastry. 1998. Output tracking control design of a helicopter model based on approximate linearization. In *Proceedings of the 37th Conference on Decision and Control*. Tampa, Florida: IEEE.

Lewis, F., S. Jagannathan, and A. Yesildirek. 1999. *Neural Network Control of Robot Manipulators and Nonlinear Systems*. London: Taylor and Francis.

Mistler, V., A. Benallegue, and N. K. M'Sirdi. 2001. Exact linearization and non- interacting control of a 4 rotors helicopter via dynamic feedback. In *10th IEEE Int. Workshop on Robot-Human Interactive Communication*. Paris.

Mokhtari, A., A. Benallegue, and Y. Orlov. 2006. Exact Linearization and Sliding Mode Observer for a Quadrotor Unmanned Aerial Vehicle. *International Journal of Robotics and Automation* 21: 39-49.

P. Castillo, R. Lozano, and A. Dzul. 2005. Stabilization of a Mini Rotorcraft Having Four Rotors. *IEEE Control System Magazine* 25: 45-55.

Prasad, J. V. R., and A. J. Calise. 1999. Adaptive nonlinear controller synthesis and flight evaluation on an unmanned helicopter. In . Kohala Coast-Island of Hawaii, USA.

Rysdyk, R., and A. J. Calise. 2005. Robust Nonlinear Adaptive Flight Control for Consistent Handling Qualities. *IEEE Transaction of Control System Technology* 13: 896-910.

Slotine, Jean-Jacques, and Weiping Li. 1991. *Applied Nonlinear Control*. Prentice Hall.

Stevens, B. L., and F. L. Lewis. 2003. *Aircraft Simulation and Control*. Wiley and Sons.

T. Madani, and A. Benallegue. 2006. Backstepping control for a quadrotor helicopter. In . Beijing, China.

Wise, K. A., J. S. Brinker, A. J. Calise, D. F. Enns, M. R. Elgersma, and P. Voulgaris. 1999. Direct Adaptive Reconfigurable Flight Control for a Tailless Advanced Fighter Aircraft. *International Journal of Robust and Nonlinear Control* 9: 999-1012.

2

Advanced Control Techniques for the Transonic Phase of a Re-Entry Flight

Gianfranco Morani, Giovanni Cuciniello,
Federico Corraro and Adolfo Sollazzo
Italian Aerospace Research Centre (CIRA)
Italy

1. Introduction

New technological developments in space engineering and science require sophisticated control systems with both high performance and reliability. How to achieve these goals against various uncertainties and off-nominal scenarios has been a very challenging issue for control system design over the last years.

Several efforts have been spent on control systems design in aerospace applications, in order to conceive new control approaches and techniques trying to overcome the inherent limitations of classical control designs.

In fact, the current industrial practice for designing flight control laws is based on Proportional Integral Derivative (PID) controllers with scheduled gains. With this approach, several controllers are designed at various points in the operative flight envelope, considering local time-invariant linear models based on small perturbations of a detailed nonlinear aircraft model. Although these techniques are commonly used in control systems design, they may have inherent limitations stemming from the poor capability of guaranteeing acceptable performances and stability for flight conditions different from the selected ones, especially when the scheduling parameters rapidly change.

This issue becomes very critical when designing flight control system for space re-entry vehicles. Indeed, space reentry applications have some distinctive features with respect to aeronautical ones, mainly related to the lack of stationary equilibrium conditions along the trajectories, to the wide flight envelope characterizing missions (from hypersonic flight regime to subsonic one) and to the high level of uncertainty in the knowledge of vehicle aerodynamic parameters.

Over the past years, several techniques have been proposed for advanced control system development, such as Linear Quadratic Optimal Control (LQOC), Eigenstructure Assignment, Robust control theory, Quantitative feedback theory (QFT), Adaptive Model Following, Feedback Linearization, Linear Parameter Varying (LPV) and probabilistic approach. Hereinafter, a brief recall of the most used techniques will be given.

Linear Quadratic Optimal Control (LQOC) allows finding an optimal control law for a given system based on a given criterion. The optimal control can be derived using Pontryagin's maximum principle and it has been commonly applied in designing Linear Quadratic Regulator (LQR) of flight control system (see Xing, 2003; Vincent et al., 1994).

The Eigenstructure Assignment consists of placing the eigenvalues of a linear system using state feedback and then using any remaining degrees of freedom to align the eigenvectors as accurately as possible (Konstantopoulos & Antsaklis, 1996; Liu & Patton, 1996; Ashari et al., 2005). Nevertheless there are several limitations, since only linear systems are considered and moreover the effects of uncertainty have been not extensively studied.
Robust analysis and control theory is a method to measure performance degradation of a control system when considering system uncertainties (Rollins, 1999; Balas, 2005). In this framework a concept of structured singular value (i.e μ-Synthesis) is introduced for including structured uncertainties into control system synthesis as well as for checking robust stability of a system.
Adaptive Model Following (AMF) technique has the advantage of strong robustness against parameter uncertainty of the system model, if compared to classical control techniques (Bodson & Grosziewicz, 1997; Kim et al., 2003). The model following approach has interesting features and it may be an important part of an autonomous reconfigurable algorithm, because it aims to emulate the performance characteristics of a target model, even in presence of plant's uncertainties.
Another powerful nonlinear design is Feedback Linearization which transforms a generic non linear system into an equivalent linear system, through a change of variables and a suitable control input (Bharadwaj et al., 1998; Van Soest et al., 2006). Feedback linearization is an approach to nonlinear control design which is based on the algebraic transformation of nonlinear systems dynamics into linear ones, so that linear control techniques can be applied.
More recently an emerging approach, named Linear Parameter Varying (LPV) control, has been developed as a powerful alternative to the classical concept of gain scheduling (Spillman, 2000; Malloy & Chang, 1998; Marcos & Balas, 2004). LPV techniques are well suited to account for on-line parameter variations such that the controllers can be designed to ensure performance and robustness in all the operative envelope. In this way a gain-scheduling controller can be achieved without interpolating between several design points.
The main effort (and also main drawback) required by the above techniques is the modelling of a nonlinear system as a LPV system. Several techniques exist but they may require a huge effort for testing controller performances on the nonlinear system. Other modelling techniques try to overcome this problem at the expense of a higher computational effort.
Finally in the last decades, a new philosophy has emerged, that is, probabilistic approach for control systems analysis and synthesis (Calafiore et al., 2007; Tempo et al., 1999; Tempo et al., 2005). In this approach, the meaning of robustness is shifted from its usual deterministic sense to a probabilistic one. The new paradigm is then based on the probabilistic definition of robustness, by which it is claimed that a certain property of a control system is "almost" robustly satisfied, if it holds for "most" instances of uncertainties. The algorithms based on probabilistic approach, usually called randomized algorithms (RAs), often have low complexity and are associated to robustness bounds which are less conservative than classical ones, obviously at the expense of a probabilistic risk.
In this chapter the results of a research activity focused on the comparison between different advanced control architectures for transonic phase of a reentry flight are reported. The activity has been carried out in the framework of Unmanned Space Vehicle (USV) program of Italian Aerospace Research Centre (CIRA), which is in charge of developing unmanned space Flying Test Beds (FTB) to test advanced technologies during flight. The first USV

Dropped Transonic Flight Test (named DTFT1) was carried out in February 2007 with the first vehicle configuration of USV program (named FTB1) (see Russo et al., 2007 for details). For this mission, a conventional control architecture was implemented. DTFT1 was then used as a benchmark application for comparison among different advanced control techniques. This comparison aimed at choosing the most suited control technique to be used for the subsequent, more complex, dropped flight test, named DTFT2, successfully carried out on April 2010. To this end, three techniques were selected after a dedicated literature survey, namely:

- μ-Control with Fuzzy Logic Gain-Scheduling
- Direct Adaptive Model Following Control
- Probabilistic Robust Control Synthesis

In the next sections, the above techniques will be briefly described with particular emphasis on their application to DTFT1 mission. In sec. 5 the performance analysis carried out for comparison among the different techniques will be presented.

2. Fuzzy scheduled MU-controller

2.1 The H∞ control problem

The H∞ Control Theory (Zhou & Doyle, 1998) rises as response to the deficiencies of the classical Linear Quadratic Gaussian (LQG) control theory of the 1960s applications. The general problem formulation is described through the following equations:

$$\begin{bmatrix} z \\ y \end{bmatrix} = P(s)\begin{bmatrix} w \\ u \end{bmatrix} = \begin{bmatrix} P_{11} & P_{12} \\ P_{21} & P_{22} \end{bmatrix}\begin{bmatrix} w \\ u \end{bmatrix}$$
$$u = K(s)y \tag{1}$$

where P is the nominal plant, u is the control variable, y is the measured variable, w is an exogenous signals (such as disturbances) and z is the error signal to be minimized. The generic control scheme is depicted in Fig 1.

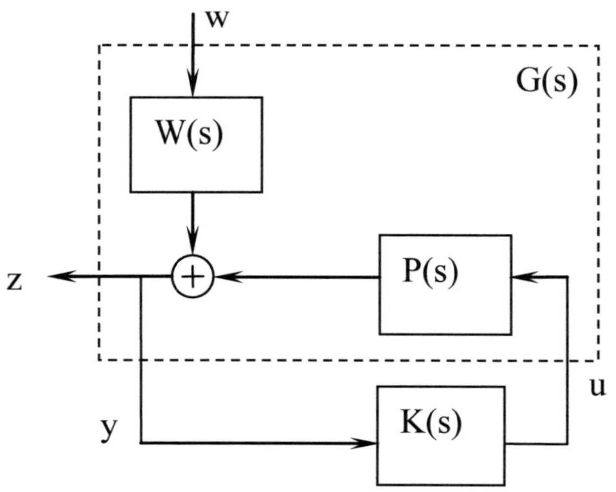

Fig. 1. Nominal Performance Scheme

It can be shown that closed-loop transfer function from w to z can be obtained via lower linear fractional transformation (Zhou & Doyle, 1998).
Therefore H_∞ control problem is to find a stabilizing controller, K, which minimizes

$$\left\|F_l(P,K)\right\|_\infty = \sup_\omega \bar{\sigma}\left(F_l(P,K)(j\omega)\right) = \gamma_M \tag{2}$$

where F_l is the lower linear fractional transformation from w to z and $\bar{\sigma}$ is the singular value of specified transfer function.
For what concerns Nominal Performance Problem, it is required that error z is kept as small as possible. To this end, a new generalized plant can be considered (see the dashed line). The weighting function penalizes the infinite-norm of new plant to achieve required performances.
In the same way, Robust Stability Problem can be solved applying Small Gain Theorem (Zhou & Doyle, 1998) to the following new generalized plant selected (see the dashed line):

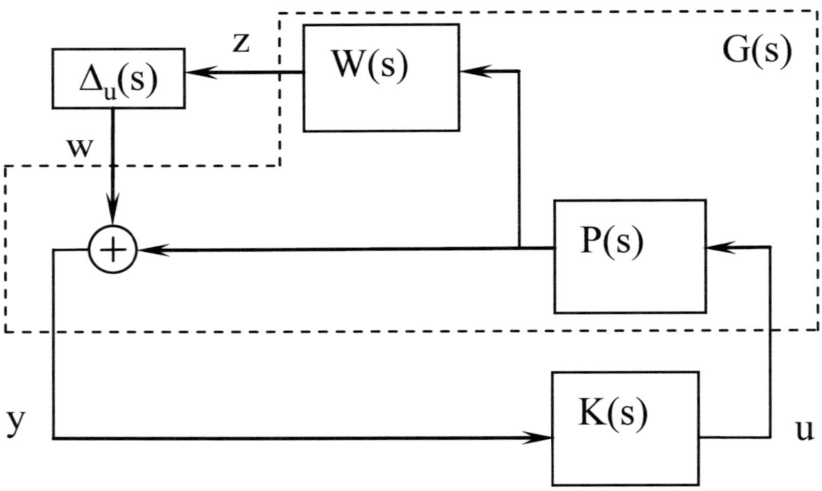

Fig. 2. Robust Stability Scheme

A more general problem to solve is Robust Performance Problem that takes into account both Nominal Performance and Robust Stability Problems. It is worth noting that a Nominal Performance Scheme allows to find a stabilizing controller that satisfies Small Gain Theorem in presence of a fictitious uncertainty block Δf(s) (with $\left\|\Delta_f(s)\right\|_\infty < 1/\gamma_M$). Hence a general scheme for Robust Performance Problem is the following one:

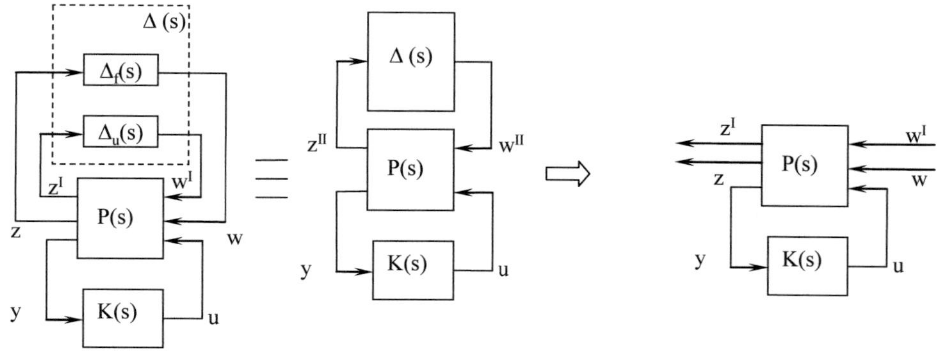

Fig. 3. Robust Performance Scheme

where:

$$\Delta(s) = \begin{bmatrix} \Delta_u(s) & 0 \\ 0 & \Delta_f(s) \end{bmatrix} \quad (3)$$

It is clear from the figures that all above problem formulations can be always rearranged to solve the same general H_∞ problem. It is worth noting that for what concerns Robust Performance Case, Δ(s) matrix has a diagonal block structure. Plant uncertainties can be structured like mixed (real and complex) uncertainties. Unfortunately H_∞ problem only deals with unstructured full complex Δ(s), so optimal (or sub-optimal) controller might be very conservative. μ-analysis and synthesis try to solve this issue by dealing with structured uncertainties.

2.2 μ-synthesis framework

The brief discussion of previous paragraph has shown how to design an H_∞ controller starting from a generalized plant. Required performances are achieved through optimal (or sub-optimal) controller by means of weighting functions of a generalized plant. In the same way, robust stability is achieved together with performances by solving Robust Performance Problem. Let $M = F_l(P,K)$, then a general scheme for μ-analysis is the following one:

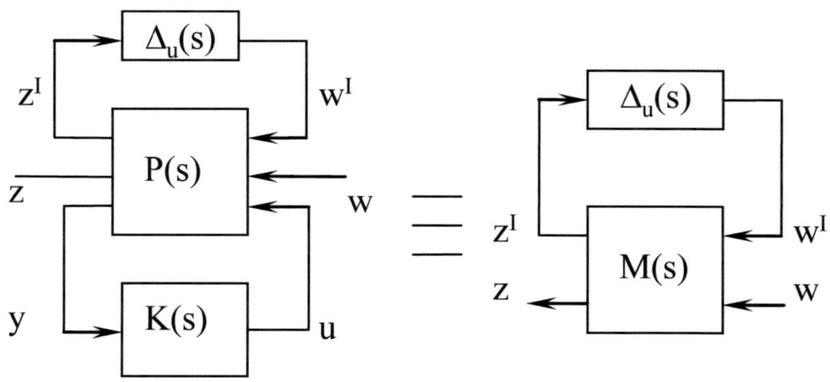

Fig. 4. μ-Analysis Scheme

Introducing structured singular value:

$$\mu_\Delta(M) := \frac{1}{\min\{\bar{\sigma}(\Delta) \ : \det(I - M\Delta) = 0\}} \tag{4}$$

(where Δ_u is the structured uncertainty mentioned earlier), for all $\Delta_u(s)$ with $\|\Delta_u(s)\|_\infty < 1/\beta$, and $\beta>0$, the loop of previous figure is internally stable and $\|F_l(M,\Delta_u)\|_\infty < \beta$ if and only if

$$\sup_{\omega \in R} \mu_\Delta(M(j\omega)) \le \beta \tag{5}$$

Therefore, given a controller K(s), μ-bound β can be numerically computed. Finally μ-synthesis framework can be represented through the following scheme.

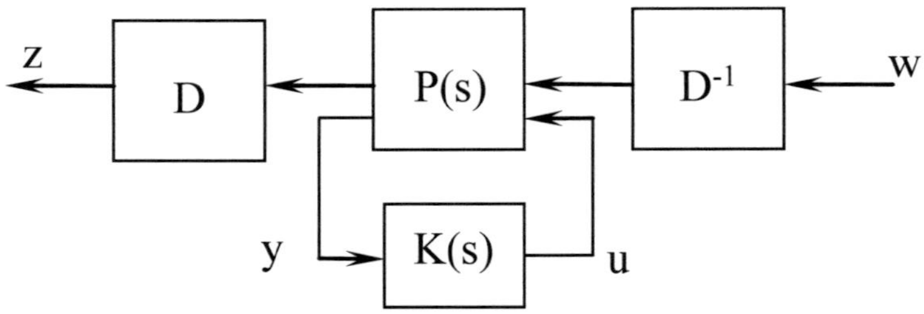

Fig. 5. μ-Synthesis via scaling

where *D* matrix allows scaling the process taking into account structured uncertainties from *z* to *w*. A commonly used methodology to solve the above problem is DK-iterations algorithm (Zhou & Doyle, 1998) that sequentially performs two parameter minimization: first minimizing over *K* with *D* fixed, then minimizing over *D* with *K* fixed, then again over *K*, and again over *D*, etc. The algorithm runs until a fixed bound is achieved and final *K* is the desired controller.

2.3 Fuzzy scheduling

Each controller developed using the technique described in the previous sections can be considered as a "local" controller, since it might not guarantee the same performances "far away" from design point (or outside a given region of flight envelope). If that region does not cover flight envelope of interest, a controller scheduling is necessary.

Many techniques and methodology have been investigated in literature (Nichols et al., 1993; Pedrycz & Peters, 1997; Hyde & Glover, 1993), but no one guarantees that scheduled controller provides robust performance to be achieved by closed loop system.

In (Pedrycz & Peters, 1997), authors present a general approach to a fuzzy interpolation of different LTI controllers. Although controllers are PID with different gains, the technique can be easily generalized to more complex LTI systems.

For what concerns application of fuzzy scheduling for the proposed DTFT application, a sort of fuzzy gain scheduling technique has been implemented using an approach similar to the one described in (Pedrycz & Peters, 1997).

Considering a system with dynamics described by the following equations:

$$\begin{aligned} \dot{x} &= f\left(x(t),u(t)\right) \\ y &= g\left(x(t),u(t)\right) \\ x &\in R^{n}, u \in R^{r}, y \in R^{m} \end{aligned} \tag{6}$$

and a family of equilibrium points (x_{ei}, y_{ei}, u_{ei}), with i=1, 2, ..., c. System linearization leads to the following linear system:

$$\begin{aligned} \dot{\tilde{x}} &= A_i\tilde{x} + B_i\tilde{u} \\ \tilde{y} &= C_i\tilde{x} + D_i\tilde{u} \end{aligned} \tag{7}$$

For each linearized model it is possible to design a local (linear) controller, L1, L2, …, Lc, on which the overall control laws will be based. Let Ω_1, Ω_2, .. Ω_c be fuzzy relations whose activation levels require specific control actions; the computations of control are then regulated by smooth, centre of gravity type of switching:

$$u = \frac{\sum_{i=1}^{c} Li \left(y * \frac{\Omega_i(x,u_{ei})}{\sum_{i=1}^{c}\Omega_i(x,u_{ei})} \right) * \Omega_i(x,u_{ei})}{\sum_{i=1}^{c}\Omega_i(x,u_{ei})} \tag{8}$$

3. Adaptive control system

Direct Adaptive Model Following (DAMF) is a Model Reference Control Strategy with strong robustness properties obtained through the use of direct adaptation of control loop gains in order to achieve a twofold objective: zero error between output of reference model and output of real plant and furthermore minimization of control effort. The proposed adaptation algorithm is based on Lyapunov theory. Hereinafter a brief mathematical description of the method, fully reported in (Kim et al., 2003), will be given. Starting from generic linear model of a plant:

$$\begin{aligned} \dot{x} &= Ax + Bu + d \\ y &= Cx \end{aligned} \tag{9}$$

where $x \in \Re^n$ is the state vector, $y \in \Re^l$ the output vector, $u \in \Re^m$ the control vector, $A \in \Re^{n\times n}$, $B \in \Re^{n\times m}$, $C \in \Re^{l\times n}$ and the term d represents the trim data, reference system dynamics are written in term of desired input-output behaviour:

$$\dot{y}_m = A_m y_m + B_m r \tag{10}$$

where y_m is the desired output for the plant, r is the reference signal, A_m and B_m represent the reference linear system dynamics. Control laws structure is defined as:

$$u = C_0\left(G_0 x + v + r + K_0 y_m\right) \tag{11}$$

where G_0, C_0 and v are adaptive control gains, while K_0 is a feed-forward gain matrix off-line computed. It is possible to demonstrate that the following adaptation rules for control laws parameters:

$$\begin{aligned} \dot{G}_0 &= -\gamma_1 B_m^T P e x^T \\ \dot{C}_0 &= -\gamma_2 C_0 B_m^T P e u^T C_0 \\ \dot{v}_0 &= -\gamma_3 B_m^T P e \end{aligned} \tag{12}$$

imply the non-positiveness of Lyapunov candidate function derivative:

$$\dot{V} = -e^T P e \leq 0 \tag{13}$$

which guarantees the asymptotical stability for the error dynamic system (Kim et al., 2003). Matrix P is the solution of Lyapunov equation:

$$A_e^T P + P A_e = -Q; \quad \text{with } Q > 0 \tag{14}$$

With reference to the implementation of adaptive technique for DTFT1 benchmark application, detailed scheme of MIMO controller is reported in Fig. 6. The design parameters of both inner and outer loops consist of a few number of matrices. First of all, Reference Dynamics are expressed by means of two matrices A_m and B_m with limitation that B_m must be chosen invertible. Desired error dynamic is regulated by means of A_e. Through this matrix it is also possible to modify system capability to reject noise and disturbances, thus defining the shape of closed loop system bandwidth. The matrix Q in Eq. (14) is used to specify tracking performance requirements of output variables. Finally, parameters $\gamma 1$, $\gamma 2$ and $\gamma 3$ are used to regulate the adaptive capability of control gains. Large values imply quick adaptivity and vice versa.

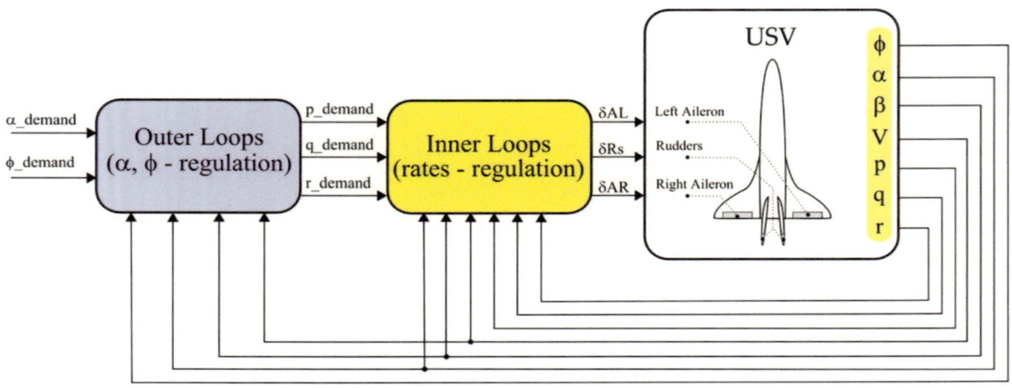

Fig. 6. The general scheme of control system architecture

Control architecture depicted in Fig. 6 is made of two MIMO control loops. The inner one is referred to the rates (p, q, r) regulation, while the outer one is used to control both angle of

attack (α) and roll angle (ϕ). Either MIMO controllers are designed with Adaptive Model Following (AMF) control technique above described.
In the following table a brief description of which variables have been used for the design of both inner and outer loops is given.

	y	x	r	u
inner loops	p, q, r	$v_{TAS}, \alpha, \beta, p, q, r$	$p_{DEM}, q_{DEM}, r_{DEM}$	$\delta_{AL}, \delta_{AR}, \delta_R$
outer loops	α, ϕ	$v_{TAS}, \alpha, \beta, \phi$	α_{DEM}, ϕ_{DEM}	$p_{DEM}, q_{DEM}, r_{DEM}$

Table 1. Controller Variables

4. Probabilistic robust controller

Within the stochastic paradigm for control system design/analysis, the meaning of robustness is moved from its well known deterministic sense to a probabilistic one. Indeed, it is claimed that a certain property of a control system is "almost" robustly satisfied, if it holds for "most" of the occurrences of uncertain variables. In other words, a risk that this property is violated by a set of uncertainties with small probability measure is considered acceptable.
Nevertheless, from a computational point of view, assessing probabilistic robustness of a given property may be more difficult than the deterministic case, since it requires the computation of a multi-dimensional probability integral. This problem is overcome by means of randomized algorithms which estimate performance probability by randomly sampling the uncertainty space, and computing bounds on the estimation error. Since estimated probability is itself a random quantity, this method always entails a certain risk of failure, i.e. there exists a nonzero probability of making an erroneous estimation. These algorithms have low complexity and are associated to robustness bounds which are less conservative than classical ones, obviously at the expense of a non deterministic result.
Randomization can be effectively used for control synthesis, by means of two different approaches. The first one aims at designing controllers that satisfy a given performance specification for most values of uncertainties, i.e. that are robust in a probabilistic sense, while the second one aims at finding a controller that maximizes the mean value of performance index, thus in the latter case the objective is to obtain a controlled system that guarantees the best performance on average (Tempo et al., 2005). For what concerns the use of this technique for DTFT1 benchmark application, the second approach has been used.
The approach used for controller synthesis was to look for a controller that (probabilistically) minimizes the mean value of the performance index, thus the objective was to obtain a controlled system that guarantees the best performance *on average*.
Performance function for the uncertain system is first defined:

$$u(\Delta) : \Delta \rightarrow \Re \tag{15}$$

the above function gives a measure of system performance for a given value of uncertainty Δ. In this application the function u is the following Boolean function which represents the "failure" of a given controller, that is,

$$u(\Delta) = \begin{cases} 1 & \text{if a given system property is not satisifed} \\ 0 & \text{otherwise} \end{cases} \tag{16}$$

Controller C_{opt} will be the one guaranteeing that the expected value of performance function u(Δ,C) is minimized:

$$C_{prob} = \arg \min_{c \in \mathbf{C}} E\left[\mathrm{u}(\Delta, C) \right] \tag{17}$$

An approximate solution can be obtained by means of randomized algorithms which are based on sampling both uncertainty set Δ and controller set C. To this end, two separate problem need to be solved: first, an estimate of the expected value is computed, then this estimate is minimized.

Computation of $E\left[\mathrm{u}(\Delta,C)\right]$ is carried out through randomization, that is, M independent, identically distributed (i.i.d.) controllers C^1, C^2, ... ,$C^M \in$ C are extracted according to their probability density function $f_C(C)$; an estimation of the minimum value is then given by:

$$\min_{i=1,2,\ldots,M} E\left[\mathrm{u}\left(\Delta, C^i\right) \right] \tag{18}$$

It is possible to demonstrate (Tempo et al., 2005) that, given $\varepsilon_1 \in [0,1]$, and $\delta \in [0,1]$, if

$$M \geq \frac{\log \frac{1}{\delta}}{\log \frac{1}{1-\varepsilon_1}} \tag{19}$$

then:

$$\mathrm{Prob}\left\{ \mathrm{Prob}\left\{ E\left[\mathrm{u}(\Delta,C)\right] \leq \min_{i=1,2,\ldots,M} E\left[\mathrm{u}\left(\Delta,C^i\right)\right] \right\} \leq \varepsilon_1 \right\} \geq 1-\delta \tag{20}$$

As it can be noticed, computation of $E\left[\mathrm{u}\left(\Delta,C^i\right)\right]$ for each C^i requires the execution of a multidimensional integral, that is very difficult in general; also this problem is solved by randomization approach.

For what concerns estimation of the expected value $E\left[\mathrm{u}\left(\Delta,C^i\right)\right]$ of performance function, N i.i.d. samples Δ^1, Δ^2, ... Δ^N are extracted from Δ, according to their density function f_Δ; performance functions $\mathrm{u}(\Delta^1,C^1)$, ... , $\mathrm{u}(\Delta^1,C^i)$ are then computed for a fixed controller $C^i \in \mathbf{C}$, and an estimation $\hat{E}_N\left[\mathrm{u}\left(\Delta,C^i\right)\right]$ of the expected value $E\left[\mathrm{u}\left(\Delta,C^i\right)\right]$ is given by:

$$\hat{E}_N\left[\mathrm{u}\left(\Delta,C^i\right)\right] = \frac{1}{N}\sum_{k=1}^{N} \mathrm{u}\left(\Delta^k,C^i\right) \tag{21}$$

It can be demonstrated (Tempo et al., 2005) that, if:

$$N \geq \frac{\log \frac{2}{\delta}}{2\varepsilon_2^2} \tag{22}$$

then:

$$\text{Prob}\left\{\left|E\left[\text{u}\left(\Delta,C^i\right)\right]-\hat{E}_N\left[\text{u}\left(\Delta,C^i\right)\right]\right|\leq\varepsilon\right\}\geq 1-\delta \tag{23}$$

In order to compute a probabilistic controller, equations (20) and (23) must be put together. To this end, it can be shown that, for any ε_1, $\varepsilon_2 \in [0,1]$ and $\delta \in [0,1]$, if

$$M \geq \frac{\log\frac{1}{\delta}}{\log\frac{1}{1-\varepsilon_1}} \quad \text{and} \quad N \geq \frac{\log\frac{2M}{\delta}}{2\varepsilon_2^2} \tag{24}$$

then

$$\text{Prob}\left\{\text{Prob}\left\{E\left[\text{u}(\Delta,C)\right]\leq \min_{i=1,2,\ldots,M}\hat{E}_N\left[\text{u}\left(\Delta,C^i\right)\right]-\varepsilon_2\right\}\leq\varepsilon_1\right\}\geq 1-\frac{\delta}{2} \tag{25}$$

The randomized probabilistic controller is given by:

$$\hat{C}_{NM}=\arg\min_{i=1,\ldots,M}\frac{1}{N}\sum_{k=1}^{N}u\left(\Delta^k,C^i\right) \tag{26}$$

Eq. (25) states that the estimated minimum $\min_{i=1,2,\ldots,M}\hat{E}_N\left[\text{u}\left(\Delta,C^i\right)\right]$ is "close" to the actual one $E\left[\text{u}\left(\Delta,C_{prob}\right)\right]$ within ε_2 in terms of probability, and this is guaranteed with an accuracy ε□ and a confidence level at least $\delta/2$.

For what concerns the implementation of the above technique to the benchmark application, a fixed control system architecture (inherited from GNC system of DTFT1) has been chosen and its parameters have been optimized according to a stochastic technique. Since controller gains are scheduled with dynamic pressure, controller design have been carried out by optimizing scheduling parameters through stochastic synthesis.

In particular, once the controller structure is defined, stochastic optimization allows selecting the optimum controller parameters also accounting for all the uncertain parameters, mainly vehicle and environment ones. For each candidate vector of control parameters, success rate is computed according to a pre-specified figure of merit and the applied uncertainties. To this end, a first test is performed considering nominal conditions, i.e. no uncertainties applied. If the considered controller passes the test in nominal conditions, uncertainty region is sampled and, for each uncertainty sample, the nonlinear test is repeated and success rate is computed. For this application, the following test success criteria have been used:

- No instability (identified as commands oscillation along the trajectory) occurring during the trajectory.
- No Out-of-Range commands deflection (Max. ± 25° for the elevons, ± 20° for the rudder);
- Satisfactory tracking performances for tracked variables (1° RMS in α and β, 3° RMS in ϕ);
- Valid aerodynamic data during the trajectory ($\alpha \in [-5 \div 18°]$, $\beta \in [-8 \div 8°]$).

5. Numerical analysis

In this section the results of a numerical analysis carried out in order to compare performance and robustness of the developed control systems will be presented. The activity has been carried out in the framework of Unmanned Space Vehicle (USV) program of Italian Aerospace Research Centre (CIRA), which is in charge of developing unmanned space Flying Test Beds (FTB) to test advanced technologies during flight. The first USV Dropped Transonic Flight Test (named DTFT1) was accomplished in February 2007 with the first vehicle configuration of USV program (named FTB1) (Russo et al., 2007). In this mission, a conventional control architecture was implemented. DTFT1 was then used as a benchmark application for comparison among different advanced control techniques. This comparison aimed at choosing the most suited control technique to be used for subsequent dropped flight test, named DTFT2, carried out on April 2010.
In the figure below the trajectory of DTFT1 mission in the plane Mach-altitude is depicted.

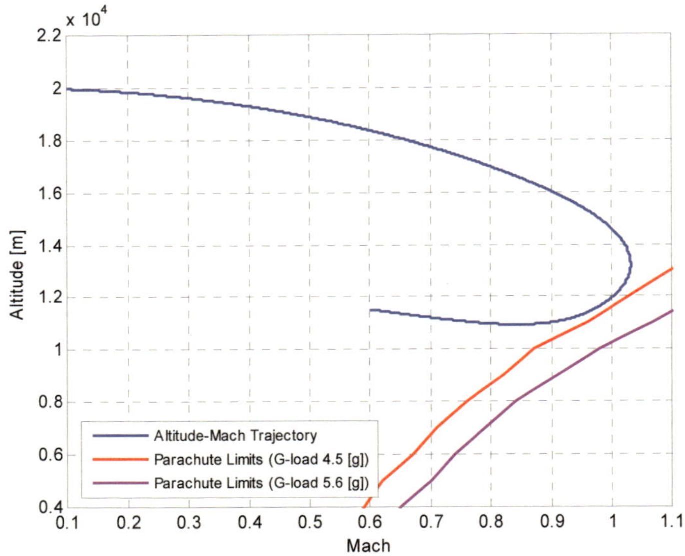

Fig. 7. Mach-Altitude Trajectory

For what concerns control performances, they are specified in the next table and they are valid only for Mach>0.7 (transonic regime):

Variable Tracked	Tracked Value [deg]	RMS Accuracy [deg]
Angle of Attack	7	1
Angle of Sideslip	0	1
Roll angle	0	3

Table 2. Tracked Variables

In order to compare the advanced controllers described in previous sections, the following scenarios have been identified:

- robustness to parametric uncertainties, namely aerodynamic ones
- robustness to initial state displacement
- robustness to navigation errors (sensor noise)
- robustness to actuator failures

Within this framework, control laws robustness is the capability of guaranteeing performances and stability of control systems in presence of above uncertainties.

Numerical evaluation reported in this section have been carried out using a complete 6 Dof model of FTB1 vehicle together with Atmospheric Model, wind model, Hydraulic Actuator System, Air Data System, Inertial Measurement System and control laws.

Parametric uncertainties have been accounted for by considering a particular aerodynamic configuration (hereafter called the worst configuration) which was identified as the aerodynamic uncertainty configuration leading to worst dynamic behaviour of FTB1 vehicle, in terms of stability, damping and control derivatives that mainly affect stability properties.

Two test cases, named C_0, and $\hat{C}_0$, have been accomplished with both nominal conditions (nominal initial state, zero navigation errors, no failure, etc.) and worst aerodynamic configuration respectively.

For what concerns robustness to initial state displacement, several off-nominal conditions in terms of Euler angles and angular rates have been considered (see the following table). Nominal DTFT1 initial state in terms of attitude, heading and angular rates is: $\phi_0 = 0$ deg, $\theta_0 = -90$ deg, $\psi = 0$ deg, $p = 0$ deg/s, $q = 0$ deg/s, $r = 0$ deg/s.

Initial State Displacement	ϕ_0 [deg]	θ_0 [deg]	ψ_0 [deg]	p_0 [deg/s]	q_0 [deg/s]	r_0 [deg/s]
I_1	-20	-89.9	0	-5	3	-3
I_2	20	-89.9	0	2	2	-2
I_3	20	-89.9	0	5	2	1
I_4	0	-85	0	-5	2	0

Table 3. Initial State Displacement

Furthermore for each case of the above table, nominal (case $C_1 - C_4$) and worst aerodynamic configuration (case $C_5 - C_8$) have been simulated.

For what concerns navigation errors, simulations have been performed with both nominal (case C_9) and worst aerodynamic configuration (case C_{10}) without any initial state have displacement.

As far as robustness to actuator failures is concerned, a rudder failure occurring after 30 s from vehicle's drop has been simulated, in particular a jam of the right rudder. It is worth noting that in this case (C_{11}) both initial state and aerodynamic configuration are nominal. All the benchmark scenarios are summarized in the following table.

Case	Aerodynamic Configuration	Initial State	Navigation Errors	Actuator Failure
C_0	Nominal	Nominal	No errors	No failure
$\hat{C}_0$	Worst-Aero 1	Nominal	No errors	No failure
C_1	Nominal	Off-nominal I_1	No errors	No failure
C_2	Nominal	Off-nominal I_2	No errors	No failure
C_3	Nominal	Off-nominal I_3	No errors	No failure
C_4	Nominal	Off-nominal I_4	No errors	No failure
C_5	Worst-Aero	Off-nominal I_1	No errors	No failure
C_6	Worst-Aero	Off-nominal I_2	No errors	No failure
C_7	Worst-Aero	Off-nominal I_3	No errors	No failure
C_8	Worst-Aero	Off-nominal I_4	No errors	No failure
C_9	Nominal	Nominal	Errors	No failure
C_{10}	Worst-Aero	Nominal	Errors	No failure
C_{11}	Nominal	Nominal	No errors	Right Rudder jamming at 30 s

Table 4. Benchmark Scenarios for Perfomance Evaluation

For what concerns robust performance indicators, tracking accuracy of trajectory has been defined as a performance parameter. Performance requirements are given in the table below:

Variable Tracked	RMS Tracking error [deg]
Angle of Attack	±1
Angle of Sideslip	±1
Roll angle	±3

Table 5. Tracking Performance requirements

In the next figures the results of scenarios C_0 and $\hat{C}_0$ are reported.

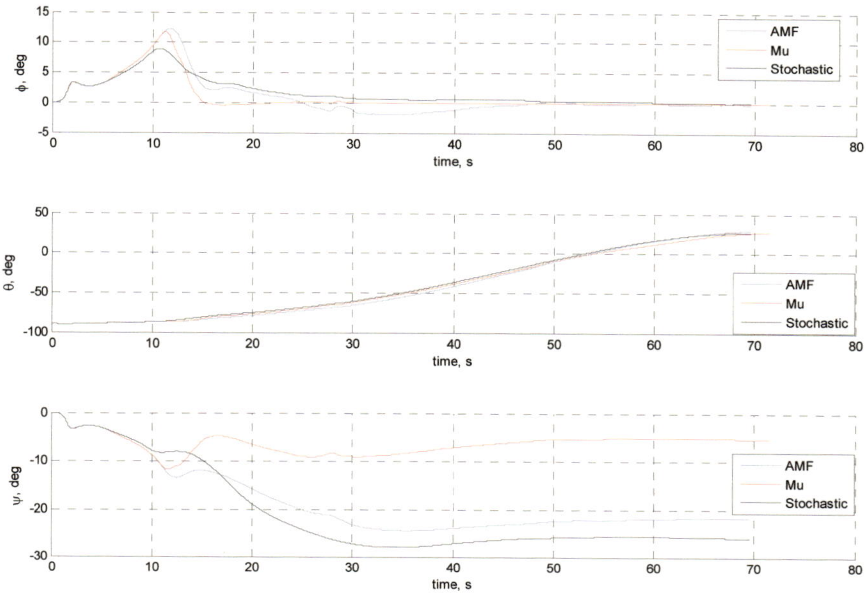

Fig. 8. Case C_0 -Euler Angles

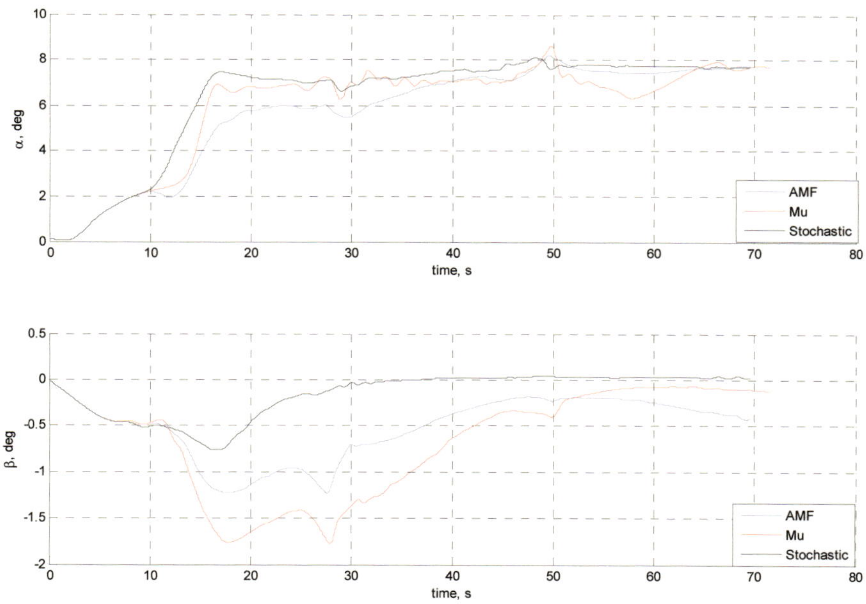

Fig. 9. Case C_0 -Incidence Angles

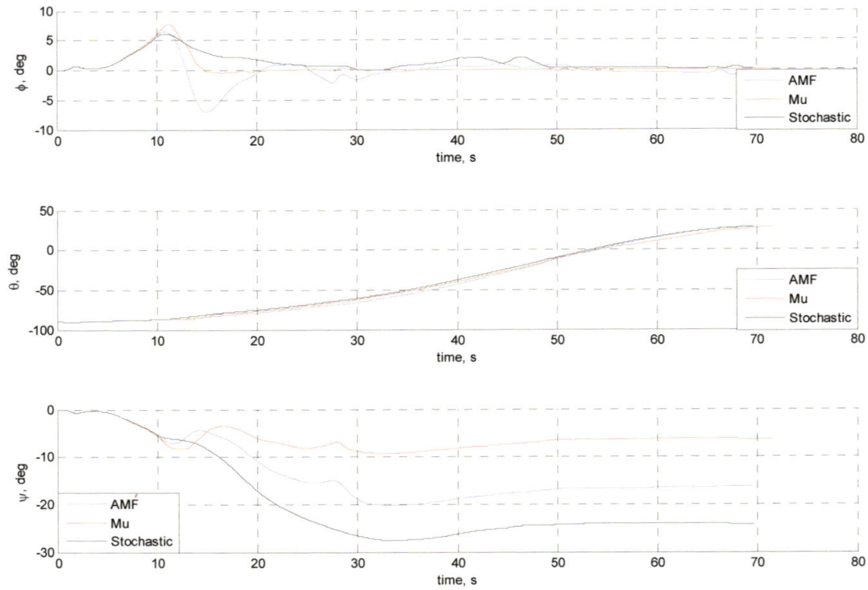

Fig. 10. Case $\hat{C}_0$ -Euler Angles

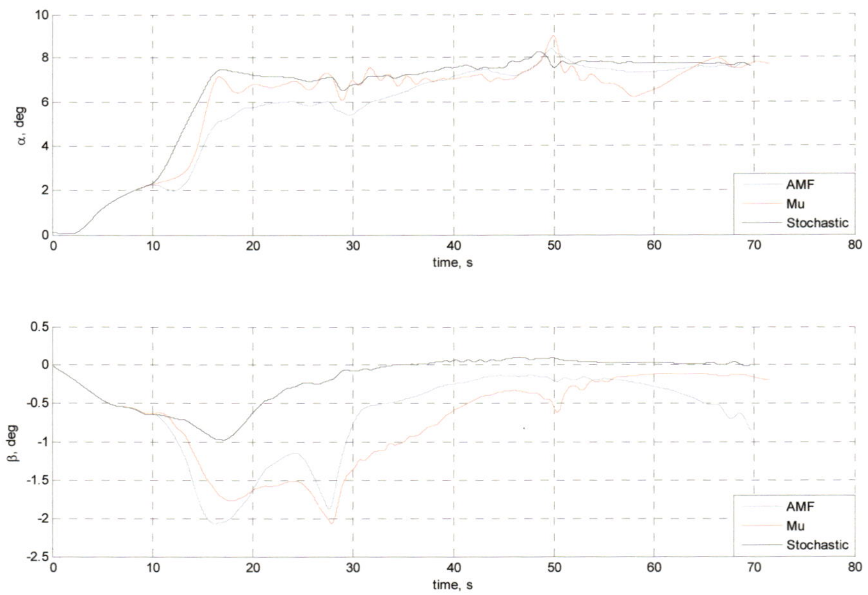

Fig. 11. Case $\hat{C}_0$ -Incidence Angles

As it can be seen from the figures, performances of all control algorithms are globally satisfactory. It is worth noting that Mu-controller shows a light unstable behaviour on sideslip angle around 50 s in worst case configuration, but it is however rapidly damped in few seconds. For what concerns uncertainties to initial state displacement, several cases have been considered with different attitude and angular velocity at vehicle drop. In the following figures, for sake of brevity, only the cases C_5 and C_8 are reported. They refer to initial state conditions I_1 and I_4 with worst aerodynamic configuration (see Table 4).

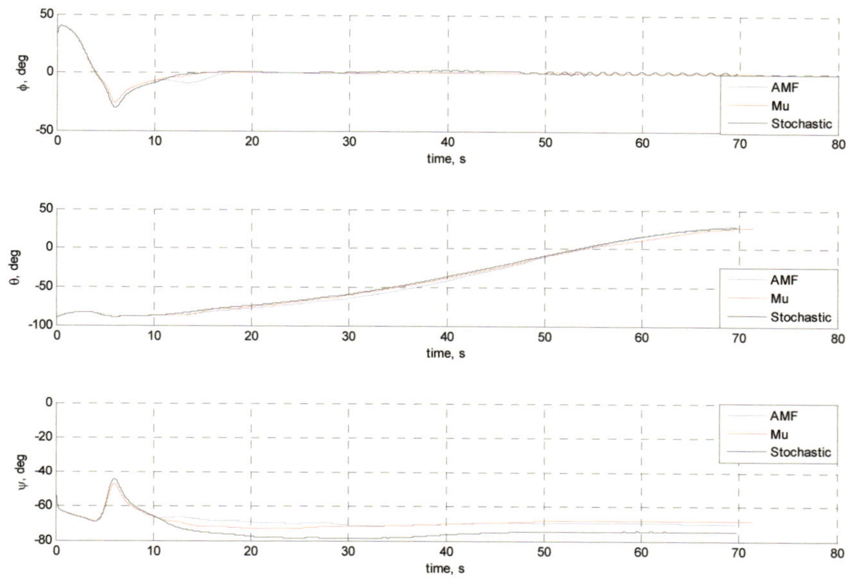

Fig. 12. Case C_5 -Euler Angles

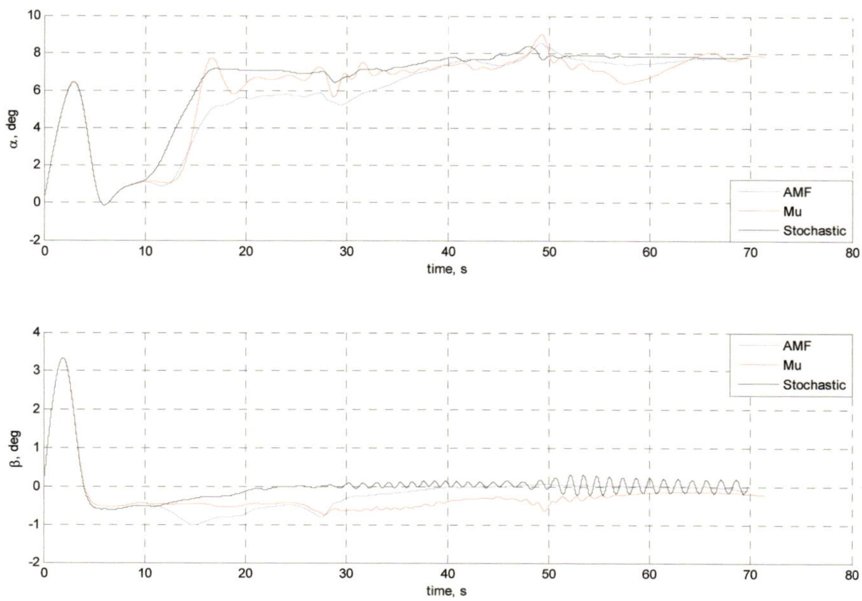

Fig. 13. Case C_5 -Incidence Angles

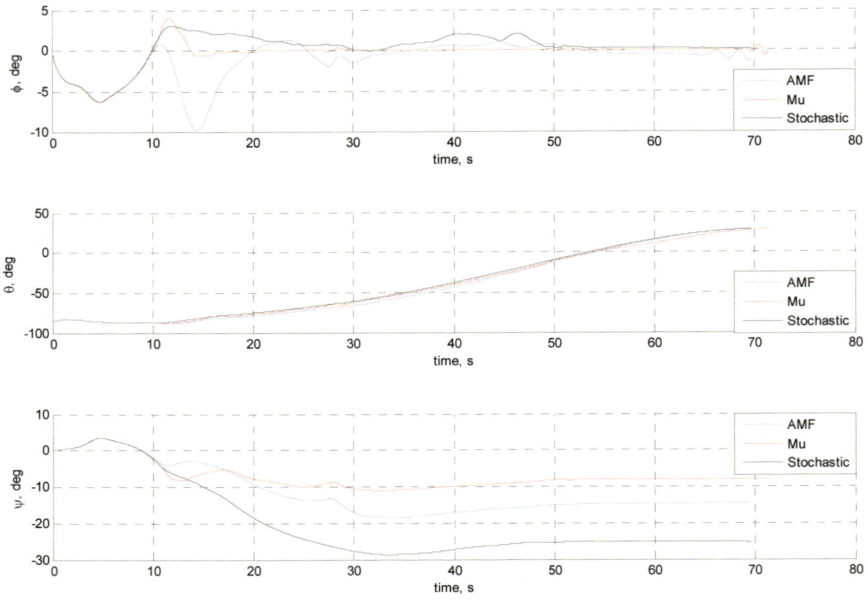

Fig. 14. Case C_8 -Euler Angles

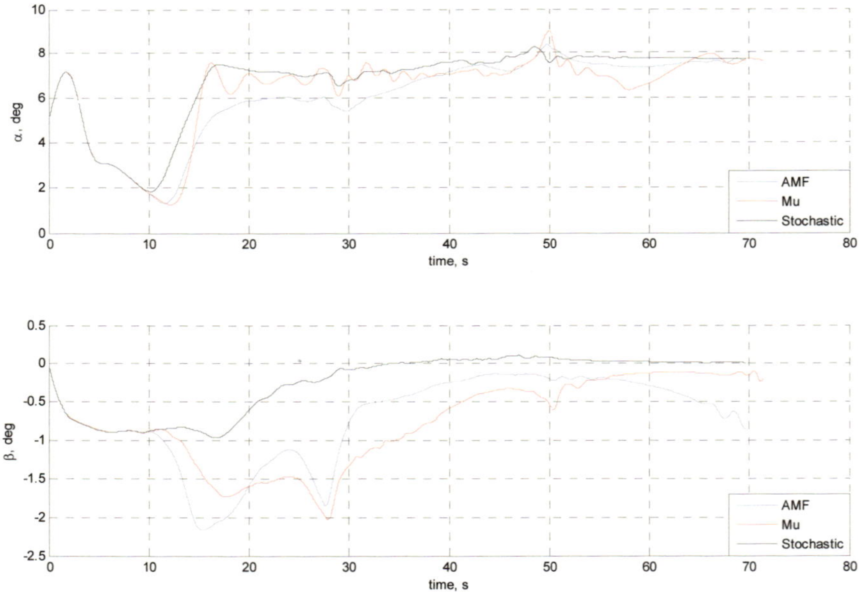

Fig. 15. Case C_8 -Incidence Angles

All controllers satisfactory work in presence of initial state displacement. In spite of a light oscillatory mode on sideslip and roll angles, stochastic controller guarantees very good performances for what concerns tracking of sideslip angle and angle of attack.

As mentioned earlier, in order to evaluate control algorithms capabilities to face disturbances such as navigation errors, two test cases have been considered, i.e. nominal and worst aerodynamic configuration (cases C_9 and C_{10}) without any initial state displacement. The comparison between controllers is reported in the following figure, only with reference to the case C_{10} for sake of simplicity.

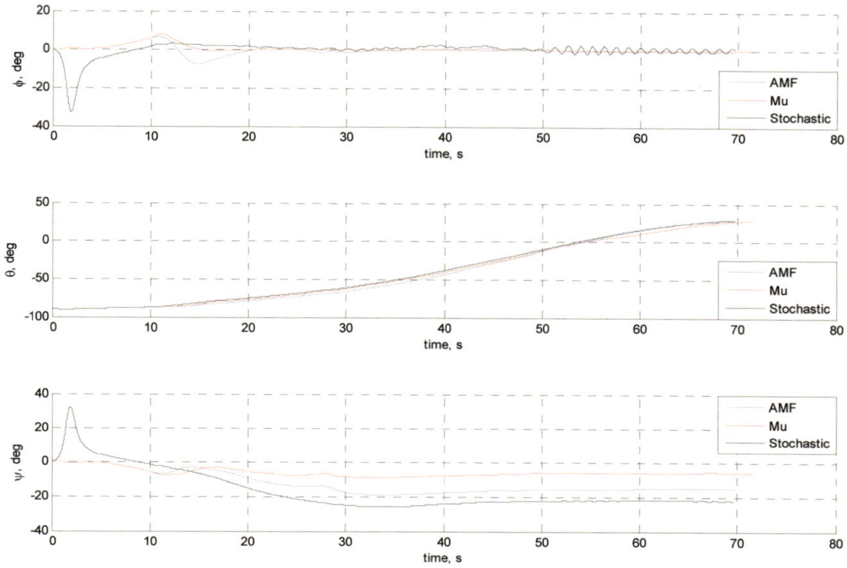

Fig. 16. Case C_{10} -Euler Angles

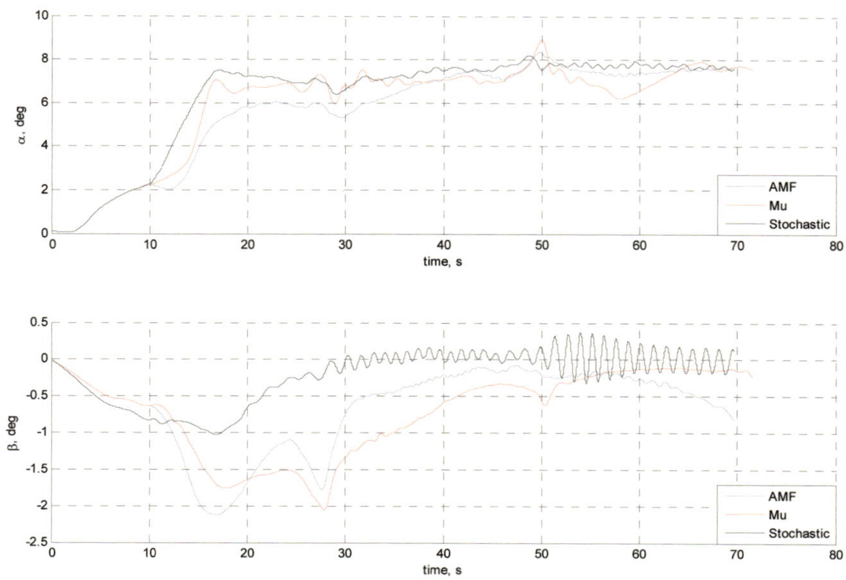

Fig. 17. Case C_{10} -Incidence Angles

Simulations show an acceptable robustness to sensor noise. A small effect on incidence angles, in terms of reduced damping, is shown by stochastic controller.
Finally algorithms robustness to an actuator failure has been evaluated. In particular a rudder jamming at t=30 s has been simulated.

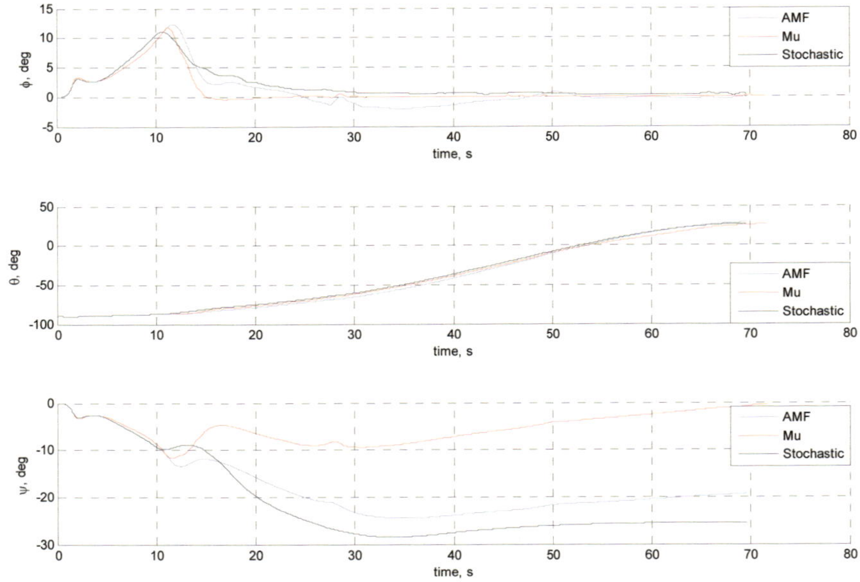

Fig. 18. Case C_{11} -Euler Angles

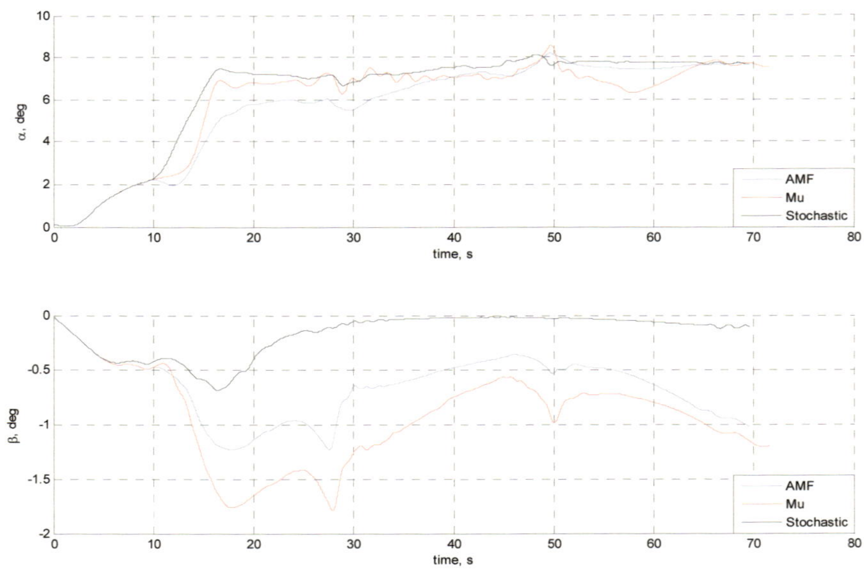

Fig. 19. Case C_{11} -Incidence Angles

The above figures show that rudder jam failure is well tolerated by all the controllers.
The following table summarizes the performances achieved by the controllers for all considered scenarios.

Case	**Adaptive Controller** AoA-AoS-Phi accuracy [deg]	**MU-Controller** AoA-AoS-Phi accuracy [deg]	**Stochastic Controller** AoA-AoS-Phi accuracy [deg]
C_0	0.87676 0.5969 0.72448	0.57129 0.88441 0.053638	0.21606 0.13512 0.61134
$\hat{C}_0$	0.90016 0.74635 0.59708	0.62284 0.91081 0.067061	0.24501 0.18806 0.8251
C_1	0.88835 0.26603 0.45265	0.5642 0.63046 0.061391	0.21624 0.066145 0.42644
C_2	0.88043 0.58375 0.75665	0.57188 0.97939 0.089979	0.21585 0.14444 0.64976
C_3	0.87282 0.62874 0.71016	0.57878 0.83847 0.038177	0.21536 0.13685 0.63915
C_4	0.86873 0.593 0.71128	0.53892 0.87025 0.05387	0.21588 0.13086 0.59952
C_5	0.91649 0.29559 0.35717	0.62481 0.40438 0.14435	0.2452 0.1115 1.1254
C_6	0.91255 0.75401 0.69885	0.62265 1.0448 0.13026	0.24415 0.20172 0.82289
C_7	0.89565 0.73756 0.66584	0.6301 0.84141 0.044825	0.24409 0.18635 0.76552
C_8	0.8919 0.74122 0.59849	0.59116 0.89455 0.10033	0.24443 0.18392 0.82521
C_9	0.87611 0.60085 0.7007	0.57183 0.88728 0.053653	0.22299 0.15436 0.61589
C_{10}	0.90059 0.74182 0.57556	0.62011 0.91444 0.061636	0.25904 0.23867 1.2075
C_{11}	0.8759 0.72793 0.80693	0.57115 1.0664 0.066715	0.20854 0.12598 0.65093

Table 6. Controller Perfromances

The comparison between the controllers has shown that:

1. Stochastic controller guarantees the best performances for what concerns the tracking of AoA and AoS and it always meets performance requirements. Anyway, it presents some light oscillations on lateral and directional dynamics in worst aerodynamic configuration.
2. μ-controller guarantees the best tracking performance in terms of roll angle, much better than required (3 deg error), but in some cases it fails to meet performance requirements on AoA and AoS (see C_6 and C_{11}).
3. Adaptive Controller guarantees almost good tracking performances even though they are never better than the other two controllers. In any case, performance requirements are always met.

Based on the above considerations, the controller obtained by means of stochastic synthesis was considered the most suited for the DTFT2 scopes, so it was selected (after a fine tuning) as a part of an advanced GNC system for DTFT2 mission, successfully carried out on April 2010.

In fact, despite the limitation of using an 'a priori' fixed control structure, control laws obtained though the stochastic synthesis have the following good features:

- excellent performances and good stability properties in spite of large uncertainties affecting the system;
- simple control structures while guaranteeing robust performances and stability as well as low computational effort and implementation simplicity.

6. Conclusions

Over the last decades many efforts have been spent to develop advanced control techniques for aerospace applications, aimed at overcoming limitations of commonly used control techniques, mainly lack of robustness against various uncertainties affecting the system to be controlled. The importance of a robust control system is readily understood when space reentry applications are considered. Indeed, these applications have some distinctive features, mainly related to the lack of stationary equilibrium conditions along the trajectories, to the wide flight envelope characterizing the missions (from hypersonic flight regime to subsonic one) and to the high level of uncertainty in the knowledge of vehicle aerodynamic parameters.

The development of an advanced control system having robustness capabilities is one of the goals of research activities carried out by Italian Aerospace Research Centre in the framework of USV program. In order to select a control strategy having the advantages above discussed, three candidate control techniques have been compared with the aim of selecting the most suited one for the second dropped flight test of USV program, named DTFT2, successfully carried out on April 2010. The three techniques are:

- μ-Controller with Fuzzy Logic Gain-Scheduling
- Direct Adaptive Model Following Control
- Probabilistic Robust Control Synthesis

In order to evaluate the robustness capabilities of proposed control algorithms, a numerical robustness analysis has been performed. Performances and stability of candidate control techniques have been evaluated in presence of several sources of uncertainties (aerodynamics, initial state, etc.) and failure scenarios.

Robustness analysis showed that all three techniques are well suited to accomplish robust control in USV DTFT1 mission, in presence of large parameters uncertainty (the vehicle mostly flies in transonic regime, where accurate aerodynamic prediction is very difficult to obtain).

Nevertheless the controller obtained by means of stochastic synthesis was selected as a part of on-board advanced GNC system for DTFT2 mission, due to its good performances and relatively simple implementation.

7. References

Ashari E. A. et al. (2005), Reconfigurable control system design using eigenstructure assignment: static, dynamic and robust approaches; International Journal of Control, vol. 78, No. 13, 10 September 2005, 1005-1016

Balas Gary J. (2005), Application of Robust Multivariable Control to Stability, Control Augmentation and Trajectory Tracking of Unmanned Space and Aerial Vehicle, CIRA Short course, 14-18 November 2005

Bharadwaj, S. Rao, Anil V., and Mease, Kenneth D. (1998), Entry Trajectory Tracking Law via Feedback Linearization, Journal of Guidance, Control, and Dynamics, Vol. 21, No. 5, September–October 1998

Bodson M., Groszkiewicz J.E. (1997), Multivariable Adaptive Algorithms for Reconfigurable Flight Control; IEEE Transactions on Control Systems Technology, vol. 5, No. 2, March 1997.

Calafiore G., Dabbene F., Tempo R. (2007), A survey of randomized algorithms for control synthesis and performance verification, Journal of Complexity 23 (2007), pp. 301-316

Hyde, A., Glover K. (1993), The Application of Scheduled H∞ Controllers to a VSTOL Aircraft, IEEE Transaction on Automatic Control, VOL. 38, NO. 7, July 1993. 1021-1039

Kim K., Lee K., Kim Y. (2003), Reconfigurable Flight Control System Design Using Direct Adaptive Method, Journal of Guidance, Control, and Dynamics, Vol. 26, N° 4, July-August 2003

Konstantopoulos K., Antsaklis P. (1996), Eigenstructure Assignment in Reconfigurable Control Systems, Technical Report of the ISIS group at the University of Notre Dame, ISIS-96-001, January, 1996

Liu, G. P. and Patton, R. J. (1996), Robust Parametric Eigenstructure Assignment, AIAA-96-3908

Malloy, D., and Chang, B.C. (1998), Stabilizing Controller Design for Linear Parameter-Varying Systems Using Parameter Feedback, Journal of Guidance, Control, and Dynamics, Vol. 21, No. 6, November–December 1998

Marcos, A., and Balas, Gary J. (2004), Development of Linear Parameter Varying Models for Aircraft, Journal of Guidance, Control, and Dynamics, Vol. 27, No. 2, March 2004

Nichols, A., Reichert, R. T., W. J. Rugh (1993), Gain Scheduling for H-infinity Controllers: A Flight Control Example, IEEE Transactions on Control Systems Technology, Vol. 1, No. 2, June 1993.

Pedrycz, W., Peters J. F. (1997), Hierarchical Fuzzy Controllers: Fuzzy Gain Scheduling, IEEE Intern. Conf. on SMC[C], 1997.1139-1143

Rollins L. (1999), Robust Control Theory, Carnegie Mellon University, 18-849b Dependable Embedded Systems, Spring 1999

Russo, G., et al. (2007), Unmanned Space Vehicle Program: DTFT in Flight Experiments, 18th ESA Symposium on European Rocket and Balloon Programmes and Related Research, Visby, Sweden, 2007.

Spillman, Mark S. (2000), Robust Longitudinal Flight Control Design Using Linear Parameter-Varying Feedback, Journal of Guidance, Control, and Dynamics, Vol. 23, No. 1, January–February 2000

Tempo R. and Dabbene F. (1999), Probabilistic Robustness Analysis and Design of Uncertain Systems, in Dynamical Systems Control, Coding, Computer Vision - New Trends, Interfaces, and Interplay, ed. by Picci G. and Gilliam D.S., Birkhauser Verlag, 1999

Tempo, R., Calafiore, G., Dabbene, F. (2005), Randomized Algorithms for Analysis and Control of Uncertain Systems ; New York: Springer-Verlag, 2005.

Van Soest, W. R., Chu, Q. P. and Mulder, J. A. (2006), Combined Feedback Linearization and Constrained Model Predictive Control for Entry Flight, Journal of Guidance, Control, and Dynamics, Vol. 29, No. 2, March–April 2006

Vincent, James H., Abbas Emami-Naeinjt and Nasser M. Khraishi (1994), Case Study Comparison of Linear Quadratic Regulator and H∞ Control Synthesis, Journal Of Guidance, Control, And Dynamics, Vol. 17, No. 5, September-October 1994

Xing, L. (2003), Comparison of Pole Assignment & LQR Design Methods for Multivariable Control for STATCOM, Thesis submitted to the Department of Mechanical Engineering, Florida State University

Zhou , K., and Doyle, John C. (1998), Essential Of Robust Control, Prentice Hal, 1998

3

Fault Tolerant Depth Control of the MARES AUV

Bruno Ferreira, Aníbal Matos and Nuno Cruz
INESC Porto and University of Porto
Portugal

1. Introduction

Control theory has been applied to several domains where practical considerations are relevant. Robotics is a notable example of this. In most cases, mobile robotic systems are governed such that their behavior obeys to a defined motion. However, during their operations, it is conceivable that faults could occur. Indeed, this assumption has to be made in order to predict a possible malfunction and to take an appropriate action according to the fault, improving the robustness and the reliability of the system. This work tackles the problem of fault detection, identification and automatic reconfiguration of an autonomous underwater vehicle (AUV). Although our emphasis will be directed to an AUV, the methods and the tools that are employed in this chapter can be easily extended to other engineering problems beyond robotics.

In this work, we will consider the MARES (Modular Autonomous Robot for Environment Sampling) (Fig. 1) Cruz & Matos (2008); Matos & Cruz (2009), a small-sized (1.5 meters long), torpedo shaped AUV weighting 32 kg, able to move at constant velocities up to 2.5 m/s. Its four thrusters provide four degrees of freedom (DOF), namely surge, heave, pitch and yaw. One of its main particularities is the capability to dive independently of the forward motion. The vertical through-hull thrusters provide heave and pitch controllability, while the horizontal ones ensure the surge and the yaw DOFs. The heave and pitch DOFs make the vertical plane control redundant when the vehicle is moving with surge velocities different from zero. In other words, the vehicle remains controllable if only one of these two DOFs is available. Such characteristic will be explored along this chapter in which the control of the nonlinear dynamics of the AUV Ferreira, Matos, Cruz & Pinto (2010); Fossen (1994) constitutes a challenging problem.

By taking advantage of the distribution of the actuators on the vehicle, it is possible to decouple the horizontal and the vertical motion. A common approach in such systems is to consider reduced models in order to simplify the analysis and the derivation of the control law (see Ferreira, Matos, Cruz & Pinto (2010); Teixeira et al. (2010) or Fossen (1994), for example). In general, for topedo-shaped vehicles, coupling effects due to composed motions (e.g., simultaneous sway and heave motions) are clearly smaller than the self effects of decomposed motion (e.g., effect of the heave motion on the heave dynamics) and can therefore be considered disturbances in the reduced model in which they are not included. Thus, a reduced model will be considered to deal with the vertical motion taking surge, heave and pitch rate as state components.

In order to make the detection and identification of possible faults, we present a method based on process monitoring by estimating relevant state variables of the system. See Frank & Ding (1997) for an overview on several techniques andZhang & Jiang (2002) for an application to a particular linear system. Wu et al. (2000) have developed an algorithm based on the two-stage

Fig. 1. MARES starting a typical mission in the ocean

Kalman filter to estimate deviations from expected input actuation for a linear system. Their approach consists in estimating the *loss of control effectiveness factors* that are added as entries of the state estimate, while guaranteeing that the corresponding estimate covariance lies in a defined interval. By imposing boundaries on the corresponding eigenvalues, it is possible to avoid impetuous corrections or to be insensitive to measurements. Inspired on the work by Zhang & Jiang (2002) and Wu et al. (2000), the present paper describes the implementation of an augmented state extended Kalman filter (ASEKF) to estimate the effectiveness of the control commands, detect and identify the possible faults.
The present work focuses on the vertical motion considering faults on the vertical thrusters. The method for accomodation of the faults consists in three main steps: fault detection, fault diagnosis and decision. Fault detection is responsible for creating a warning whenever an abrupt or an incipient fault happens, while fault diagnosis distinguishes and identifies the fault. In the presence of faults, a decision must be taken, adopting a suitable control law to stabilize the vertical motion. In the presence of faults in one of the vertical thrusters, the heave motion will no longer be controllable. Consequently, a control law derived for normal operation could be inadequate or even turn the feedback system unstable when such a fault occurs. An algorithm has to be developed in order to make the behavior of the robotic system tolerant to faults.
Making use of the pitch angle controllability, we will derive two control laws to drive the vehicle to a depth reference, possibly time variant. To achieve so, we make use of the Lyapunov theory, adopting the backstepping method Khalil (2002). Nevertheless, the presence of biases in steady state shifts the error at equilibrium away from zero. Those biases are commonly induced by unmodeled, neglected effects or external disturbances whose values are hard to observe or to estimate. The introduction of an integral term, under some assumptions, would solve the problem allowing the error to converge to zero as time goes to infinity. Based on the conditional integrators, extended by Singh & Khalil (2005) to more general control framework beyond sliding mode control Seshagiri & Khalil (2005), we present a control law that makes it possible to achieve asymptotic regulation of the vehicle depth error

when operating with only one thruster. The method considers an integral component in the control law derived in the backstepping first step.
Beyond the derivation and the particularization of the mathematical tools used here, we demonstrate our approach by illustrating the work with real experiments, voluntarily inducing faults on the system, and analyze the behavior of the dynamic system under such conditions.
The organization of the chapter is as follow: the section 2 describes the MARES AUV and presents the main mathematical models under consideration here, namely the kinematic and the dynamic models. In section 3, we introduce the relevant concepts and formulate the ASEKF from a reduced model of the vertical dynamics and describe how to identify the fault. In section 4, we derive the control law to drive the depth error to zero making use of the tools described above. Finally, in 5, our solution is demonstrated through real experiments.

2. MARES

The MARES autonomous underwater vehicle was developed in 2006 at the Faculty of Engineering of the University of Porto (FEUP). Typical operations have been performed in the ocean and fresh water, collecting relevant data for surveys and environmental monitoring during several tens of missions to date. Its configuration was specially designed to dive vertically in the water column while its horizontal motion is controlled independently, resulting in truly decoupled vertical and horizontal motions. Such characteristic is particularly appreciated in missions where the operation area is restricted or precise positioning is required. Parallel to the missions to collect data, the MARES AUV has also been used as a testbed for intensive research being performed in several problems related to robotics, specially on localization and control. Thus, besides the typical applications, several missions have been conducted to test and to verify implemented algorithms.
Before presenting our method to detect a possible fault and to control the vehicle under such situation, let us first introduce the kinematics and dynamics concepts and equations. We assume an inertial earth-fixed frame $\{I\} = \{x_I, y_I, z_I\}$, where $x_I, y_I, z_I \in \mathbb{R}^3$ are orthonormal vectors (in the marine literature, they are often made coincident with north, east and down directions, respectively), and a body-fixed frame $\{B\} = \{x_B, y_B, z_B\}$, where $x_B, y_B, z_B \in \mathbb{R}^3$ are orthonormal vectors frequently refered to as surge, sway and heave directions, respectively (see Fig. 2). The absolute position and orientation of the vehicle is expressed in the inertial frame $\{I\}$ through the vector $\eta_c = [\eta_1, \eta_2]^T = [x, y, z, \phi, \theta, \psi]^T$, where $\eta_2 = [\phi, \theta, \psi]^T$ is the vector of euler angles with respect to x_I, y_I and z_I, and $[x, y, z]$ are the coordinate of the frame $\{B\}$ expressed in $\{I\}$. The vehicle's velocity, expressed in the body frame $\{B\}$, is given by $\nu_c = [u, v, w, p, q, r]^T$, where p, q and r are the angular velocities along x_B, y_B and z_B, respectively. The velocities in both referentials are related through the kinematic equation Fossen (1994)

$$\dot{\eta}_c = J(\eta_2)\nu_c, \tag{1}$$

where $\dot{\eta}_c$ denotes the time derivative of η and $J(\eta_2) = block\,diag[J_1, J_2]$ represents the rotation matrix, with $J_1, J_2 \in \mathbb{R}^{3\times3}$. Although this transformation is common in the literature to map vectors from a referential frame to another, it is not the only one. An alternative can be found in quaternions (see Zhang (1997), for an introduction and useful results), avoiding the singularity problems of the matrix J_2. However, in this chapter, we will assume that the values of the angles that make the matrix J_2 singular (and J, consequently) are not reached. Moreover, since the water currents present in the ocean and in the rivers do not influence the development of the present work, they will not be considered for simplicity.

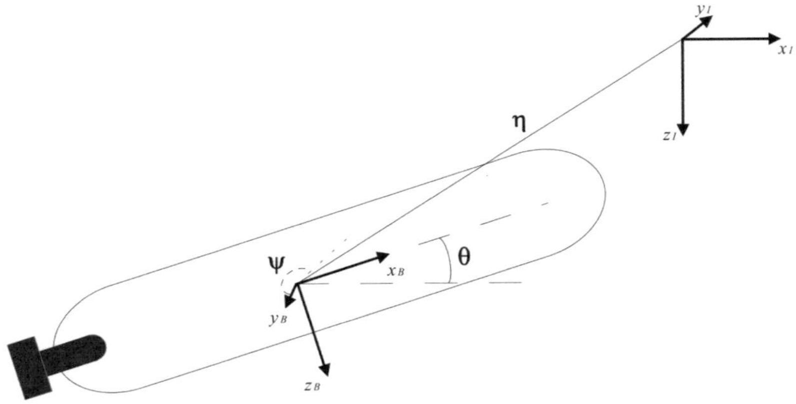

Fig. 2. Body frame and inertial frame referentials

As it is well known, rigid bodies moving in the three dimensional space are governed by nonlinear equations. For the particular case of underwater vehicles, such equations include the effect of the added mass, viscous damping, restoring and actuation forces and moments. Following the notation in Fossen (1994), the nonlinear second order, six dimensions equation is written as:

$$M_c\dot{\nu}_c = -C_c(\nu_c)\nu_c - D_c(\nu_c)\nu_c - g_c(\eta_c) + \tau_c \tag{2}$$

where $M_c \in \mathbb{R}^{6\times 6}$ is the sum of the body inertia and added mass matrices, $C_c \in \mathbb{R}^{6\times 6}$ results from the sum of the Coriolis and centrifugal terms from body inertia and added mass, $D_c \in \mathbb{R}^{6\times 6}$ is the viscous damping matrix, $g_c \in \mathbb{R}^6$ is the vector of restoring forces and moments and $\tau_c \in \mathbb{R}^6$ is the vector of actuation forces and moments.

Such system is complex and the task of deriving control laws that ensure stability is not trivial, having led to order reduction in several works (see Ferreira, Matos, Cruz & IEEE (2010); Fossen (1994); Teixeira et al. (2010), for example). By taking advantage of the body shape symmetries and of the configuration of the actuators on the body, it is usual to decouple the complex motion in more elementary ones. However, this has consequences since some cross-coupling terms are eliminated but their influence is often small and can be neglected or considered disturbances. Such approach has been implemented in the MARES AUV and the corresponding performances were already demonstrated in Ferreira, Matos, Cruz & Pinto (2010). The current thruster configuration on the MARES makes it possible to decouple the motion into the vertical and the horizontal plane. Since the roll angle is stable (and $\phi \approx 0$) two reduced order models are extracted. See Ferreira, Matos, Cruz & Pinto (2010), for further details.

3. Fault detection and identification

Under normal operation, the vertical thrusters of MARES provide the capability to control almost independently the pitch and heave degrees of freedom (DOF). In the same way, the horizontal thrusters make possible the control on the surge and yaw DOFs. As it will be exposed later, the vehicle remains controllable if one of the vertical thrusters fails. As an aside, note that the same is not verified if one the two horizontal thrusters fails since the surge and the yaw motion can no longer be decoupled. The derivation of the control laws is left to the next section.

Several approaches for fault identification and detection have been proposed, commonly based on observation of the residuals Frank & Ding (1997) either by using state observers or by using accurate models. Our approach makes use of an extended Kalman filter to detect the faults, indirectly exploiting the residuals to estimate actuation bias variables additionally incorporated in the state. A similar approach was already carried out by Zhang & Jiang (2002) where they used a two stage Kalman filter to identify the faults on actuation of a linear system. The method presented here makes it possible not only to detect faults but also to identify the faulty actuator. By taking advantage of the cyclic predictions and corrections, the main idea behind our approach is to estimate the biases on the actuation (or deviation from the nominal value, commonly referred to as loss of control effectiveness factor in Wu et al. (2000); Zhang & Jiang (2002) whose values should theoretically equal zero when no fault is occuring.

3.1 Vertical plane dynamics

The implementation of the extended Kalman filter assumes the use of a reasonably accurate dynamics model that recreates mathematically the behavior of the system for the prediction step. As it was pointed out in the previous section, the use of the complete model of the vehicle dynamics is complex and computationally expensive. Thus, from 2, we derive the reduced model for the vertical motion, considering that cross-terms are negligible:

$$M\dot{\nu} = -C(\nu)\nu - D(\nu)\nu - g(\eta_2) + P_f \mathbf{f_t}(t) \tag{3}$$

where $\nu = [u, w, q]^T$ and

$$M = \begin{bmatrix} m - X_{\dot{u}} & 0 & -X_{\dot{q}} \\ 0 & m - Z_{\dot{w}} & -Z_{\dot{q}} \\ -M_{\dot{u}} & -M_{\dot{w}} & -M_{\dot{q}} \end{bmatrix}, \; C(\nu) = \begin{bmatrix} 0 & -mq & -Z_{\dot{w}}w - Z_{\dot{q}}q \\ -mq & 0 & X_{\dot{u}}u + X_{\dot{q}}q \\ Z_{\dot{w}}w + Z_{\dot{q}}q & -X_{\dot{u}}u - X_{\dot{q}}q & 0 \end{bmatrix},$$

$$D(\nu) = -\begin{bmatrix} X_{|u|u}|u| & 0 & X_{|q|q}|q| \\ 0 & Z_{|w|w}|w| & Z_{|q|q}|q| \\ M_{|u|u}|u| & M_{|w|w}|w| & M_{|q|q}|q| \end{bmatrix}, \; g(\eta_2) = -\begin{bmatrix} (W - B)\sin\theta \\ (B - W)\cos\theta \\ -z_{CB}B\sin\theta \end{bmatrix},$$

$$P_f = \begin{bmatrix} 1 & 1 & 0 & 0 \\ 0 & 0 & 1 & 1 \\ 0 & 0 & x_{ts} & x_{tb} \end{bmatrix}, \; \mathbf{f_t}(t) = \begin{bmatrix} f_p(t) \\ f_r(t) \\ f_s(t) \\ f_b(t) \end{bmatrix}.$$

The authors recommend Fossen (1994); Triantafyllou & Hover (2002) for details about the parameters above and their derivation, and Hoerner (1993); White (2008) for further details. Notice that $\mathbf{f_t}(t)$ is the vector of forces applied by the thrusters that are generated according to a given control law, and f_p, f_r, f_s and f_b are scalars that represents the force applied by the port, starboard, stern and bow thrusters, respectively. We assume that such forces can be directly measured during operation. The inclusion of the surge velocity is required in this reduced order model due to the nonnegligible influence it has on the vertical plane dynamics. The parameters used in the reduced model are listed in the table 1.

3.2 Augmented state extended Kalman filter formulation

Our final goal in this section is to detect and to identify a fault occuring on one of the vertical thrusters. To this end, one aims to quantify the loss of control effectiveness of the referred actuators: The effective force applied by the vertical thrusters may differ from the commanded one. We will consider that f_s and f_b are the commanded forces, which may not correspond

Parameter	Value	units	Description
m	$3.20 \cdot 10^{1}$	kg	Vehicle's mass
W	$3.14 \cdot 10^{2}$	N	Vehicle's weight
B	$3.16 \cdot 10^{2}$	N	Vehicle's bouyancy
z_{CB}	$-4.40 \cdot 10^{-3}$	m	z_B of CB w.r.t CG
$X_{\dot{u}}$	$-1.74 \cdot 10^{0}$	kg	Added mass longitudinal term
$X_{\dot{q}}$	$-3.05 \cdot 10^{-2}$	$kg \cdot m$	Added mass cross-term
$Z_{\dot{w}}$	$-4.12 \cdot 10^{1}$	kg	Added mass heave term
$Z_{\dot{q}}$	$-1.23 \cdot 10^{-1}$	$kg \cdot m$	Added mass cross-term
$M_{\dot{u}}$	$-3.05 \cdot 10^{-2}$	$kg \cdot m$	Added mass cross-term
$M_{\dot{w}}$	$-1.23 \cdot 10^{-1}$	$kg \cdot m$	Added mass cross-term
$M_{\dot{q}}$	$-6.07 \cdot 10^{0}$	$kg \cdot m^2$	Added mass pitch term
$X_{\lvert u \rvert u}$	$-1.04 \cdot 10^{1}$	$kg \cdot m^{-1}$	Drag longitudinal term
$X_{\lvert q \rvert q}$	$4.84 \cdot 10^{-2}$	$kg \cdot m$	Drag cross term
$Z_{\lvert w \rvert w}$	$-1.16 \cdot 10^{2}$	$kg \cdot m^{-1}$	Drag heave term
$Z_{\lvert q \rvert q}$	$-5.95 \cdot 10^{0}$	$kg \cdot m$	Drag cross-term
$M_{\lvert u \rvert u}$	$-2.11 \cdot 10^{-1}$	kg	Drag cross-term
$M_{\lvert w \rvert w}$	$-8.26 \cdot 10^{0}$	kg	Drag cross-term
$M_{\lvert q \rvert q}$	$-1.56 \cdot 10^{1}$	$kg \cdot m^2$	Drag pitch term
x_{ts}	$-3.21 \cdot 10^{-1}$	m	x_B of stern vertical thruster w.r.t CG
x_{tb}	$5.34 \cdot 10^{-1}$	m	x_B of bow vertical thruster w.r.t CG

Table 1. Reduced model terms

to the effective applied force. Like in many other problems in robotics, it is often difficult or even impossible to measure such forces. Measuring relative or absolute motion variables then becomes an alternative and the choice of the state to be observed is directly influenced by the variables that can be actually measured. Therefore, we propose the following model for the fault free ideal system:

$$\begin{aligned} \dot{\mathbf{x}} = \begin{bmatrix} \dot{\mathbf{x}}_1 \\ \dot{\mathbf{x}}_2 \end{bmatrix} = \begin{bmatrix} \dot{z} \\ \dot{\theta} \\ \dot{w} \\ \dot{q} \end{bmatrix} &= A_l(\mathbf{x})\mathbf{x} + f_u(\mathbf{x}, u) + f(\mathbf{x}, u, \mathbf{f_v}) + w^x(t) \\ y = h(x) + v_n(t), & \end{aligned} \tag{4}$$

where $w^x \in \mathbb{R}^4$ is a zero-mean Gaussian noise vector with autocorrelation matrix $Q^w(t)$, $\mathbf{x_1} = [z, \theta]^T$, $\mathbf{x_2} = [w, q]^T$, A_l and $f_u(\cdot)$ are easily derived from the kinematics model in 1 as

$$A_l(\mathbf{x}) = \begin{bmatrix} 0 & 0 & \cos\theta & 0 \\ 0 & 0 & 0 & 1 \\ 0 & 0 & 0 & 0 \\ 0 & 0 & 0 & 0 \end{bmatrix}, \quad f_u(\mathbf{x}, u) = \begin{bmatrix} -u\sin\theta \\ 0 \\ 0 \\ 0 \end{bmatrix},$$

assuming $\phi = 0$, and $f(\cdot)$ results from the dynamics model 3 as

$$f(\mathbf{x}, u, \mathbf{f_v}) = \begin{bmatrix} 0 \\ 0 \\ \dot{w} \\ \dot{q} \end{bmatrix} = S\dot{\nu} = SM^{-1}\big(-C(\nu)\nu - D(\nu)\nu - g(\mathbf{x_2}) + P_v\mathbf{f_v}(t)\big),$$

$$S = \begin{bmatrix} 0 & 0 & 0 \\ 0 & 0 & 0 \\ 0 & 1 & 0 \\ 0 & 0 & 1 \end{bmatrix}, \quad P_v = \begin{bmatrix} 0 & 0 \\ 1 & 1 \\ x_{ts} & x_{tb} \end{bmatrix}, \quad \mathbf{f_v} = \begin{bmatrix} f_s(t) \\ f_b(t) \end{bmatrix}.$$

Regarding the output y of 4, the dimension of the function $h(\mathbf{x})$ depends on the measurements and consequently on the on-board sensors. Here, we will assume we are able to observe the depth z, the pitch angle θ and the pitch rate q. Thus it results

$$h(\mathbf{x}) = C_h\mathbf{x}, \quad C_h = \begin{bmatrix} 1 & 0 & 0 & 0 \\ 0 & 1 & 0 & 0 \\ 0 & 0 & 0 & 1 \end{bmatrix}.$$

The vector $v_n \in \mathbb{R}^3$ is the output noise, assumed to be zero-mean, Gaussian noise with autocorrelation matrix R^v.

Notice that $\mathbf{x_2}$ is the vector containing the last two entries of the velocity vector ν, i.e., $\nu = [u, \mathbf{x_2}]$. For simplicity of notation, in the expressions above we wrote ν instead of $[u, \mathbf{x_2}]$. Recall that we assumed that u is a measured variable, or at least, it can be accurately estimated. Indeed, it could be included in the state in 4 but the complexity of this latter would increase without advantages in the approach.

In order to model the possible loss of control effectiveness, let us define $\gamma = [\gamma_s, \gamma_b]^T$ as the vector of loss of control effectiveness factors, adopting the same notation as in Wu et al. (2000). Introducing these multiplicative factors in 4, the augmented state model results in

$$\begin{aligned} \dot{\mathbf{x}} &= A_l(\mathbf{x})\mathbf{x} + f_u(\mathbf{x}, u) + f(\mathbf{x}, u, \mathbf{f_v}) + E(\mathbf{f_v})\gamma + w^x(t) \\ \dot{\gamma} &= w^\gamma(t) \\ y &= C_h x + v_n(t), \end{aligned} \tag{5}$$

where $w^\gamma \in \mathbb{R}^2$ is a zero-mean, Gaussian noise vector with autocorrelation matrix Q^γ, uncorrelated with w^x, and

$$E(\mathbf{f_v}) = SM^{-1}P_v\,\mathrm{diag}(\mathbf{f_v}).$$

As it can be seen in 5, γ is assumed to be driven only by the noise w^γ. This comes from the fact that, in real scenarios, it is impossible to predict how the fault and, consequently, how γ evolve. In such situation, the most appropriate is to model the evolution with a noise vector w^γ with a sufficiently large autocorrelation (see Wu et al. (2000)), whose entries can play an important role in the design of the augmented state estimator, as it will be seen later on.

Making $s = [\mathbf{x}^T, \gamma^T]^T$, we rewrite 5 on the form

$$\begin{aligned} \dot{s} &= A_s(s)s + f_{us}(s, u) + f_s(s, u, \mathbf{f_v}) + E_s(\mathbf{f_v})s + w_s(t) \\ y &= C_s s + v_n(t). \end{aligned} \tag{6}$$

where

$$A_s = \begin{bmatrix} A_l & 0_{4\times2} \\ 0_{2\times4} & 0_{2\times2} \end{bmatrix}, \; f_{us} = \begin{bmatrix} f_u \\ 0_{2\times1} \end{bmatrix}, \; f_s = \begin{bmatrix} f \\ 0_{2\times1} \end{bmatrix},$$

$$E_s = \begin{bmatrix} 0_{4\times4} & E \\ 0_{2\times4} & 0_{2\times2} \end{bmatrix}, \; w_s = \begin{bmatrix} w^x \\ w^\gamma \end{bmatrix} \; C_s = \begin{bmatrix} C_h & 0_{3\times2} \end{bmatrix}.$$

The discrete time representation of 6 follows

$$\begin{aligned} s_{k+1} &= A_{sk}(s_k)s_k + f_{usk}(s_k,u_k) + f_{sk}(s_k,u_k,\mathbf{f_v}_k) + E_{sk}(\mathbf{f_v}_k)s_k + w_{sk} \\ y_{k+1} &= C_s s_{k+1} + v_{k+1}, \end{aligned} \tag{7}$$

where β_k represents the discrete time equivalent vector, or matrix, β at time t_k.
We assume that the process noise w_{sk} and the output noise v_k are uncorrelated, i.e., $E\{w_{sk}v_k^T\} = 0$. The autocorrelation of the process noise and of the output are respectively given by

$$E\{w_{sk}w_{sk}^T\} = Q_k = \begin{bmatrix} Q_k^x & 0 \\ 0 & Q_k^\gamma \end{bmatrix}, \quad E\{v_k v_k^T\} = R_k. \tag{8}$$

The formulation of a Kalman filter assumes the use of a model of the process which is a mathematical representation of the dynamics. However, the mathematical translation of the dynamics of a given system may be inaccurate or may not describe entirely its behavior. This is the case in hydrodynamics, where the models are complex, difficult to extract. Moreover, there is no complete theory that allows for determining an accurate model and calculations of parameters mostly rely on empirical or semi-empirical formulas.
Hence, we define $\hat{\beta}$ as the estimate of the generic vector, or matrix, β. The augmented state extended Kalman filter formulation follows now directly from Gelb (1974). During the prediction stage, the state estimate and the covariance matrix evolve according to

$$\hat{s}_{k+1|k} = \hat{A}_{\hat{s}k}(\hat{s}_k)\hat{s}_k + \hat{f}_{usk}(\hat{s}_k,\hat{u}_k) + \hat{f}_{sk}(\hat{s}_k,\hat{u}_k,\mathbf{f_v}_k) + \hat{E}_{\hat{s}k}(\mathbf{f_v}_k)\hat{s}_k \tag{9}$$

$$P_{k+1|k} = F_k P_k F_k^T + Q_k, \tag{10}$$

where F_k stands for the Jacobian of $\dot{s}$ evaluated at $\hat{s}_k$:

$$F_k = \frac{\partial \dot{s}}{\partial s}_{|s=\hat{s}_k}.$$

The so-called Kalman gain and the updates of the estimate and of the covariance matrix are respectively given by

$$K_{k+1} = P_{k+1|k}C_s^T(C_s P_{k+1|k} C_s^T + R_k)^{-1} \tag{11}$$

$$\hat{s}_{k+1|k+1} = s_{k+1|k} + K_{k+1}(y_{k+1} - C_s s_{k+1|k}) \tag{12}$$

$$P_{k+1|k+1} = (I - K_{k+1}C_s)P_{k+1|k}. \tag{13}$$

From the state estimate, it is now possible to extract the vector γ_k, whose entries constitute the base to determine whether a fault has occured or not.
As it was stated earlier, the autocorrelation matrix Q_k^γ can play a significant role to avoid divergence or guarantee faster convergence of the estimate of the loss of control effectiveness

factors. For the sake of clarity, from 10 and 13 we can decompose P_k as

$$P_k = \begin{bmatrix} P_k^x & P_k^{\gamma x} \\ P_k^{x\gamma} & P_k^{\gamma} \end{bmatrix}$$

Thus, from 10 and 13, we can conclude

$$P_{k+1|k}^{\gamma} = P_{k|k}^{\gamma} + Q_k^{\gamma}$$

$$P_{k+1|k+1}^{\gamma} \leq P_{k+1|k}^{\gamma}$$

where we used the fact that $P_{k+1|k} > 0$. Hence, the autocorrelation matrix Q_k^{γ} can be set such that P^{γ} lies in an interval, preventing state corrections to be excessive, when P^{γ} is too large, or to be insufficient with slow convergence, when P^{γ} is too small. Taking the eigenvalues λ^{γ} of P^{γ} as measures, we propose the following function

$$Q^{\gamma} = \begin{cases} diag(q_{\gamma}, q_{\gamma}) & , \text{ if } \max(\lambda_1^{\gamma}, \lambda_2^{\gamma}) < \lambda_{max} \\ 0 & , \text{ if } \max(\lambda_1^{\gamma}, \lambda_2^{\gamma}) \geq \lambda_{max} \end{cases} \quad (14)$$

where q_{γ} is the autocorrelation of $\gamma_i, i = 1, 2$ and λ_{max} is a preset maximum constant.

3.3 Fault identification

The loss control effectiveness factors provide an estimate of the performances of the actuators. Ideally, a fault would be identified whenever the absolute value of one of the factors would rise above a preset threshold. However, model uncertainties will be directly reflected in these factors. Even in normal operation, with the actuators working perfectly, the loss of control effectiveness factors may diverge from zero, reflecting, for example, the effect of a damping force greater than the modeled. As these errors are frequently commited on the overall model, their effects are verified on all actuators either by increasing or decreasing the loss of control effectiveness factors. Hence, for the present case, a reasonable measure of the malfunction of one of the thrusters is given by the difference of the corresponding loss of control effectiveness factor estimate. On the other hand, taking a decision about the malfunction of a given thruster should also be based on the confidence of the factor estimate, which can be indirectly taken from the eigenvalues λ^{γ} of P_k^{γ}, avoiding taking decisions on transient state, while considerable corrections on the state are being performed. Thus, we propose the following measure for fault detection:

$$\delta = \frac{|\gamma_s - \gamma_b|}{f_{\lambda}(\lambda_1^{\gamma} + \lambda_2^{\gamma})}. \quad (15)$$

where f_{λ} is a monotically increasing function of its argument.
Whenever δ is greater than a preset threshold, a fault is detected and the identification is made according to the greater λ, i.e., if $\gamma_s > \gamma_b$ then the stern thruster is faulty and vice-versa.

4. Control of MARES

In the presence of a faulty vertical thruster, the reconfiguration of the actuation is required. Otherwise, keeping the same actuation will likely lead to instability or to other pratical problems such as thruster dammage or large battery consumption, for example. Therefore, the control law for normal operation could be inadequate and another control law must take

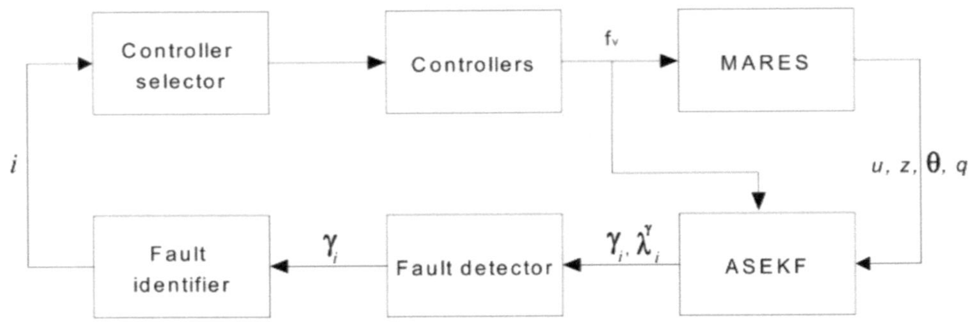

Fig. 3. Operation of the fault overall fault detection and recovery algorithm

over. In this section, we first start by deriving such controller and present the main concepts behind the derivation of the control for normal operation in order to make the result section clear.

4.1 Control under fault

We consider now the scenario in which only one of the vertical thrusters is available to control the motion of the vehicle. Under such situation, the heave DOF is no longer controllable but the depth is still controllable by manipulating pitch. Based on the Lyapunov theory we will derive a controller that makes it possible to control the vehicle's depth, while assuming that the absolute value of the surge velocity is sufficiently large to compensate the vehicle's flotation. The derivation of the controller employs the well know backstepping method as well as conditional integrators to achieve asymptotic regulation.
As the final goal in this section is to control the vehicle depth, we will assume that roll angle is null ($\phi = 0$), resulting:

$$\dot{z} = -u \sin\theta + w \cos\theta \tag{16}$$

Let us introduce the error variable $e_z = z - z_d$, which we want to drive to zero, and the quadratic Lyapunov function:

$$V_1 = \frac{1}{2} e_z^2, \tag{17}$$

whose time derivative results

$$\dot{V}_1 = e_z \dot{e}_z = e_z(-u \sin\theta + w \cos\theta - \dot{z}_d). \tag{18}$$

Although u, θ and z are measured by sensors or estimated, it is hard to accurately compute w due to model uncertainties and measurement noise. Thus we will assume that it constitutes a disturbance acting on the system, shifting the equilibrium point $e_z = 0$ to an uncertain value. Throughout the following developments, we will consider that the surge velocity is maintained constant in order to simplify our approach. Indeed, in most missions the surge velocity is intended to be constant along the trajectory. Moreover, the limited actuation on the vertical thruster makes the pitch angular velocity to lie in a bounded interval. Hence, from the vertical dynamics, we can assume that there exists an upper bound on the absolute value of $w \in [-w_{max}, w_{max}]$.

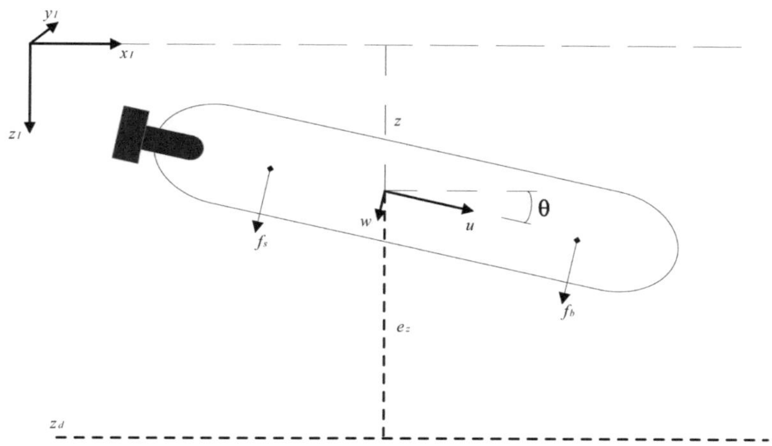

Fig. 4. Depth control by actuating either f_s or f_b

Inspired by Singh & Khalil (2005), let us suppose we are able to handle θ directly through the virtual control law

$$\theta = \theta_d(e_z) = \arcsin\Big[-\frac{1}{u}(\dot{z}_d - \alpha(e_z)\varphi(\frac{e_z}{\mu}))\Big], \tag{19}$$

where $\varphi(\cdot) : \mathbb{R} \to \mathbb{R}$ is the continuous infinitely differentiable sigmoid function

$$\varphi(x) = \frac{2}{1+e^{-\varsigma x}} - 1, \tag{20}$$

which verifies $x\varphi(x) > 0,\ x \neq 0$, and $\alpha(\cdot) > 0$ is a continuous function left to be determined later. Of course, handling θ directly and instantaneously is not realistic and such assumption will be lifted next. In opposition to Singh & Khalil (2005), we have selected a sigmoid function φ instead of a saturated linear function due to the differentiability characteristic. Let us take take $\varsigma = 2$, which will make $\varphi(\cdot)$ equal to the hyperbolic tangent function.
Assuming z_d sufficiently smooth, $u > 0$ and imposing

$$-\frac{1}{u}(\dot{z}_d - \alpha(e_z)\varphi(\frac{e_z}{\mu})) \leq 1, \tag{21}$$

the time derivative of the Lyapunov function in 18 results

$$\begin{aligned} \dot{V}_1 &= e_z\Big(-\alpha(e_z)\varphi(\frac{e_z}{\mu}) + w\cos\theta\Big) \\ &\leq -e_z\alpha(e_z)\varphi(\frac{e_z}{\mu}) + |e_z|w_{max}, \end{aligned} \tag{22}$$

where we used the fact that $|\cos\theta| \leq 1$ and w be bounded. By choosing appropriately $\alpha(\cdot)$ and $\varphi(\cdot)$, the system can now be made pratically stable (see Singh & Khalil (2005)). Hence, let us define $\varepsilon \in (0,1)$ and take $\mu = \frac{1}{\tanh^{-1}(\varepsilon)}$, then choosing $\alpha(e_z) = K_z,\ K_z \in (\frac{w_{max}}{\varepsilon}, u + \dot{z}_d]$, such that it satisfies 21, it comes

$$\dot{V}_1 < 0, \ ||e_z|| > 1$$
$$\dot{V}_1 \leq -e_z\alpha(e_z)\varphi(\frac{e_z}{\mu}) + |e_z|w_{max}, \ ||e_z|| \leq 1.$$

Thus, the system is made pratically stable and the invariant set for which the error tends to can be made arbitrarily small by handling μ. However, a too small μ induces chattering phenomena which are intended to be minimal.
Taking into account that V_1 is a strictly increasing function of e_z from the last inequalities, we can state that the error enters a positively invariant set $\Omega = \{e_z \leq 1\}$. However, due to non-null disturbances considered above, asymptotic stability can not be achieved. Therefore, following the same idea as in Singh & Khalil (2005), the conditional integrator is now introduced to obtain asymptotic convergence to the origin $e_z = 0$. Modifiyng the control law in 19 to include an integral component, it results

$$\theta_d(e_z) = \arcsin\Big[-\frac{1}{u}(\dot{z}_d - \alpha(e_z)\varphi(\frac{e_z+\sigma}{\mu}))\Big], \tag{23}$$

where

$$\dot{\sigma} = -\gamma\sigma + \mu\varphi(\frac{e_z+\sigma}{\mu}), \ \ \gamma > 0, \ \ \sigma(t_0) = 0.$$

Since $|\phi(x)| < 1, \ \forall x \in \mathbb{R}$, it is easy to check that $\sigma \leq \frac{\mu}{\gamma}$. In order to guarantee convergence to zero, one has to set γ and μ such that the maximum absolute value of the integral satisfies$|\sigma| > \mu|\varphi^{-1}(\frac{w_{max}}{K_z})|$. Although conservative, this will allow the integral component to compensate the disturbance effect. By applying theorem 1 in Singh & Khalil (2005), convergence to $e_z = 0$ as $t \to \infty$ is ensured.
So far, we have considered that we are able to handle θ directly, which is not true, as it was stated before. Thus, based on the backstepping method Khalil (2002), let us introduce the new error variable $e_\theta = \theta - \theta_d$ and the new augmented Lyapunov function as follows

$$V_2 = V_1 + \frac{1}{2}e_\theta^2, \tag{24}$$

whose time derivative results

$$\dot{V}_2 = \dot{V}_1 + e_\theta(\dot{\theta} - \dot{\theta}_d), \tag{25}$$

with

$$\dot{\theta}_d = -\frac{\ddot{z}_d - K_z\frac{\partial}{\partial t}\varphi(\frac{e_z+\sigma}{\mu})}{\left(u^2 - (\dot{z}_d - K_z\varphi(\frac{e_z+\sigma}{\mu}))^2\right)^{1/2}}.$$

Then by imposing

$$\dot{\theta} = q_d = \dot{\theta}_d - K_\theta e_\theta, \ \ K_\theta > 0, \tag{26}$$

the time derivative of the augmented Lyapunov function satisfies $\dot{V}_2 \leq \dot{V}_1 - K_\theta e_\theta^2$. Taking into account the previous result about the convergence of e_z to zero and the fact that V_1 is a class $\mathcal{K}_\infty$ function, we can deduce that $V_2 \to 0$ as $t \to \infty$.
Nevertheless, we are not able to handle $\dot{\theta}$ directly and, as it can be seen from 1 and 2, a last step is required. Hence, we define $e_q = q - q_d = S_q\nu - q_d$, with $S_q = [0, 0, 1]$, as the pitch rate

error variable as well as the new augmented Lyapunov function:

$$V_3 = V_2 + \frac{1}{2}e_q^2. \tag{27}$$

Considering 3, the time derivative results

$$\begin{aligned} \dot{V}_3 &= \dot{V}_2 + e_q(S_q\dot{\nu} - q_d) \\ &= \dot{V}_2 + e_q(S_q M^{-1}(-C(\nu)\nu - D(\nu)\nu - g(\eta_2) + P_i f_{pi}) - \dot{q}_d). \end{aligned} \tag{28}$$

where P_i and f_{pi}, $i = \{s, b\}$, are given as functions of the actuator configuration. When the vehicle is operating with only one thruster, either stern or bow thruster, P_i and f_{pi} are respectively given by

$$P_s = \begin{bmatrix} 1 & 1 & 0 \\ 0 & 0 & 0 \\ 0 & 0 & x_{ts} \end{bmatrix}, \quad P_b = \begin{bmatrix} 1 & 1 & 0 \\ 0 & 0 & 0 \\ 0 & 0 & x_{tb} \end{bmatrix}, \quad f_{ps} = \begin{bmatrix} f_p \\ f_r \\ f_s \end{bmatrix}, \quad f_{pb} = \begin{bmatrix} f_p \\ f_r \\ f_b \end{bmatrix}.$$

Note that P_i takes the form $P = [P_h | P_{vi}]$, where $P_h \in \mathbb{R}^{3\times 2}$ is the submatrix composed by the first two columns of P_i and $P_{vi} \in \mathbb{R}^{3\times 1}$ is the last column of P_i. Further, let us decouple the input vector into $f_{pi} = [f_h^T, f_i^T]^T$, where $f_h \in \mathbb{R}^2$ is composed by the first two entries of f_{pi} and f_i is the last entry of this latter, which we can manipulate directly. By considering the decoupled form of P_i, we can rewrite 28 as

$$\dot{V}_3 = \dot{V}_2 + e_q(S_q M^{-1}(-C(\nu)\nu - D(\nu)\nu - g(\eta_2) + P_h f_h + P_{vi} f_i) - \dot{q}_d).$$

Clearly $S_q M^{-1} P_h f_h = 0$, which means that the horizontal thrusters have no direct influence on the pitch dynamics (see the entries of P_s and P_b).
Finally, defining the proportional gain $K_q > 0$ and choosing the control law

$$f_i = (S_q M^{-1} P_{vi})^{-1}(S_q M^{-1}(C(\nu)\nu + D(\nu)\nu + g(\eta)) + \dot{q}_d - K_q e_q), \; i = \{s, b\}, \tag{29}$$

the time derivative of the Lyapunov function 28 becomes

$$\dot{V}_3 = \dot{V}_2 - K_q e_q^2. \tag{30}$$

Therefore, the convergence of the error e_z to zero is then guaranteed by setting the input f_i according to the control law 29. Note that Equation 29 gives the two control laws for either actuating with only stern or bow thruster, being different on the entries of P_{vi} only.

4.2 Control without fault

Under normal operation, the two through-hull thrusters provide controllability on the heave and the pitch DOFs. We will not give emphasis to the derivation of this controller since it was previously derived in Ferreira, Matos, Cruz & Pinto (2010). We aim at exposing the main concepts that led to the control law, in order to better understand the results of the next section. The controller was derived using common backstepping with no integral terms.
In opposition to the previous subsection, the errors considered for the control with the two thrusters are bidimensional vectors. Naturally, the error vector for vertical position comes

$$e'_p = \begin{bmatrix} z - z_d \\ \theta - \theta_d \end{bmatrix},$$

assuming that $\theta, \theta_d \neq \pi/2$. Following the same method as previously, a first Lyapunov function is defined as a quadratic function of the error e_p and its time derivative is made negative definite by adequately choosing $\dot{z}$ and $\dot{\theta}$ as virtual control variables to achieve asymptotic stability.
A new augmented Lyapunov function is then introduced by adding a quadratic term of the error

$$e'_v = \begin{bmatrix} w - w_d \\ q - q_d \end{bmatrix},$$

and, from the reduced dynamics model 3, the control law for the two vertical thrusters $f_{pv} = \chi(e'_p, e'_v, u, \eta_2) \in \mathbb{R}^2$ is determined such that the time derivative of the augmented Lyapunov function is made negative definite.

5. Experiments and results

To validate the method described in the previous sections, several experimental tests were conducted. The results will be presented in a decoupled way in order to facilitate the exposition and the analysis. First, the results obtained from the ASEKF will be exposed for the system under normal operation with induced faults. Then the performances of the controller derived in the previous section will be presented.

5.1 Fault detection

Several tests were performed in order to verify the behavior of the fault detection and identification algorithm. Under normal operation, we have intentionally induced faults in the thrusters with the aim of analyzing the behavior of our approach. The following graphs expose the evolution of some of the most relevant variables referred in section 3. The model errors and uncertainties were not corrected so that we could observe behaviors similar to those occuring with real faults. Although not explicitly written, the units of the loss of control effectiveness factors, their difference, the eigenvalues of the corresponding covariance matrix as well as the fault measure are dimensionless, while angles and linear distances are expressed in radians and meters, respectively.
The fault in one of the thruster can be simulated by a conteracting force in the same axis of force application, with an opposite direction. The following tests were carried out such that a force with an opposite direction was applied in an axis near the axis of force application along all the operation or along part of it. The fault measure computed and exposed in the graphs below was set, according to 15, equal to

$$\delta = \frac{|\gamma_s - \gamma_b|}{\sqrt{\lambda_1^\gamma + \lambda_2^\gamma}}.$$

Fig. 5 presents the evolution of the state estimate in normal operation without conteracting forces. It can be seen in Fig. 5(a), however, that the loss of control effectiveness factors are non-null, in opposition to what would be expected. Indeed, such behavior is due to errors of model parameters, neglected dynamics effects, discretization errors as well as linearization in the extended Kalman filter formulation. Nevertheless, it can be seen they are limited and their differences in Fig. 5(b) are confined to a well defined interval. This effect is unavoidable since deviations of the mathematical model will be directly reflected in the loss of control effectiveness factors.
On the other hand, the eigenvalues of the submatrix P^γ (Fig. 5(c)) are monotically increasing along the operation due to the reduced actuation, and consequently poor observability. In

fact, the vertical thrusters are mainly compensating the difference between the weight and the bouyancy of the vehicle, which is less than 1.5 Newton, well below the maximum actuation (24N in each thruster). Although it is not shown in the graph, the eigenvalues would stabilize after some more time of operation.

In Fig. 6, the graphs show the evolution of the variables during the operation for which 1.5N and 2.5N of force are opposing the vertical bow and stern thrusters, respectively. It can be seen that the eigenvalues (Fig. 6(c)) reach their "steady stade" after 40 seconds. In this case, the controller compensates the effects of the opposition force by increasing the actuation, which, in turn, makes the observability better than in the previous case (compare Fig. 6(c) with Fig. 5(c)). Regarding the loss of control effectiveness factors, one can observe that the bow thruster is more affected by the added forces than the stern one when compared with Fig. 5(a). This behavior is explained by the fact that the opposition forces were applied asymetrically with respect to the center of gravity of the vehicle.

In Fig. 7 and Fig. 8, we present the same variables as in the previous figures, but now with a disturbance occuring at time $t = 60s$, approximately. The graphs in Fig. 7 are related to the disturbance applied in the stern thruster while the graphs in Fig. 8 corresponds to the disturbance in the bow thruster. Comparing the results in Fig. 5 and Fig. 6 with those in Fig. 7 and Fig. 8, the occurence of faults in the last two experiences is evident through simple analysis of Fig. 7(b) and Fig. 8(b).

By comparing the results exposed in Fig. 5(d), 6(d), 7(d), 8(d), we can conclude that setting the threshold $\delta_{th} = 7$ would adequatly detect the occurence of a fault, while ensuring a sufficiently large margin to avoid false dectections. The choice of such threshold has to be made according to the sensitivity desired for the detections of faults which must be mainly related to the accuracy of the model.

The identification is made according to the stated before, being the thruster with bigger $|\gamma|$ the one that is identified as faulty. In both situations (Fig. 7(a) and Fig. 8(a)), the faulty thruster is easily identifiable. Still, practical considerations have to be made at this stage: One can observe that δ reaches larger values in the initial transient instants due to small eigenvalues, which could lead to false detections (depending on the threshold). To avoid so, a possibility would be integrating the value of δ when it is above the threshold, while using a forgetting factor, and defining another threshold for the value of the integral. Nonetheless, the simple method with the threshold defined above would be sufficient for the present work, as it can be seen from the experiments.

On the other hand, the relation between the actuation amplitude and the eigenvalues of the matrix P^{γ} is now notable in Fig. 7(c) and Fig. 8(c). With the application of the disturbance, the controller increases the actuation and, as a consequence, the observability is made greater (see 7(c) and 8(c)).

5.2 Control

The control laws derived in section 4 were implemented in MARES and tested in real conditions. The results of the operations are shown in this section, being the analysis of the performances of the controllers the main topic. The vehicle missions were programed such that it navigates at a constant depth reference with a given constant orientation. We must highlight that several unconsidered disturbances have acted on the vehicle during operations: The vehicle was subject to more bouyancy than the assumed in the mathematical model; the fedback depth measurement is actually performed in the nose of MARES instead of the vehicle's center of gravity. Such disturbances induce undesired effects on the controllers. However, the following figures show the robustness of our approach. The variables shown in the graphs have the following units correspondence: Depth is expressed in meters, the pitch angle is expressed in radians while the surge velocity is expressed in meters per second.

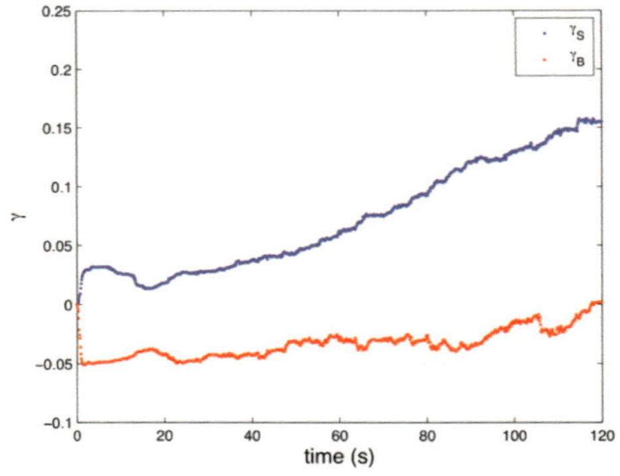

(a) Control effectiveness factors

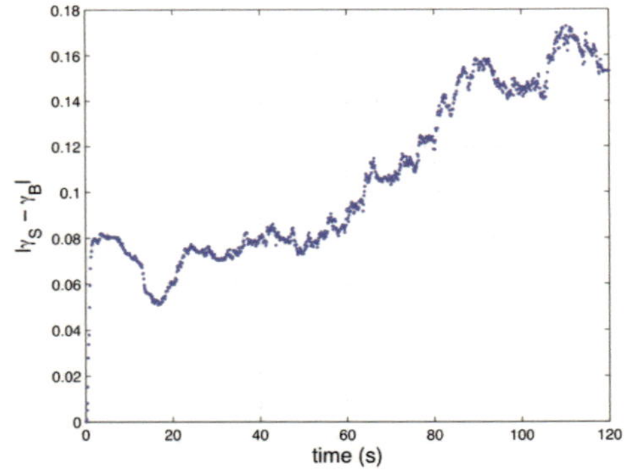

(b) Difference of control effectiveness factors

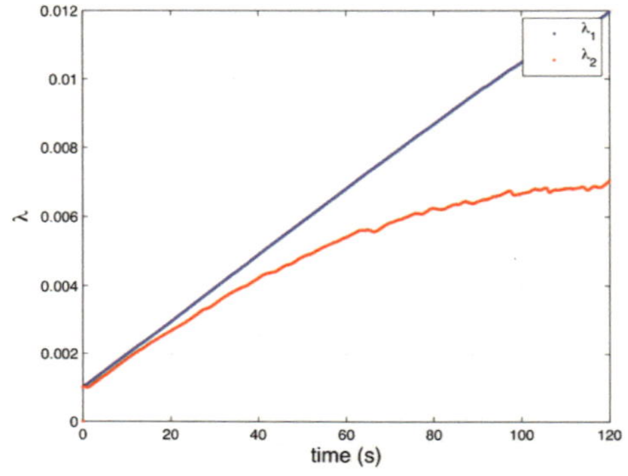

(c) Eigenvalues

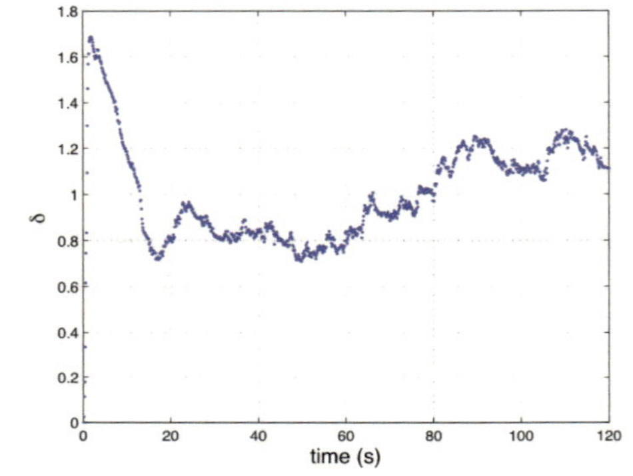

(d) Fault measure

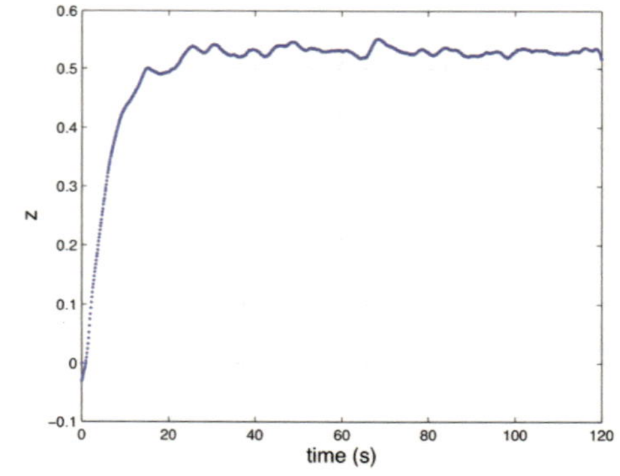

(e) Depth

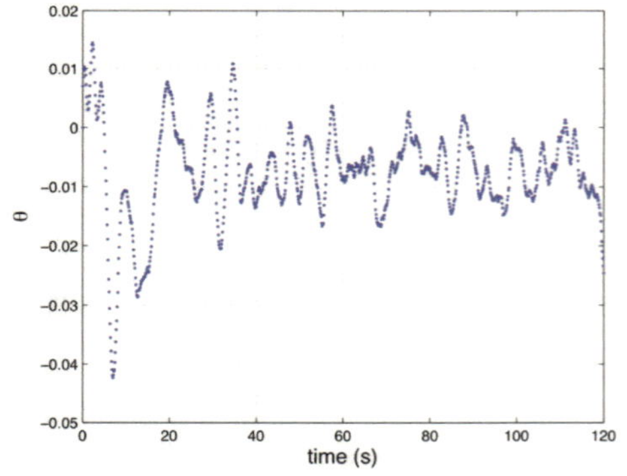

(f) Pitch angle

Fig. 5. MARES under normal operation (no faults)

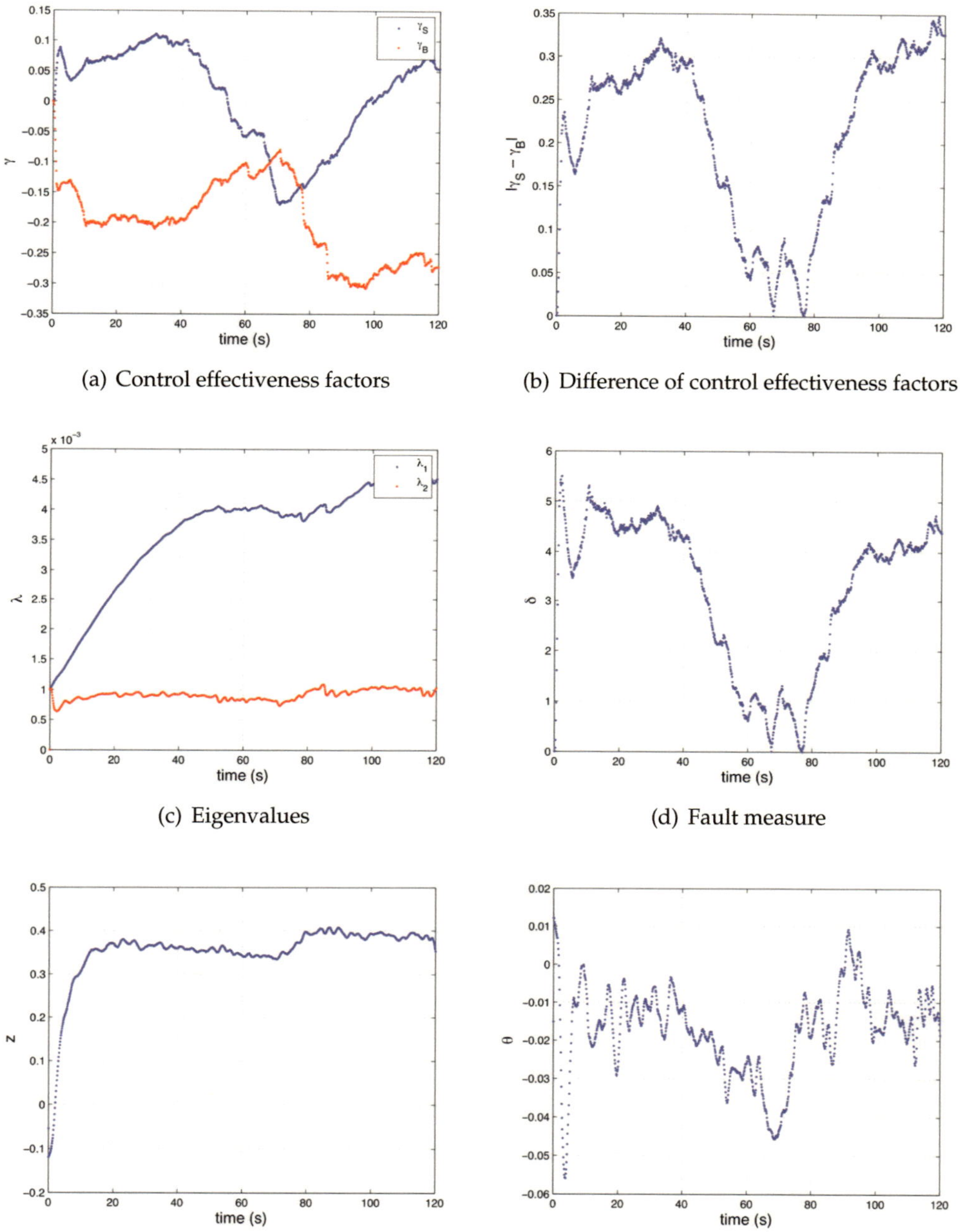

(a) Control effectiveness factors

(b) Difference of control effectiveness factors

(c) Eigenvalues

(d) Fault measure

(e) Depth

(f) Pitch angle

Fig. 6. MARES with constant disturbances: bouyancy added in the nose and the tail asymetrically

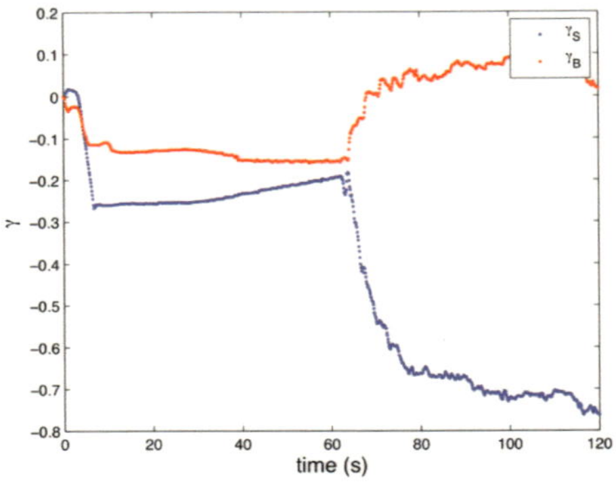

(a) Control effectiveness factors

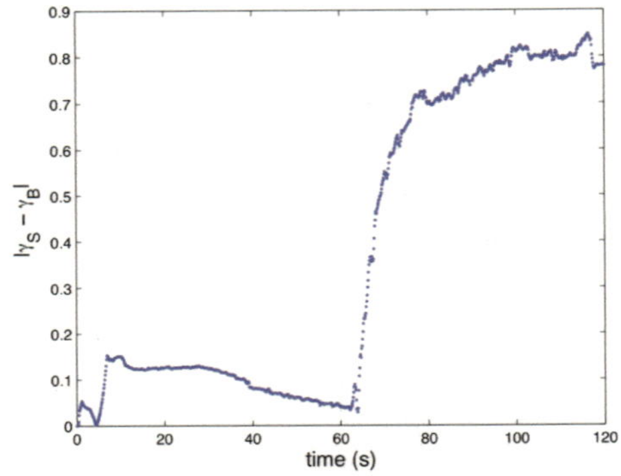

(b) Difference of control effectiveness factors

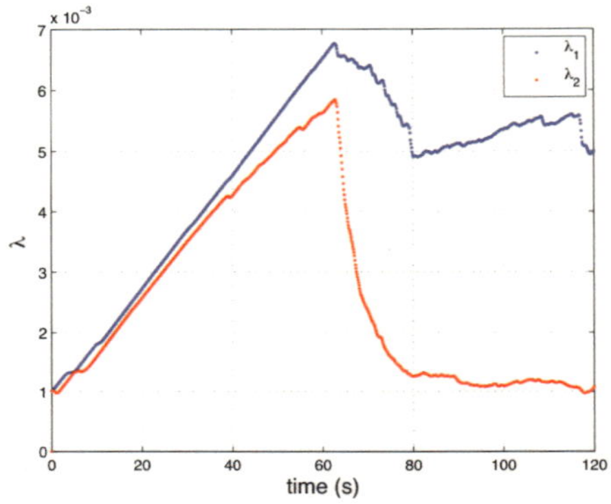

(c) Eigenvalues

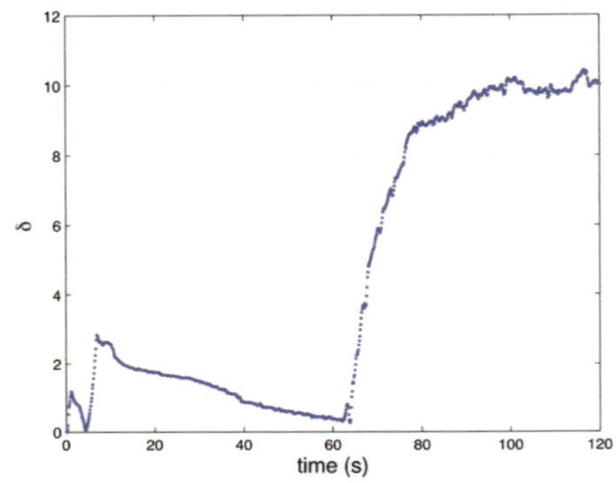

(d) Fault measure

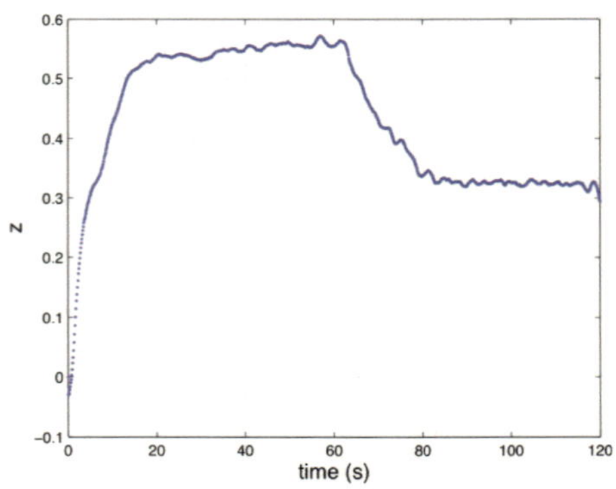

(e) Depth

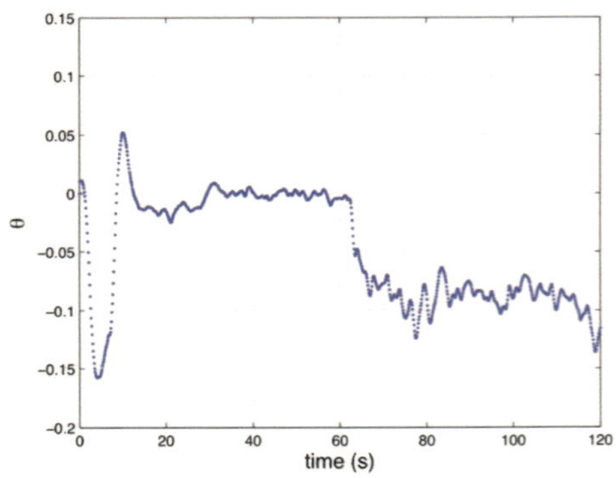

(f) Pitch angle

Fig. 7. MARES with disturbance in the stern thruster at $t = 60s$ approximately

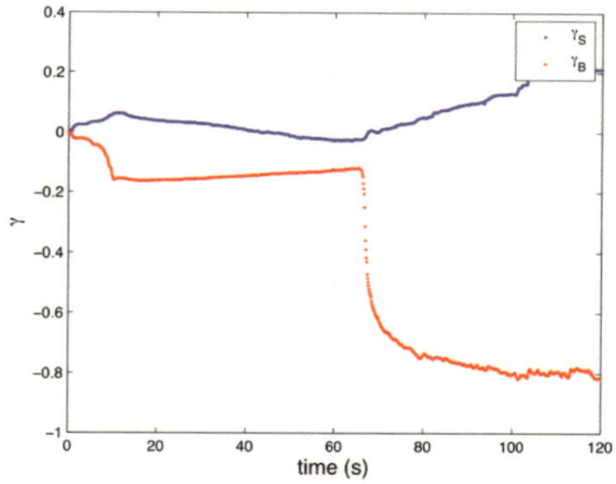

(a) Control effectiveness factors

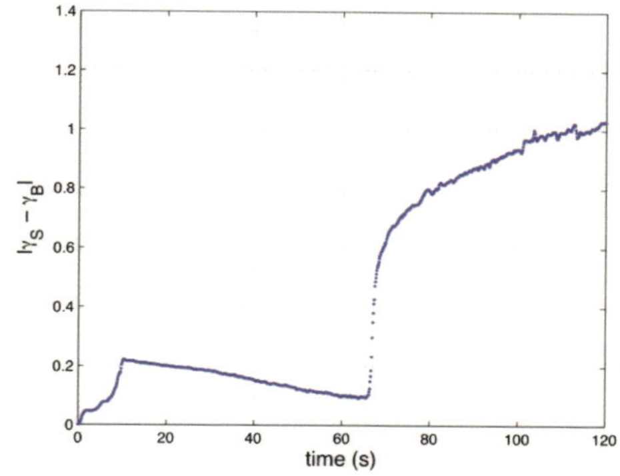

(b) Difference of control effectiveness factors

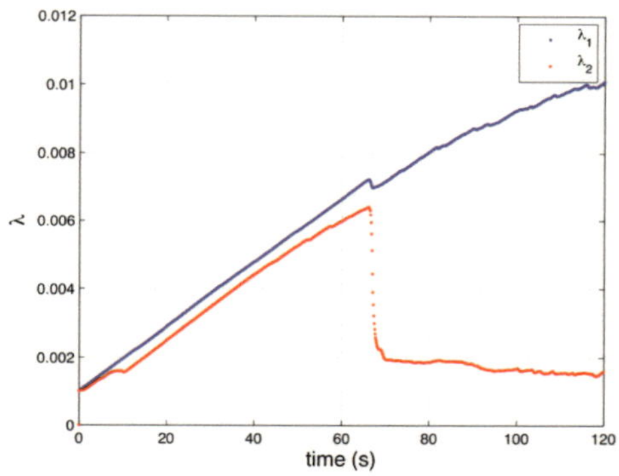

(c) Eigenvalues

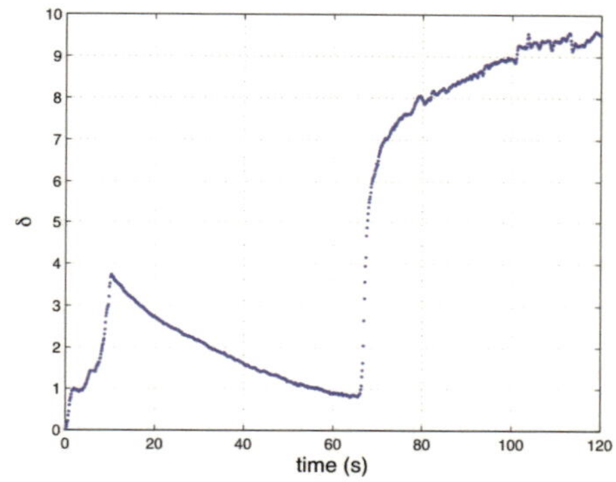

(d) Fault measure

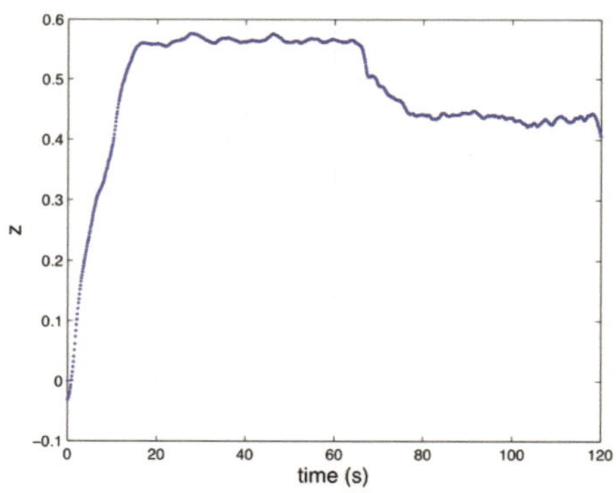

(e) Depth

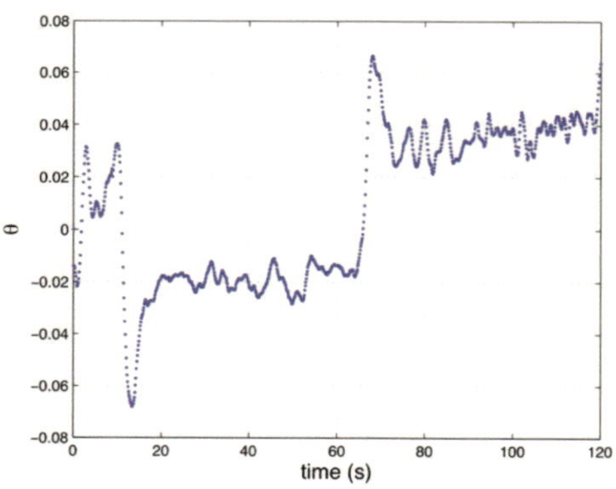

(f) Pitch angle

Fig. 8. MARES with disturbance in the bow thruster at $t = 60s$ approximately

Figures 9 to 11 show the results obtained during the operation of the controllers. The missions were set so that the vehicle starts diving with the two vertical thrusters simultaneously controlling pitch and depth, with surge velocity $u = 0$. At time $t = 20s$, a fault is simulated

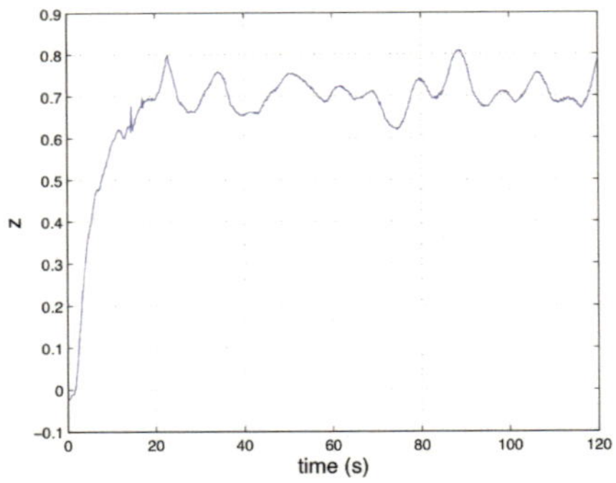

(a) Depth

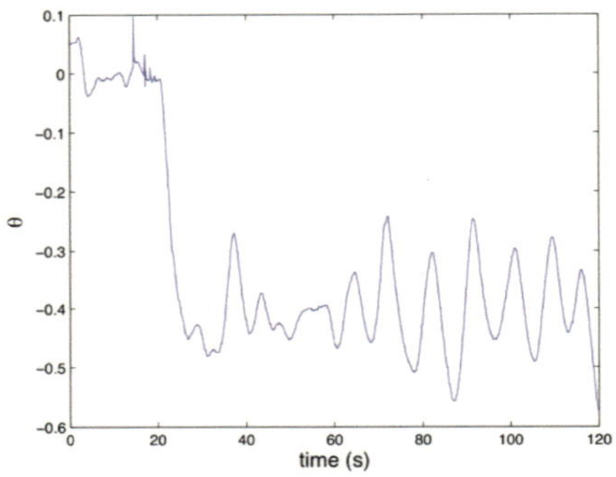

(b) Pitch angle

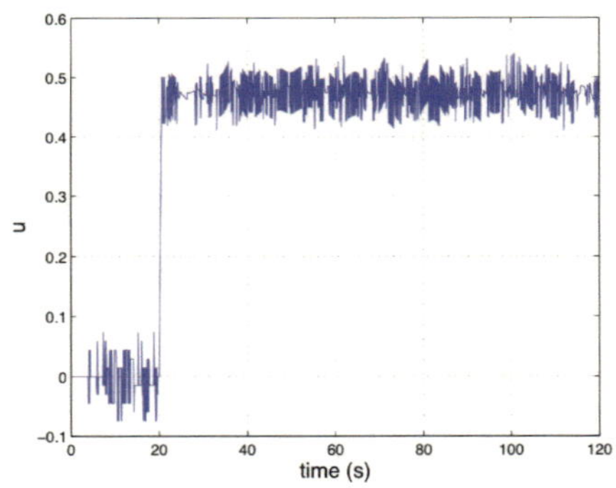

(c) Surge velocity

Fig. 9. MARES controlling depth with bow and horizontal thrusters only ($z_d = 0.7$ m)

and one of the controllers for degraded mode with only one vertical thruster starts operating. The figures show the results for different surge velocities and for the two controllers.
The Fig. 9 shows the variables directly measured from the sensors for a mission with a depth reference $z_d = 0.7$m, only with bow thruster. One can verify that the depth (9(a)) is reasonably close the reference and the small oscillation is due to natural disturbances that the vehicle

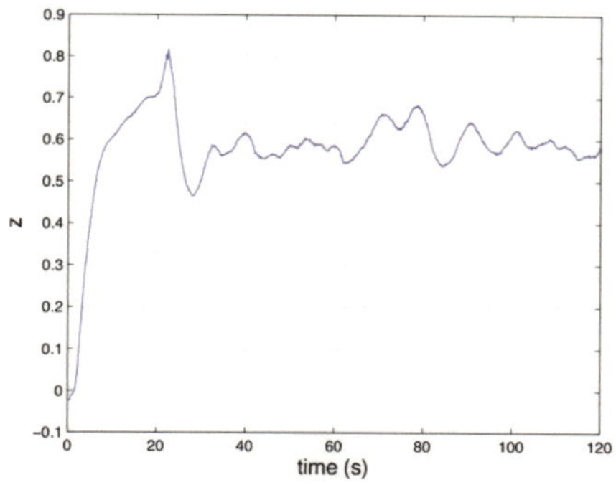

(a) Depth

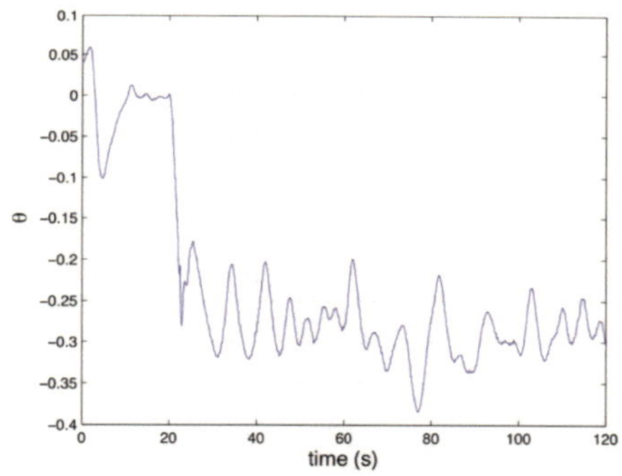

(b) Pitch angle

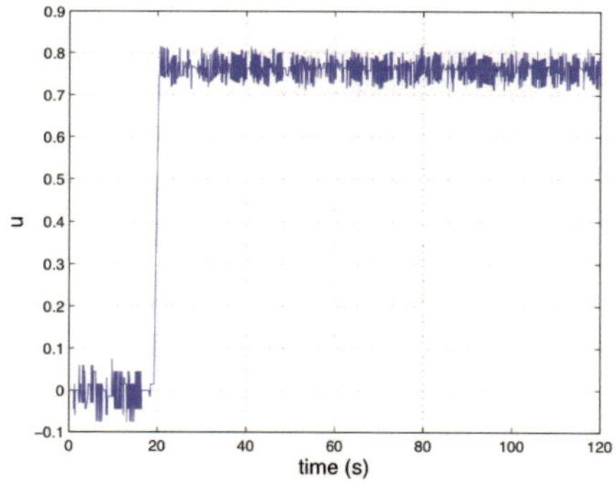

(c) Surge velocity

Fig. 10. MARES controlling depth with bow and horizontal thrusters only ($z_d = 0.6$ m)

founds in practical operations. Moreover, the commanded forces are affected by delays and thruster model is subject to uncertainties, which certainly influence the behavior.
On the other hand, the Fig. 11(a) shows that the oscillation amplitude is bigger than in Fig. 9(a) and 9(a), having been induced by more disturbances. Moreover, for stern-only control,

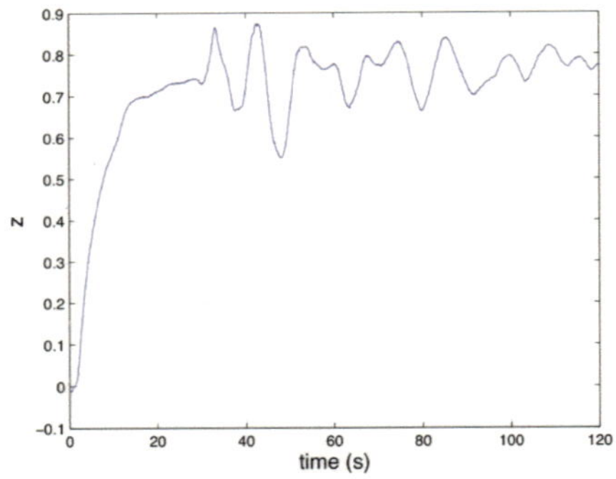

(a) Depth

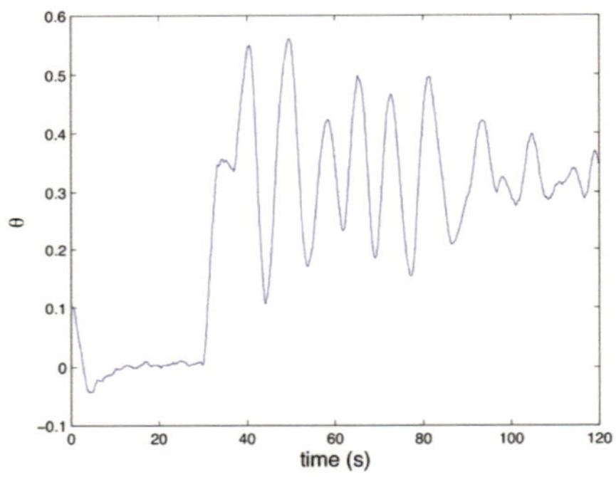

(b) Pitch angle

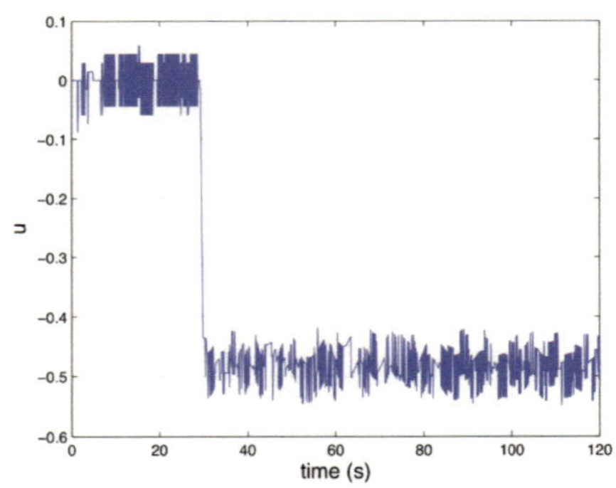

(c) Surge velocity

Fig. 11. MARES controlling depth with stern and horizontal thrusters only ($z_d = 0.8$ m)

more actuation is needed for the same pitch angle, in steady state. Such behavior is originated by the smaller distance at which the stern thruster is placed relatively to the center of gravity, comparing to the bow thruster (see Table 1).
Figures 9 to 11 show the robustness of the control laws, independently of the surge velocity. Even in the presence of unconsidered disturbances naturally induced by the environment, the graphs 9(a), 10(a) and 11(a) demonstrate that the controllers provide satisfactory results in real operation.

6. Conclusions

In this work, we have presented a complete method for fault detection, identification and automatic reconfiguration of the MARES AUV. Although we have focused on this particular system, the concepts and ideas can easily be extended to other problems, even beyond robotics.
Based on the dynamics and kinematics models, we have formulated a filter to estimate possible actuation biases. The augmented state extended Kalman filter was chosen to handle the problem, contemplating a reduced order model to simplify the analysis and the formulation. Bias variables were introduced in the state to be estimated as loss of control effectiveness factors whose values reflect the commanded underactuation/overactuation. Along with these estimated variables, the eigenvalues of the corresponding submatrix in the state estimate covariance matrix were taken to define a fault measure. Such measure was then used to generate a fault detection warning through comparison of its value with a given threshold. Finally, the faulty thruster is identified through analysis of the biases amplitude.
When a fault occurs and the corresponding thruster is set off, a suitable control law has to take over to ensure that the on-going mission succeed. To achieve so, we have defined two control laws for which we have based the derivation on Lyapunov theory and on backstepping method and further applied conditional integrators in order to drive the vehicle depth to a given reference with a null error in steady state.
At last, we demonstrated the performances of the developed method through real experiments in which we verified the operation of both estimator and controllers. Even in the presence of unconsidered disturbances, naturally induced by the environment, we have demonstrated that the controllers provide satisfactory results for several surge velocities and different thruster configurations.

7. Acknowledgement

Bruno Ferreira was supported by the Portuguese Foundation for Science and Technology through the Ph.D. grant SFRH/BD/60522/2009.

8. References

Cruz, N. A. & Matos, A. C. (2008). The MARES auv, a modular autonomous robot for environment sampling, *Oceans 2008, Vols 1-4*, Oceans-IEEE, IEEE, New York, pp. 1996–2001.
Ferreira, B., Matos, A., Cruz, N. & IEEE (2010). Single beacon navigation: Localization and control of the mares auv, *Oceans 2010*, Oceans-IEEE, IEEE, New York.
Ferreira, B., Matos, A., Cruz, N. & Pinto, M. (2010). Modeling and control of the MARES autonomous underwater vehicle, *Marine Technology Society Journal* 44(2): 19–36.
Fossen, T. I. (1994). *Guidance and Control of Ocean Vehicles*, John Whiley & Sons Ltd.
Frank, P. M. & Ding, X. (1997). Survey of robust residual generation and evaluation methods in observer-based fault detection systems, *Journal of Process Control* 7(6): 403–424.

Gelb, e. A. (1974). *Optimal applied estimation*, MIT Press.
Hoerner, S. F. (1993). *Fluid dynamics drag: theoretical, experimental and statistical information*, Published by the author, Bakerfield, CA.
Khalil, H. (2002). *Nonlinear Systems*, 3 edn, Prentice Hall, Michigan State University, East Lansing.
Matos, A. & Cruz, N. (2009). MARES - navigation, control and on-board software, *in* A. V. Inzartsev (ed.), *Underwater Vehicles*, InTech.
Seshagiri, S. & Khalil, H. K. (2005). Robust output feedback regulation of minimum-phase nonlinear systems using conditional integrators, *Automatica* 41(1): 43–54.
Singh, A. & Khalil, H. K. (2005). Regulation of nonlinear systems using conditional integrators, *International Journal of Robust and Nonlinear Control* 15(8): 339–362. 930VR
Teixeira, F. C., Aguiar, A. P. & Pascoal, A. (2010). Nonlinear adaptive control of an underwater towed vehicle, *Ocean Engineering* 37(13): 1193–1220.
Triantafyllou, M. S. & Hover, F. S. (2002). Maneuvering and control of marine vehicles, *Technical report*, Massachussets Institute of Technology.
White, F. M. (2008). *Fluid Mechanics*, 5 edn, McGraw Hill, Boston.
Wu, N. E., Zhang, Y. M. & Zhou, K. M. (2000). Detection, estimation, and accommodation of loss of control effectiveness, *International Journal of Adaptive Control and Signal Processing* 14(7): 775–795.
Zhang, F. Z. (1997). Quaternions and matrices of quaternions, *Linear Algebra and Its Applications* 251: 21–57.
Zhang, Y. M. & Jiang, J. (2002). Active fault-tolerant control system against partial actuator failures, *IEEE Proceedings-Control Theory and Applications* 149(1): 95–104.

4

Robust Control Design for Automotive Applications: A Variable Structure Control Approach

Benedikt Alt and Ferdinand Svaricek
University of the German Armed Forces Munich
Germany

1. Introduction

The steady rise in fuel prices and the increased awareness on climate issues led and still lead to considerable efforts in the development of automotive engines and drivetrains (Guzzella & Sciarretta (2005)). Thus, fuel savings and emission reduction are of general interest and obviously as important as improved riding comfort or driveability.

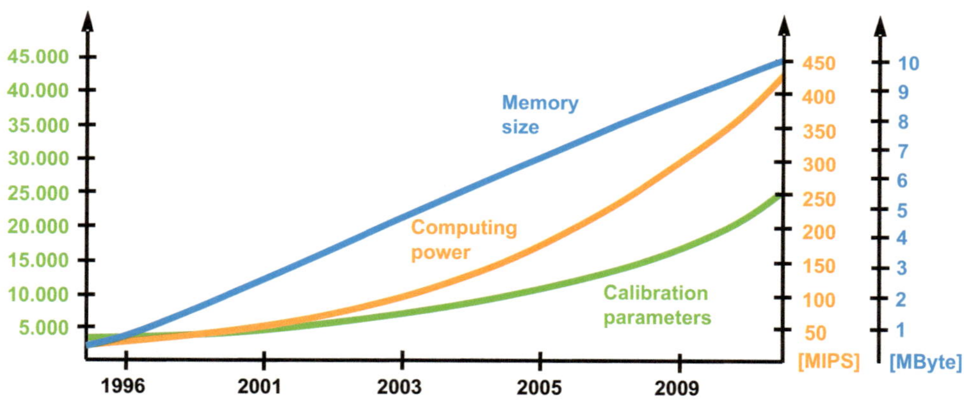

Fig. 1. Evolution of memory size, computing power and number of calibration parameters of an automotive engine control unit from 1996 to 2009 (ETAS GmbH (2010))

However, it is hard to find a suitable trade-off between all of these requirements and many resulting solutions lead to increased complexity of the vehicle systems. This is in particular true for common automotive combustion engines where the number of free calibration parameters of the corresponding electronic control unit (ECU) software has been increased up to five times during the last fifteen years (see Figure 1). From today's state of the art it takes up to five calibration engineers one whole year to finish all the calibration work on a series-production engine (Reif (2007)). Consequently, this time consuming calibration results in considerable development cost. Since the complexity of future drivetrains (e.g. battery electric vehicles or hybrid electric vehicles) will be drastically increased (Ehsani et al. (2010)) an ongoing rise on development cost is inevitable. However, with this effect cars may become unaffordable to many customers in the near future. Thus, novel control design strategies have to be introduced such that today's and future calibration work is minimized.

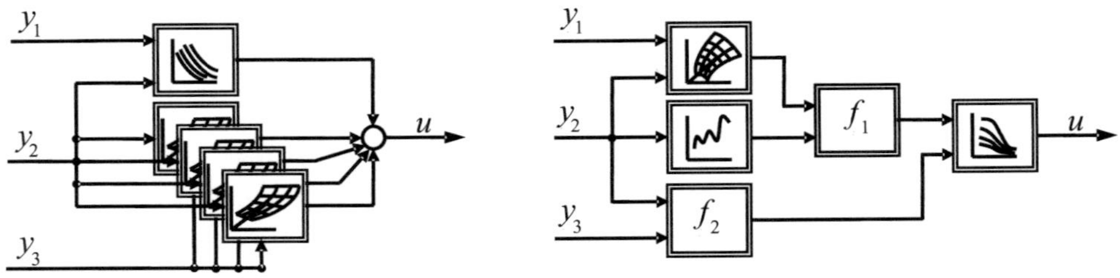

Fig. 2. Comparison of black box approach (left) and model based approach (right) for control design in ECU software development

Since several years common black box control design approaches are more and more replaced with model based design strategies (see Figure 2). Here, the corresponding control parameters are referred to a single subsystem of the plant and no longer to the entire process. Thus, each parameter has a clear physical meaning and any model uncertainties or unknown load torque disturbances can be systematically incorporated within the control design process. With this strategy a considerable reduction of calibration efforts can be achieved (Schopp et al. (2010)). However, the efforts for the design of a suitable process model have to be taken into account as well, since it is not easy to find a trade-off between model accuracy and complexity. Thus, it becomes clear that model based control design strategies are not the unique solution to minimize the development cost on ECU software. Often the desired reduction of efforts is less than expected. To overcome this major drawback a combination of model based and robust control design strategies is proposed since it is the best way to reduce the modeling and calibration efforts similarly (Alt (2010)).

Among robust control design methods the class of variable structure controllers (especially sliding mode controllers (SMCs)) is well known for their low burden on model accuracy. Regarding the operating range of a common combustion engine it is well known that the operating range of sliding mode control is enlarged compared to conventional solutions with gain scheduling techniques and heuristically tuned PI or PID controllers even if simple linear system models are used for control design (Edwards & Spurgeon (1998)). Hence, the total number of required operating points can be considerably reduced thus leading to less calibration efforts (see Figure 3). Moreover, sliding mode control shows good robustness properties against a wide class of model uncertainties and external disturbances including environmental influences, aging and tolerance effects (Hung et al. (1993); Utkin (1977)).

Due to its discontinuous nature a high frequency oscillation may arise and deteriorate the performance of closed-loop systems with SMCs (Utkin et al. (2009)). These so called chattering effects take usually place if the plant includes actuator dynamics which cannot be neglected (e.g. electromechanical actuators) or if the discretization effects affect the overall system behaviour. To alleviate the chattering phenomenon several control design approaches have been investigated. Among these control design methods second order sliding modes (SOSM) controllers attract great attention since they guarantee excellent robustness properties and even better accuracy compared to conventional SMCs (Alt et al. (2009a); Bartolini et al. (1998); Butt & Bhatti (2009); Khan et al. (2001); Levant (1993)).

In this contribution a SOSM based control strategy will be applied to a typical automotive control design task, namely the idle speed control (ISC) of a spark ignition (SI) engine (Alt

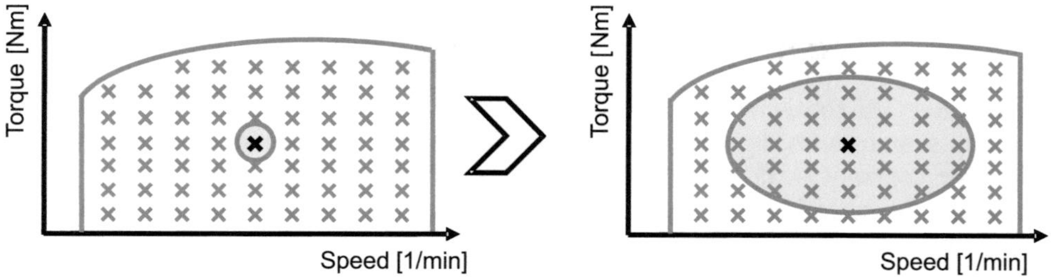

Fig. 3. Operating ranges of heuristically tuned PI or PID controller (left) and SMC controller (right)

et al. (2009b)). For this purpose a short introduction on SMCs and SOSM based controllers will be given. Here, the robustness properties will be analyzed and it will be shown how chattering effects can be alleviated efficiently. Then, the ISC control design task will be outlined and a corresponding simulation model will be introduced and validated on a research vehicle. Finally, the SOSM based control design approach will be applied to the ISC problem. Since the control parameters remain fixed no gain scheduling technique is necessary. Thus, the overall design and calibration efforts are considerably reduced compared to the series-production solution. However, representative nonlinear simulation and experimental results show impressively that the proposed controller is still able to satisfy all current ISC design requirements.

2. Sliding mode control and second order sliding mode control

Sliding mode control theory has attracted great interest among scientists and control engineers within the last decades. The resulting control laws can be applied but are not restricted to affine nonlinear single input single output (SISO) systems

$$\begin{aligned} \dot{\mathbf{x}}(t) &= \mathbf{f}(\mathbf{x}(t)) + \mathbf{g}(\mathbf{x}(t))\, u(t) + \mathbf{z}(\mathbf{x}(t)) \\ y(t) &= h(\mathbf{x}(t)) \end{aligned} \tag{1}$$

where $\mathbf{x} = \mathbf{x}(t) \in \mathbb{R}^n$, $u = u(t) \in \mathbb{R}$ and $y = y(t) \in \mathbb{R}$. The system nonlinearities $\mathbf{f} \in \mathbb{R}^n$, $\mathbf{g} \in \mathbb{R}^n$ and $h \in \mathbb{R}$ are considered to be sufficiently smooth (Bartolini et al. (1998)). The discontinuous structure of these sliding mode controllers allows to switch between different system structures (or components) such that a new type of system motion, called sliding mode, exists in a dedicated manifold $\sigma(\mathbf{x}) = 0$. In particular the corresponding system trajectory moves onto this sliding manifold in finite time which leads to better system performance than the asymptotic behaviour of e.g. linear control systems. After reaching the manifold $\sigma(\mathbf{x}) = 0$ the system motion is uniquely characterized from the design of the sliding manifold and independent to any of the corresponding subsystems. Thus, once the system trajectory reached the sliding manifold its motion is insensitive to model uncertainties and disturbances that satisfy the so-called matching conditions (see Drazenovic (1969)). Here, the term matching conditions means that all these model uncertainties and disturbances enter the system through the control channel.

Regarding the overall control gain of the sliding mode control law the aforementioned robustness properties are easy to understand. As soon as the system trajectory reaches the sliding manifold the corresponding sliding variable $\sigma(\mathbf{x})$ is equal to zero. Since $\sigma(\mathbf{x})$ appears in the denominator of the overall control gain $k = \frac{u}{\sigma}$ this variable is drastically increased. In practice that means that the discontinuous control law acts directly with its maximum but finite control input if the system motion on the sliding manifold is affected. Due to that high gain effect the robustness properties of the sliding mode control system are similar to a closed-loop system with high-gain control law (Khalil (1996)). On the contrary to this class of nonlinear controllers the corresponding sliding mode control input doesn't suffer from unrealistic large control efforts. Instead it is well known that this control input is bounded by a finite value as shown in Figure 4. In the remainder of this section the following second

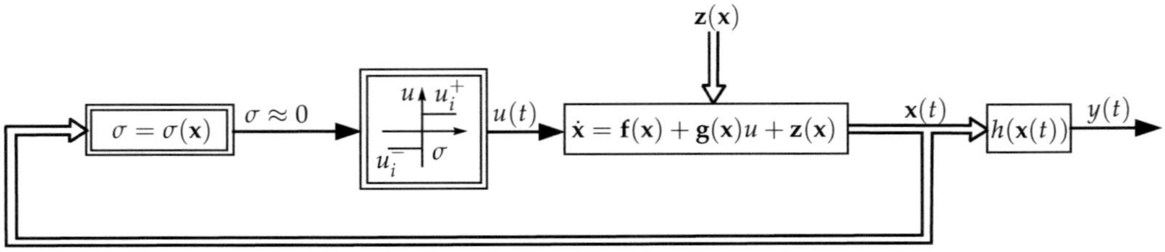

Fig. 4. Nonlinear single input single output (SISO) system with sliding mode controller, high gain effect with good robustness properties against matched model uncertainties and external disturbances after the system trajectory has reached the sliding manifold

order system

$$\begin{aligned} \dot{x}_1 &= x_2 \,, \\ \dot{x}_2 &= a^2 u \end{aligned} \tag{2}$$

with $a > 0$ is considered to explain the design of first order and second order sliding mode control laws. First, a so-called first order sliding mode control law (Perruquetti & Barbot (2002)) is given that guarantees the existence and the reachability (Edwards & Spurgeon (1998)) of the sliding motion in the entire state space:

$$u_{smc} = -\delta \left|x_1\right| \operatorname{sgn}\left(\sigma\left(x_1, x_2\right)\right) = \begin{cases} \delta \left|x_1\right| & \text{for } \sigma\left(\mathbf{x}\right) < 0 \\ -\delta \left|x_1\right| & \text{for } \sigma\left(\mathbf{x}\right) > 0 \end{cases} . \tag{3}$$

As soon as the system trajectory reaches the sliding manifold $\sigma(\mathbf{x}) = 0$ the control input u_{smc} shows a switching effect with infinite frequency. Of course, this infinite fast switching effect cannot occur in practical applications since each actuator has a limited bandwith and the corresponding control laws are calculated with finite sampling rates. Thus, the intended ideal sliding motion is also not realizable and the system trajectory oscillates around the given manifold as shown in Figure 5. These so-called chattering effects have to be alleviated in practical applications since chattering may lead to high power loss or even damages on the actuators or the overall system (Utkin et al. (2009)). Thus, the alleviation of chattering effects has been also intensively studied in the last decades (Bartolini et al. (1998); Hung et al. (1993); Utkin (1977); Utkin et al. (2009); Young et al. (1999)). Here, the so-called boundary layer approach (Edwards & Spurgeon (1998)) represents an efficient solution for many practical applications. However, it is well known that this alleviation approach suffers from reduced robustness properties since the system trajectory is no longer able to reach the sliding manifold exactly. Instead it can only be guaranteed that the trajectory moves within a dedicated boundary layer around the sliding manifold.

Another interesting approach for the alleviation of chattering effects can be found within the class of second order sliding mode (SOSM) controllers (Bartolini et al. (1998); Levant (1993);

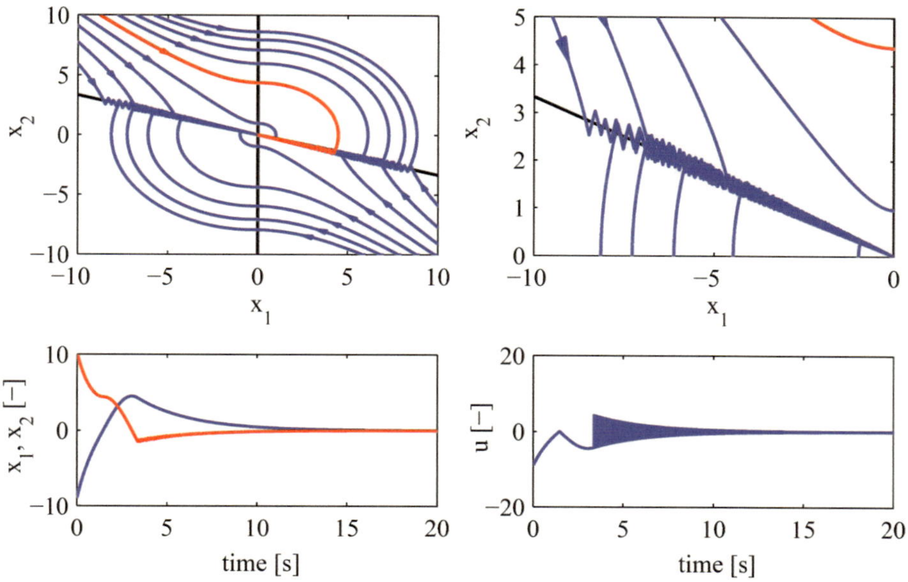

Fig. 5. Nonlinear simulation results for example in (2) with first order sliding mode control law in (3), sampling time $t_s = 20$ ms, phase portrait of closed-loop system (up, left), zoom-in of phase portrait (up, right), system states x_1 (blue) and x_2 (red), (low, left) and control input u (low, right)

Utkin et al. (2009)). The corresponding design of this specific control law which is referred to super twisting algorithm (STA) is briefly discussed in the remainder of this section. For this purpose the class of affine nonlinear SISO systems is considered as already introduced in (1). Additionally, it is assumed that the system trajectory should reach the sliding manifold $\dot{\sigma} = \sigma = 0$ in finite time and that the relative degree of this system is one, i.e. the control input appears in the first time derivative of the sliding variable $\sigma(\mathbf{x})$. Although this assumption looks restrictive it has been shown in Alt (2010) that many systems in the field of automotive, electric drive or robotic systems fulfill this requirement. Finally, the first and second order time derivatives of $\sigma(\mathbf{x})$ have to be calculated for the following control design steps:

$$\begin{aligned} \dot{\sigma} &= \frac{\partial}{\partial t}\sigma + \frac{\partial}{\partial \mathbf{x}}\sigma\left(\mathbf{f}(\mathbf{x}) + \mathbf{g}(\mathbf{x})\,u + \mathbf{z}(\mathbf{x})\right), \\ \ddot{\sigma} &= \underbrace{\frac{\partial}{\partial t}\dot{\sigma} + \frac{\partial}{\partial \mathbf{x}}\dot{\sigma}\left(\mathbf{f}(\mathbf{x}) + \mathbf{g}(\mathbf{x})\,u + \mathbf{z}(\mathbf{x})\right)}_{\phi(\mathbf{x})} + \underbrace{\frac{\partial}{\partial u}\dot{\sigma}}_{\gamma(\mathbf{x})}\,\dot{u}\,. \end{aligned} \tag{4}$$

From $\dot{\sigma}$ and $\ddot{\sigma}$ it can be clearly seen that the lumped model uncertainties and external disturbances $\mathbf{z}(\mathbf{x})$ appear within $\phi(\mathbf{x})$ and $\gamma(\mathbf{x})$. However, no detailed knowledge of these nonlinear relationships is required for the following control design steps. Instead it turned out to be sufficient to introduce dedicated lower and upper bounds $|\phi(\mathbf{x})| < \Phi$ and $0 < \Gamma_m < \gamma(\mathbf{x}) < \Gamma_M$ on $\phi(\mathbf{x})$ and $\gamma(\mathbf{x})$, respectively to cope with the matched model uncertainties and external disturbances where $\Phi, \Gamma_m, \Gamma_M \in \mathbb{R}^+$. Thus, the robustness properties are considered to be similar to those of a closed-loop system with first order sliding mode control law.

For a better general understanding the reduction of the chattering effects can be related to the additional integrator within the well-known form of the super twisting algorithm (Fridman & Levant (2002)) control law

$$\begin{aligned} u_{sta} &= u_{sta,1} + u_{sta,2} \,, \\ \dot{u}_{sta,1} &= \begin{cases} -u_{sta} & \text{for } |u_{sta}| > 1 \\ -W \mathrm{sgn}(\sigma) & \text{for } |u_{sta}| \leq 1 \,, \end{cases} \\ u_{sta,2} &= \begin{cases} -\lambda \, |\sigma_0|^{\rho} \, \mathrm{sgn}(\sigma) & \text{for } |\sigma| > \sigma_0 \\ -\lambda \, |\sigma|^{\rho} \, \mathrm{sgn}(\sigma) & \text{for } |\sigma| \leq \sigma_0 \,. \end{cases} \end{aligned} \tag{5}$$

Thus, the discontinuous first order sliding mode control law in (3) is replaced by a continuous alternative. However, the resulting implementation and calibration efforts are increased with regards to practical applications.

For the calculation of the control gains W, λ and ρ the first order time derivative $\dot{u}_{sta}$ of the control variable from (5) has to be inserted in the right hand side of the second order time derivative $\ddot{\sigma}$ in (4) where $|u_{sta}| \leq 1$ and $|\sigma| \leq \sigma_0$:

$$\ddot{\sigma} = \phi(\mathbf{x}) - \gamma(\mathbf{x}) \left(W \mathrm{sgn}(\sigma) + \rho\lambda \frac{\dot{\sigma}}{|\sigma|^{1-\rho}} \right) . \tag{6}$$

Considering the lower and upper bounds Φ, Γ_m and Γ_M of $\phi(\mathbf{x})$ and $\gamma(\mathbf{x})$, the right hand side of $\ddot{\sigma}$ turns from an ordinary differential equation into a differential inclusion (Emelyanov et al. (1996); Levant (1993)):

$$\ddot{\sigma} \in [\Gamma_m W - \Phi, \Gamma_M W - \Phi] - [\Gamma_m, \Gamma_M] + \rho\lambda \frac{\dot{\sigma}}{|\sigma|^{1-\rho}} . \tag{7}$$

With regards to the calibration of the control gains W, λ and ρ it can be clearly seen from (7) that no unique bounds can be given such that the system trajectory reaches $\dot{\sigma} = \sigma = 0$ in finite time. However, with some further dedicated assumptions some more conservative bounds (Fridman & Levant (2002)) on W, λ and ρ can be introduced to satisfy this stringent condition:

$$\begin{aligned} W &> \frac{\Phi}{\Gamma_m} \,, \\ \lambda^2 &\geq \frac{4\Phi}{\Gamma_m^2} \frac{\Gamma_M (W + \Phi)}{\Gamma_m (W - \Phi)} \,, \\ 0 &< \rho \leq 0.5 \,. \end{aligned} \tag{8}$$

Here, it has to be noted that the assumptions on these conservative bounds for deriving W, λ and ρ may vary from reference to reference (see Levant (1993; 1998)). In practice, these sufficient conditions on W, λ and ρ are often used to simplify the heuristic calibration process (Bartolini et al. (1999)).

Finally, the introductory example in (2) is considered to show the efficiency of the super twisting algorithm in terms of chattering alleviation purposes. The corresponding simulation results are depicted in Figure 6 and it can be clearly seen that the system trajectories reach the sliding manifold $\dot{\sigma} = \sigma = 0$ in finite time. Additionally, the chattering effects are considerably reduced.

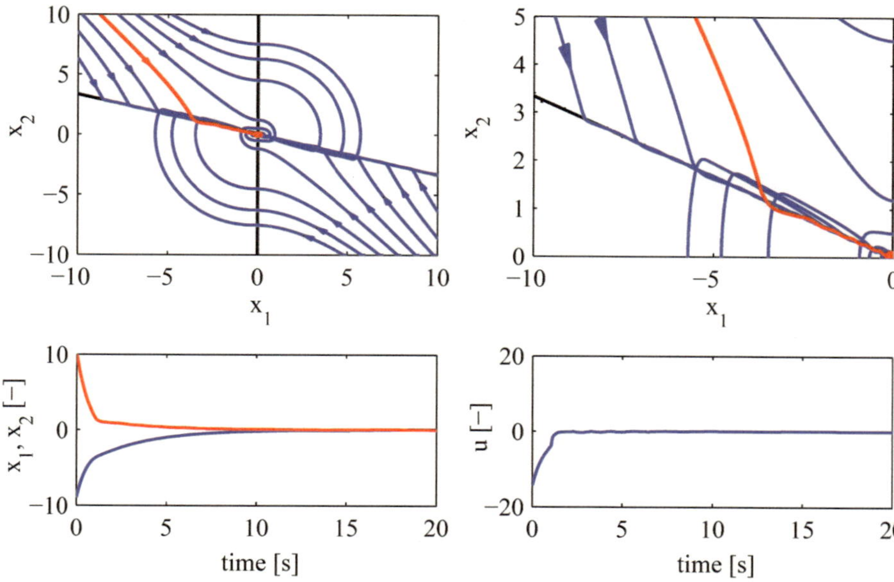

Fig. 6. Nonlinear simulation results for example in (2) with second order sliding mode control law in (5), sampling time $t_s = 20$ ms, phase portrait of closed-loop system (up, left), zoom-in of phase portrait (up, right), system states x_1 (blue) and x_2 (red), (low, left) and control input u (low, right)

3. Nonlinear engine model

In this section a mathematical model of the spark ignition (SI) engine is briefly discussed. In the remainder of this contribution this engine model will basically be used as a nonlinear simulation model and thus as virtual engine test rig. It incorporates both the overall system dynamics of the engine and the torque structure of current engine management systems. For modeling purposes of the engine a continuous time mean value modeling approach turned out to be sufficient at idle condition (Guzzella & Sciarretta (2005)). This means, that all internal processes of the engine are spread out over one combustion period and differences from cylinder to cylinder are neglected. Thus, it is sufficient to take only the electronic throttle with its position controller, the intake manifold and the rotational dynamics of the crankshaft into account:

$$\begin{aligned} \dot{\alpha}_{thr} &= -\frac{1}{\tau_{thr}}\alpha_{thr} + \frac{1}{\tau_{thr}}\alpha_{thr,u} \,, \\ \dot{p}_{im} &= \frac{R\theta_{im}}{V_{im}}(\dot{m}_{thr} - \dot{m}_{cc}) \,, \\ \dot{N} &= \frac{30}{\pi J}(T_{ind} - T_{loss} - T_{load}) \,, \end{aligned} \tag{9}$$

where τ_{thr} represents the time constant of the closed loop behaviour of the electronic throttle. The variables $\dot{m}_{thr} = \dot{m}_{thr}(p_{im}, \alpha_{thr,u})$ and $\dot{m}_{cc} = \dot{m}_{cc}(p_{im}, N)$ denote the air mass flow rates into the intake manifold and the combustion chamber, respectively. For the calculation of the indicated torque $T_{ind} = T_{ind}(\dot{m}'_{cc}, T_{ign,u}(t - \tau_d))$ per combustion cycle the air mass flow rate into the combustion chamber has to be related to the crank-angle domain based software features of the electronic control unit:

$$\dot{m}'_{cc} = \frac{120}{N_{cc}N}\dot{m}_{cc} \,. \tag{10}$$

Additionally, the physical actuator inputs (throttle position $\alpha_{thr,u}$ and ignition setting $\alpha_{ign,SP}$) are transformed into torque demands $T_{air,u}$ and $T_{ign,u}$ on the air path and on the ignition path, respectively. In general the torque demand $T_{ign,u}$ is considered as only control input acting directly on the indicated torque T_{ind} and hence on the engine speed N. The remaining control input $T_{air,u}$ on the air path influences however the maximum brake torque $T_{bas} = T_{bas}(\dot{m}_{cc}, N)$. Thus both control inputs affect also the torque reserve

$$T_{res} = T_{bas} - T_{ind} \,. \tag{11}$$

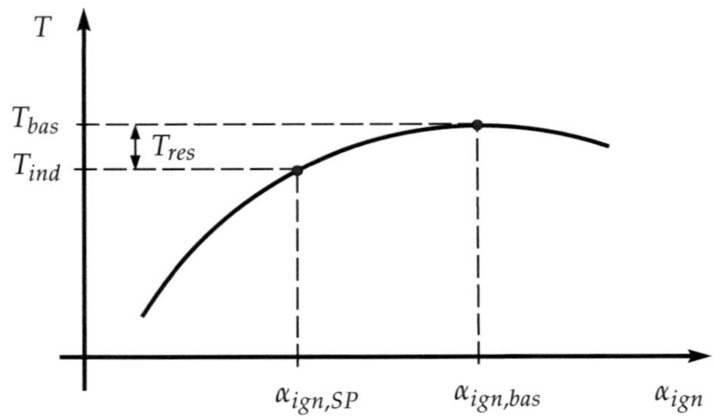

Fig. 7. Engine torque over spark ignition setting α_{ign} with fixed intake manifold mass flow $\dot{m}_{thr} = \dot{m}_{thr}(p_{im}, \alpha_{thr,u})$, this characteristic is also known as spark sweep

As seen in Figure 7 the torque reserve T_{res} represents the amount of torque that is available on the ignition path. Hence there exists a unidirectional coupling between the torque demands on the air and the ignition path and the system outputs because the air path is able to adjust the dynamic actuator constraints on the ignition path. With equations (9), (10), (11) and the ECU related software structure from Alt (2010) a nonlinear state space representation can be derived, where $\mathbf{x} = [\alpha_{thr}\ p_{im}\ N]^T$, $\mathbf{u} = \left[T_{ign,u}\ T_{air,u}\right]^T$ and $\mathbf{y} = [N\ T_{res}]^T$:

$$\begin{aligned} \begin{bmatrix} \dot{x}_1 \\ \dot{x}_2 \\ \dot{x}_3 \end{bmatrix} &= \begin{bmatrix} f_1(x_1, x_2, x_3, u_2) \\ f_{21}(x_2, x_3) + f_{22}(x_1) \\ f_{31}(x_2, x_3) + f_{32}(u_1) \end{bmatrix} , \\ \begin{bmatrix} y_1 \\ y_2 \end{bmatrix} &= \begin{bmatrix} x_3 \\ h_{21}(x_2, x_3, u_2) - h_{22}(x_2, x_3, u_1) \end{bmatrix} . \end{aligned} \tag{12}$$

The structure of the overall nonlinear engine model is shown in Figure 8. Here, it can be clearly seen that there exists a unidirectional coupling between the control inputs $T_{ign,u}$, $T_{air,u}$ and the outputs N and T_{res}. In the remainder of this paper the nonlinear model (12) is used as a

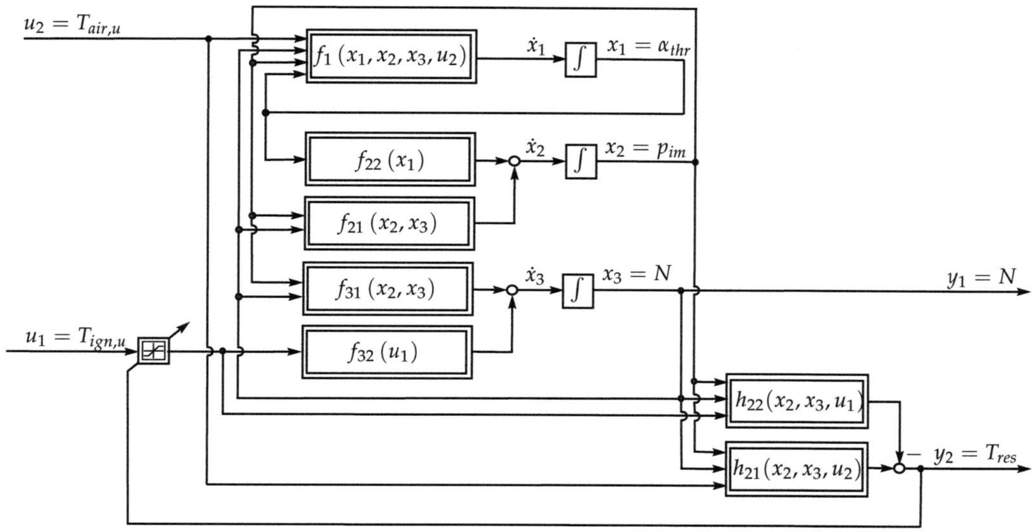

Fig. 8. Structure of nonlinear engine model

virtual test rig for the simulation studies. To show the performance of the proposed modeling approach a validation process has been carried out on a series-production vehicle with a 2.0l SI engine and a common rapid control prototyping system. Since the validation should cover the whole idle operating range different engine speed setpoints have to be considered. In Figure 9 and 10 two representative examples are shown where the corresponding engine speed setpoint $N_{SP} = 800\ 1/\text{min}$ is situated in the middle of the idle operating range. For identification purposes a step in the torque demand $T_{air,u}$ on the air path and a step in the torque demand $T_{ign,u}$ on the ignition path are applied to the system. In the first case the maximum torque T_{bas} of the engine is increased while the indicated torque T_{ind} remains nearly the same. Due to the unidirectional coupling the engine speed N is not affected. In the second case the engine speed N and the torque reserve are both affected due to the step demand on the control input $T_{ign,u}$. From both Figures it can be also seen that there exists a good matching between the outputs of the simulation model and the real plant measurements.

4. Idle speed control design

In this section a decoupling controller is proposed that will be able to hold the engine speed N and the torque reserve T_{res} at their reference values N_{SP} and $T_{res,SP}$, respectively. Whenever the engine runs at idle condition and the reference value of the torque reserve $T_{res,SP}$ is greater than zero, this ISC controller will be active. The corresponding control structure is shown in Figure 11. Here, it can be seen that the novel ISC controller includes two individual feedback controllers and a decoupling compensation.

First, the design of the decoupling compensation is shown which will improve the driver's impression on the engine quality. In particular he should not registrate any influence on the engine speed N when changes in the reference value of the torque reserve $T_{res,SP}$ occur. As seen in (12) the unilateral coupling between the control inputs $T_{ign,u}$, $T_{air,u}$ and the outputs N and T_{res} has to be taken into account such that any influence on the engine speed N vanishes. This decoupling compensation is based on a linear time invariant (LTI) model that can either

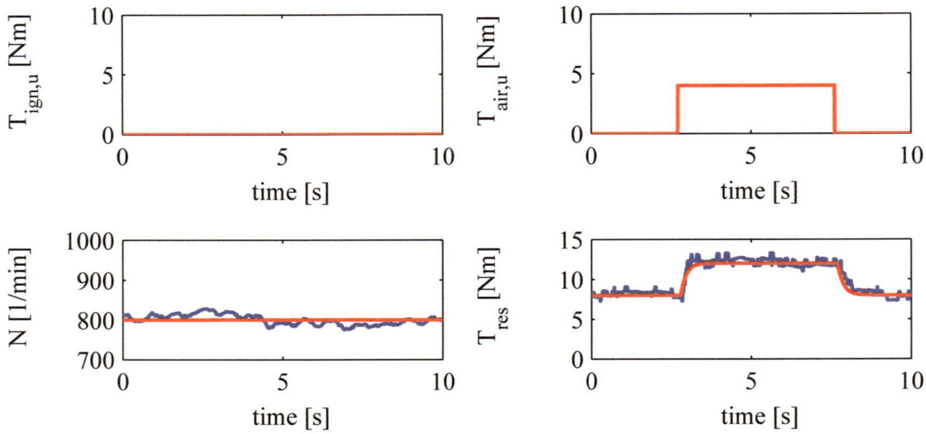

Fig. 9. Experimental results for validation of the nonlinear engine model, step on the air path torque demand: Control input $T_{ign,u}$ (up, left), control input $T_{air,u}$ (up, right), engine speed N (low, left), torque reserve T_{res} (low, right), experimental results (blue), simulation results (red)

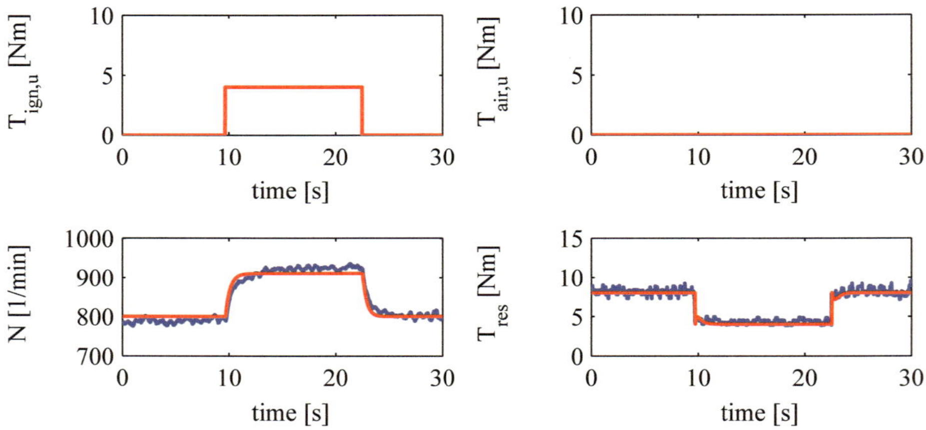

Fig. 10. Experimental results for validation of the nonlinear engine model, step on the ignition path torque demand: Control input $T_{ign,u}$ (up, left), control input $T_{air,u}$ (up, right), engine speed N (low, left), torque reserve T_{res} (low, right), experimental results (blue), simulation results (red)

be derived using analytical linearization or by system identification methods (Ljung (1999)). In many automotive control problems the latter techniques are more common since often no detailed nonlinear mathematical model is available. Instead test rig measurements are easily accessible. For this reason the remainder of the work is also based on identification methods. The resulting LTI models are generally valid in the neighbourhood of given operating points. Here, the required test rig measurements are taken from the validated nonlinear simulation model of (12) for the sake of simplicity. The aforementioned operating point with its reference values for the engine speed $N_{SP,0} = 800\,1/\text{min}$ and the torque reserve $T_{res,SP,0} = 8$ Nm represents a good choice for the following control design steps since it is situated in the middle

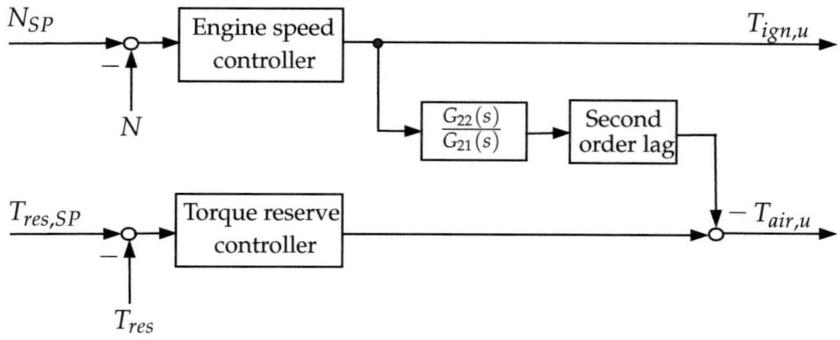

Fig. 11. Block diagram of the decoupling controller at idle condition

of the range at idle condition. If the behaviour of the nonlinear engine model at this operating point has to be described with a LTI model it is clear that the unidirectional coupling structure is still conserved. Hence, the LTI model can be written as

$$\begin{aligned} N(s) &= G_{12}(s)T_{ign,u}(s) , \\ T_{res}(s) &= G_{21}(s)T_{air,u}(s) + G_{22}(s)T_{ign,u}(s) . \end{aligned} \tag{13}$$

The operating point dependent continuous time transfer functions $G_{12}(s)$, $G_{21}(s)$ and $G_{22}(s)$ are calculated from various step responses using MATLAB's System Identication Toolbox (Ljung (2006)):

$$\begin{aligned} G_{12}(s) &= \frac{246.4}{s + 2.235} , \\ G_{21}(s) &= \frac{4.618}{s + 4.625} , \\ G_{22}(s) &= \frac{-26.14s - 91.07}{s^2 + 45.03s + 90.9} . \end{aligned} \tag{14}$$

The parameters of $G_{12}(s)$, $G_{21}(s)$ and $G_{22}(s)$ are calculated numerically using a maximum likelihood criterion. That means the underlying identification algorithm is based on continuous time low order transfer functions and it includes an iterative estimation method that minimizes the prediction errors. From the LTI model in (13) it can be seen that the transfer function

$$G_{Ds}(s) = \frac{G_{22}(s)}{G_{21}(s)} \tag{15}$$

helps to compensate the influence of the torque demand $T_{ign,u}$ on the torque reserve T_{res} efficiently. Hence, the decoupled system with its inputs $T_{ign,u}$ and $T_{air,u}$ can be controlled by two feedback controllers which are designed independently of each other. Since the dynamics of the air path are generally much slower than the dynamics on the ignition path a second order lag is additionally introduced to smooth the transient behaviour of the decoupling compensation in (15), see Figure 11. The corresponding damping of this filter and its natural frequency have to be determined experimentally.

For the design of both feedback controllers linear control theory would be generally sufficient as shown in current series-production applications or even in Kiencke & Nielsen (2005). Nevertheless, it is well known that classical linear controllers often do their job only in the neighbourhood of an operating point and the control parameters have to be scheduled over the entire operating range. This leads to time-consuming calibration efforts. In this work the potential of sliding mode control theory will be particularly analyzed with regards to reduced calibration efforts. Hence, both feedback controllers are designed using a second order sliding modes (SOSM) control design approach that has already been introduced in Section 2. This so-called super twisting algorithm (STA) has been developed to control systems with relative degree one in order to avoid chattering effects. Furthermore, it does not need any information on the time derivative of the sliding variable. For these reasons the super twisting algorithm has become very popular in recent years and it has been adopted to many real world control applications so far (Alt et al. (2009a); Butt & Bhatti (2009); Perruquetti & Barbot (2002)). In the following steps the control law for the engine speed N is derived while the engine runs at idle and the condition $T_{res} > 0$ holds true. This control law includes two major parts:

$$
\begin{aligned}
u_N &= u_{N,1} + u_{N,2}\,, \\
\dot{u}_{N,1} &= \begin{cases} -u_{N,1} & \text{for } |u_{N,1}| > 1 \\ -W_{N,1}\text{sgn}(\sigma_N) & \text{for } |u_{N,1}| \leq 1\,, \end{cases} \\
u_{N,2} &= -\lambda_{N,1}\,|\sigma_N|^{\rho_{N,1}}\,\text{sgn}(\sigma_N)\,,
\end{aligned}
\tag{16}
$$

where $\sigma_N = 0$ with $\sigma_N = N - N_{SP}$ represents the engine speed related sliding manifold.

For the application of the super twisting algorithm it has to be guaranteed that the considered system has relative degree one. For this purpose the time derivative

$$
\dot{\sigma}_N = f_{31}(x_2, x_3) + f_{32}(u_1) - \dot{N}_{SP}
\tag{17}
$$

is calculated using the nonlinear model in (12). Here, it can be clearly seen that the control input u_1 appears in $f_{32}(u_1)$ and thus in the first time derivative of σ_N. Thus, the aforementioned relative degree one condition is fulfilled for this case and the super twisting algorithm can be applied. For the calibration of the control gains $W_{N,1}, \lambda_{N,1}$ and $\rho_{N,1}$ sufficient conditions for finite time convergence to the sliding surface $\sigma_N = 0$ are derived in Levant (1993). Here, it is shown that starting from an initial value $\sigma_{N,0}$ at an arbitrary time instant $t_{N,0}$ the variable σ_N converges to $\sigma_N = 0$ if the following sufficient conditions (Fridman & Levant (2002); Levant (1993; 1998)) on $W_{N,1}, \lambda_{N,1}$ and $\rho_{N,1}$ are satisfied:

$$
\begin{aligned}
W_{N,1} &> \frac{\Phi_{N,1}}{\Gamma_{N,m1}}\,, \\
\lambda_{N,1}^2 &\geq \frac{4\Phi_{N,1}}{\Gamma_{N,m1}^2}\,\frac{\Gamma_{N,M1}(W_{N,1} + \Phi_{N,1})}{\Gamma_{N,m1}(W_{N,1} - \Phi_{N,1})}\,, \\
0 &< \rho_{N,1} \leq 0.5\,.
\end{aligned}
\tag{18}
$$

Here, the variables $\Gamma_{N,m1}$ and $\Gamma_{N,M1}$ denote lower and upper limitations of the nonlinear relationship $f_{31}(x_2, x_3) - \dot{N}_{SP}$, where

$$0 < \Gamma_{N,m} \leq f_{31}(x_2, x_3) - \dot{N}_{SP} \leq \Gamma_{N,M} \,. \tag{19}$$

Additionally, the variable $\Phi_{N,1}$ represents an upper bound for all effects which appear in case of model uncertainties due to the inversion of $f_{32}(u_1)$:

$$|f_{32}(f_{32}^*(u_1))| \leq \Phi_{N,1} \,. \tag{20}$$

Here, $f_{32}^*(u_1)$ denotes the nominal value of $f_{32}(u_1)$. Hence, the design of the engine speed controller is complete. The design of the torque reserve controller runs similarly to (16). The corresponding control law includes also an integral and a nonlinear part:

$$\begin{aligned} u_{Tres} &= u_{Tres,1} + u_{Tres,2} \,, \\ \dot{u}_{Tres,1} &= \begin{cases} -u_{Tres,1} & \text{for } |u_{Tres,1}| > 1 \\ -W_{Tres,1}\text{sgn}(\sigma_{Tres}) & \text{for } |u_{Tres,1}| \leq 1 \,, \end{cases} \\ u_{Tres,2} &= -\lambda_{Tres,1} \, |\sigma_{Tres}|^{\rho_{Tres,1}} \, \text{sgn}(\sigma_{Tres}) \,. \end{aligned} \tag{21}$$

where $\sigma_{Tres} = 0$ with $\sigma_{Tres} = T_{res} - T_{res,SP}$ represents the torque reserve related sliding manifold.

For the application of the super twisting algorithm it has to be again guaranteed that the considered system has relative degree one. For this purpose the time derivative

$$\dot{\sigma}_{Tres} = \frac{\partial h_2}{\partial x_2} f_{21}(x_2, x_3) + \frac{\partial h_2}{\partial x_2} f_{22}(x_1) - \dot{T}_{res,SP} \,. \tag{22}$$

is calculated using the nonlinear relationship from (12) while the corresponding time derivative of T_{res} is simplified to

$$\dot{T}_{res} \approx \frac{\partial h_2}{\partial x_2} \dot{x}_2 \,. \tag{23}$$

From (22) it can be clearly seen that the state x_1 appears in the nonlinear relationship $\frac{\partial h_2}{\partial x_2} f_{22}(x_1)$ and thus in the first time derivative of σ_{Tres}. However, to satisfy the relative degree one condition the dynamics of the subordinated electronic throttle control loop $\dot{x}_1 = f_1(x_1, x_2, x_3, u_2)$ in (12) have to be neglected for the following control design steps. This assumption is justified since the time lag of the subordinated throttle control loop is ten times smaller than the remaining ones of the SI engine model. With this simplification the state $x_1 = \alpha_{thr}$ is assumed to be equal to the control input $\alpha_{thr,SP}$ of the subordinated closed-loop system.

Under these conditions the time derivative of the torque reserve related sliding surface is given with

$$\dot{\sigma}_{Tres} = \frac{\partial h_2}{\partial x_2} f_{21}(x_2, x_3) + \frac{\partial h_2}{\partial x_2} f_{22}(f_{22}^{*(-1)}(u_2)) - \dot{T}_{res,SP} \,. \tag{24}$$

With this assumption the corresponding system fulfills the relative degree one condition. Thus, the super twisting algorithm can be also applied to the torque reserve controller.

Regarding the control gains $W_{Tres,1}$, $\lambda_{Tres,1}$ und $\rho_{Tres,1}$ it has to be guaranteed similar to the engine speed controller that starting from an initial value $\sigma_{Tres,0}$ at an arbritary time instant

$t_{Tres,0}$ the sliding variable σ_{Tres} converges to $\sigma_{Tres} = 0$ in finite time. For this purpose the following sufficient conditions (Fridman & Levant (2002); Levant (1993; 1998)) have to be fulfilled:

$$\begin{aligned} W_{Tres,1} &> \frac{\Phi_{Tres,1}}{\Gamma_{Tres,m1}}\,, \\ \lambda^2_{Tres,1} &\geq \frac{4\Phi_{Tres,1}}{\Gamma^2_{Tres,m1}} \frac{\Gamma_{Tres,M1}(W_{Tres,1} + \Phi_{Tres,1})}{\Gamma_{Tres,m1}(W_{Tres,1} - \Phi_{Tres,1})}\,, \\ 0 &< \rho_{Tres,1} \leq 0.5\,. \end{aligned} \tag{25}$$

Here, the variables $\Gamma_{Tres,m1}$ and $\Gamma_{Tres,M1}$ denote lower and upper limitations of the nonlinear relationship $\frac{\partial h_2}{\partial x_2} f_{21}(x_2, x_3) - \dot{T}_{res,SP}$:

$$0 < \Gamma_{Tres,m} \leq \frac{\partial h_2}{\partial x_2} f_{21}(x_2, x_3) - \dot{T}_{,res,SP} \leq \Gamma_{Tres,M}\,. \tag{26}$$

The variable $\Phi_{Tres,1}$ represents similar to $\Phi_{N,1}$ an upper bound for all effects which appear due to possible model uncertainties that are related to the inversion of $f_{22}(u_2)$:

$$\left| \frac{\partial h_2}{\partial x_2} f_{22}(f_{22}^{*(-1)}(T_{air,u})) \right| \leq \Phi_{Tres,1}\,. \tag{27}$$

Here, $f^*_{22}(u_2)$ denotes the nominal value of $f_{22}(u_2)$. Note that, in practice the engine speed control and the torque control loops are affected by model uncertainties and external disturbances leading to imperfect decoupling properties of the multivariable system. Nevertheless, it is well known from literature (Alt et al. (2009a); Bartolini et al. (1999); Levant (1993; 1998)) that the sliding surfaces $\sigma_N = 0$ and $\sigma_{Tres} = 0$ can still be reached in this case. Thus, the engine speed control and the torque reserve control loops are supposed to be robust against any disturbances due to improper decoupling. Finally, it has been shown in Alt (2010) that this multivariable control design approach leads to better performance and less calibration efforts than a similar approach without decoupling compensation.

5. Nonlinear simulation and experimental results

This section illustrates the efficiency and the robustness properties of the proposed decoupling controller. For this purpose some representative nonlinear simulation and experimental results are shown. All the simulations are based on the nonlinear engine model of Alt (2010) with a controller sampling time of $t_s = 10$ ms. The experimental results include representative field test data with a 2.0l series-production vehicle and a common rapid control prototyping system.

In the first scenario the disturbance rejection properties of the closed-loop system are evaluated. For this purpose an additional load torque of $T_{load} = 8$ Nm (e.g. power steering) is applied to the engine at $t_1 = 4$ s and removed again at $t_2 = 9$ s. From Figure 12 it can be seen that due to this load torque the engine speed N and the torque reserve T_{res} drop below their reference values while the corresponding transients stay below $\Delta N = 40$ 1/min and $\Delta T_{res} = 8$ Nm, respectively. However, the proposed idle speed controller steers both variables back to their reference values $N_{SP} = 800$ 1/min and $T_{res,SP} = 8$ Nm within less than 2 s.

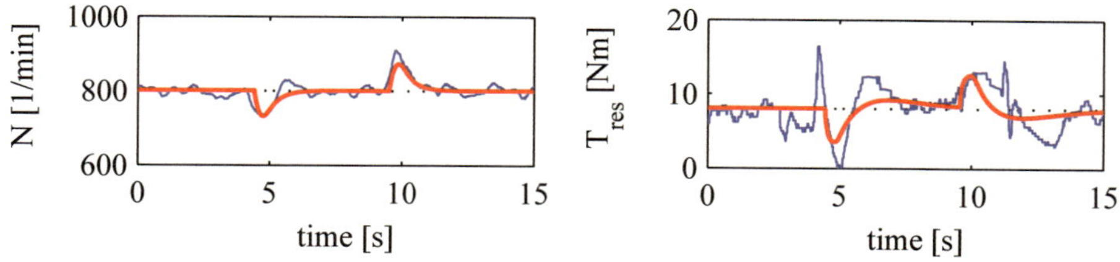

Fig. 12. Nonlinear simulation and experimental results for super twisting algorithm based decoupling controller, disturbance rejection properties: Engine speed N (left), torque reserve T_{res} (right), experimental results (blue), simulation results (red)

When disabling the load torque similar effects take place. Considering the engine speed N it can also clearly be seen that there exists a good matching between the nonlinear simulation data and the experimental measurements. For the torque reserve T_{res} this matching is less perfect since this variable is much more prone to unmodelled dynamics and tolerance effects that have not been considered in the nonlinear simulation model. This effect will be further evaluated in Section 6. In a second representative scenario the engine speed reference value

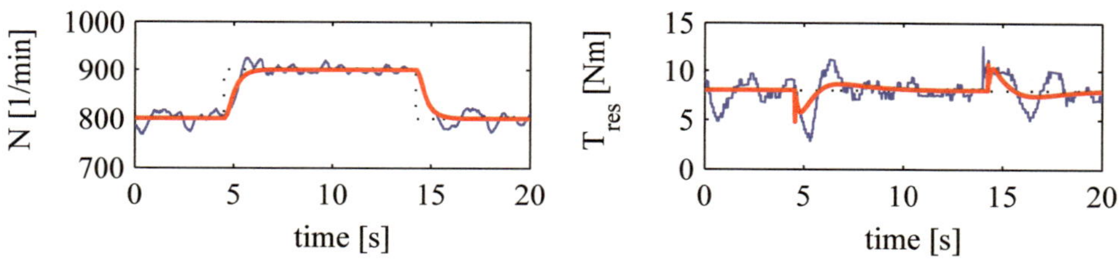

Fig. 13. Nonlinear simulation and experimental results for super twisting algorithm based decoupling controller, tracking of an engine speed reference step profile: Engine speed N (left), torque reserve T_{res} (right), experimental results (blue), simulation results (red)

N_{SP} is increased at $t_1 = 4$ s and lowered again at $t_2 = 14$ s. The corresponding simulation results are shown in Figure 13. Regarding the step response of the engine speed N it can be clearly seen that no overshoot occurs and the settling times are within less than 2 s and thus reasonable small. Additionally, the torque reserve T_{res} shows only small deviations due to the

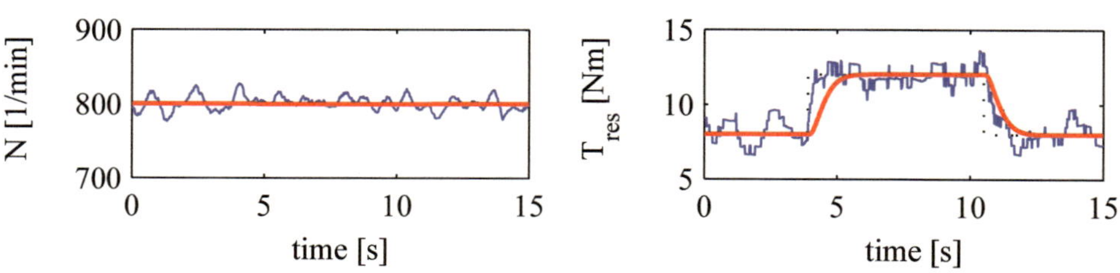

Fig. 14. Nonlinear simulation and experimental results for super twisting algorithm based decoupling controller, tracking of a torque reserve reference step profile: Engine speed N (left), torque reserve T_{res} (right), experimental results (blue), simulation results (red)

step changes on the engine speed N and it returns to its reference value $T_{res,SP}$ within a short settling time.
Similar results can be seen from Figure 14 where the torque reserve reference value $T_{res,SP}$ is increased at $t_1 = 3$ s and lowered again at $t_2 = 14$ s. During these changes on the torque reserve T_{res} the minimization of any effects on the engine speed N is considered as most important design criteria since this behaviour would affect the driver's comfort. From Figure 14 it can be clearly seen that the proposed idle speed controller is able to fulfill this requirement as specified. As known from existing series-production ISC controllers this overall performance can not be achieved using classical linear control design approaches without gain scheduling. Finally, the step response of the torque reserve T_{res} is also without any overshoot and faster than that for the engine speed N.

6. Robustness analysis

After the first experimental studies the robustness properties of the closed-loop system have to be analyzed in detail. For the sake of simplicity this analysis will be performed using the validated nonlinear simulation model from Alt (2010). Here, a representative disturbance rejection scenario is used to illustrate the major effects of model uncertainties on the closed-loop system performance. This simulation scenario includes an external load torque disturbance of $T_{load} = 10$ Nm which is applied to the engine at $t_1 = 10$ s and removed again at $t_2 = 20$ s. The overall robustness analysis covers variations of $\pm$ 10 % in up to 19 different characteristic maps of the nonlinear simulation model. In particular, the system nonlinearities f_1, f_{21}, f_{22}, f_{31}, f_{32} and $h_2 = h_2(h_{21}, hh22)$ are varied one after another using multiplicative uncertainty functions:

$$\begin{aligned}
f_1 &= d_{1,\pm} \cdot f_{1,nom} \quad \text{mit} \quad d_{1,\pm} \in [0.9, 1.1]\,,\\
f_{21} &= d_{21,\pm} \cdot f_{21,nom} \quad \text{mit} \quad d_{21,\pm} \in [0.9, 1.1]\,,\\
f_{22} &= d_{22,\pm} \cdot f_{22,nom} \quad \text{mit} \quad d_{22,\pm} \in [0.9, 1.1]\,,\\
f_{31} &= d_{31,\pm} \cdot f_{31,nom} \quad \text{mit} \quad d_{31,\pm} \in [0.9, 1.1]\,,\\
f_{32} &= d_{32,\pm} \cdot f_{32,nom} \quad \text{mit} \quad d_{32,\pm} \in [0.9, 1.1]\,,\\
h_2 &= d_{2,\pm} \cdot h_{2,nom} \quad \text{mit} \quad d_{2,\pm} \in [0.9, 1.1]\,.
\end{aligned} \tag{28}$$

Furthermore, the intake-to-torque-production delay τ_d has been increased up to 4 times to cope with any signal communication problems

$$\tau_d = \tau_{d,nom} + \Delta\tau_d \quad \text{with} \quad \Delta\tau_d = 20 \text{ ms}. \tag{29}$$

All these nonlinear simulation results are depicted in Figure 15.
In a second step all resulting deviations $\text{dev}(N)$ and $\text{dev}(T_{res})$ on the nominal behaviour of the engine speed and the torque reserve are scaled with the reference values of the operating point ($N_{SP,0} = 800$ 1/min, $T_{res,SP,0} = 8$ Nm):

$$\text{dev}(N(t)) = \frac{|\max(\Delta N_{\pm}(t)) - \Delta N_{nom}(t)|}{N_{SP}} \cdot 100\,, \tag{30}$$

$$\text{dev}(T_{res}(t)) = \frac{|\max(\Delta T_{res\pm}(t)) - \Delta T_{res,nom}(t)|}{T_{res,SP}} \cdot 100\,, \tag{31}$$

where $\Delta N_{nom}(t) = |N_{SP}(t) - N_{nom}(t)|$ and $\Delta T_{res,nom}(t) = |T_{res,SP}(t) - T_{res,nom}(t)|$ represent the resulting errors to the corresponding reference values N_{SP} and $T_{res,SP}$ while the engine

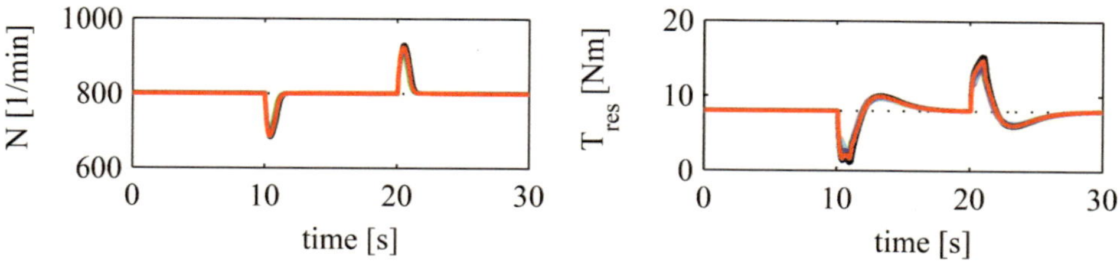

Fig. 15. Simulation results of engine speed N (left) and torque reserve T_{res} (right) for load torque disturbance rejection scenario with multiplicative variations on the system nonlinearities and the intake-to-torque production delay

operates in nominal condition. In Figure 16 the calculated deviations $\mathrm{dev}(N)$ and $\mathrm{dev}(T_{res})$ are shown for all 20 variations with strongest impact $\max(\Delta N_{\pm}(t)) = |N_{SP}(t) - N_{\pm}(t)|$ and $\max(\Delta T_{res,\pm}(t)) = |T_{res,SP}(t) - T_{res,\pm}(t)|$ on the closed-loop system.

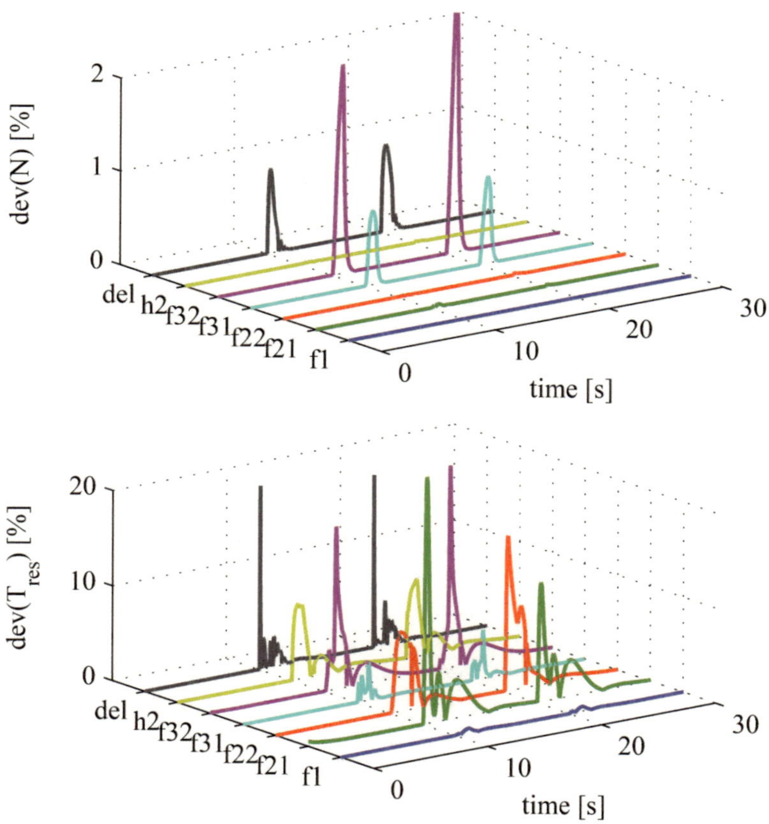

Fig. 16. Simulation results of engine speed deviations $\mathrm{dev}(N)$ (up) and torque reserve deviations $\mathrm{dev}(T_{res})$ (low) for load torque disturbance rejection scenario with multiplicative variations on the system nonlinearities and the intake-to-torque production delay

From Figure 16 it can be seen that the engine speed deviation $\mathrm{dev}(N)$ is bounded with about 1 % while the deviation $\mathrm{dev}(T_{res})$ on the torque reserve is bounded with about 15 %. This large peak deviation on the torque reserve seems to be not reasonable since the impact on the system parameters is bounded with only 10 %. However, it has to be noted that the calibration of the controllers allows to find a trade off between the accuracy on N and T_{res} and thus to penalize the engine speed error more than the torque reserve error. Since the comfort and the driver's impression on the engine quality are mainly affected by deviations on the engine speed it becomes clear that large control errors on N should be more penalized than deviations on T_{res}. Keeping this effect in mind it can be anyhow summarized that the proposed control framework shows good robustness properties despite any uncertainties in the system parameters, e.g. aging, tolerance effects or environmental influences.

7. Conclusion and future work

The paper deals with the idle speed control problem which represents an interesting multivariable control design application in the field of modern automotive spark ignition engines. In idle condition the engine speed and the torque reserve should be held at their reference values. The key design requirements include the decoupling of the underlying multivariable system and the improvement of the robustness properties against unknown load torque disturbances and tolerance effects. In the first step a nonlinear engine model is introduced that includes both the main dynamics of the engine internal processes and also the major parts of the torque structure of current engine management systems. The resulting nonlinear simulation model is validated on a series-production vehicle and it is used as a virtual engine test rig. Then, a decoupling control framework is introduced that is able to hold the idle engine speed and the torque reserve at their reference values despite external load torque disturbances or even uncertainites in the system parameters or the intake-to-torque-production delay.

The multivariable control framework consists of two independent feedback controllers and a decoupling compensation. Each of these two controllers is based on a second order sliding modes control design method that is also known as super twisting algorithm. The decoupling compensation is based on an identified linear time invariant model of the plant that is valid around a given operating point which is situated in the middle of the idle operating range. Here, the required LTI model is deduced from test rig measurements using system identification methods. The efficiency of the proposed control framework is shown by nonlinear simulation results. It can be seen that the controller shows good performance for the large signal behaviour although it is only designed for the neigbourhood of the given operating point. Nonlinear simulation and experimental results show as well that the proposed controller is able to handle a wide operating range at idle condition while the control gains remain unchanged. Hence, the proposed control framework is easier to calibrate since the number of control parameters is severly reduced compared to classical series-production control design methods using gain scheduling techniques. The efficiency and the robustness properties against system uncertainties and variations in the intake-to-torque production delay are evaluated by extended simulation studies. Current research includes the application of this second order sliding modes based multivariable design approach in the field of other automotive control design tasks (i.e. hybrid electric vehicles) and aerospace applications (i.e. smart aeroengines).

8. Acknowledgment

This work has been supported by IAV GmbH Gifhorn in Germany. The authors express their gratitude to Jan Peter Blath and Matthias Schultalbers for their support.

9. References

Alt, B. (2010). *Modellbasierte Regelung ausgewählter Komponenten im Antriebsstrang eines Kraftfahrzeugs mit Ottomotor*, PhD thesis, Universität der Bundeswehr München, Germany.

Alt, B., Blath, J., Svaricek, F. & Schultalbers, M. (2009a). Control of idle engine speed and torque reserve with higher order sliding modes, *Proceedings of the Multi-Conference on Systems and Control, Saint Petersburg, Russia*.

Alt, B., Blath, J., Svaricek, F. & Schultalbers, M. (2009b). Multiple sliding surface control of idle engine speed and torque reserve with dead start assist control, *IEEE Transactions on Industrial Electronics*, Vol. 56: 3580 – 3592.

Bartolini, G., Ferrara, A., Levant, A. & Usai, E. (1999). On second order sliding mode controllers, *in* K. Young & U. Özgüner (eds), *Variable structure systems, sliding mode and nonlinear control*, Springer, London, Berlin, Heidelberg.

Bartolini, G., Ferrara, A. & Usai, E. (1998). Chattering avoidance by second order sliding mode control, *IEEE Transactions on Automatic Control*, Vol. 43, No. 2: 241 – 246.

Butt, Q. & Bhatti, A. (2009). Estimation of gasoline-engine parameters using higher order sliding mode, *IEEE Transactions on Industrial Electronics*, Vol. 55, No. 11: pp. 3891 – 3898.

Drazenovic, B. (1969). The invariance conditions in variable structure systems, *Automatica*, Vol. 5, No. 3: pp. 287 – 295.

Edwards, C. & Spurgeon, S. (1998). *Sliding mode control, theory and applications*, Taylor and Francis Ltd. London, UK.

Ehsani, M., Gao, Y. & Emadi, A. (2010). *Modern electric, hybrid electric and fuel cell vehicles*, CRC Press, Boca Raton, London, New York.

Emelyanov, S., Korovin, S. & Levant, A. (1996). Higher-order sliding modes in control systems, *Computational Mathematics and Modeling*, Vol. 7: pp. 294 – 318.

ETAS GmbH (2010).
URL: *http://www.etas.com/de/products/applications calibrating automotive electronics.php*

Fridman, L. & Levant, A. (2002). Higher-order sliding modes, *in* W. Perruquetti & J. Barbot (eds), *Sliding mode control in engineering*, Marcel Dekker Inc., New York, Basel.

Guzzella, L. & Sciarretta, A. (2005). *Vehicle propulsion systems, Introduction to modeling and optimization*, Springer, Berlin, Heidelberg, New York.

Hung, J., Gao, W. & Hung, J. (1993). Variable structure control: a survey, *IEEE Transactions on Industrial Electronics*, Vol. 40, No. 1: 2 – 22.

Khalil, H. (1996). *Nonlinear Systems, 3rd edition*, Prentice Hall, Upper Saddle River, NJ, USA.

Khan, M., Spurgeon, S. & Bhatti, A. (2001). Robust speed control of an automotive engine using second order sliding modes, *Proceedings of the European Control Conference 2001, Porto, Portugal*.

Kiencke, U. & Nielsen, L. (2005). *Automotive control systems for engine, driveline and vehicle - 2nd edition*, Springer, Berlin, Heidelberg, New York.

Levant, A. (1993). Sliding order and sliding accuracy in sliding mode control, *International Journal of Automatic Control*, Vol. 58, No. 6: 1247 – 1263.

Levant, A. (1998). Robust exact differentiation via sliding mode technique, *Automatica,* Vol. 34: 379 – 384.

Ljung, L. (1999). *System identification - theory for the user*, Prentice Hall, Upper Saddle River, NJ, USA.

Ljung, L. (2006). *System identification toolbox for use with Matlab*, The Mathworks Inc., Natick, MA, USA.

Perruquetti, W. & Barbot, J. (2002). *Sliding mode control in engineering*, Marcel Dekker Inc. New York, Basel.

Reif, K. (2007). *Automobilelektronik*, Vieweg & Teubner, Wiesbaden, Germany.

Schopp, G., Burkhardt, T., Dingl, J., Schwarz, R. & Eisath, C. (2010). Funktionsentwicklung und Kalibration für aufgeladene Motoren - Modellbasiert vom Konzept bis zur Serie, *in* R. Isermann (ed.), *Elektronisches Management motorischer Antriebe*, Vieweg & Teubner, Wiesbaden, Germany.

Utkin, V. (1977). Variable structure systems with sliding modes, *IEEE Transactions on Automatic Control,* Vol. 22: 212 – 222.

Utkin, V., Guldner, J. & Shi, J. (2009). *Sliding mode control in electromechanical systems*, CRC Press, Boca Raton, London, New York.

Young, K., Utkin, V. & Özgüner, . (1999). A control engineer's guide to sliding mode control, *IEEE Transactions on Control Systems Technology,* Vol. 7, No. 3: pp. 328 – 342.

5

Robust Active Suspension Control for Vibration Reduction of Passenger's Body

Takuma Suzuki and Masaki Takahashi
Keio University
Japan

1. Introduction

An automotive performance has improved from the demand of ride comfort and driving stability. Many research have proposed various control system design methods for active and semi-active suspension systems. These studies evaluated the amount of reduced vibration in the vehicle body, i.e., the vertical acceleration in the center-of-gravity (CoG) of the vehicle's body (Ikeda *et al.*, 1999; Kosemura *et al.*, 2008; Itagaki *et al.*, 2008). However, any passengers always do not sit in the CoG of the vehicle body. In the seated position that is not the CoG of the vehicle body, vertical acceleration is caused by vertical, roll and pitch motion of the vehicle. In nearly the resonance frequency of the seated human, the passenger's vibration becomes larger than the seated position's vibration of the vehicle body due to the seated human dynamics.

The seated human dynamics and human sensibility of vibration are cleared by many researchers. So far some human dynamics model has been proposed (Tamaoki *et al.*, 1996, 1998, 2002; Koizumi *et al.*, 2000). Moreover, some of them are standardized in ISO (ISO-2631-1, 1997; ISO-5982, 2001). At the research as for automotive comfort with the passenger-vehicle system, M.Oya et al. proposed the suspension control method considering the passenger seated position in the half vehicle model (Oya *et al.*, 2008). G.J. Stein et al. evaluated passenger's head acceleration at some vehicle velocities and some road profiles (Guglielmino *et al.*, 2008). There are few active suspension control design methods which are positively based on a passenger's dynamics and the seating position. These methods can be expected to improve the control performance.

In this paper, new active suspension control method is developed to reduce the passenger's vibration. Firstly, a vehicle and passenger model including those dynamics at seated position is constructed. Next, a generalized plant that uses the vertical acceleration of the passenger's head as one of the controlled output is constructed to design the linear H_∞ controller. In this paper, this proposed method defines as "Passenger Control". "Passenger Control" means passenger's vibration control. Moreover, in an active suspension control, it is very important to reduce the vibration at the condition of the limited actuating force. Then, we design two methods which are "Vehicle CoG Control", and "Seat Position Control", and compare the proposed method with two methods. "Vehicle CoG Control" means vibration control of vehicle. "Seat Position Control" means vibration control of seat position. Finally, several simulations are carried out by using a full vehicle model which has active suspension system. From the result, it was confirmed that in nearly the resonance frequency of a passenger's head in the vertical direction, "Passenger Control" is effective in reducing a passenger's vibration better than "Vehicle CoG Control" and "Seat Position Control". The numerical simulation results show that the proposed method has the highest control performance which is vibration reduction of the passenger's head per generated force by the active suspension. Moreover, the results show that the proposed method has robustness for the difference in passenger's vibration characteristic.

2. Modeling

2.1 Modeling of the vehicle

Figure 1 shows a full vehicle model which is equipped with an active suspension between each wheel and the vehicle body. The weight of the vehicle body is supported by the spring. We assume that a vehicle model is a generic sedan car as shown in Table 1. The equations of motion which are, bounce, roll, pitch and each unsprung motion are as follows:

$$M_b \ddot{z}_{cg} = \sum_{i=1}^{4} f_{si}\,, \tag{1}$$

$$I_r \ddot{\phi} = \frac{T_f}{2}\left(f_{s1} - f_{s2}\right) + \frac{T_r}{2}\left(f_{s3} - f_{s4}\right) + M_b g H_r \phi\,, \tag{2}$$

$$I_p \ddot{\theta} = -L_f\left(f_{s1} + f_{s2}\right) + L_r\left(f_{s3} + f_{s4}\right) + M_b g H_p \theta\,, \tag{3}$$

$$M_{ti} \ddot{z}_{ui} = -F_{di} + K_t z_{ti} \quad (i = 1, \cdots, 4)\,. \tag{4}$$

where H_r is the distance from a roll center to the CoG of the vehicle body, and H_p is the distance from a pitch center to the CoG of the vehicle body. These parameters are constant. The spring coefficients of each wheel are different from each other, and were set to $K_{1,2} = K_f$, $K_{3,4} = K_r$, z_{si} means a suspension stroke of each wheel, z_{ti} means deformation of the each tire.

$$\begin{aligned}
z_{s1} &= z_{cg} + T_f/2\phi - L_f\theta - z_{u1} \\
z_{s2} &= z_{cg} - T_f/2\phi - L_f\theta - z_{u2} \\
z_{s3} &= z_{cg} + T_r/2\phi + L_r\theta - z_{u3} \\
z_{s4} &= z_{cg} - T_r/2\phi + L_r\theta - z_{u4} \\
z_{ti} &= z_{ri} - z_{ui} \quad (i = 1, \cdots, 4)
\end{aligned} \tag{5}$$

The spring and damping forces which act between the wheels and the vehicle body, are given by the following equation.

$$F_{di} = -K_i z_{si} - C_i(t)\dot{z}_{si} \quad (i = 1, \cdots, 4) \tag{6}$$

2.2 Modeling of the passenger

Various models of a seated human have been proposed so far. In this paper, the passenger's motion is expressed to the seated human model shown in Fig. 2. Therefore, it is easy to understand the passenger's motion. To the seated position, Ps, the body part has three

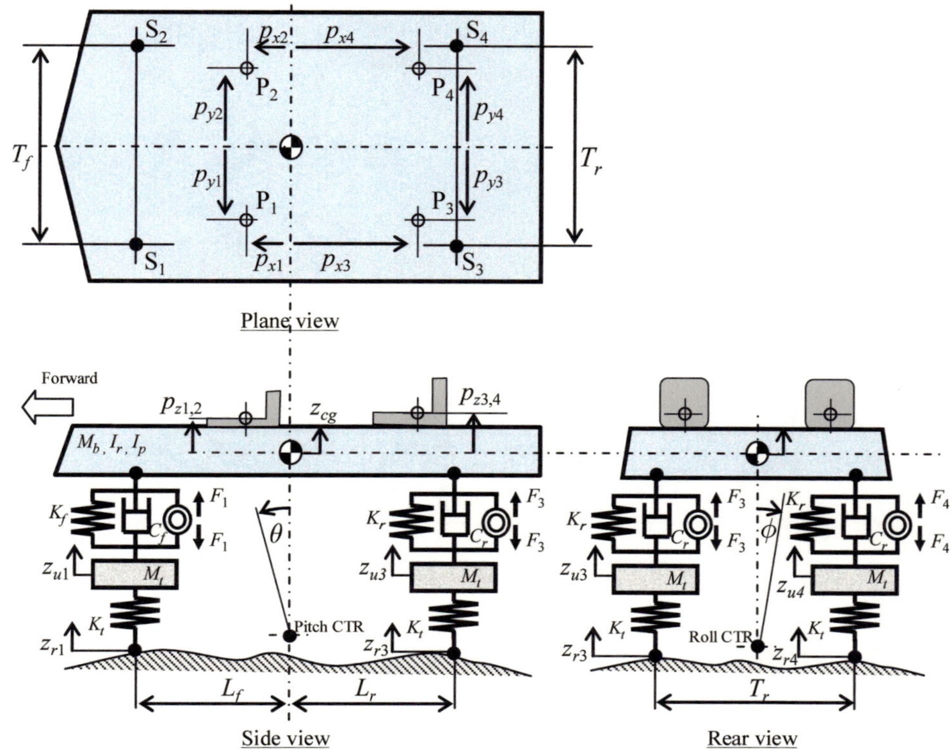

Fig. 1. Full vehicle model

Symbol	Value		Symbol	Value	
M_b	1900	kg	T_f	1.53	m
M_t	50	kg	T_r	1.50	m
I_r	600	kgm^2	H_r	0.45	m
I_p	3000	kgm^2	H_p	0.53	m
K_f	33×10^3	N/m	$p_{x1,2}$	0.04	m
K_r	31×10^3	N/m	p_{y1}	0.4	m
K_t	260×10^3	N/m	p_{y2}	-0.4	m
L_f	1.34	m	$p_{z1,2}$	-0.045	m
L_r	1.46	m			

Table 1. Specification of vehicle mode

degree of freedom (DOF) which is longitudinal, lateral, and vertical motions. The head has 3 DOF. First, the head moves up and down to the body parts. Second, the head rotates around the point, Pp, at the pitch direction. Third, the head rotates around the point, Pr, at the roll direction. Thus the passenger model has a total of 6 DOF. Between the each part, it has a spring and a damper. The equation of motion of the passenger model is as follows.

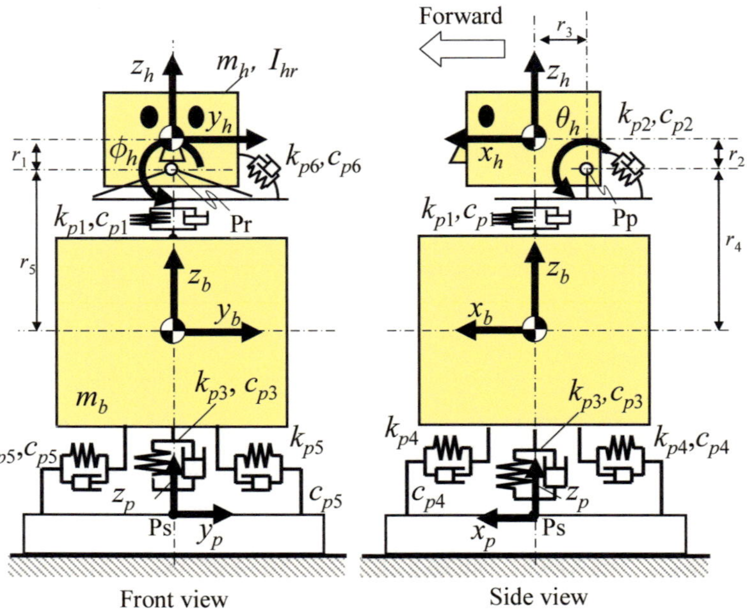

Fig. 2. Passenger model

i	k_{pi} [N/m]	c_{pi} [N/m/s]	r_{pi} [m]
1	40000	2000	0.1
2	15*	0.9**	0.1
3	96000	1120	0.05
4	22500	600	0.2
5	2000	400	0.3
6	20*	1.2**	0.3

Table 2. Specification of passenger model (*:Nm/rad, **:Nm/rad/s)

$$m_{bh}\ddot{x}_b = 2k_{p4}\left(x_p - x_b\right) + 2c_{p4}\left(\dot{x}_p - \dot{x}_b\right) + \left(-c_{p2}\dot{\theta}_h - k_{p2}\theta_h\right)/r_4 \tag{7}$$

$$m_{bh}\ddot{y}_b = 2k_{p5}\left(y_p - y_b\right) + 2c_{p5}\left(\dot{y}_p - \dot{y}_b\right) - \left(-c_{p6}\dot{\phi}_h - k_{p6}\phi_h\right)/r_5 \tag{8}$$

$$\begin{aligned} m_b\ddot{z}_b &= k_{p3}\left(z_p - z_b\right) + c_{p3}\left(\dot{z}_p - \dot{z}_b\right) - \left[k_{p1}\left(z_b - z_h\right) + c_{p1}\left(\dot{z}_b - \dot{z}_h\right)\right] \\ &+ \left(-c_{p2}\dot{\theta}_h - k_{p2}\theta_h\right)/r_3 \end{aligned} \tag{9}$$

$$m_h\ddot{z}_h = k_{p1}\left(z_b - z_h\right) + c_{p1}\left(\dot{z}_b - \dot{z}_h\right) - \left(-c_{p2}\dot{\theta}_h - k_{p2}\theta_h\right)/r_3 \tag{10}$$

$$
\begin{aligned}
I_{hr}\ddot{\phi}_h &= -c_{p6}\dot{\phi}_h - k_{p6}\phi_h \\
&\quad + m_{hbh}r_1\left[2k_{p5}\left(y_p - y_b\right) + 2c_{p5}\left(\dot{y}_p - \dot{y}_b\right) - \left(-c_{p6}\dot{\phi}_h - k_{p6}\phi_h\right)/r_5\right]
\end{aligned} \tag{11}
$$

$$
\begin{aligned}
I_{hp}\ddot{\theta}_h &= -c_{p2}\dot{\theta}_h - k_{p2}\theta_h + \left[k_{p1}\left(z_b - z_h\right) + c_{p1}\left(\dot{z}_b - \dot{z}_h\right)\right]r_3 \\
&\quad - m_{hbh}r_2\left[2k_{p4}\left(x_p - x_b\right) + 2c_{p4}\left(\dot{x}_p - \dot{x}_b\right) + \left(-c_{p2}\dot{\theta}_h - k_{p2}\theta_h\right)/r_4\right]
\end{aligned} \tag{12}
$$

Where,

$$
m_{bh} = m_b + m_h, \quad m_{hbh} = \frac{m_b}{m_b + m_h}.
$$

Each parameter of the passenger model is set to m_b = 45 kg, m_h = 7.5 kg, I_{hr} = 8.3×10^{-2} kgm^2, I_{hp} = 5.0 kgm^2, and I_{hp} = 5.5×10^{-2} kgm^2 based on the adult male's height and weight data. In addition, the acceleration of the passenger's head is derived from a geometric relation.

$$
\begin{aligned}
\ddot{x}_h &= \ddot{x}_b + \ddot{\theta}_h/r_2 \\
\ddot{y}_h &= \ddot{y}_b + \ddot{\phi}_h/r_1
\end{aligned} \tag{13}
$$

As shown in Table 2, the spring, the damper, and length were adjusted to conform the passenger model and an experimental data which was reported in previous research shown in Figs. 3 and 4 (Tamaoki *et al.*, 1996, 1998). The results shown in Figs. 3(c) and 4(c) demonstrate that the gain characteristics of the model were nearly equal to the experimental ones. However, as shown in Figs. 3(b) and 4(b), there were some differences in the high-frequency band for the phase properties. To reduce these differences, the passenger model must be made more complex, but this necessitates the use of a higher order control system. Because the purpose of our controller is to reduce the vertical vibration of the passengers in comparison with the lateral vibration, we designed it using this passenger model.

2.3 Vehicle-passenger model

In this section, the passenger for the vehicle model was assumed to sit in the front-left seat in designing the control system to reduce passenger vibration and motion. The vehicle-passenger model is shown in Fig. 5. The passenger model is set to the vehicle model in a front-left seat to design the controller. The translational motion of the position of the seat and the motion of the vehicle have the following relation;

$$
\begin{aligned}
x_{p1} &= \left(H_p + p_{1z}\right)\theta \\
y_{p1} &= -\left(H_r + p_{1z}\right)\phi \\
z_{p1} &= z_{cg} + p_{1y}\phi - p_{1x}\theta
\end{aligned} \tag{14}
$$

The equation of state of the vehicle-passenger model is defined as the following equation.

$$
\dot{x}(t) = A_p x(t) + B_{p1}w(t) + B_{p2}u(t) \tag{15}
$$

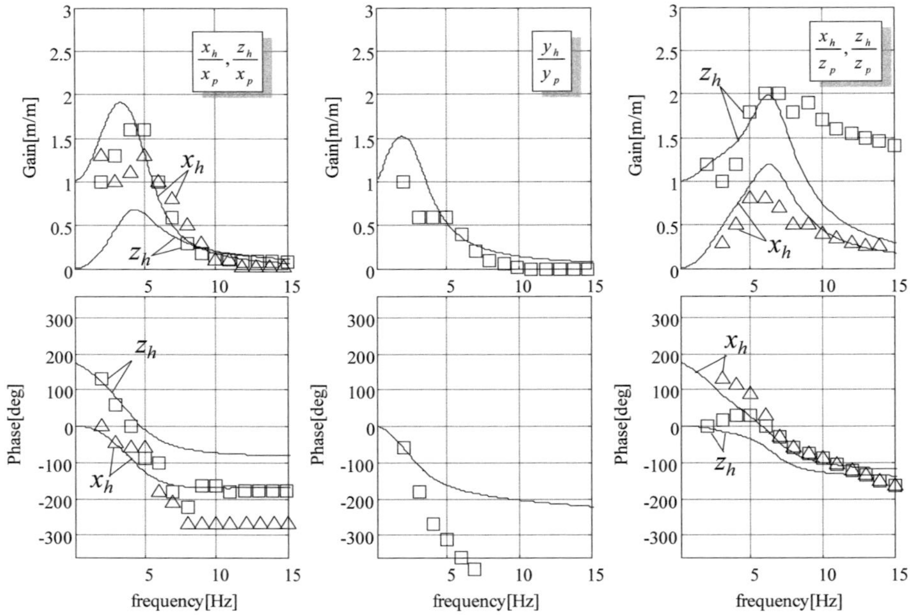

(a) x_p input (b) y_p input (c) z_p input

Fig. 3. Transfer function from seat to the head (Translational motion, Dot: Experiment (Tamaoki *et al.*, 1998), Line: model)

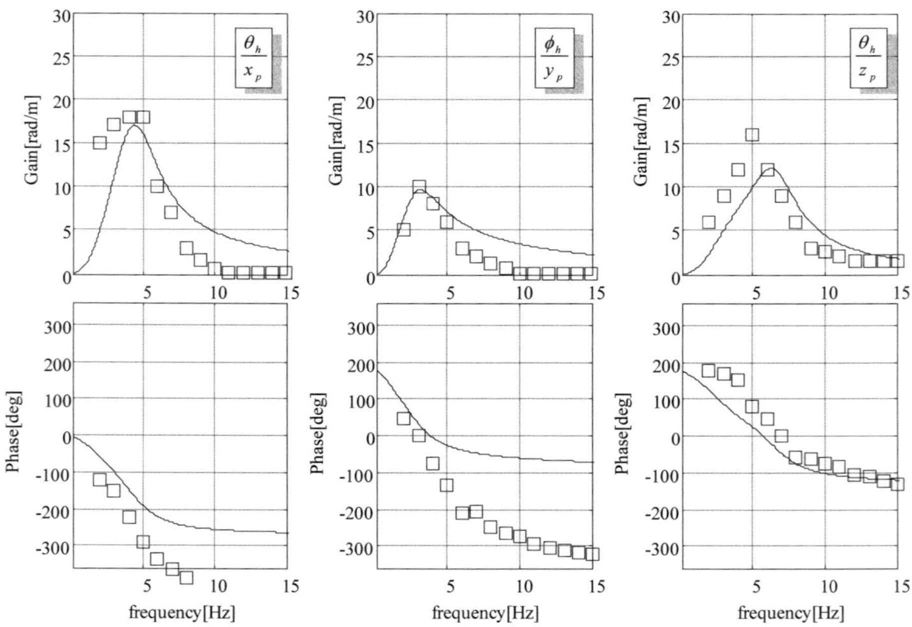

(a) x_p input (b) y_p input (c) z_p input

Fig. 4. Transfer function from seat to the head (Rotational motion, Dot: Experiment (Tamaoki *et al.*, 1996), Line: Model)

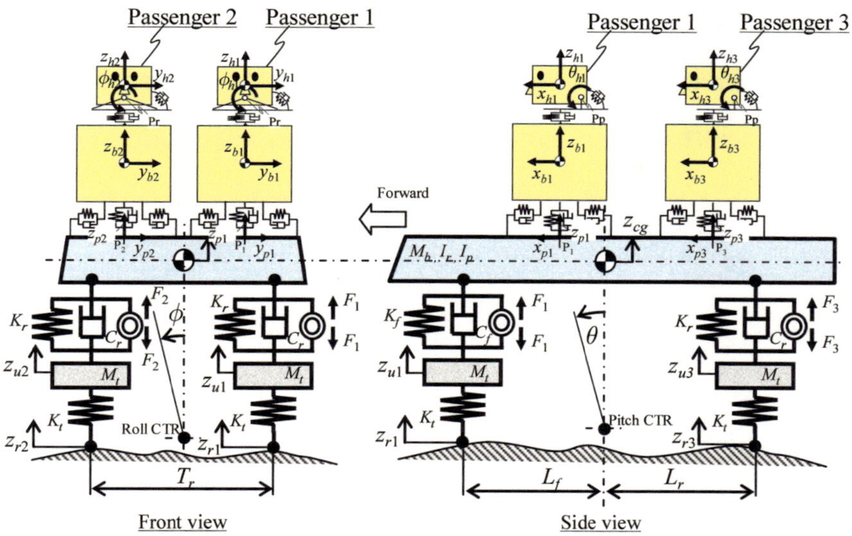

Fig. 5. Vehicle-passenger model

Where,

$$
\begin{aligned}
x(t) &= \left[z_{u1}\, z_{u2}\, z_{u3}\, z_{u4}\, z_{cg}\, \phi\, \theta\, \dot{z}_{u1}\, \dot{z}_{u2}\, \dot{z}_{u3}\, \dot{z}_{u4}\, \dot{z}_{cg} \dot{\phi}\, \dot{\theta}\, x_{b1}\, y_{b1}\, z_{b1}\, z_h\, \phi_{h1}\, \theta_{h1}\, \dot{x}_{b1}\, \dot{y}_{b1}\, \dot{z}_{b1}\, \dot{z}_h\, \dot{\phi}_{h1}\, \dot{\theta}_{h1} \right]^T \\
w(t) &= \left[z_{r1}\, z_{r2}\, z_{r3}\, z_{r4} \right]^T \\
u(t) &= \left[F_1\, F_2\, F_3\, F_4 \right]^T
\end{aligned}
.
$$

3. Design of controller

3.1 Disturbance accommodating control

We found that feedforward control of disturbance information in the finite frequency range and feedback control improve performance (Okamoto *et al.*, 2000). The power spectral density of the actual velocity of disturbances had flat characteristics in a low frequency, and decreased according to frequency at a region of high frequency. We assumed that it regarded as the colored noise formed by shaping filter which has a transfer function with low-pass characteristics. This filter of the each wheel is based on the road condition which defined by ISO (ISO-8608, 1995). The filter is as follows:

$$
\begin{aligned}
& Q_{di} \begin{cases} \dot{x}_{di}(t) = A_{di} x_{di}(t) + B_{di} w_{gi}(t) \\ w_i(t) = C_{di} x_{di}(t) \qquad (i = 1, \cdots, 4) \end{cases} \\
& \frac{w_i(s)}{w_{gi}(s)} = \frac{\varpi_d^{\,2}}{s^2 + 2\xi_d \varpi_d s + \varpi_d^{\,2}} \qquad (i = 1, \cdots, 4)
\end{aligned}
\quad . \tag{16}
$$

where, w_{gi} is road input of the each wheel, w_i is road input of the vehicle-passenger model of the generalized plant to design the controller as shown in Fig. 6. It was referred to as $\varpi_d = 50 \times 2\pi$ and $\xi_d = 0.706$.

3.2 Disturbance accommodating H_∞ control.

The feedforward control of disturbances resulted in worse accuracy outside the assumed frequency (Okamoto *et al.*, 2000). Furthermore, because each resonance frequency of the vehicles, passenger, and tire differs, the control system design considering each resonance frequency is needed. Therefore, the control system was designed by using the H_∞ method in the control theory.

We integrated each state variable of the road disturbance model and frequency weights for controlled values. The frequency weights are as follows:

$$Q_{wi}\begin{cases}\dot{x}_{wi}(t)=A_{wi}x_{wi}(t)+B_{wi}z_{pi}(t)\\ z_{gi}(t)=C_{wi}x_{wi}(t) \qquad (i=1,\cdots,4)\end{cases} \\ \frac{z_{gi}(s)}{z_{pi}(s)}=K_{wi}W_i(s) \qquad (i=1,\cdots,4) \quad . \tag{17}$$

where, z_{pi} is controlled value of the vehicle-passenger model, z_{gi} is controlled value of the generalized plant. Figure 6 shows a block diagram of the generalized plant to design the controller, and the state-space form of the generalized plant is as follows:

$$\begin{aligned}\dot{x}_g(t)&=A_gx_g(t)+B_{g1}w_g(t)+B_{g2}u(t)\\ z_g(t)&=C_{g1}x_g(t) \qquad\qquad +D_{g12}u(t) \quad .\\ y_g(t)&=C_{g2}x_g(t)+D_{g21}w_g(t)\end{aligned} \tag{18}$$

H_∞ norm of the transfer function from disturbance $w_g(t)$ to controlled value $z(t)$ is expressed by the following equation.

$$\left\|G_{z_gw_g}\right\|_\infty=\sup_w\frac{\left\|z_g\right\|_2}{\left\|w_g\right\|_2} \tag{19}$$

$$\min_u\left\|G_{z_gw_g}\right\|_\infty=:\gamma^* \tag{20}$$

where, γ^* is a minimum of H_∞ norm of the generalized plant realized with H_∞ controller. The controller is the following equation (Glover & Doyle, 1988).

$$\begin{aligned}\dot{x}_k(t)&=A_kx_k(t)+B_ky(t)\\ u(t)&=C_kx_k(t)\end{aligned} \tag{21}$$

The measured outputs, $y(t)$, are four vertical accelerations of the wheel position of the vehicle body. The controlled values, $z(t)$, are vertical acceleration of the passenger's head, vertical velocity of the sprung, tire deformation, and actuating force. Frequency weight W_i, shown in Fig. 7 was determined by trial and error.

A bandpass filter, W_1, that had a peak frequency equal to the resonance frequency of the passenger's head was used based on sensitivity curves (ISO-2631-1, 1997), such as that being standardized by ISO and shown in Fig. 8. In order to prevent the increase of response in each resonance, a low pass filter W_2 and a bandpass filter W_3 are used. Moreover, to prevent steady control input and minimize energy consumption, a high pass filter, W_4, was used.

We compare the proposal method and two generalized control methods to verify the control performance. As one of the generalized control methods, the controller in which the one of the controlled values is vertical acceleration of the body CoG (Vehicle CoG Control), is designed. Another is that one of the controlled values is vertical acceleration of a seated position (Seat Position Control). The design of two generalized control methods are changed the controlled value z_1 into the vertical acceleration of CoG of the vehicle body and seated position, respectively. Frequency weights, $W_1(s)$, $W_2(s)$, $W_3(s)$, $W_4(s)$, $K_{w2} = 400$, $K_{w3} = 5000$, and $K_{w4} = 1.31$, use the same value also in the three methods. The following section describes K_{w1}.

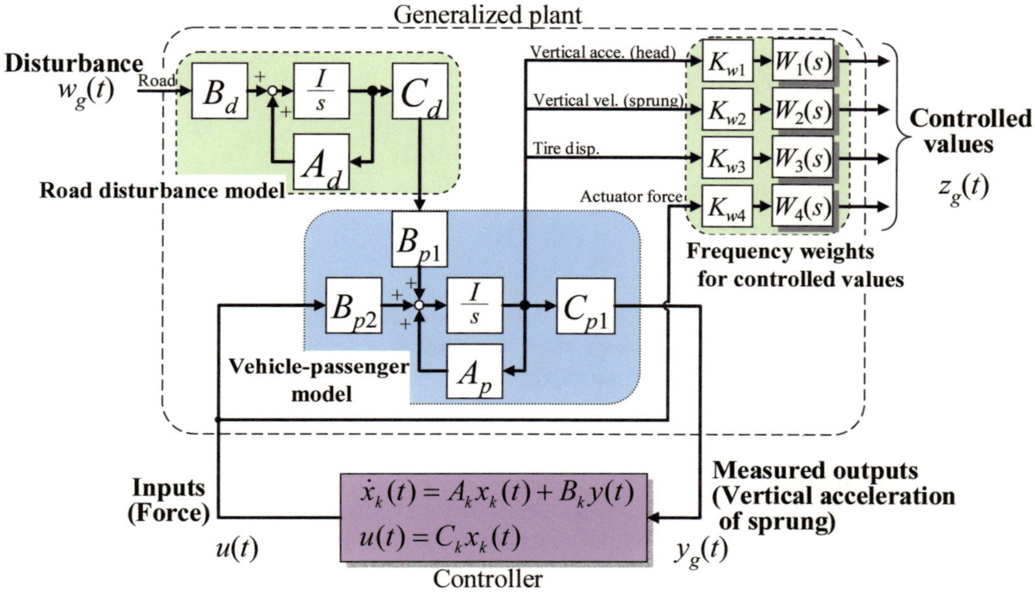

Fig. 6. Generalized plant for "Passenger Control"

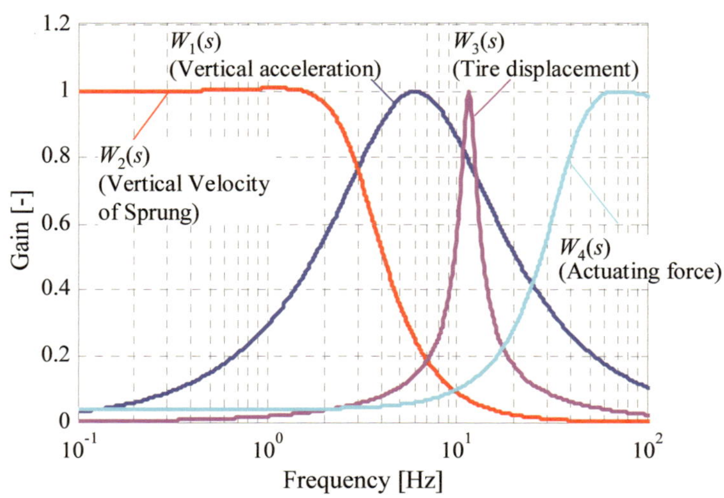

Fig. 7. Frequency weights for controlled value

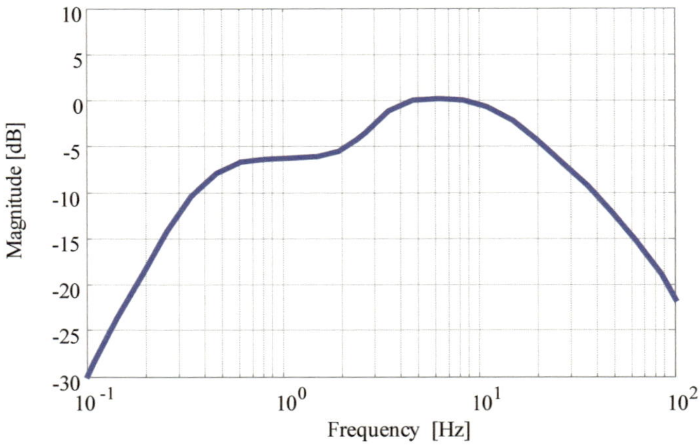

Fig. 8. Sensitivity curve of vertical vibration (ISO-2631-1, 1997)

4. Simulation

In this section, two kinds of numerical simulations were carried out. One is to verify control performance in comparison with other methods. Another is to verify robustness for the difference in passenger's vibration characteristic.

4.1 Assumption

We verified the effectiveness of the proposed method by using the vehicle-passenger model with H_∞ controller. We used MATLAB (The Math Work Inc.) for the calculations, and the Runge-Kutta method for the differential equations. The computational step size is 1 ms. In addition, it assumes that we perform the evaluation in an ideal condition, and the model to design the controller and the model for evaluation are same models.

4.2 Driving condition

It assumes that the PSD characteristic of the road surface is C class defined by ISO (ISO-8608, 1995). The vehicle speed is 16.6 m/s (60 km/h). The vehicle runs the straight for 10 seconds, and the input of the road surface to each wheel is independent. Figure 9 shows the PSD of the road disturbance. Figure 10 shows the road displacement.

4.3 Design of the frequency weight K_{w1}

In each method, if the evaluation function of acceleration is raised, it is clear that each acceleration set as the controlled value is reduced, and the actuating force increases. To set the same actuating force, frequency weight K_{w1} of each method was adjusted so that RMS value of the actuating force of the four wheels sets to 1000 N. The each frequency weight, K_{w1}, of "Vehicle CoG Control", "Seat Position Control" and "Passenger Control" is 244, 315, and 78 respectively.

4.4 Difference of vehicle-passenger model

In the numerical simulation, there are some diffidence in the vehicle-passenger model as shown in Table 3. In sections 4.5.1 and 4.5.2, passenger models sit in the front-left seat and front-right seat. In section 4.5.2, some specifications of the passenger model are different from the generalized plant to design controller.

4.5 Results

4.5.1 Comparison with the "Vehicle CoG Control" and "Seat Position Control"

Figure 11 shows the time histories of the vehicle and the passenger 1's vertical acceleration for 3 second. In this paper, passenger 1 sits a front-left seat, and passenger 2 sits a front-right. In the acceleration of the vehicle body, it was confirmed that there is few differences among the three methods. On the other hand, in the acceleration of the passenger's head, the proposed method is the smallest, and it was confirmed that the proposed method is effective for the passenger's vibration reduction.

The actuating force of each wheel in each method is shown in Fig. 12. In the Vehicle CoG Control, the actuating force of all wheels is generated in the same direction. In the other method, the actuating force of the left/right wheel is generated in a different direction. Therefore, the vertical accelerations of the seated position and the passenger's head are reduced by controlling the roll motion of the vehicle body.

Figure 13 shows the Lissajous figure of lateral and vertical accelerations of the seated position, the passenger's body and the head part respectively. This figure is seen from the front of vehicle. In upper-right figure of Fig. 13 (c), the proposed method has control effect which vertical acceleration of the passenger 1's head is reduced in comparison with "Vehicle CoG Control". Moreover, the proposed method has not only the vibration reduction effect of the passenger 1's head, but also the vertical acceleration reduction effect of the passenger 1's body.

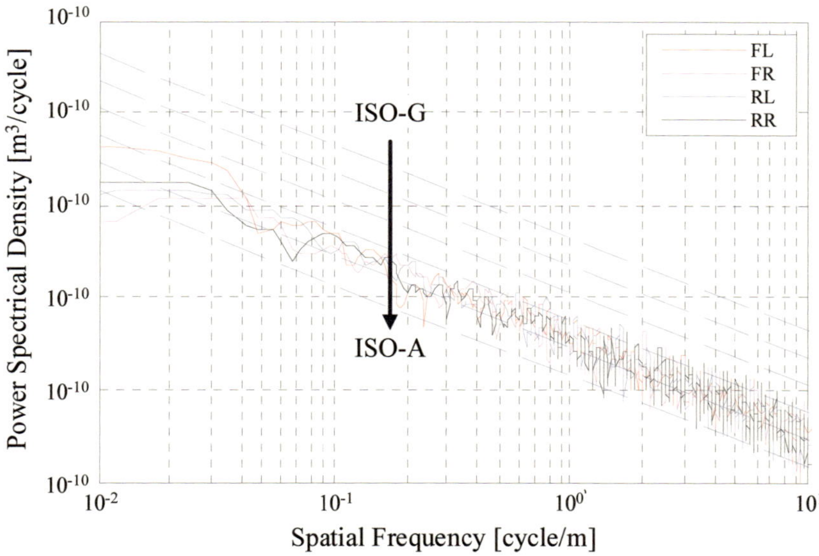

Fig. 9. PSD of road surface profile

Figure 14 shows power spectrum density (PSD) of the vertical acceleration of the passenger's head in each method, and actuating force. In the frequency band of 4-7 Hz with resonance of a passenger's head, although the proposed method has the vibration reduction effect better than other methods. On the other hand, PSD of the actuating force does not necessarily have the highest value in the frequency band. In this frequency band, the proposed method can reduce the passenger's vibration by the limited actuating force.

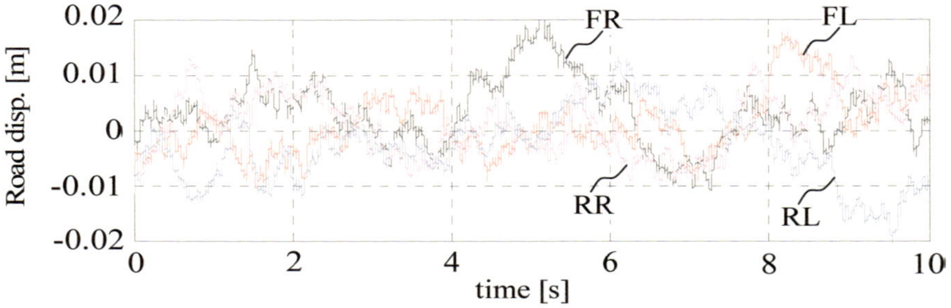

Fig. 10. Road displacement

<table>
<tr><th rowspan="2" colspan="2"></th><th rowspan="2">Generalized plant to design controller</th><th colspan="2">Simulation model</th></tr>
<tr><th>Section 4.5.1</th><th>Section 4.5.2</th></tr>
<tr><td colspan="2">Vehicle</td><td>Fig. 1, Table 1</td><td>←</td><td>←</td></tr>
<tr><td colspan="2">Passenger</td><td>Fig. 2, Table 2</td><td>←</td><td>Table 4</td></tr>
<tr><td></td><td>Seated position</td><td>•Front-left</td><td>•Front-left
•Front-right</td><td>←</td></tr>
</table>

Table 3. Vehicle-passenger model

In each frequency band, the sensitivity of the vertical acceleration for the human is defined by sensitivity curves (ISO-2631, 1997). In this paper, we estimate the root mean square (RMS) value which is added the sensitivity compensation expressed by a high order transfer function (Rimel & Mansfield, 2007). Figure 15 shows the ratio of the RMS value of each vertical acceleration to those values of "Vehicle CoG Control". In the passenger 1, it was confirmed that the proposed method can reduce the RMS value of the passenger 1's head (head 1). Moreover, in the passenger 2, it was confirmed that the RMS value of the passenger 2's head (head 2) is not increased by the proposed method, and the proposed method had the vibration reduction effect equivalent to the generalized control methods.

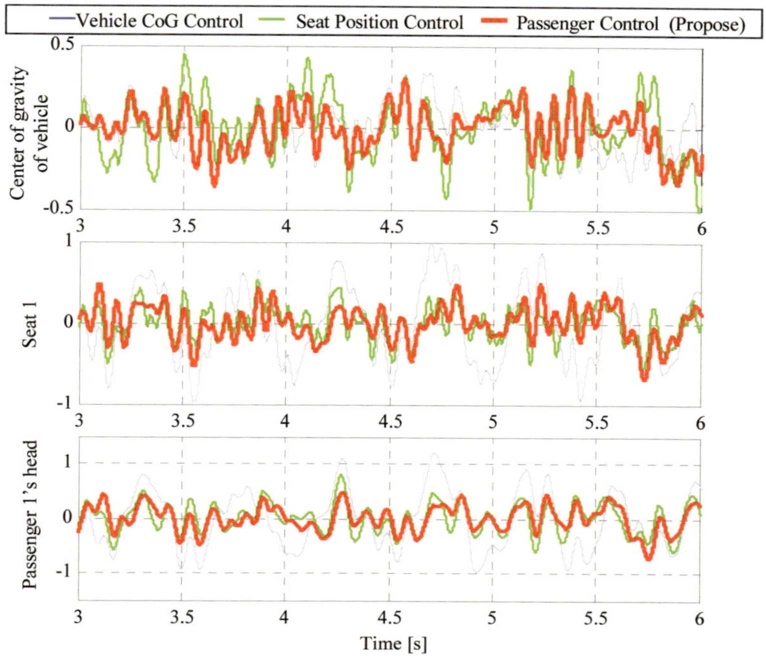

Fig. 11. Vehicle and passenger's behavior (Vertical acceleration, unit : m/s^2)

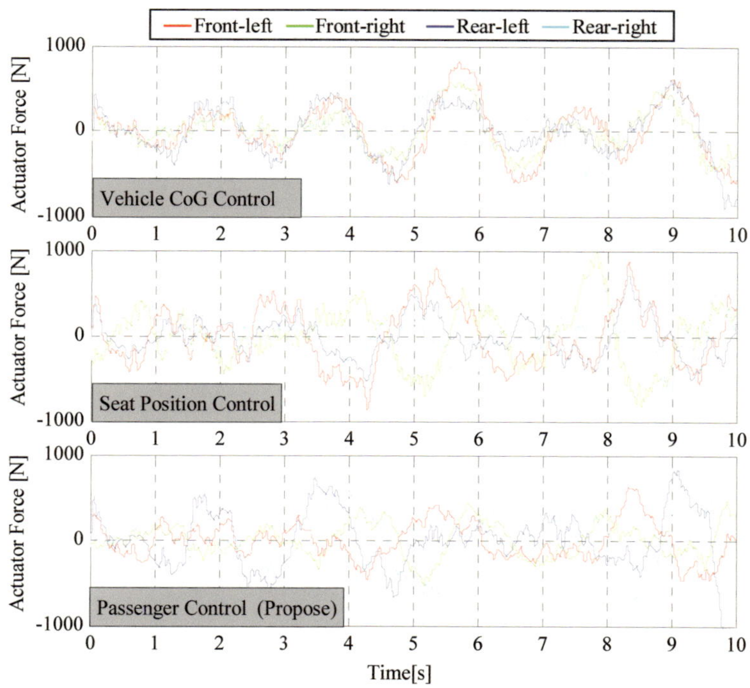

Fig. 12. Actuating force

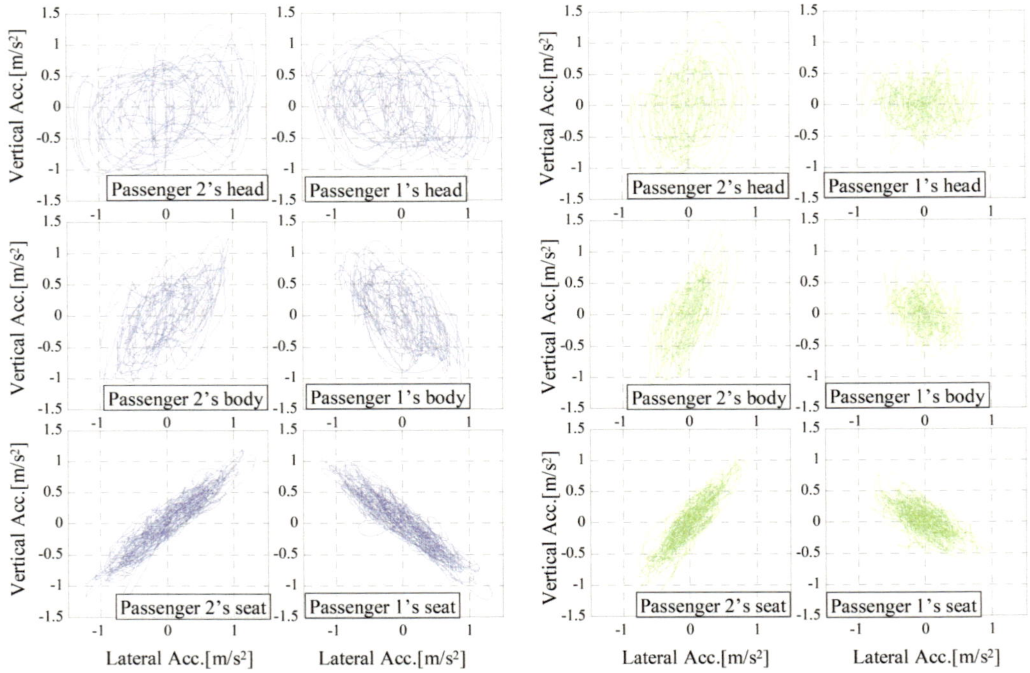

(a) Vehicle CoG Control (b) Seat Position Control

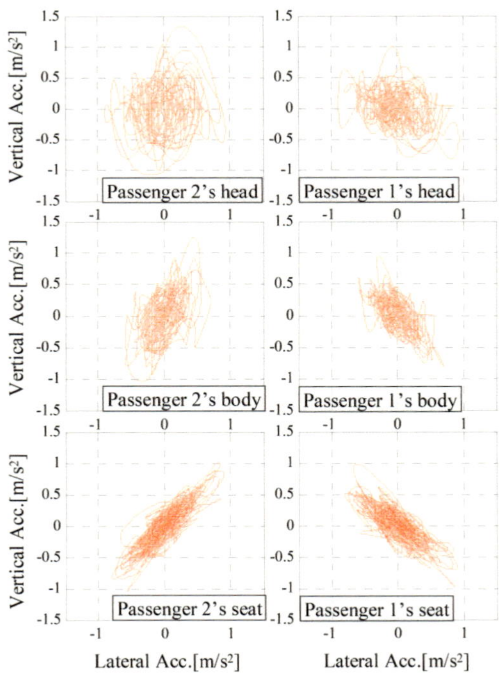

(c) Passenger Control (Propose)

Fig. 13. Lissajous figure (Lateral and vertical acceleration)

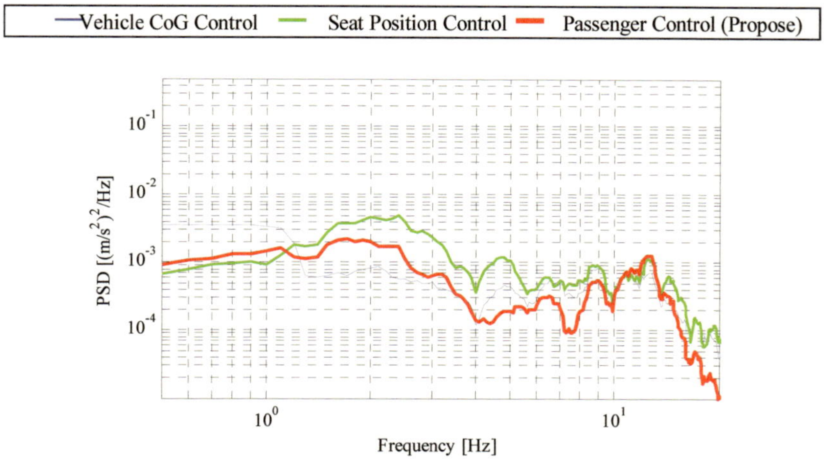

(a) Vertical acceleration (Center of gravity of vehicle)

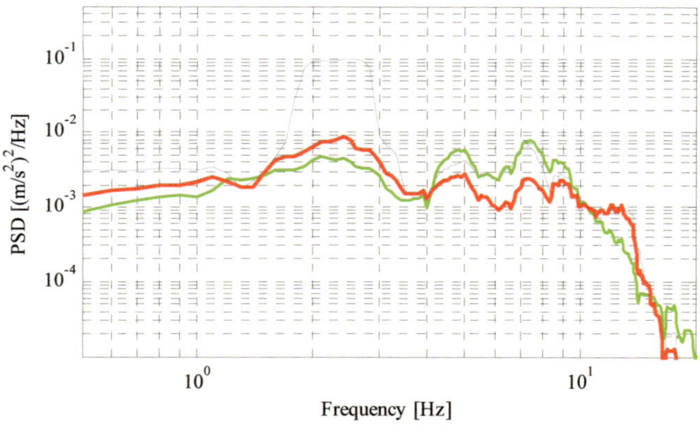

(b) Vertical acceleration (Passenger1's head)

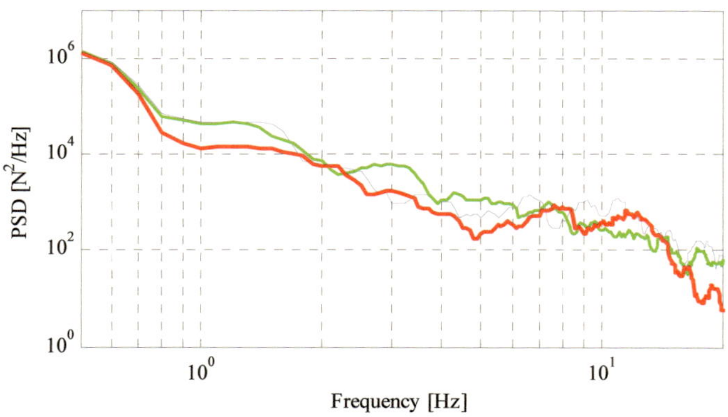

(c) Actuator force

Fig. 14. Power spectral density

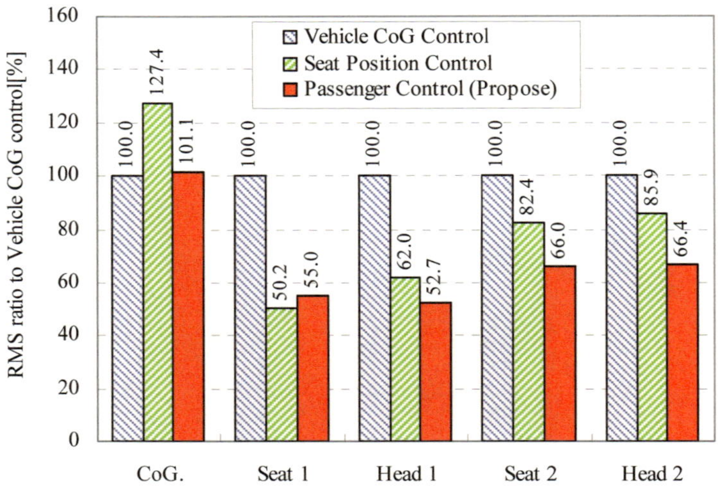

Fig. 15. RMS value of vertical acceleration

From these results, it was confirmed that the proposed method can effectively reduce passenger's vibration by using H_∞ control which including the dynamics of human body and seated position. By means of setting the passenger' motion to one of the amounts of evaluation function, the proposed method can directly control the passenger's vibration.

4.5.2 Comparison with the different passenger model

In this section, the robust performance against the difference in a passenger's vibration characteristic is verified. In previous research, there are many reports about seated human

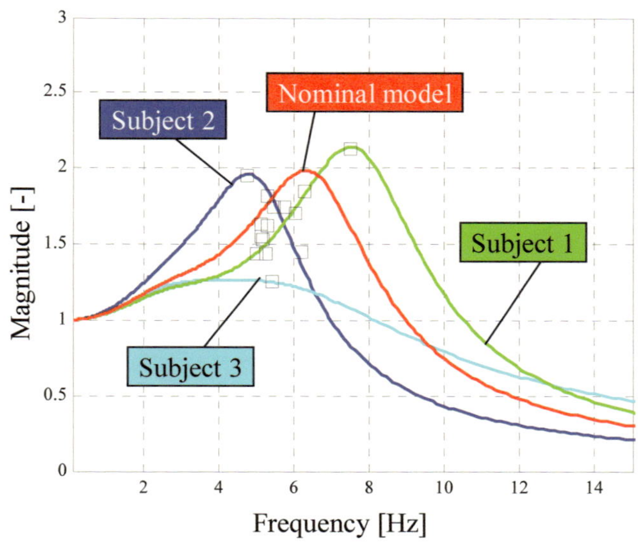

Fig. 16. Frequency response from seat to the head (Vertical motion, dot : Experiment (Varterasian & Thompson, 1977), Line : Model)

dynamics. Varterasian & Thompson reported the seated human dynamics from a large person to a small person(Varterasian & Thompson, 1977). Robust performance is verified by supposition that such person sits in the vehicle. Figure 16 shows the frequency response from vertical vibration of seat to vertical vibration of the head. Dot is 15 subjects' resonance peak. In this section, three outstanding subjects' data of their report is modeled in the vibration characteristic of vertical direction. The damper and spring were adjusted to conform the passenger model and an experimental data. The characteristic of the passenger model of three outstanding subjects are shown in Table 4.

	k_{p3} [N/m]	c_{p3} [N/m/s]
Nominal model	960000	1120
Subject 1	1320000	1150
Subject 2	576000	960
Subject 3	960000	2550

Table 4. Difference of specifications

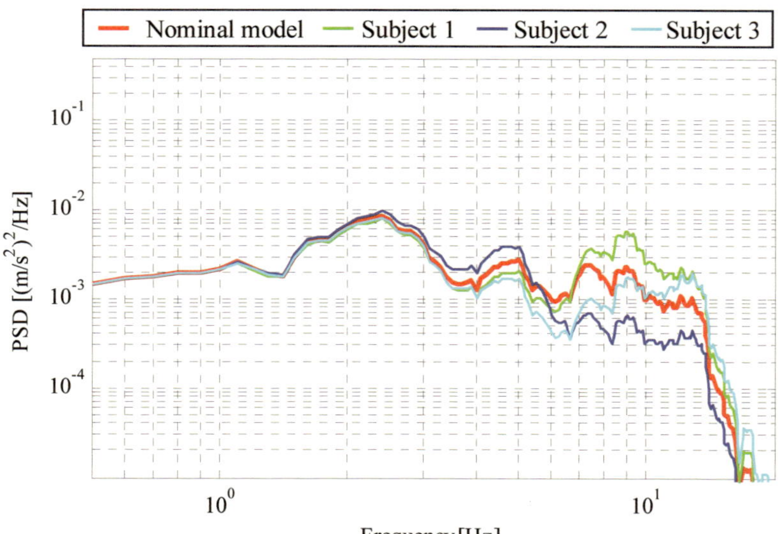

Fig. 17. PSD of vertical acceleration (Passenger 1's head)

The numerical simulation is carried out on the same road surface conditions as the section 4.5.1. Figure 17 shows PSD of the vertical acceleration of the passenger 1's head and Fig. 18

shows RMS value. In PSD of 7 Hz or more, RMS value of vertical acceleration of subject 1's head becomes higher than the nominal model. Moreover, RMS of subject 1 is the highest. On the other hand, RMS of subjects 2 and 3 is reduced in comparison with the nominal model. The physique of subject 1 differs from other subjects. When such a person sits, the specified controller should be designed. From these results, the proposed method has robustness for the passenger of the general physique.

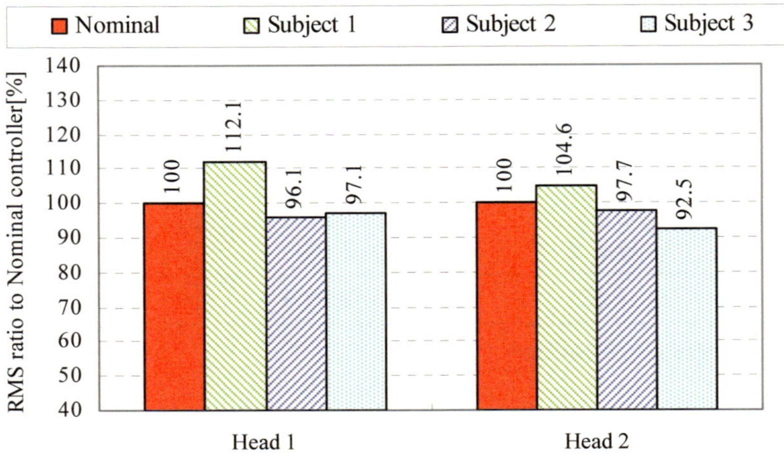

Fig. 18. RMS value of vertical acceleration of passenger 1's head

5. Conclusion

This study aims at establishing a control design method for the active suspension system in order to reduce the passenger's vibration. In the proposed method, a generalized plant that uses the vertical acceleration of the passenger's head as one of the controlled output is constructed to design the linear H_∞ controller. In the simulation results, when the actuating force is limited, we confirmed that the proposed method can reduce the passenger's vibration better than two methods which are not include passenger's dynamics. Moreover, the proposed method has robustness for the difference in passenger's vibration characteristic.

6. Acknowledgment

This work was supported in part by Grant in Aid for the Global Center of Excellence Program for "Center for Education and Research of Symbiotic, Safe and Secure System Design" from the Ministry of Education, Culture, Sport, and Technology in Japan.

7. References

Ikeda, S.; Murata, M.; Oosako, S. & Tomida, K. (1999). Developing of New Damping Force Control System -Virtual Roll Damper Control and Non-liner H_∞ Control-, *Transactions of the TOYOTA Technical Review*, Vol.49. No.2, pp.88-93

Kosemura, R.; Takahashi, M. & and Yoshida, K. (2008). Control Design for Vehicle Semi-Active Suspension Considering Driving Condition, *Proceedings of the Dynamics and Design Conference 2008*, 547, Kanagawa, Japan, September, 2008

Itagaki, N.; Fukao, T.; Amano, M.; Ichimaru, N.; Kobayashi, T. & Gankai, T. (2008). Semi-Active Suspension Systems based on Nonlinear Control, *Proceedings of the 9th International Symposium on Advanced Vehicle Control 2008*, pp. 684-689, Kobe, Japan, October, 2008

Tamaoki, G.; Yoshimura, T. & Tanimoto, Y. (1996). Dynamics and Modeling of Human Body Considering Rotation of the Head, *Proceedings of the Dynamics and Design Conference 1996*, 361, pp. 522-525, Fukuoka, Japan, August, 1996

Tamaoki, G.; Yoshimura, T. & Suzuki, K. (1998). Dynamics and Modeling of Human Body Exposed to Multidirectional Excitation (Dynamic Characteristics of Human Body Determined by Triaxial Vibration Test), *Transactions of the Japan Society of Mechanical Engineers, Series C*, Vol.64, No.617, pp. 266-272

Tamaoki, G. & Yoshimura, T. (2002). Effect of Seat on Human Vibrational Characteristics, *Proceedings of the Dynamics and Design Conference 2002*, 220, Kanazawa, Japan, October, 2002

Koizumi, T.; Tujiuchi, N.; Kohama, A. & Kaneda, T. (2000). A study on the evaluation of ride comfort due to human dynamic characteristics, *Proceedings of the Dynamics and Design Conference 2000*, 703, Hiroshima, Japan, October, 2000 ISO-2631-1 (1997). Mechanical vibration and shock–Evaluation of human exposure to whole-body vibration -, *International Organization for Standardization* ISO-5982 (2001). Mechanical vibration and shock –Range of idealized value to characterize seated body biodynamic response under vertical vibration, *International Organization for Standardization*

Oya, M.; Tsuchida, Y. & Qiang, W. (2008). Robust Control Scheme to Design Active Suspension Achieving the Best Ride Comfort at Any Specified Location on Vehicles, *Proceedings of the 9th International Symposium on Advanced Vehicle Control 2008*, pp.690-695, Kobe, Japan, October, 2008

Guglielmino, E.; Sireteanu, T.; Stammers, C. G.; Ghita, G. & Giuclea, M. (2008). *Semi-Active Suspension Control -Improved Vehicle Ride and Road Friendliness*, Springer-Verlag, ISBN- 978-1848002302, London

Okamoto, B. and Yoshida, K. (2000). Bilinear Disturbance-Accommodating Optimal Control of Semi-Active Suspension for Automobiles, *Transactions of the Japan Society of Mechanical Engineers, Series C*, Vol.66, No.650, pp. 3297-3304

Glover, K. & Doyle, J.C. (1988). State-space Formula for All Stabilizing Controllers that Satisfy an H_∞-norm Bound and Relations to Risk Sensitivity, *Journal of the Systems and Control letters*, 11, pp.167-172

ISO-8608 (1995). Mechanical vibration -Road surface profiles - Reporting of measured data, *International Organization for Standardization*

Rimel, A.N. & Mansfield, N.J. (2007). Design of digital filters for Frequency Weightings Required for Risk Assessment of workers Exposed to Vibration, *Transactions of the Industrial Health*, Vol.45, No.4, pp. 512-519

Varterasian, H. H. & Thompson, R. R. (1977). The Dynamic Characterristics of Automobiles Seats with Human Occupants, *SAE Paper*, No. 770249

6

Modelling and Nonlinear Robust Control of Longitudinal Vehicle Advanced ACC Systems

Yang Bin[1], Keqiang Li[2] and Nenglian Feng[1]
[1]*Beijing University of Technology*
[2]*Tsinghua University*
China

1. Introduction

Safety and energy are two key issues to affect the development of automotive industry. For the safety issue, the vehicle active collision avoidance system is developing gradually from a high-speed adaptive cruise control (ACC) to the current low-speed stop and go (SG), and the future research topic is the ACC system at full-speed, namely, the advanced ACC (AACC) system. The AACC system is an automatic driver assistance system, in which the driver's behavior and the complex traffic environment ranging are taken into account from high-speed to low-speed. By combining the function of the high-speed ACC and low-speed SG, the AACC system can regulate the relative distance and the relative velocity adaptively between two vehicles according to the driving condition and the external traffic environment. Therefore, not only can the driver stress and the energy consumption caused by the frequent manipulation and the traffic congestion both be reduced effectively at the urban traffic environment, but also the traffic flow and the vehicle safety will be improved on the highway.

Taking the actual traffic environment into account, the velocity of vehicle changes regularly in a wide range and even frequently under SG conditions. It is also subject to various external resistances, such as the road grade, mass, as well as the corresponding impact from the rolling resistance. Therefore, the behaviors of some main components within the power transmission show strong nonlinearity, for instance, the engine operating characteristics, automatic transmission switching logic and the torque converter capacity factor. In addition, the relative distance and the relative velocity of the inter-vehicles are also interfered by the frequent acceleration/deceleration of the leading vehicle. As a result, the performance of the longitudinal vehicle full-speed cruise system (LFS) represents strong nonlinearity and coupling dynamics under the impact of the external disturbance and the internal uncertainty. For such a complex dynamic system, many effective research works have been presented. J. K. Hedrick et al. proposed an upper+lower layered control algorithm concentrating on the high-speed ACC system, which was verified through a platoon cruise control system composed of multiple vehicles [1-3]. K. Yi et al. applied some linear control methods, likes linear quadratic (LQ) and proportional–integral–derivative (PID), to design the upper and lower layer controllers independently for the high-speed ACC system [4]. In ref.[5], Omae designed the model matching control (MMC) vehicle high-speed ACC system based on the H-*infinity* (H_{inf}) robust control method. To achieve a tracking control between the relative distance and the relative velocity of the inter-vehicles, A. Fritz proposed a nonlinear vehicle model for the high-speed ACC system with four state variables in refs.[6, 7], and designed a variable structure control (VSC) algorithm based on the feedback linearization. In ref. [8], J.E. Naranjo used the fuzzy theory to design a coordinate control algorithm between the throttle actuator and the braking system. It has been verified on an ACC and SG cruise system. Utilizing the model predictive control (MPC) method, D. Coron designed an ACC control system for a SMART Car [9]. G. N. Bifulco applied the human artificial intelligence to study an ACC control algorithm with anthropomorphic function [10]. U. Ozguner investigated the impact of inter-vehicles communications on the performance of vehicle cruise control system [11]. J. Martinez, et al. proposed a reference model-based method, which has been applied to the ACC and SG system, and achieved an expected tracking performance at full-speed condition [12]. Utilizing the idea of hierarchical design method, P. Venhovens proposed a low-speed SG cruise control system, and it has been verified on a BMW small sedan [13]. Y. Yamamura developed an SG control method based on an existing framework of the ACC control system, and applied it to the SG cruise control [14]. Focusing on the low-speed condition of the heavy-duty vehicles, Y. Bin et al. derived a nonlinear model [15, 16] and applied the theory of nonlinear disturbance decoupling (NDD) and LQ to the low-speed SG system [17, 18].

In the previous research works, the controlled object (i.e. the dynamics of the controlled vehicle) was almost simplified as a linear model without considering its own mass, gear position and the uncertainty from external environment (likes, the change of the road grade). Furthermore, the analysis of the disturbance from the leading vehicle's acceleration/ deceleration was not paid enough consideration, which has become a bottleneck in limiting the enhancement of the control performance. To summarize, based on a detailed analysis of the impact from the practical high/low speed operating condition, the uncertainty of complex traffic environment, vehicle mass, as well as the change of gear shifting to the vehicle dynamic, an innovative LFS model is proposed in this study, in which the dynamics of the controlled vehicle and the inter-vehicles are lumped together within a more accurate and reasonable mathmatical description. For the uncertainty, strong nonlinearity and the strong coupling dynamics of the proposed model, an idea of the step-by-step transformation and design is adopted, and a disturbance decoupling robust control (DDRC) method is proposed by combining the theory of NDD and VSC. On the basis of this method, it is possible to weaken the matching condition effectively within the invariance of VSC, and decouple the system from the external disturbance completely while with a simplified control structure. By this way, an improved AACC system for LFS based on the DDRC method is designed. Finally, a simulation in view of a typical vehicle moving scenario is conducted, and the results demonstrate that the proposed control system not only achieves a global optimization by means of a simplified control structure, but also exhibits an expected dynamic response, high tracking accuracy and a strong robustness regarding the external disturbance from the leading vehicle's frequent acceleration/deceleration and the internal uncertainty of the controlled vehicle.

2. LFS model

The LFS is composed of a leading vehicle and a controlled vehicle, and the block diagram is shown in Figure 1. The controlled vehicle is a heavy-duty truck, whose power transmission is composed of an engine, torque converter, automatic transmission and a final drive. The brake system is a typical one with the assistance of the compressed air. On-board millimetric wave radar is used to detect the information from the inter-vehicles (i.e., the relative distance and the relative velocity), which is installed in the front-end frame bumper of the controlled vehicle.

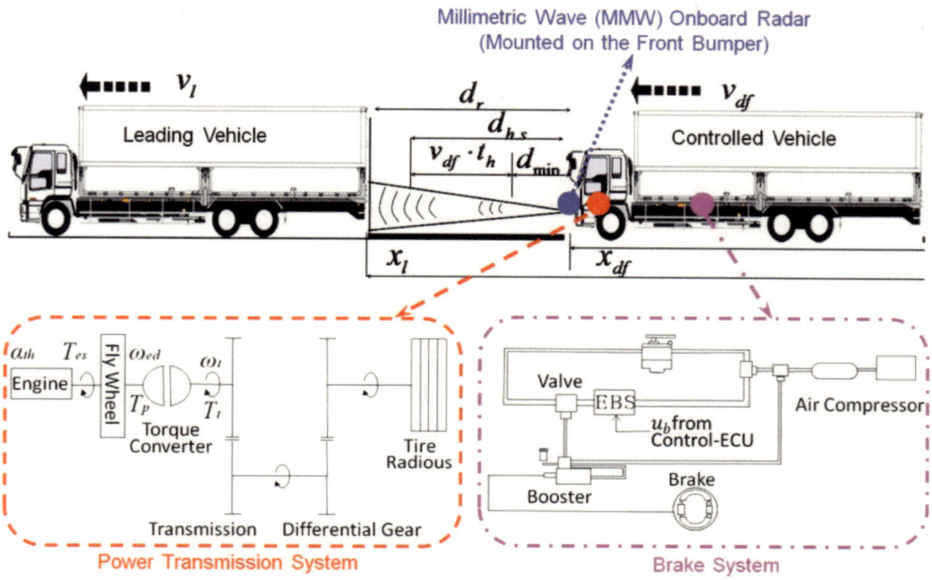

Fig. 1. Block diagram of LFS

x_l, x_{df}, v_l, v_{df} are absolute distance (m) and velocity (m/s) between the leading vehicle and the controlled vehicle, respectively. $d_r=x_l-x_{df}$ is an actual relative distance between the two vehicles. Desired relative distance can be expressed as $d_{h,s}=d_{min}+v_{df}t_h$, where, d_{min}=5m, t_h=2s. $v_r=v_l-v_{df}$ is an actual relative velocity. The purpose of LFS is to achieve the tracking of the inter-vehicles relative distance/relative velocity along a desired value. Therefore, a dynamics model of LFS at low-speed condition has been derived in ref. [15], which consists of two parts. The first part is the longitudinal dynamics model of the controlled vehicle, in which the nonlinearity of some main components, such as the engine, torque converter, etc, is taken into account. However, this model is only available at the following strict assumptions:

- the vehicle moves on a flat straight road at a low speed (<7m/s)
- assume the mass of vehicle body is constant
- the automatic transmission gear box is locked at the first gear position
- neglect the slip and the elasticity of the power train

The second part is the longitudinal dynamics model of the inter-vehicles, in which the disturbance from frequent accelartion/deccelartion of the leading vehicle is considered.

In general, since the mass, road grade and the gear position of the automatic transmission change regularly under the practical driving cycle and the traffic environment, the longitudinal dynamics model of the controlled vehicle in ref. [15] can only be used in some way to deal with an ideal traffic environment (i.e., the low-speed urban condition). In view of the uncertainties above, in this section, a more accurate longitudinal dynamics model of the controlled vehicle is derived for the purpose for high-speed and low-speed conditions (that is, the full-speed condition). After that, it will be integrated with a longitudinal dynamics model of the inter-vehicles, and an LFS dynamics model for practical applications can be obtained in consideration of the internal uncertainty and the external disturbance. It is a developed model with enhanced accuracy, rather than a simple extension in contrast with ref. [15].

2.1 Longitudinal dynamics model of the controlled vehicle

Based on the vehicle multi-body dynamics theory [19], modeling principles, and the above assumptions, two nominal models of the longitudinal vehicle dynamics are derived firstly according to the driving/braking condition:
The driving condition:

$$\dot{\boldsymbol{X}} = \boldsymbol{F}_{av}(\boldsymbol{X}) + \boldsymbol{G}_{av}(\boldsymbol{X})\alpha_{th} \Rightarrow \begin{bmatrix} \dot{x}_1 \\ \dot{x}_2 \end{bmatrix} = \begin{bmatrix} f_{av1}(\boldsymbol{X}) \\ f_{av2}(\boldsymbol{X}) \end{bmatrix} + \begin{bmatrix} g_{av1}(\boldsymbol{X}) \\ g_{av2}(\boldsymbol{X}) \end{bmatrix} \alpha_{th} \tag{1}$$

where two state variables are $x_1=\omega_t$ (turbine speed (r/min)) and $x_2=\omega_{ed}$ (engine speed (r/min)); a control variable is a_{th} (percentage of the throttle angle (%)); definitions of nonlinear items $f_{av1}(\boldsymbol{X})$, $f_{av2}(\boldsymbol{X})$, $g_{av1}(\boldsymbol{X})$ and $g_{av2}(\boldsymbol{X})$ are presented in Appendix (1).
The braking condition:

$$\dot{\boldsymbol{X}} = \boldsymbol{F}_{dv}(\boldsymbol{X}) + \boldsymbol{G}_{dv}(\boldsymbol{X})u_b \Rightarrow \begin{bmatrix} \dot{x}_1 \\ \dot{x}_2 \\ \dot{x}_3 \end{bmatrix} = \begin{bmatrix} f_{dv1}(\boldsymbol{X}) \\ f_{dv2}(\boldsymbol{X}) \\ f_{dv3}(\boldsymbol{X}) \end{bmatrix} + \begin{bmatrix} g_{dv1}(\boldsymbol{X}) \\ g_{dv2}(\boldsymbol{X}) \\ g_{dv3}(\boldsymbol{X}) \end{bmatrix} u_b \tag{2}$$

where $x_3=a_b$ is a braking deceleration (m/s^2); u_b is a control variable of the desired input voltage of EBS (V); definitions of nonlinear items $f_{dv1}(\boldsymbol{X})$~$f_{dv3}(\boldsymbol{X})$ and $g_{dv1}(\boldsymbol{X})$~$g_{dv3}(\boldsymbol{X})$ are presented in Appendix (2).
As mentioned earlier, models (1) and (2) are available based upon some strict assumptions. In view of the actual driving condition and complex traffic environment, some uncertainties which this heavy-duty vehicle may possibly encounter can be presented as follows:

1. variation of the mass $10{,}000\text{kg} \le M \le 25{,}000\text{kg}$
2. variation of the road grade $-3° \le \varphi_s \le 3°$
3. gear position shifting of the automatic transmission $i_{g1}=3.49$, $i_{g2}=1.86$, $i_{g3}=1.41$, $i_{g4}=1$, $i_{g5}=0.7$, $i_{g6}=0.65$.
4. mathematical modeling error from the engine, torque converter and the heat fade efficiency of the braking system.

Considering the uncertainties above, two longitudinal dynamics models of the controlled vehicle differ from Eqs. (1) and (2) are therefore expressed as
Driving condition:

$$\dot{\boldsymbol{X}} = \left[\boldsymbol{F}_{av}(\boldsymbol{X}) + \Delta\boldsymbol{F}_{av}(\boldsymbol{X})\right] + \left[\boldsymbol{G}_{av}(\boldsymbol{X}) + \Delta\boldsymbol{G}_{av}(\boldsymbol{X})\right]\alpha_{th} \tag{3}$$

Braking condition:

$$\dot{\boldsymbol{X}} = \left[\boldsymbol{F}_{dv}(\boldsymbol{X}) + \Delta\boldsymbol{F}_{dv}(\boldsymbol{X})\right] + \left[\boldsymbol{G}_{dv}(\boldsymbol{X}) + \Delta\boldsymbol{G}_{dv}(\boldsymbol{X})\right]u_b \tag{4}$$

where $\Delta\boldsymbol{F}_{av}(\boldsymbol{X}),\Delta\boldsymbol{G}_{av}(\boldsymbol{X}),\Delta\boldsymbol{F}_{dv}(\boldsymbol{X}),\Delta\boldsymbol{G}_{dv}(\boldsymbol{X})$ are system uncertain matrixes relative to the nominal model. They are influenced by various factors, and are described as

$$\Delta\boldsymbol{F}_{av}(\boldsymbol{X})=\begin{bmatrix}\Delta f_{av1}\\ \Delta f_{av2}\end{bmatrix},\ \Delta\boldsymbol{G}_{av}(\boldsymbol{X})=\begin{bmatrix}\Delta g_{av1}\\ \Delta g_{av2}\end{bmatrix},\Delta\boldsymbol{F}_{dv}(\boldsymbol{X})=\begin{bmatrix}\Delta f_{dv1}\\ \Delta f_{dv2}\\ \Delta f_{dv3}\end{bmatrix},\ \Delta\boldsymbol{G}_{dv}(\boldsymbol{X})=\begin{bmatrix}\Delta g_{dv1}\\ \Delta g_{dv2}\\ \Delta g_{dv3}\end{bmatrix}.$$

In terms of multiple factors of the uncertain matrixes, it is difficult to estimate the upper and lower boundaries of Eqs. (3) and (4) precisely by using the mathematical analytic method. Therefore, a simulation model of the heavy-duty vehicle is created at first by using the MATLAB/Simulink software, which will be used to estimate the upper and lower boundaries of the uncertain matrixes. To determine the upper and lower boundaries, an analysis on extreme driving/breaking conditions of models (3) and (4) is.

At first, the analysis of Eq. (3) indicates that with the increase of the mass M, road grade φ_s and the gear position, the item of $f_{av1}(\boldsymbol{X})$ converges reversely to its minimum value relative to the nominal condition (at a given ω_t, ω_{ed}). Similarly, the extreme operating condition for the maximum value of $f_{av1}(\boldsymbol{X})$ can be obtained. The analysis above can be applied equally to other items of Eq. (3), and can be summarized as the following two extreme conditions:

(a1) If the vehicle mass is M=10,000kg, the road grade is φ_s=-3° and the automatic transmission is locked at the first gear position, then the upper boundary of Δf_{av1} can be estimated.

(a2) If the vehicle mass is M=25,000kg, the road grade is φ_s=-3° and the automatic transmission is shifted to the third gear position (supposing that the gear position can not be shifted up to the sixth gear position, since it should be subject to a known gear shift logic under a given actual traffic condition), then the lower boundary of Δf_{av1} can be estimate.

On the analysis of Eq. (4), two extreme conditions corresponding to the upper and lower boundaries can also be obtained:

(b1) If the vehicle mass is M=10,000kg, the road grade is φ_s=-3°, the braking deceleration is a_b=0m/s^2 and the gear position is locked at the first gear position, then the upper boundary of Δf_{dv1} can be estimated.

(b2) If the vehicle mass is M=25,000kg, the road grade is φ_s=3°, the braking deceleration is a_b=-2m/s^2 (assuming it as the maximum braking deceleration commonly used) and the gear position is locked at the third gear position, then the lower boundary of Δf_{dv1} can be estimated.

By the foregoing analysis, the extreme and nominal operating conditions will be simulated respectively by using the simulation model of the heavy-duty vehicles. In order to activate entire gear positions of the automatic transmission, the vehicle is accelerated from 0m/s to the maximum velocity by inputting a engine throttle percentage of 100%. After that, the throttle angle percentage is closed to 0%, and the velocity is slowed down gradually in the following two patterns:

1. according to the requirement of (b1) condition, the vehicle is slowed down until stop by the use of the engine invert torque and the road resistance.
2. according to the requirement of (b2) condition, the vehicle is slowed down until stop through an adjoining of a deceleration a_b=-2m/s^2 generated by the EBS, as well as the sum of the engine invert torque and the road resistance.

According to the above extreme conditions (a1), (a2), (b1), (b2), the variation range of each uncertainty can be obtained by simulation, as shown in Figures 2 and 3. For removing the influence from the gear position, the *x*-coordinates in Figures 2 and 3 have been transferred into a universal scale of the engine speed.

For instance (see solid line in Figure 2), during the increase of the engine speed in condition (a1), the upper boundary of the item Δf_{av1} increases gradually, while the items Δf_{av2}, Δg_{av2} change trivially. As to the increase of the engine speed in condition (a2) (see dashed line in Figure 2), the lower boundary of the item Δf_{av1} increases rapidly at the beginning, and then drops slowly. The minimum value appears approximately at the slowest speed of the engine (i.e., the idle condition). The items Δf_{av2}, Δg_{av2} decrease during the engine speed increases.

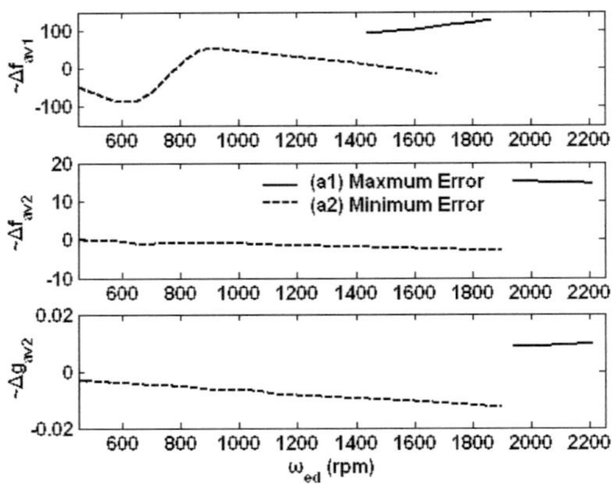

Fig. 2. Changes of uncertain items under driving condition

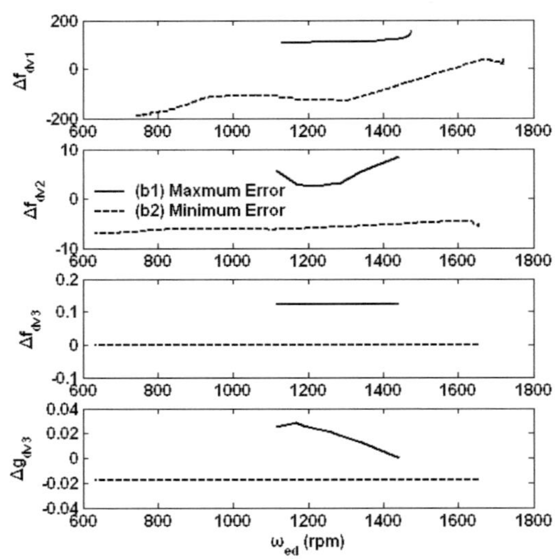

Fig. 3. Changes of uncertain items under braking condition

From the above simulation results, it is easy to calculate the upper and lower boundaries of the uncertain matrixes in Eqs. (3) and (4):

Driving condition:

$$-86 \le \Delta f_{av1} \le 127, \ -2.75 \le \Delta f_{av2} \le 15, \ \Delta g_{av1} = 0, \ -0.0127 \le \Delta g_{av2} \le 0.001.$$

Braking condition:

$$-188 \le \Delta f_{dv1} \le 155, \ -7 \le \Delta f_{dv2} \le 8.45, \ 0 \le \Delta f_{dv3} \le 0.124, \ \Delta g_{dv1} = \Delta g_{dv2} = 0,$$
$$-0.0174 \le \Delta g_{dv3} \le 0.029$$

where a unit of $\Delta f_{a*}, \Delta f_{d*}$ is $r / \min^2$, two units of $\Delta g_{a*}, \Delta g_{d*}$ are $r / \left(\min^2 \cdot \%\right)$ and $m / \left(s^3 \cdot V\right)$, respectively.

To verify the proposed models, some profiles are prepared in Figure 4 according to the aforementioned extreme conditions. They include the throttle angle percentage, EBS desired braking voltage and the road grade containing two values of $\pm 3°$. Figures 5 and 6 are the

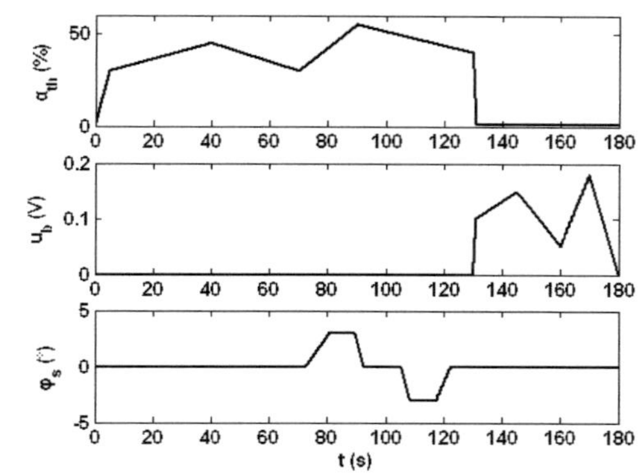

Fig. 4. Profiles of throttle angle percentage, EBS desired braking voltage and road grade

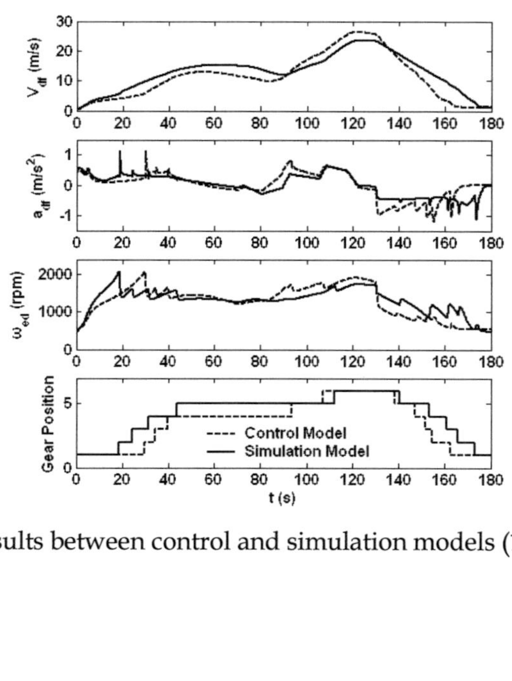

Fig. 5. Comparison results between control and simulation models (10,000kg)

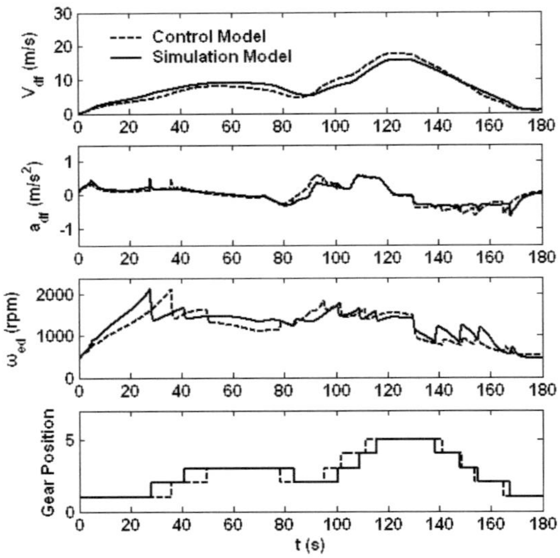

Fig. 6. Comparison results between control and simulation models (25,000kg)

comparison results corresponding to 10,000kg and 25,000kg, respectively. The dashed lines and the solid lines are the results of the control models (3) and (4) and the simulation models, respectively. It can be seen from the comparison results that the control models (3) and (4) are able to approximate the simulation models very closely, even in the case of a wide variation ranges of the velocity (0m/s~28m/s), mass (10,000kg~25,000kg) and the gear positions of the automatic transmission (1~6 gears). Because the models (3) and (4) only present the longitudinal dynamics of the controlled vehicle, the inter-vehicles dynamics has to be considered furthermore such that a completed dynamics model of the LFS at full-speed can be obtained.

2.2 Longitudinal dynamic model of the inter-vehicles

For the purposes of vehicular ACC or SG cruise control system design, many well-known achievements on the operation policy for the inter-vehicles relative distance and velocity have been intense studied [20, 21]. Focusing on the AACC system, the operation policy for the inter-vehicles relative distance and relative velocity should be determined so as to

- maintain desirable spacing between the vehicles
- ensure string stability of the convoy

Inspired by previous research [1], [2], [7] on the design of upper level controller, the operation policy of inter-vehicles relative distance and relative velocity can be defined as

$$\begin{aligned} \varepsilon_d &= d_{h,s} - d_r = \left(d_{\min} + v_{df} t_h\right) - \left(x_l - x_{df}\right) \\ \varepsilon_v &= a_{df} t_h - v_r = a_{df} t_h - \left(v_l - v_{df}\right) \end{aligned} . \tag{5}$$

where a_{df} is a controlled vehicle acceleration (m/s²); ε_d is a tracking error of the longitudinal relative distance (m); ε_v is a tracking error of the longitudinal relative velocity (m/s).

As the illustration of the vehicle longitudinal AACC system (see Figure 1), it should be noted that an item $a_{df}t_h$ is introduced to define the inter-vehicles relative velocity ε_v so as to fit the dynamical process from one stable state to another one. In contrast to Eq. (5), conventional operation policy of inter-vehicles relative velocity is often defined as $\varepsilon_v = v_l - v_{df}$, which only focuses on the static situation of invariable velocity following. However, on account of the dynamic situation of acceleration/deceleration, the previously investigation [15, 16] has demonstrated that it is dangerous and uncomfortable for the AACC system to track a vehicle in front still adopted conventional operation policy. Therefore, an item of $a_{df}t_h$ is proposed to capture accurately the human driver's longitudinal behavior aiming at this situation. Generally, Eq. (5) can be regarded as the dynamical operation policy.

The accuracy of Eq. (5) is validated by the following experimental tests, which is carried out under complicated down-town traffic conditions in terms of five skillful adult drivers (including four males and one female). Two cases including an acceleration tracking and a deceleration approaching are considered. In the case of acceleration tracking, the driver is closing up a leading vehicle without initial error of relative distance and relative velocity. Then, the driver adjusts his/her velocity to the one of the vehicle in front. The headway distance aimed at by the driver during the tracking is essentially depending on the driver's desire of safety. In the case of deceleration approaching, the driver is closing down a leading vehicle with constant velocity. The driver brakes to reestablish the minimal headway distance, and then follow the leading vehicle with the same velocity. The experimental data presented in Figure 7 are the mean square value of five drivers' results. The comparison results confirm that Eq. (5) shows a sufficient agreement with practical driver manipulation, which can be adopted in the design of vehicle longitudinal AACC system.

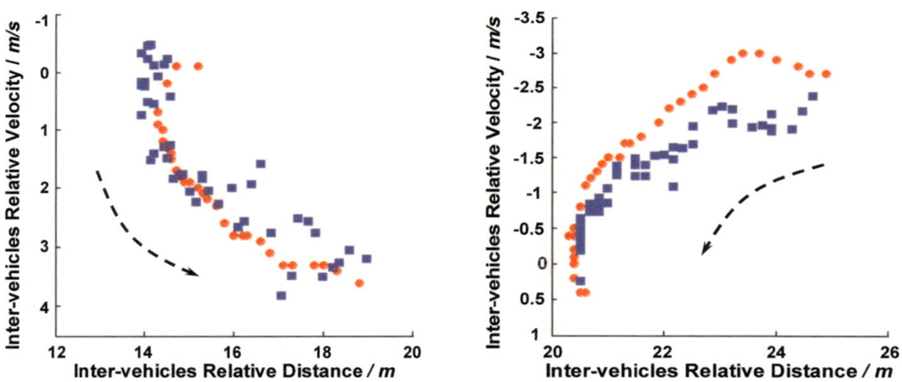

■Operation Policy •Experimental Data

(a) Acceleration tracking condition (b) Deceleration approaching condition

Fig. 7. Comparison results between experimental data and operation policy

By virtue of the operation policy (5), the mathematical model of inter-vehicles longitudinal dynamics is created

$$\begin{aligned} \dot{\varepsilon}_d &= \varepsilon_v = a_{df}t_h - \left(v_l - v_{df}\right) \\ \dot{\varepsilon}_v &= \dot{a}_{df}t_h - \left(\dot{v}_l - \dot{v}_{df}\right) \end{aligned} \quad . \tag{6}$$

where $\dot{v}_l$ is a leading vehicle acceleration (m/s²), which is generally limited within an extreme acceleration/deceleration condition, i.e., $-2\text{m}/\text{s}^2 \le \dot{v}_l \le 2\text{m}/\text{s}^2$.

Although the inter-vehicles dynamics is considered in Eq. (6), the dynamics of the controlled vehicle that has great impact on the performance of entire system has been ignored instead. Actually, two aforementioned models are interrelated and coupled mutually in the LFS. To overcome the disadvantages of the existing independent modeling method, a more accurate model will be proposed in the following to describe the dynamics of the LFS reasonably. In this model, the longitudinal vehicle dynamics models (3) and (4) with uncertainty and the longitudinal inter-vehicles dynamic model (6) are both taken into account. As a result, a control system can be designed on this platform, and an optimal tracking performance with better robustness can also be achieved.

2.3 LFS dynamics model

Firstly, take the time derivative of the state variable $\dot{\omega}_t$ in Eq. (3), and obtain $\ddot{\omega}_t$. After that, $\dot{\omega}_t, \ddot{\omega}_t$ are substituted into Eq. (6) by virtue of the relationship $a_{df} = \alpha_n \dot{\omega}_t = \dfrac{2\pi r_t}{60 i_g i_0}\dot{\omega}_t$. Finally, an LFS dynamics model for the driving condition is derived according to Eqs. (3) and (6). It is a combination of the dynamics between the controlled vehicle and the inter-vehicles, as well as the uncertainty from actual driving conditions.

$$\dot{\boldsymbol{X}} = \left[\boldsymbol{F}_a(\boldsymbol{X}) + \Delta\boldsymbol{F}_a(\boldsymbol{X})\right] + \left[\boldsymbol{G}_a(\boldsymbol{X}) + \Delta\boldsymbol{G}_a(\boldsymbol{X})\right]\alpha_{th} + \boldsymbol{P}_a(\boldsymbol{X})w$$
$$\Leftrightarrow \begin{cases} \dot{\varepsilon}_d = f_{a1} \\ \dot{\varepsilon}_v = \left(f_{a2} + \Delta f_{a2}\right) + \left(g_{a1} + \Delta g_{a1}\right)\alpha_{th} + p_{a1}\dot{v}_l \\ \dot{\omega}_t = f_{a3} + \Delta f_{a3} \\ \dot{\omega}_{ed} = \left(f_{a4} + \Delta f_{a4}\right) + \left(g_{a2} + \Delta g_{a2}\right)\alpha_{th} \end{cases} \tag{7}$$

where $\boldsymbol{X} = \begin{bmatrix}\varepsilon_d & \varepsilon_v & \omega_t & \omega_{ed}\end{bmatrix}^T$ is a vector of the state variables, $w = \dot{v}_l$ is a disturbance variable, and α_{th} is a control variable. The definition of each item in Eq. (7) can be referred to Appendix (2).

Similarly, an LFS dynamics model for the braking condition is achieved:

$$\dot{\boldsymbol{X}} = \left[\boldsymbol{F}_d(\boldsymbol{X}) + \Delta\boldsymbol{F}_d(\boldsymbol{X})\right] + \left[\boldsymbol{G}_d(\boldsymbol{X}) + \Delta\boldsymbol{G}_d(\boldsymbol{X})\right]u_b + \boldsymbol{P}_d(\boldsymbol{X})w$$
$$\Leftrightarrow \begin{cases} \dot{\varepsilon}_d = f_{d1} \\ \dot{\varepsilon}_v = \left(f_{d2} + \Delta f_{d2}\right) + \left(g_{d1} + \Delta g_{d1}\right)u_b + p_{d1}\dot{v}_l \\ \dot{\omega}_t = f_{d3} + \Delta f_{d3} \\ \dot{\omega}_{ed} = f_{d4} + \Delta f_{d4} \\ \dot{a}_b = \left(f_{d5} + \Delta f_{d5}\right) + \left(g_{d2} + \Delta g_{d2}\right)u_b \end{cases} \tag{8}$$

where $\boldsymbol{X} = \begin{bmatrix}\varepsilon_d & \varepsilon_v & \omega_t & \omega_{ed} & a_b\end{bmatrix}^T$ is a vector of the state variables, u_b is a control variable. The definition of each item in Eq. (8) can be referred to Appendix (4).

According to the analysis of the extreme driving/braking conditions in 2.1, an approximate ranges of the upper and lower boundaries regarding uncertain items in Eqs. (7) and (8) can be calculated through simulation.

Driving condition:

$$-104 \le \Delta f_{a2} \le 203, -0.031 \le \Delta g_{a1} \le 0.0027$$

Braking condition:

$$-192 \le \Delta f_{d2} \le 174, -0.0153 \le \Delta g_{d1} \le 0.022$$

where an unit of Δf_* is m/s^2, units of $\Delta g_{a1}, \Delta g_{d1}$ are $m/(s^2 \cdot \%)$ and $m/(s^2 \cdot V)$, respectively.
The analysis of the dynamics models (7) and (8) indicates that the LFS is an uncertain affine nonlinear system, in which the strong nonlinearities and the coupling properties caused by the disturbance and the uncertainty are represented. These complex behaviors result in more difficulties while implementing the control of the LFS, since the state variables ε_d, ε_v are influenced significantly by the nonlinearity, uncertainty, as well as the disturbance from the leading vehicle's acceleration/deceleration. However, because the longitudinal dynamics of the controlled vehicle and the inter-vehicles can be described and integrated into a universal frame of the state space equation accurately, this would be helpful for the purpose of achieving a global optimal and a robust control for the LFS.
The LFS AACC system intends to implement the accurate tracking control of the inter-vehicles relative distance/relative velocity under both high-speed and crowded traffic environments. Thus, the system should be provided with strong robustness in view of the complex external disturbance and the internal uncertainty, as well as the capability to eliminate the impact from the system's strong nonlinearity at low-speed. Focusing on the LFS, refs. [22-27] presented an NDD method to eliminate the disturbance effectively, which was, however, limited to some certain affine nonlinear systems. Utilizing the invariance of the sliding mode in VSC, the control algorithm proposed in refs. [28, 29] can implement the completely decoupling of all state variables from the disturbance and the uncertainty. But, it is not a global decoupling algorithm, and should also be submitted to some strict matching conditions. Refs. [30-34] studied the input-output linearization on an uncertain affine nonlinear system, but did not discuss the disturbance decoupling problem. On a nonlinear system with perturbation, ref. [35] gave the necessary and sufficient condition for the completely disturbance decoupling problem, but did not present the design of the feedback controller. To avoid the disadvantages of those control algorithms mentioned above, a DDRC method combining the theory of NDD and VSC is proposed in regard to the complex dynamics of the LFS.

3. DDRC method

The basic theory of DDRC method is inspired by the idea of the step-by-step transformation and design. First, on account of a certain affine nonlinear system with disturbance, the NDD theory based on the differential geometry is used to implement the disturbance decoupling and the input-output linearization. Hence, a linearized subsystem with partial state variables is given, in which the invariance matching conditions of the sliding mode can be discussed easily via VSC theory, and then a VSC controller can be deduced. Finally, two methods will be integrated together such that a completely decoupling of the system from the external disturbance, and a weakened invariance matching condition with a simplified control system structure are obtained.

3.1 NDD theory on certain affine nonlinear system

At first, consider a certain dynamics model of the LFS, where uncertain items of $\Delta \boldsymbol{F}_a(\boldsymbol{X})$, $\Delta \boldsymbol{G}_a(\boldsymbol{X})$, $\Delta \boldsymbol{F}_d(\boldsymbol{X})$ and $\Delta \boldsymbol{G}_d(\boldsymbol{X})$ are considered as zero. Hence, a certain affine nonlinear system can be simplified as

$$\begin{cases} \dot{\boldsymbol{X}} = \boldsymbol{F}(\boldsymbol{X}) + \boldsymbol{G}(\boldsymbol{X})u + \boldsymbol{P}(\boldsymbol{X})w \\ y = h(\boldsymbol{X}) \end{cases} \tag{9}$$

where $\boldsymbol{X} \in R^n$ and $u, w, y \in R$ are system state variable, control variable, disturbance variable and output variable, respectively, $\boldsymbol{F}$, $\boldsymbol{G}$, $\boldsymbol{P}$, h are differentiable functions of $\boldsymbol{X}$ with corresponding dimensions.

The basic theory of NDD is trying to seek a state feedback, and construct a closed-loop system as follows

$$\begin{cases} \dot{\boldsymbol{X}} = \boldsymbol{F}(\boldsymbol{X}) + \boldsymbol{G}(\boldsymbol{X})\alpha(\boldsymbol{X}) + \boldsymbol{G}(\boldsymbol{X})\beta(\boldsymbol{X})v + \boldsymbol{P}(\boldsymbol{X})w = \tilde{\boldsymbol{F}}(\boldsymbol{X}) + \tilde{\boldsymbol{G}}(\boldsymbol{X})v + \boldsymbol{P}(\boldsymbol{X})w \\ y = h(\boldsymbol{X}) \end{cases} \tag{10}$$

If there is an invariant distribution $\Delta(\boldsymbol{X})$ that exists over $\tilde{\boldsymbol{F}}(\boldsymbol{X}), \tilde{\boldsymbol{G}}(\boldsymbol{X})$, and satisfies

$$span\{\boldsymbol{P}\} \subset \Delta(\boldsymbol{X}) \subset \Omega^{\perp}(\boldsymbol{X}) \tag{11}$$

where

$$\Omega(\boldsymbol{X}) = \begin{bmatrix} dh(\boldsymbol{X}) & dL_{\boldsymbol{F}}h(\boldsymbol{X}) & \cdots & dL_{\boldsymbol{F}}^{r-1}h(\boldsymbol{X}) \end{bmatrix}^T .$$

Then, the output y can be decoupled from the disturbance w, and we have a r-dimension coordinate transformation

$$\boldsymbol{Z} = \boldsymbol{\psi}(\boldsymbol{X}) = [z_1, \cdots, z_r]^T = \left[h(\boldsymbol{X}), \cdots, L_{\boldsymbol{F}}^{r-1}h(\boldsymbol{X})\right]^T \tag{12}$$

as well as an n-r-dimension coordinate transformation

$$\boldsymbol{\mu} = \boldsymbol{\phi}(\boldsymbol{X}) = \left[\mu_1(\boldsymbol{X}), \cdots, \mu_{n-r}(\boldsymbol{X})\right]^T \tag{13}$$

where $\boldsymbol{\mu}$ satisfies

$$d\mu_i(\boldsymbol{X})\boldsymbol{G}(\boldsymbol{X}) = 0, \quad \forall \boldsymbol{X} \in U, \quad i = 1, \cdots, n-r \tag{14}$$

In this way, the original closed-loop system (9) can be modified as a following form over the new coordinate

$$\begin{aligned} \dot{z}_i &= z_{i+1} \qquad 1 \le i \le r-1 \\ \dot{z}_r &= v \end{aligned} \tag{15}$$

$$\dot{\boldsymbol{\mu}} = \boldsymbol{Q}(\boldsymbol{Z}, \boldsymbol{\mu}) + \boldsymbol{K}(\boldsymbol{Z}, \boldsymbol{\mu})w \tag{16}$$

Obviously, Eq. (15) is a linearized decoupling subsystem, while Eq. (16) is a nonlinear internal dynamic subsystem subject to the disturbance. The invariant distribution $\Delta(\boldsymbol{X})$ is defined as $\left[\tilde{\boldsymbol{F}},\Delta\right](\boldsymbol{X})\subset\Delta(\boldsymbol{X})$, L is a Lie derivative, defined as $L_{\boldsymbol{F}}\boldsymbol{G}=\left(\frac{\partial \boldsymbol{G}}{\partial \boldsymbol{X}}\right)\boldsymbol{F}$, r is a relative degree, defined as $L_{G}L_{F}^{r-1}h(\boldsymbol{X})\neq 0$ [36], $*^{\perp}$ is an orthogonal of" $*$ "[37]. Eq. (10) is a necessary and sufficient condition of the disturbance decoupling problem, which can be expressed in the equivalent form

$$\Omega(\boldsymbol{X})\boldsymbol{P}(\boldsymbol{X})=\boldsymbol{0} \tag{17}$$

State feedback is

$$u=\alpha(\boldsymbol{X})+\beta(\boldsymbol{X})v=\frac{-L_{F}^{r}h(\boldsymbol{X})+v}{L_{G}L_{F}^{r-1}h(\boldsymbol{X})} \tag{18}$$

If the disturbance w is measurable, the following state feedback can be considered

$$u=\alpha(\boldsymbol{X})+\beta(\boldsymbol{X})v+\gamma(\boldsymbol{X})w \ =\frac{-L_{F}^{r}h(\boldsymbol{X})+v-L_{P}L_{F}^{r-1}h(\boldsymbol{X})w}{L_{G}L_{F}^{r-1}h(\boldsymbol{X})} \tag{19}$$

In this way, a weakened necessary and sufficient condition of the disturbance decoupling problem is achieved as

$$\Omega(\boldsymbol{X})\left[\boldsymbol{G}(\boldsymbol{X})\gamma+\boldsymbol{P}(\boldsymbol{X})\right]=\boldsymbol{0} \tag{20}$$

As a result, some existing linear control methods (likes, LQ, pole placement) can be used to implement the pole placement over the linearized decoupling subsystem. In the following, the NDD theory is used to discuss the VSC problem of the affine nonlinear systems under the impact of the uncertainty.

3.2 VSC of uncertain affine nonlinear systems based on NDD

Considering Eqs. (7) and (8) with uncertainty, they can be simplified as a more general forms for the analysis, i.e.,

$$\begin{cases}\dot{\boldsymbol{X}}=\left[\boldsymbol{F}(\boldsymbol{X})+\Delta\boldsymbol{F}(\boldsymbol{X})\right]+\left[\boldsymbol{G}(\boldsymbol{X})+\Delta\boldsymbol{G}(\boldsymbol{X})\right]u+\left[\boldsymbol{P}(\boldsymbol{X})+\Delta\boldsymbol{P}(\boldsymbol{X})\right]w\\ y=h(\boldsymbol{X})\end{cases} \tag{21}$$

where $\boldsymbol{F}$, $\boldsymbol{G}$, $\boldsymbol{P}$, h indicate the certain part of the system, and they are defined as Eq. (8), $\Delta\boldsymbol{F}$, $\Delta\boldsymbol{G}$, $\Delta\boldsymbol{P}$ indicate the uncertain part correspondingly.

At first, take first derivative of the output variable $y=h(\boldsymbol{X})$:

$$\begin{aligned}\dot{z}_1&=\frac{dy}{dt}\\&=\frac{\partial h(\boldsymbol{X})}{\partial \boldsymbol{X}}\left[\boldsymbol{F}(\boldsymbol{X})+\boldsymbol{G}(\boldsymbol{X})u+\boldsymbol{P}(\boldsymbol{X})w\right]+\frac{\partial h(\boldsymbol{X})}{\partial \boldsymbol{X}}\left[\Delta\boldsymbol{F}(\boldsymbol{X})+\Delta\boldsymbol{G}(\boldsymbol{X})u+\Delta\boldsymbol{P}(\boldsymbol{X})w\right]\\&=\left[L_{F}h(\boldsymbol{X})+L_{G}h(\boldsymbol{X})u+L_{P}h(\boldsymbol{X})w\right]+\left[L_{\Delta F}h(\boldsymbol{X})+L_{\Delta G}h(\boldsymbol{X})u+L_{\Delta P}h(\boldsymbol{X})w\right]\end{aligned} \tag{22}$$

Obviously, if

$$L_{\Delta F}h(\boldsymbol{X}) = L_{\Delta G}h(\boldsymbol{X}) = L_{\Delta P}h(\boldsymbol{X}) = 0 \tag{23}$$

then according to the definition of the relative degree and Eq. (17), Eq. (22) becomes

$$\dot{z}_1 = L_F h(\boldsymbol{X}) = z_2 \tag{24}$$

Differentiate Eq. (24) again yields

$$\begin{aligned}\dot{z}_2 &= \frac{dL_F h(\boldsymbol{X})}{dt} \\ &= \frac{\partial L_F h(\boldsymbol{X})}{\partial \boldsymbol{X}}\left[\boldsymbol{F}(\boldsymbol{X}) + \boldsymbol{G}(\boldsymbol{X})u + \boldsymbol{P}(\boldsymbol{X})w\right] + \frac{\partial L_F h(\boldsymbol{X})}{\partial \boldsymbol{X}}\left[\Delta \boldsymbol{F}(\boldsymbol{X}) + \Delta \boldsymbol{G}(\boldsymbol{X})u + \Delta \boldsymbol{P}(\boldsymbol{X})w\right] \\ &= \left[L_F^2 h(\boldsymbol{X}) + L_G L_F h(\boldsymbol{X})u + L_P L_F h(\boldsymbol{X})w\right] + \left[L_{\Delta F} L_F h(\boldsymbol{X}) + L_{\Delta G} L_F h(\boldsymbol{X})u + L_{\Delta P} L_F h(\boldsymbol{X})w\right]\end{aligned} \tag{25}$$

which in turn deduces

$$L_{\Delta F} L_F h(\boldsymbol{X}) = L_{\Delta G} L_F h(\boldsymbol{X}) = L_{\Delta P} L_F h(\boldsymbol{X}) = 0 \tag{26}$$

By the definition of relative degree and Eq. (17), Eq. (25) becomes

$$\dot{z}_2 = L_F^2 h(\boldsymbol{X}) = z_3 \tag{27}$$

After differentiating r times, we find that

$$\begin{aligned}\dot{z}_r &= L_G L_F^{r-1} h(\boldsymbol{X})u + L_F^r h(\boldsymbol{X}) + L_{\Delta F} L_F^{r-1} h(\boldsymbol{X}) + L_{\Delta G} L_F^{r-1} h(\boldsymbol{X})u + L_{\Delta P} L_F^{r-1} h(\boldsymbol{X})w \\ &= \left[L_{\Delta F} L_F^{r-1} h(\boldsymbol{X}) + \alpha(\boldsymbol{X}) L_{\Delta G} L_F^{r-1} h(\boldsymbol{X})\right] + \left[1 + \beta(\boldsymbol{X}) L_{\Delta G} L_F^{r-1} h(\boldsymbol{X})\right]v + L_{\Delta P} L_F^{r-1} h(\boldsymbol{X})w\end{aligned} \tag{28}$$

Based on the above proof, the disturbance decoupling problem of uncertain affine nonlinear systems can be solved, if there exist VSC matching conditions such that

$$\text{(c1) } L_{\Delta F} L_F^i h(\boldsymbol{X}) = 0, L_{\Delta G} L_F^i h(\boldsymbol{X}) = 0, L_{\Delta P} L_F^i h(\boldsymbol{X}) = 0,\ L_{\Delta P} L_F^i h(\boldsymbol{X}) = 0, (0 \le i \le r-2)$$

$$\text{(c2) } \|\Delta \boldsymbol{F}(\boldsymbol{X})\| \le f_m, \|\Delta \boldsymbol{G}(\boldsymbol{X})\| \le g_m, \|\Delta \boldsymbol{P}(\boldsymbol{X})\| \le p_m,\ \|w\| \le w_m$$

where $\|\bullet\|$ is a norm of the vector or matrix of "•", that is $\left\|\left(a_{ij}\right)_{n\times n}\right\| = \max\limits_{1\le i\le n} \sum\limits_{j=1}^{n} \left|a_{ij}\right|$; f_m, g_m, p_m, w_m are perturbation boundaries of the corresponding given matrixes.

Summing up the definition of the relative degree, matching conditions (c1) and the coordinate transformation **Z**=*ψ*(**X**), we obtain a closed-loop system over the new coordinate by substituting the state feedback (18) or (19) into Eq. (21), which has the form

$$\begin{aligned}\dot{z}_i &= z_{i+1}, \quad 1 \le i \le r-1, \\ \dot{z}_r &= \left[L_{\Delta F} L_F^{r-1} h(\boldsymbol{X}) + \alpha(\boldsymbol{X}) L_{\Delta G} L_F^{r-1} h(\boldsymbol{X})\right] + \left[1 + \beta(\boldsymbol{X}) L_{\Delta G} L_F^{r-1} h(\boldsymbol{X})\right]v + L_{\Delta P} L_F^{r-1} h(\boldsymbol{X})w\end{aligned} \tag{29}$$

$$\dot{\boldsymbol{\mu}} = \boldsymbol{Q}(\boldsymbol{Z},\boldsymbol{\mu}) + \Delta \boldsymbol{Q}(\boldsymbol{Z},\boldsymbol{\mu}) + \Delta \boldsymbol{R}(\boldsymbol{Z},\boldsymbol{\mu})v + \left[\boldsymbol{K}(\boldsymbol{Z},\boldsymbol{\mu}) + \Delta \boldsymbol{K}(\boldsymbol{Z},\boldsymbol{\mu})\right]w \tag{30}$$

It can be noticed from Eq. (29) that for the state variables z_i of the first r-1 dimensions, the linearization and the disturbance decoupling have been achieved, except for the remaining z_r (Eq. (30)). By virtue of the invariance of the sliding mode in VSC [28], it will be used in consequence to eliminate the disturbance and the uncertainty on z_r.
Based on the VSC theory [28], a switching function is designed easily by taking advantage of the linearized decoupling subsystem (29) over the new coordination

$$S_Z = S(\boldsymbol{Z}) = \boldsymbol{C}[z_1 \cdots z_r]^T \tag{31}$$

where $\boldsymbol{C}=[c_1,\ldots,c_{r-1},1]$ is a normal constant coefficient matrix to be determined. Once the system is controlled towards the sliding mode, it satisfies

$$S_Z = \boldsymbol{C}[z_1 \cdots z_r]^T = 0 \tag{32}$$

yielding the following reduced-order equation

$$\dot{z}_i = z_{i+1}, \qquad 1 \le i \le r-1 \tag{33}$$

Clearly, a desired dynamic performance of each state variable in Eq. (33) can be achieved by configuring the coefficient $\boldsymbol{C}$.
As the desired dynamic performance of the sliding mode has already been achieved, an appropriate VSC law is to be defined so as to ensure the desired sliding mode occurring within a finite time. It is convenient to differentiate the switching function (31), and derive the following equation in terms of Eq. (29)：

$$\dot{S}_Z = \boldsymbol{A}_{S_Z} + \Delta\boldsymbol{A}_{S_Z} + \left(\boldsymbol{B}_{S_Z} + \Delta\boldsymbol{B}_{S_Z}\right)v + \Delta\boldsymbol{C}_{S_Z} w \tag{34}$$

where

$$\boldsymbol{A}_{S_Z} = \sum_{i=1}^{r-1} c_i \dot{z}_i, \Delta\boldsymbol{A}_{S_Z} = \boldsymbol{\varphi}\cdot\Delta\boldsymbol{F} + \alpha\boldsymbol{\varphi}\cdot\Delta\boldsymbol{G},\ \boldsymbol{B}_{S_Z} = 1, \Delta\boldsymbol{B}_{S_Z} = \beta\boldsymbol{\varphi}\cdot\Delta\boldsymbol{G},$$

$$\Delta\boldsymbol{C}_{S_Z} = \boldsymbol{\varphi}\cdot\Delta\boldsymbol{P},\ \boldsymbol{\varphi} = \frac{\partial L_{\boldsymbol{F}}^{r-1} h(\boldsymbol{X})}{\partial \boldsymbol{X}}.$$

Considering an VSC law below

$$v = -\boldsymbol{B}_{S_Z}^{-1}\left[\boldsymbol{A}_{S_Z} + a_s S_Z + b_s \operatorname{sgn}(S_Z)\right] \qquad (a_s \ge 0, b_s > 0) \tag{35}$$

an inequality below can be derived from the matching condition (c2), Eqs. (31) and (34).

$$\begin{aligned}
S_Z\dot{S}_Z &= S_Z\left\{\Delta\boldsymbol{A}_{S_Z} + \Delta\boldsymbol{C}_{S_Z} w - a_s S_Z - b_s \operatorname{sgn}(S_Z) - \Delta\boldsymbol{B}_{S_Z}\boldsymbol{B}_{S_Z}^{-1}\left[\boldsymbol{A}_{S_Z} + a_s S_Z + b_s \operatorname{sgn}(S_Z)\right]\right\} \\
&\le \|S_Z\|\|\Delta\boldsymbol{A}_{S_Z}\| + \|S_Z\|\|\Delta\boldsymbol{C}_{S_Z} w\| - a_s\|S_Z\|^2 - b_s\|S_Z\| + \|\Delta\boldsymbol{B}_{S_Z}\|\left(\|\boldsymbol{B}_{S_Z}^{-1}\boldsymbol{A}_{S_Z}\|\|S_Z\| + a_s\|S_Z\|^2 + b_s\|S_Z\|\right) \\
&\le \|S_Z\|\|\boldsymbol{\varphi}\|\left(f_m - \|\alpha\| g_m\right) + \|S_Z\|\|\boldsymbol{\varphi}\| p_m w_m - a_s\|S_Z\|^2 \\
&\quad - b_s\|S_Z\| + \|\beta\boldsymbol{\varphi}\| g_m\left(\|\boldsymbol{B}_{S_Z}^{-1}\boldsymbol{A}_{S_Z}\|\|S_Z\| + a_s\|S_Z\|^2 + b_s\|S_Z\|\right) \\
&= -\|S_Z\|^2 a_s\left(1 - \|\beta\boldsymbol{\varphi}\| g_m\right) - \|S_Z\|\left\{b_s\left(1 - \|\beta\boldsymbol{\varphi}\| g_m\right)\right. \\
&\quad \left. - \left[\|\varphi\|\left(f_m - \|\alpha\| g_m + p_m w_m\right) + \|\beta\boldsymbol{\varphi}\|\|\boldsymbol{B}_{S_Z}^{-1}\boldsymbol{A}_{S_Z}\| g_m\right]\right\}
\end{aligned} \tag{36}$$

It is noticed from Eq. (36) that if the perturbation boundary g_m of uncertain part $\Delta\boldsymbol{G}$ satisfies

$$g_m < \|\beta\boldsymbol{\varphi}\|^{-1} \tag{37}$$

then defining

$$b_s > \frac{\|\boldsymbol{\varphi}\|\left(f_m - \|\alpha\| g_m + p_m w_m\right) + \|\beta\boldsymbol{\varphi}\| \left\|\boldsymbol{B}_{S_Z}^{-1} \boldsymbol{A}_{S_Z}\right\| g_m}{1 - \|\beta\boldsymbol{\varphi}\| g_m} \tag{38}$$

may lead to the following inequality:

$$S_Z \dot{S}_Z < 0 \tag{39}$$

Namely, the convergence condition of the sliding mode is achieved.

From the above verification, the desired sliding mode is achievable under the VSC law (35), as long as the matching condition (c2) and the constraints (38) are satisfied. Since Eqs. (31) and (35) are the switching function and the control law over the new coordinate **X**, they should be transferred back to the original coordinate **Z** by adopting the inverse transformation $\boldsymbol{Z}=\psi(\boldsymbol{X})$. Finally, the DDRC law can be achieved by substituting the VSC law over the original coordinate into the disturbance decoupling state feedback control law (Eq. (18) or Eq. (19)).

To summarize, for an uncertain affine nonlinear system, if the disturbance decoupling condition (17) or (20) and the matching conditions of (c1) and (c2) hold respectively for the certain part and the uncertain part, the DDRC method with the combination of NDD and VSC theories can be figured out as the following design procedure:

Step 1. According to the NDD theory of affine nonlinear systems, the feedback control law (Eq. (18) or (19)) and the coordinate transformation (Eqs. (12) and (13)) are derived to transfer the original system into the linearized decoupling normal form (Eq. (15)) over the new coordinate.

Step 2. Give the VSC matching conditions (c1) and (c2) for the uncertain part of the affine nonlinear systems.

Step 3. Utilize the linearized decoupling normal form (Eq. (15)) over the new coordinate to design the switching function (Eq. (31)), and determine its coefficients accordingly.

Step 4. Design the VSC law (Eq. (35)) based on the perturbation boundary (37) of the uncertainty part, and the convergence condition of the sliding mode (39).

Step 5. Define the coordinate transformation (12) to transfer the switching function (Eq. (31)) and the VSC law (Eq. (35)) from the new coordinate **Z** back to the original coordinate **X**.

Step 6. Substitute the VSC law (Eq. (18) or (19)) over the original coordinate into the feedback control law, and yield the DDRC method.

A block diagram of the closed-loop system for the aforementioned DDRC method is shown in Figure 8, which includes two feedback loops. The nonlinear loop (i.e., the NDD loop) is used to achieve the disturbance decoupling and the partial linearization, regarding the system output y from the disturbance w. On the other hand, the linear loop (i.e., the VSC loop) is used to restrain the system's uncertainty and regulate the closed-loop dynamic performance.

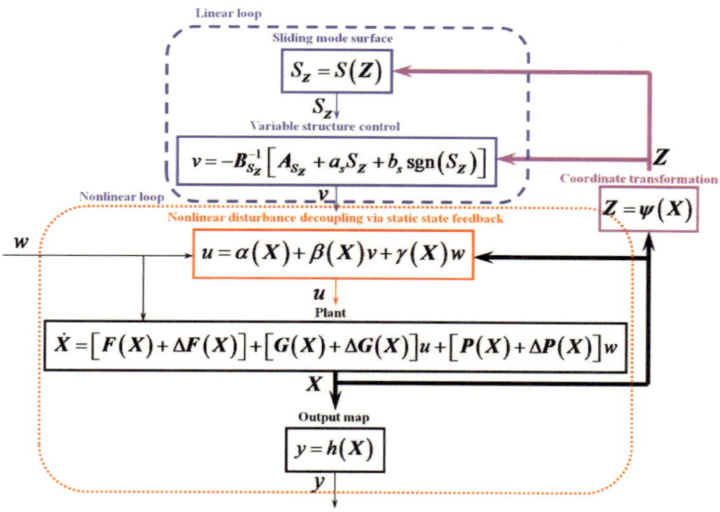

Fig. 8. Block diagram of closed-loop system for DDRC method

4. LFS AACC system

In this section, the proposed DDRC method will be used to design the LFS AACC system with respect to the driving and the braking conditions.

4.1 LFS AACC system for driving condition

Recall the procedure in 2.2, the disturbance decoupling problem on the LFS dynamics model without the impact of the uncertainty is considered (i.e., for the uncertain items of Eq. (7) let $\Delta F_a(X)=0$, $\Delta G_a(X)=0$). On the purpose of LFS AACC system, the following affine nonlinear system with the output variable is defined:

$$\begin{cases} \dot{X} = F_a(X) + G_a(X)\alpha_{th} + P_a(X)w \\ y = h(X) = \varepsilon_d \end{cases} \tag{40}$$

By adopting the NDD theory of certain affine nonlinear system, the relative degree of system (40) is calculated as

$$L_{F_a}h(X) = \varepsilon_v, L_{G_a}h(X) = [1 \quad 0 \quad 0 \quad 0]G_a = 0, L_{G_a}L_{F_a}h(X) = [0 \quad 1 \quad 0 \quad 0]G_a \neq 0.$$

Obviously, the relative degree is $r = 2$, which results in the following matrix

$$\Omega_a(X) = \begin{bmatrix} dh(X) \\ dL_{F_a}h(X) \end{bmatrix} = \begin{bmatrix} 1 & 0 & 0 & 0 \\ 0 & 1 & 0 & 0 \end{bmatrix} \tag{41}$$

Then, it is easy to verify that

$$\Omega_a(X)P_a = \begin{bmatrix} 1 & 0 & 0 & 0 \\ 0 & 1 & 0 & 0 \end{bmatrix} P_a = \begin{bmatrix} 0 \\ p_{a1} \end{bmatrix} \neq \mathbf{0} \tag{42}$$

That is to say, the disturbance decoupling from system (40) can not be achieved by the state feedback (18), because the necessary and sufficient condition (17) is not satisfied. Thus, one can turn to the state feedback (19) with measurable disturbance. Note that if

$$\gamma_a = -\left(\frac{p_{a1}}{g_{a1}}\right) \tag{43}$$

then the necessary and sufficient condition (20) is satisfied, i.e.,

$$\Omega_a(\boldsymbol{X})[G_a\gamma_a + \boldsymbol{P}_a] = \begin{bmatrix} 1 & 0 & 0 & 0 \\ 0 & 1 & 0 & 0 \end{bmatrix}[G_a\gamma_a + \boldsymbol{P}_a] = \boldsymbol{0} \tag{44}$$

By Eq. (19), the decoupling state feedback is obtained as

$$\alpha_{th} = \alpha_a(\boldsymbol{X}) + \beta_a(\boldsymbol{X})v_{ua} + \gamma_a(\boldsymbol{X})w = \frac{-f_{a2}(\omega_t,\omega_{ed}) + v_{ua} - p_{a1}w}{g_{a1}(\omega_t,\omega_{ed})} \tag{45}$$

and the corresponding coordinate transformation with r=2 dimensions is

$$\boldsymbol{Z}_a = \boldsymbol{\psi}_a(\boldsymbol{X}) = \begin{bmatrix} z_{a1} \\ z_{a2} \end{bmatrix} = \begin{bmatrix} h(\boldsymbol{X}) \\ L_{\mathbf{F}_a}h(\boldsymbol{X}) \end{bmatrix} = \begin{bmatrix} \varepsilon_d \\ \varepsilon_v \end{bmatrix} \tag{46}$$

where

$$L_{\mathbf{F}_a}^2 h(\boldsymbol{X}) = f_{a2}, L_{\mathbf{P}_a}L_{\mathbf{F}_a}h(\boldsymbol{X}) = p_{a1}, L_{\mathbf{G}_a}L_{\mathbf{F}_a}h(\boldsymbol{X}) = g_{a1}.$$

Additionally, in order to complete the coordinate transformation, the remaining n-r=2 dimensional coordinates μ_{a1}, μ_{a2} should satisfy the following condition:

$$\frac{\partial \mu_{ai}}{\partial \boldsymbol{X}}\boldsymbol{G}_a = \begin{bmatrix} \frac{\partial \mu_{ai}}{\partial \varepsilon_d} & \frac{\partial \mu_{ai}}{\partial \varepsilon_v} & \frac{\partial \mu_{ai}}{\partial \omega_t} & \frac{\partial \mu_{ai}}{\partial \omega_{ed}} \end{bmatrix}\begin{bmatrix} 0 \\ g_{a1} \\ 0 \\ g_{a2} \end{bmatrix} = 0 \qquad (i = 1,2) \tag{47}$$

The purpose is to ensure the diffeomorphism relationship of the coordinate transformation between the original and the new one (in other words, it is a one-to-one continuous coordinate transformation between the original and the new one, the same is for the inverse transformation). Obviously, one solution of the partial differential Eq. (47) is

$$\begin{aligned} \mu_{a1} &= \omega_t \\ \mu_{a2} &= -\varepsilon_v + \alpha_n t_h\left(b\omega_t\omega_{ed} + c\omega_{ed}^2 + d\frac{\omega_t^3}{\omega_{ed}}\right) \end{aligned} \tag{48}$$

Hence, the transformation of the remaining 2 dimensional coordinates is

$$\boldsymbol{\mu}_a = \boldsymbol{\phi}_a(\boldsymbol{X}) = \begin{bmatrix} \mu_{a1}(\boldsymbol{X}) \\ \mu_{a2}(\boldsymbol{X}) \end{bmatrix} \tag{49}$$

Up to now, the decoupling state feedback (Eq. (45)) and the coordinate transformation (Eqs. (46) and (49)) have been obtained for the certain part of the LFS dynamics model under the driving condition.

Further consideration on the uncertain part of model (7) will be continued. On the basis of the design procedure (Step2) in 3.2, the matching conditions (c1) and (c2) have to be verified at first, and

$$\begin{aligned} L_{\Delta F_a} h(X) &= [1 \quad 0 \quad 0 \quad 0] \Delta F_a = 0 \\ L_{\Delta G_a} h(X) &= [1 \quad 0 \quad 0 \quad 0 \quad 0] \Delta G_a = 0 \end{aligned} \tag{50}$$

It should be noticed from 1.2 and 1.3 that the uncertain items $\Delta F_a(X)$, $\Delta G_a(X)$ and the disturbance w are subject to the following limited upper boundaries:

$$\begin{aligned} \|\Delta F_a(X)\| &\le f_{am} = 203 \\ \|\Delta G_a(X)\| &\le g_{am} = 0.031 \\ \|w\| &\le w_{am} = 2 \end{aligned} \tag{51}$$

By substituting the decoupling state feedback $u=a_{th}$ (Eq. (45)) into model (7), and making use of the coordinate transformations (46) and (49), a linearized subsystem below can be achieved, in which the certain part is completely decoupled from the disturbance.

$$\begin{cases} \begin{bmatrix} \dot{z}_{a1} \\ \dot{z}_{a2} \end{bmatrix} = \overbrace{\begin{bmatrix} 0 & 1 \\ 0 & 0 \end{bmatrix} \begin{bmatrix} z_{a1} \\ z_{a2} \end{bmatrix} + \begin{bmatrix} 0 \\ 1 \end{bmatrix} v_{ua}}^{\text{Certain part}} + \overbrace{\begin{bmatrix} 0 \\ \Delta f_{a2} - \dfrac{f_{a2}}{g_{a1}} \Delta g_{a1} \end{bmatrix} + \begin{bmatrix} 0 \\ \dfrac{\Delta g_{a1}}{g_{a1}} \end{bmatrix} v_{ua} + \begin{bmatrix} 0 \\ -\dfrac{p_{a1}}{g_{a1}} \Delta g_{a1} \end{bmatrix} w}^{\text{Part of uncertain and disturbance}} \\ y = \varepsilon_d \end{cases} \tag{52}$$

Besides, a nonlinear dynamic internal subsystem without separating from the disturbance and the uncertainty is yielded

$$\dot{\mu}_a = Q_a(Z_a, \mu_a) + \Delta Q_a(Z_a, \mu_a) + [K_a(Z_a, \mu_a) + \Delta K_a(Z_a, \mu_a)] w \tag{53}$$

where

$$Q_a(Z_a, \mu_a) = \begin{bmatrix} a\mu_{a1}^2 + \dfrac{\mu_{a2} + z_{a2}}{\alpha_n t_h} + l \\ -\left(2a\mu_{a1} + \dfrac{1}{t_h}\right)\left(\alpha_n t_h a \mu_{a1}^2 + \mu_{a2} + z_{a2} + \alpha_n t_h l\right) \end{bmatrix}, \quad K_a(Z_a, \mu_a) = \begin{bmatrix} 0 \\ -p_{a1} \end{bmatrix}.$$

Based on the analysis of the extreme operating conditions in 2.1, it can be noticed that the items ΔQ_a, ΔK_a are constants with limited upper boundaries.

For the certain part of Eq. (52), it is clear that the state variables z_{a1}, z_{a2} have been completely decoupled from the disturbance w. In order to enhance the system's robustness from the remaining uncertain part and the disturbance within the linearized decoupling subsystem (52), we may design the following switching function over the new coordinate by making use of Eq. (52).

$$S_{a\mathbf{Z}} = \mathbf{C}_a \begin{bmatrix} z_{a1} \\ z_{a2} \end{bmatrix} \tag{54}$$

where $\mathbf{C}_a$=[c_{a1} 1] is a coefficient matrix to be determined. Once the system is controlled towards the sliding mode, it obeys

$$S_{a\mathbf{Z}} = c_{a1}z_{a1} + z_{a2} = 0 \Rightarrow z_{a2} = -c_{a1}z_{a1} \tag{55}$$

and the order of Eq. (52) can be reduced to

$$\dot{z}_{a1} = z_{a2} \tag{56}$$

Clearly, the disturbance and the uncertainty have been separated from Eq. (56). In this way, substituting Eq. (56) into Eq. (55) yields

$$\dot{z}_{a1} + c_{a1}z_{a1} = 0 \tag{57}$$

By the Laplace transform, an eigenvalue equation of Eq. (57) is obtained as

$$s + c_{a1} = 0 \tag{58}$$

To achieve a desired dynamic performance and a stable convergence of the sliding mode, the coefficient c_{a1} can be determined by employing the pole assignment method. That is, the eigenvalue of Eq. (58) should be assigned strictly in the negative half plane. Without loss of generality, it can be chosen herein as c_{a1}=1.

The VSC law is designed below by the procedure (Step4) of 3.2, in order to guarantee that the desired sliding mode occurs within a finite time. First, a VSC law is obtained on the basis of Eq. (35):

$$v_{ua} = -\mathbf{B}_{S_{aZ}}^{-1}\left[\mathbf{A}_{S_{aZ}} + a_{as}S_{a\mathbf{Z}} + b_{as}\operatorname{sgn}(S_{a\mathbf{Z}})\right] \tag{59}$$

where $\mathbf{A}_{S_{aZ}} = c_{a1}\dot{z}_{a1}, \mathbf{B}_{S_{aZ}} = 1$. For determining the coefficients a_{as}, b_{as}, the perturbation boundary of g_{am} should be verified such that

$$g_{am} < \left\|\beta_a\boldsymbol{\varphi}_a\right\|^{-1} \tag{60}$$

where $\boldsymbol{\varphi}_a$=[0 1 0 0]. According to Eq. (45) and the analysis of 3.2, it is easy to obtain

$$\left\|\beta_a\boldsymbol{\varphi}_a\right\|^{-1} = \left(\max\left|\frac{1}{g_{a1}(\omega_t,\omega_{ed})}\right|\right)^{-1} = 0.98 \tag{61}$$

Clearly, the condition of Eq. (60) is satisfied. Then, the parameter b_{as} will be determined by the inequality (38). Recalling the analysis results of 3.1, $\|\alpha_a\| = \max\left|-\frac{f_{a2}(\omega_t,\omega_{ed})}{g_{a1}(\omega_t,\omega_{ed})}\right| = 16.33$ is given. On this basis, it is reasonable to suppose that the absolute value of the extreme relative velocity tracking error is max$|\varepsilon_v|$=35m/s. It can be presented as a scenario that the leading vehicle moves forward with a maximum velocity 35m/s relative to the statical controlled vehicle (assuming this given value is an actual maximum velocity). The values above will be substituted into the right hand side of the inequality (38), and we have

$$\frac{\|\boldsymbol{\varphi}_a\|\left(f_{am}-\|\alpha_a\|g_{am}\right)+\|\beta_a\boldsymbol{\varphi}_a\|\left\|\boldsymbol{B}_{S_{aZ}}^{-1}\boldsymbol{A}_{S_{aZ}}\right\|g_{am}}{1-\|\beta_a\boldsymbol{\varphi}_a\|g_{am}}=210.25 \tag{62}$$

Then, the parameter b_{as}=250 can be determined, and a_{as}=10 is achieved separately by the condition of a_{as}>0.

By the procedure (Step5) in 3.2, the coordinate transformations $\boldsymbol{Z}_a=\psi_a(\boldsymbol{X})$ and $\boldsymbol{\mu}_a=\phi_a(\boldsymbol{X})$ will be used to transfer the new coordinates ($\boldsymbol{Z}_a$, $\boldsymbol{\mu}_a$) back to the original coordinate $\boldsymbol{X}$. In this way, the switching function over the original coordinate becomes

$$S_{aZ}=c_{a1}z_{a1}+z_{a2}\overset{Z_a=\psi_a(X)}{\Rightarrow}S_{aX}=c_{a1}\varepsilon_d+\varepsilon_v \tag{63}$$

the VSC law (57) over the original coordinate has the form

$$v_{ua}=-\left[c_{a1}\varepsilon_v+a_{as}S_{aX}+b_{as}\,\mathrm{sgn}\left(S_{aX}\right)\right] \tag{64}$$

With substitutions of S_{aX} and v_{ua} into Eq. (45), a AACC system based on the DDRC method is finally obtained as

$$\alpha_{th}=\frac{-f_{a2}\left(\omega_t,\omega_{ed}\right)}{g_{a1}\left(\omega_t,\omega_{ed}\right)}-\frac{\left[c_{a1}\varepsilon_v+a_{as}\left(c_{a1}\varepsilon_d+\varepsilon_v\right)+b_{as}\,\mathrm{sgn}\left(S_{aX}\right)\right]+p_{a1}w}{g_{a1}\left(\omega_t,\omega_{ed}\right)} \tag{65}$$

The control laws designed above only satisfy the convergence stability and the robustness of the linearized decoupling subsystem. In order to ensure the stability of the total system, the stability of the remaining nonlinear internal dynamic subsystem has to be verified, so that the problem of tracking control can be solved completely. Based on ref. [38], the study on the stability of nonlinear internal dynamic subsystem can be turned into the study on its zero dynamics correspondingly. Therefore, let $\Delta\boldsymbol{Q}_a=\Delta\boldsymbol{K}_a=0$, i.e., ignore the tiny impact of the uncertain part. Then the zero dynamics of the nonlinear internal dynamic subsystem (53) owing to z_{a1}, z_{a2}, w=0 is obtained as follows

$$\begin{cases}\dot{\mu}_{a1}=a\mu_{a1}^2+\dfrac{\mu_{a2}}{\alpha_n t_h}+l\\[2ex]\dot{\mu}_{a2}=-\left(2a\mu_{a1}+\dfrac{1}{t_h}\right)\left(\alpha_n t_h a\mu_{a1}^2+\mu_{a2}+\alpha_n t_h l\right)\end{cases} \tag{66}$$

To verify the asymptotic stability of Eq. (66) at the equilibrium point (z_{a1}, z_{a2}, μ_{a1}, μ_{a2})=0, a candidate Lyapunov function is chosen:

$$V_a\left(\mu_{a1},\mu_{a2}\right)=-\left(\alpha_n\mu_{a1}+\mu_{a2}\right)^2 \tag{67}$$

The time derivative with respect to the Lyapunov function Eq. (67) is

$$\begin{aligned}\frac{dV_a}{dt} &= -2(\alpha_n\mu_{a1}+\mu_{a2})(\alpha_n\dot{\mu}_{a1}+\dot{\mu}_{a2}) \\ &= 2a\mu_{a1}\left(\alpha_n t_h a\mu_{a1}^2+\mu_{a2}+\alpha_n t_h l\right)\times\left[\alpha_n\omega_t+\alpha_n t_h\left(b\omega_t\omega_{ed}+c\omega_{ed}^2+d\frac{\omega_t^3}{\omega_{ed}}\right)\right]\end{aligned} \tag{68}$$

Because $a,b,c,d,\alpha_n,t_h,l,\omega_t,\omega_{ed}>0$ and $\dot{\omega}_t>0$ under the driving condition of the vehicle acceleration, we have

$$\alpha_n t_h a\mu_{a1}^2+\mu_{a2}+\alpha_n t_h l=\alpha_n t_h\dot{\mu}_{a1}=\alpha_n t_h\dot{\omega}_t>0,\ (\text{if } \dot{\omega}_t\neq 0) \tag{69}$$

In addition, it is easy to verify

$$\alpha_n\omega_t+\alpha_n t_h\left(b\omega_t\omega_{ed}+c\omega_{ed}^2+d\frac{\omega_t^3}{\omega_{ed}}\right)>0 \quad (\text{if } \omega_t\neq\omega_{ed}\neq 0) \tag{70}$$

Therefore, $\frac{dV_a}{dt}>0$. The following inequality is satisfied:

$$V_a\frac{dV_a}{dt}<0 \qquad (\text{is } \omega_t\neq\omega_{ed}\neq 0 \text{ and } \dot{\omega}_t\neq 0) \tag{71}$$

The zero dynamics is asymptotically stable.

4.2 LFS AACC system for braking condition

The design of LFS AACC system under the braking condition is similar to under the driving condition. Regarding the purpose of the LFS AACC system (8), the output can be defined as $y=h(\mathbf{X})=\varepsilon_d$. Then, the relative degree is obtained as r=2, and the decoupling state feedback is achieved according to Eq. (19) as

$$u_b=\alpha_d(\mathbf{X})+\beta_d(\mathbf{X})v_{ud}+\gamma_d(\mathbf{X})w=\frac{-f_{d2}(\omega_t,\omega_{ed},a_b)+v_{ud}-p_{d1}w}{g_{d1}(\omega_t,a_b)} \tag{72}$$

The corresponding coordinate transformation is given as

$$\mathbf{Z}_d=\boldsymbol{\psi}_d(\mathbf{X})=\begin{bmatrix}z_{d1}\\ z_{d2}\end{bmatrix}=\begin{bmatrix}h(\mathbf{X})\\ L_{\mathbf{F}_d}h(\mathbf{X})\end{bmatrix}=\begin{bmatrix}\varepsilon_d\\ \varepsilon_v\end{bmatrix} \tag{73}$$

$$\boldsymbol{\mu}_d=\boldsymbol{\phi}_d(\mathbf{X})=\begin{bmatrix}\mu_{d1}(\mathbf{X})\\ \mu_{d2}(\mathbf{X})\\ \mu_{d3}(\mathbf{X})\end{bmatrix}=\begin{bmatrix}\omega_t\\ \omega_{ed}\\ \varepsilon_v-\alpha_n t_h d_d a_b\end{bmatrix} \tag{74}$$

Taking further account of the influence from system's uncertainty, we have

$$\begin{aligned}L_{\Delta\mathbf{F}_d}h(\mathbf{X})&=[1\ \ 0\ \ 0\ \ 0]\Delta\mathbf{F}_d=0\\ L_{\Delta\mathbf{G}_d}h(\mathbf{X})&=[1\ \ 0\ \ 0\ \ 0\ \ 0]\Delta\mathbf{G}_d=0\end{aligned} \tag{75}$$

That is to say, the matching condition (c1) is satisfied with respect of uncertain items $\Delta F_d(X)$, $\Delta G_d(X)$. Besides, on the analyses of 2.2 and 2.3, the uncertain items $\Delta F_d(X)$, $\Delta G_d(X)$ and the disturbance w are subject to the following limited upper boundaries:

$$\begin{aligned}\left\|\Delta F_d(X)\right\| &\le f_{dm} = 192\\ \left\|\Delta G_d(X)\right\| &\le g_{dm} = 0.029\end{aligned} \tag{76}$$

By substituting the decoupling state feedback (72) into model (8), and making use of the coordinate transformations (73) and (74), a linearized subsystem (77) can be achieved, in which the certain part is completely decoupled from the disturbance.

$$\begin{cases}\begin{bmatrix}\dot z_{d1}\\ \dot z_{d2}\end{bmatrix} = \overbrace{\begin{bmatrix}0 & 1\\ 0 & 0\end{bmatrix}\begin{bmatrix}z_{d1}\\ z_{d2}\end{bmatrix} + \begin{bmatrix}0\\ 1\end{bmatrix}v_{ud}}^{\text{Certain part}} + \overbrace{\begin{bmatrix}0\\ \Delta f_{d2} - \dfrac{f_{d2}}{g_{d1}}\Delta g_{d1}\end{bmatrix} + \begin{bmatrix}0\\ \dfrac{\Delta g_{d1}}{g_{d1}}\end{bmatrix}v_{ud} + \begin{bmatrix}0\\ -\dfrac{p_{d1}}{g_{d1}}\Delta g_{d1}\end{bmatrix}w}^{\text{Part of uncertain and disturbance}}\\ y = \varepsilon_d\end{cases} \tag{77}$$

Additionally, a nonlinear internal dynamic subsystem with the influence of the disturbance and uncertainty is presented

$$\dot{\mu}_d = Q_d(Z_d, \mu_d) + \Delta Q_d(Z_d, \mu_d) + \left[K_d(Z_d, \mu_d) + \Delta K_d(Z_d, \mu_d)\right]w \tag{78}$$

where

$$Q_d(Z_d,\mu_d) = \begin{bmatrix} a_d\mu_{d1}^2 + b_d\mu_{d1}\mu_{d2} + c_d\mu_{d2}^2 + \dfrac{-\mu_{d3} + z_{d2}}{\alpha_n t_h} - \vartheta_d \\ e_d\mu_{d1}^2 + f_d\mu_{d1}\mu_{d2} + g_d\mu_{d2}^2 + h_d\mu_{d2} + i_d \\ -\left(2\dfrac{a_d}{d_d}\mu_{d1} + \dfrac{b_d}{d_d}\mu_{d2} + \dfrac{1}{t_h d_d}\right)\times\left(a_d\mu_{d1}^2 + b_d\mu_{d1}\mu_{d2} + c_d\mu_{d2}^2 + \dfrac{-\mu_{d3} + z_{d2}}{\alpha_n t_h} - \vartheta_d\right) \\ -\left(\dfrac{b_d}{d_d}\mu_{d1} + 2\dfrac{c_d}{d_d}\mu_{d2}\right)\times\left(e_d\mu_{d1}^2 + f_d\mu_{d1}\mu_{d2} + g_d\mu_{d2}^2 + h_d\mu_{d2} + i_d\right)\end{bmatrix}$$

$$K_d(Z_d,\mu_d) = \begin{bmatrix}0\\ 0\\ -p_{d1}\end{bmatrix}.$$

According to the analysis of 2.1, items ΔQ_d, ΔK_d are the constants with limited upper boundaries.

Similarly, the VSC law can be designed as

$$v_{ud} = -B_{S_{dZ}}^{-1}\left[A_{S_{dZ}} + a_{ds}S_{dZ} + b_{ds}\,\mathrm{sgn}(S_{dZ})\right] \tag{79}$$

where the sliding mode surface is

$$S_{dZ} = c_{d1}z_{d1} + z_{d2} \tag{80}$$

By ignoring the tedious calculation process, the parameters are given directly as $\boldsymbol{A}_{S_{dz}} = c_{d1}\dot{z}_{d1}, \boldsymbol{B}_{S_{dz}} = 1,\ c_{d1} = 1,\ a_{ds} = 10,\ \ b_{ds} = 185$. By transferring v_{ud} back to the original coordinate and substituting it into Eq. (72), the AACC law is finally obtained as

$$u_b = -\frac{f_{d2}(\omega_t, \omega_{ed}, a_b)}{g_{d1}(\omega_t, a_b)} - \frac{\left[c_{d1}\varepsilon_v + a_{ds}(c_{d1}\varepsilon_d + \varepsilon_v) + b_{ds}\,\mathrm{sgn}(S_{dX})\right] + p_{d1}w}{g_{d1}(\omega_t, a_b)} \tag{81}$$

where $S_{aX} = c_{d1}\varepsilon_d + \varepsilon_v$, which is a sliding mode surface over the original coordinate.
The remaining nonlinear internal dynamic subsystem (78) should be verified as well to ensure the stability of the total system. At first, if z_{d1}, z_{d2}, w=0 and the impact of uncertain items $\Delta\boldsymbol{Q}_d$, $\Delta\boldsymbol{K}_d$ can be neglected, then the zero dynamics becomes

$$\begin{cases} \dot{\mu}_{d1} = a_d\mu_{d1}^2 + b_d\mu_{d1}\mu_{d2} + c_d\mu_{d2}^2 - \dfrac{\mu_{d3}}{\alpha_n t_h} - \vartheta_d \\ \dot{\mu}_{d2} = e_d\mu_{d1}^2 + f_d\mu_{d1}\mu_{d2} + g_d\mu_{d2}^2 + h_d\mu_{d2} + i_d \\ \dot{\mu}_{d3} = -\left(2\dfrac{a_d}{d_d}\mu_{d1} + \dfrac{b_d}{d_d}\mu_{d2} + \dfrac{1}{t_h d_d}\right)\times\left(a_d\mu_{d1}^2 + b_d\mu_{d1}\mu_{d2} + c_d\mu_{d2}^2 - \dfrac{\mu_{d3}}{\alpha_n t_h} - \vartheta_d\right) \\ \quad -\left(\dfrac{b_d}{d_d}\mu_{d1} + 2\dfrac{c_d}{d_d}\mu_{d2}\right)\times\left(e_d\mu_{d1}^2 + f_d\mu_{d1}\mu_{d2} + g_d\mu_{d2}^2 + h_d\mu_{d2} + i_d\right) \end{cases} \tag{82}$$

Then, a candidate Lyapunov function is chosen as

$$V_d(\mu_{d1}, \mu_{d2}, \mu_{d3}) = \mu_{d2} + \left(\frac{2\pi\alpha_n t_h d_d r_t}{60 i_g i_0}\mu_{d1} + \int \mu_{d3}dt\right) \tag{83}$$

Since z_{d2}=0, it is easy to obtain

$$V_d(\mu_{d1}, \mu_{d2}, \mu_{d3}) = \mu_{d2} = \omega_{ed} > 0 \qquad (\text{if } \omega_{ed} \neq 0) \tag{84}$$

The time derivative with respect to the Lyapunov function Eq. (83) is

$$\frac{dV_d}{dt} = \dot{\mu}_{d2} + \left(\frac{2\pi\alpha_n t_h d_d r_t}{60 i_g i_0}\dot{\mu}_{d1} + \mu_{d3}\right) = \dot{\mu}_{d2} = \dot{\omega}_{ed} \tag{85}$$

For the braking condition, the engine operates under the decelerating mode, hence

$$\frac{dV_d}{dt} = \dot{\omega}_{ed} < 0 \ \ (\text{if } \dot{\omega}_{ed} \neq 0) \tag{86}$$

Assembling Eqs. (84) and (85), the following inequality is hold

$$V_d\frac{dV_d}{dt} < 0 \ \ (\text{if } \omega_{ed} \neq 0 \text{ and } \dot{\omega}_{ed} \neq 0) \tag{87}$$

Thus, the zero dynamics of the nonlinear internal dynamic subsystem (78) is asymptotically stable as long as z_{d1}, z_{d2}, w=0.

5. Simulation and analysis

Base on above analysis of the control system under the driving/braking conditions, the LFS AACC system applying the DDRC method can be designed as the block diagram in Figure 9. The system consists of three parts: the controlled object of a convoy with two vehicles, DDRC system, and the input/output signals.

In order to verify the control performance of the LFS AACC system, a typical driving cycle of the leading vehicle's aceeleration/deceleration, velocity, as well as the road grade are given in Figure 10. The road grade changes from 0°~$+3^o$ to 0°~-3° in a period of 80s~90s and 110s~120s, respectively. Furthermore, the conditions from the high-speed to low-speed SG, and two cases of mass equaling 10,000kg and 25,000kg are included. The initial errors at 0s for the inter-vehicles relative distance and relative velocity are set to 0m and 0m/s, respectively. Table 1 and the solid lines in Figures 11 and 12 are the coefficients and the simulation results, respectively for the proposed control system. In contrast, the coefficients and some simulation results of an upper LQ+lower PID hierarchical control system proposed in ref. [1] are also presented respectively in Table 2 and by the dotted lines in Figures 11 and 12. The comparison results of the throttle angle, desired input voltage of EBS, engine speed, automatic transmission gear position, relations of relative distance/relative velocity tracking error verses time scale, as well as the phase chart of the relative distance/relative velocity tracking error are shown in Figures 11 and 12 in sequences of (a)~(f).

Driving condition	c_{a1}=1	a_{as}=10	b_{as}=250
Braking condition	c_{d1}=1	a_{ds}=10	b_{ds}=185

Table 1. Control parameters of DDRC system

Conditions	Upper layer LQ parameters		Lower layer PID parameters
	Q	R	P, I, D
Driving	[7 0,0 4]	10	800, 560, 15
Braking			350, 150, 20

Table 2. Control parameters of hierarchical control system

As illustrated by Figures 11 (a)~(d), for the proposed control system, the throttle angle and the EBS desired input voltage exhibit smooth response characteristic, rapid convergence and small oscillation, even at the moment of gear switching. However, for the hierarchical control system, it shows intense and long time oscillations especially at low-speed condition (shown as dashed border subfigures inside the Figures 11 (a) and (b)), which have impacts on the vehicle's comfortability severely.

This is because the small parameters are adopted by the proposed control system as the consequence of applying DDRC method (shown as Tables 1), and thus the unmodeled high frequency oscillation can be effectively eliminated, in contrast with the hierarchical control system adopting large parameters (shown as Tables 2). Moreover, during the time period of 0s ~ 73s and 130s ~ 200s in Figures 11(e) and 12(e), the simulation results of the proposed control system indicate that the errors of the relative distance and the relative velocity are constrained within the range of ± 0.02m and -0.05m/s~0.02m/s, respectively. The tracking accuracy of the proposed control system is enhanced and almost frees from the disturbance of the leading vehicle's acceleration/deceleration. However, for the hierarchical control system, it is affected obviously by the change of the leading vehicle's acceleration/deceleration, and touches the maximum value of ± 0.1m. Finally, the comparison between (e) and (f) in Figures 11 and 12 demonstrates a superior robustness for the proposed control system in spite of the uncertainties caused by the road grade, gear position and the vehicle mass. Particularly, while the road grade changes between ± 3° in the time period of 80s~120s, the tracking error of the relative distance and the relative velocity for the proposed control system are less than ± 0.05m and -0.04m/s ~ 0.02m/s, in contract to larger than ± 0.15m and ± 0.05m/s of the hierarchical control system.

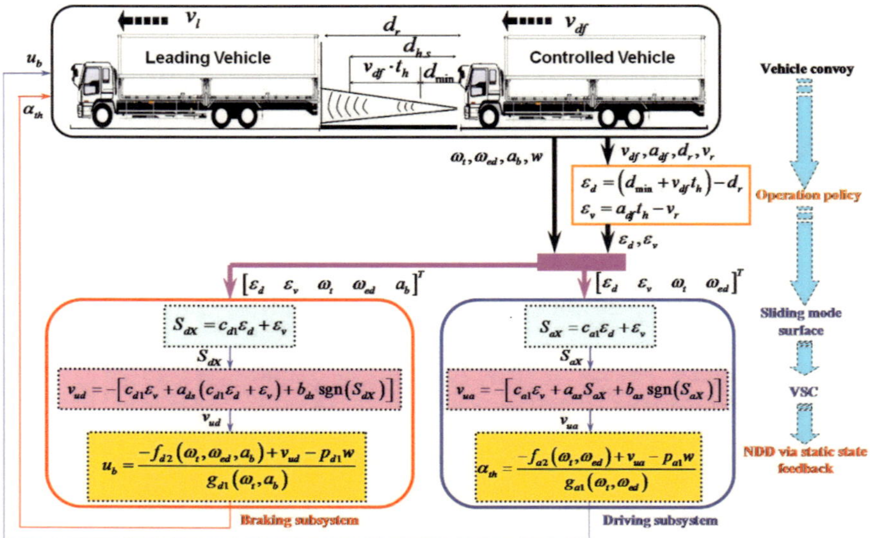

Fig. 9. LFS ACC system using the DDRC method

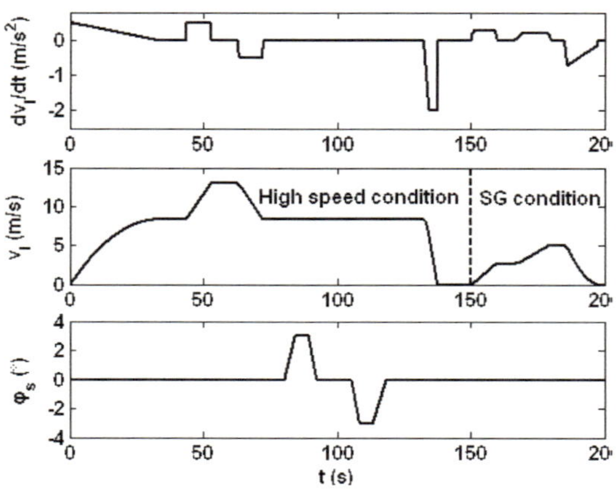

Fig. 10. Profile of leading vehicle driving cycle and road grade

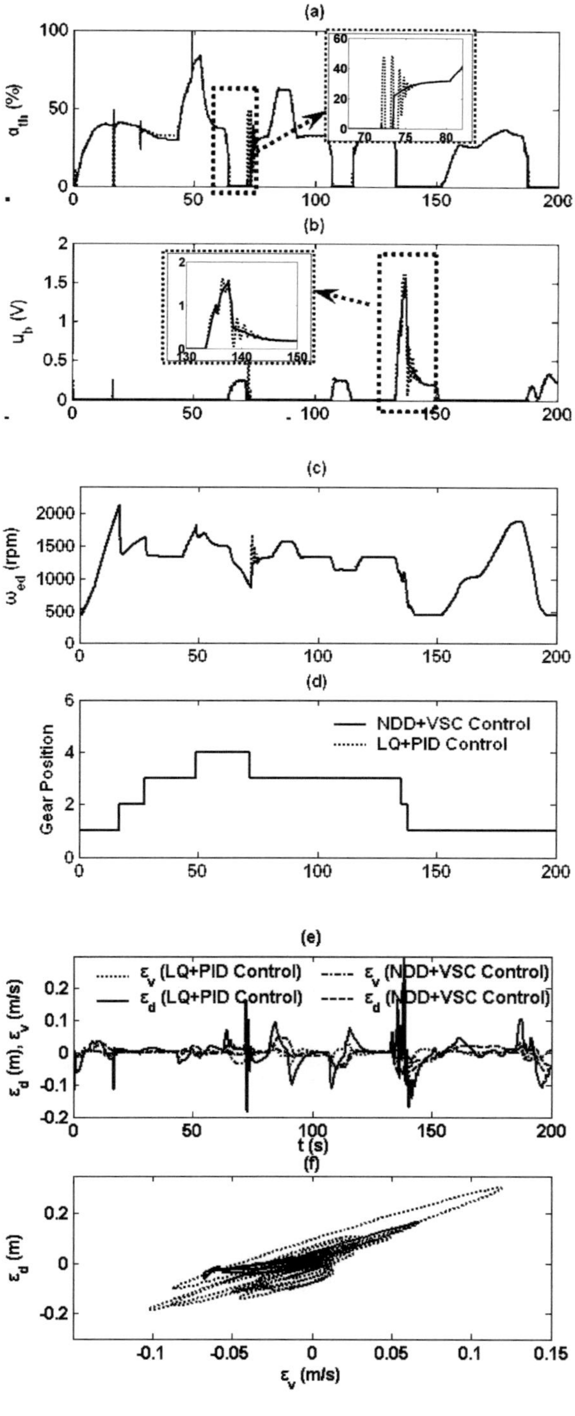

Fig. 11. Simulation results (mass is 10,000 kg)

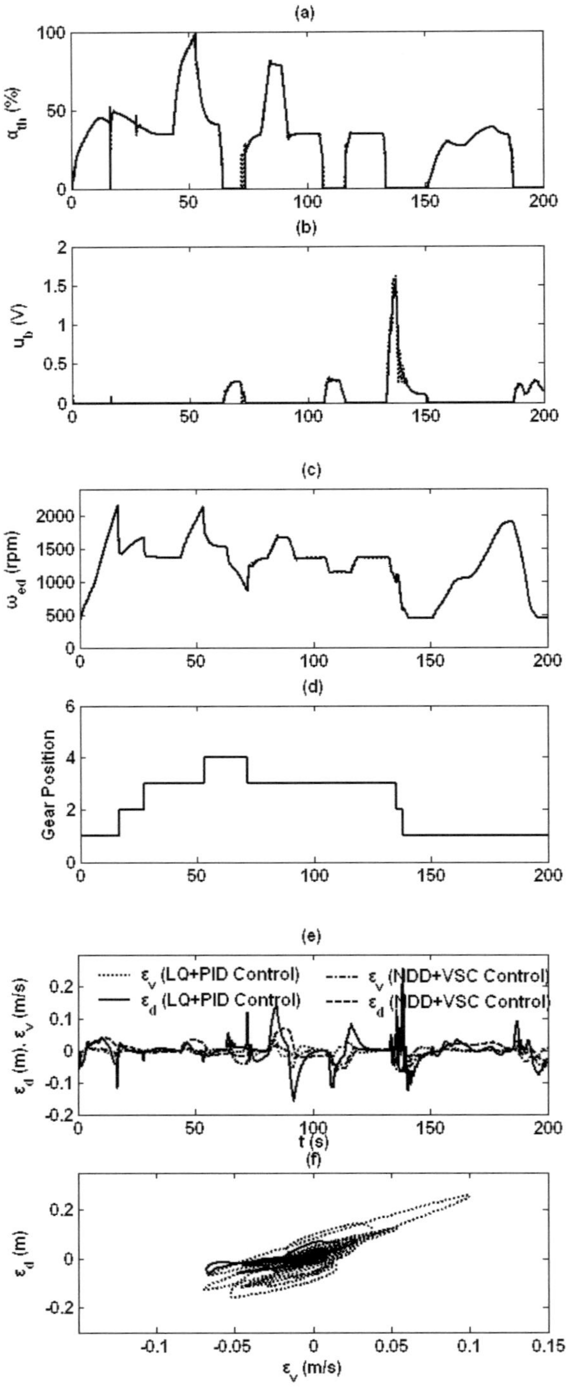

Fig. 12. Simulation results (mass is 25,000 kg)

From above analysis and the simulation results, it seems that the influence of nonlinearity, external disturbance and the variable uncertainties have been eliminated by adopting the proposed DDRC method for the LFS AACC system, and it results in a significant improvement of the tracking accuracy, robustness, as well as the response characteristics of the actuator system (i.e., the throttle angle and the EBS desired input voltage). In addition, the control structure and the parameters are simplified, and easy to determine in comparison with the hierarchical control algorithm.

6. Conclusion

In this study, an LFS nonlinear dynamics model is proposed by integrating the dynamics of the inter-vehicles and the controlled vehicle. Then, a DDRC method is developed, and used to design the LFS AACC system. Finally, the control performance is verified by the numerical simulation under a typical driving cycle. The simulation results confirm the followings:

1. The proposed LFS model not only can describe the vehicle's strong nonlinearity at low-speed conditions and the uncertainty induced by the complex traffic environment and the road condition, but also is able to express the strong coupling characteristics due to frequent change of the leading vehicle's acceleration/deceleration at high-speed condition. Particularly, the dynamics of the inter-vehicles and the controlled vehicle are lumped together within a universal state space equation.
2. The tracking accuracies at high-speed and low-speed SG condition, as well as the robustness to the external disturbance and the model parameter uncertainty have been improved simultaneously, because the DDRC method is applied in the design of the LFS ACC system.
3. The actuators' high frequency oscillation caused by the unmodeled part has been restrained through using small parameters, and this leads to a control system with simplified structure.

7. Appendixes

Appendix 1. Definition of the matrix items in Eq. (1)

$$f_{av1}(\mathbf{X}) = \left\{ \left[\tau_1\left(\frac{\omega_t}{\omega_{ed}}\right)^2 + \tau_2\left(\frac{\omega_t}{\omega_{ed}}\right) + \tau_3 \right]\left(\frac{\omega_{ed}}{60}\right)^2 \frac{t_1\left(\frac{\omega_t}{\omega_{ed}}\right) + t_2}{g_g} \frac{i_{g_1} i_0 \eta_k}{r_t} - \frac{\eta_1\left(\frac{\omega_t}{\eta_2}\right)^2 + M(\eta_r + \sin\varphi_s) g_g}{g_g} \right\} \frac{i_{g_1} i_0}{\sigma_C}$$

$$f_{av2}(\mathbf{X}) = \left\{ (k_2\omega_{ed} + k_4) g_g - \left[\tau_1\left(\frac{\omega_t}{\omega_{ed}}\right)^2 + \tau_2\left(\frac{\omega_t}{\omega_{ed}}\right) + \tau_3 \right]\left(\frac{\omega_{ed}}{60}\right)^2 \right\} \frac{\eta_2}{I_e}$$

$$g_{av1}(\mathbf{X}) = 0$$

$$g_{av2}(\mathbf{X}) = (k_1\omega_{ed} + k_3)\frac{g_g \eta_2}{I_e}$$

Appendix 2. Definition of the matrix items in Eq. (2)

$$f_{dv1}(\mathbf{X})=\left\{\left[\tau_{d1}\left(\frac{\omega_t}{\omega_{ed}}\right)^2+\tau_{d2}\left(\frac{\omega_t}{\omega_{ed}}\right)+\tau_{d3}\right]\left(\frac{\omega_{ed}}{60}\right)^2\frac{i_{g_1}i_0\eta_k}{r_t}+Ma_b-\frac{\eta_1\left(\frac{\omega_t}{\eta_2}\right)^2+M(\eta_r+\sin\varphi_s)g_g}{g_g}\right\}\frac{i_{g_1}i_0}{\sigma_C}$$

$$f_{dv2}(\mathbf{X})=\left\{(k_2\omega_{ed}+k_4)g_g-\left[\tau_{d1}\left(\frac{\omega_t}{\omega_{ed}}\right)^2+\tau_{d2}\left(\frac{\omega_t}{\omega_{ed}}\right)+\tau_{d3}\right]\left(\frac{\omega_{ed}}{60}\right)^2\right\}\frac{\eta_2}{I_e}$$

$$f_{dv3}(\mathbf{X})=-\frac{1}{t_r}a_b$$

$$g_{dv1}(\mathbf{X})=g_{dv2}(\mathbf{X})=0$$

$$g_{dv3}(\mathbf{X})=-\frac{k_b+k_a a_b^2 v}{t_r}$$

Appendix 3. Definition of the matrix items in Eq. (7)

$$f_{a1}(\varepsilon_v)=\varepsilon_v$$

$$f_{a2}(\omega_t,\omega_{ed})=\left[\alpha_n t_h\left(2a\omega_t+b\omega_{ed}+3d\frac{\omega_t^2}{\omega_{ed}}\right)+\alpha_n\right]\dot{\omega}_t+\alpha_n t_h\left(b\omega_t+2c\omega_{ed}-d\frac{\omega_t^3}{\omega_{ed}^2}\right)$$
$$\left(e\omega_t^2+f\omega_t\omega_{ed}+g\omega_{ed}^2+h\omega_{ed}+i\right)$$

$$f_{a3}(\omega_t,\omega_{ed})=a\omega_t^2+b\omega_t\omega_{ed}+c\omega_{ed}^2+d\frac{\omega_t^3}{\omega_{ed}}-\vartheta$$

$$f_{a4}(\omega_t,\omega_{ed})=e\omega_t^2+f\omega_t\omega_{ed}+g\omega_{ed}^2+h\omega_{ed}+i$$

$$g_{a1}(\omega_t,\omega_{ed})=\alpha_n t_h\left(b\omega_t+2c\omega_{ed}-d\frac{\omega_t^3}{\omega_{ed}^2}\right)(j\omega_{ed}+k)$$

$$g_{a2}(\omega_{ed})=(j\omega_{ed}+k)$$

$$p_{a1}=-1$$

$$\Delta f_{a2}(\omega_t,\omega_{ed})=\left[\alpha_n t_h\left(2a\omega_t+b\omega_{ed}+3d\frac{\omega_t^2}{\omega_{ed}}\right)+\alpha_n\right]\Delta f_{av1}+\alpha_n t_h\left(b\omega_t+2c\omega_{ed}-d\frac{\omega_t^3}{\omega_{ed}^2}\right)\Delta f_{av2}$$

$$\Delta g_{a1}(\omega_t,\omega_{ed})=\alpha_n t_h\left(b\omega_t+2c\omega_{ed}-d\frac{\omega_t^3}{\omega_{ed}^2}\right)\Delta g_{av2}$$

$$\Delta f_{a3}=\Delta f_{av1},\Delta f_{a4}=\Delta f_{av2},\Delta g_{a2}=\Delta g_{av2}$$

where $a,b,c,d,e,f,g,h,i,j,k,\vartheta$ are constant coefficients, their specific values can be referred to ref. [16].

Appendix 4. Definition of the matrix items in Eq. (8)

$$f_{d1}(\varepsilon_v)=\varepsilon_v$$
$$f_{d2}(\omega_t,\omega_{ed},a_b)=\left[\alpha_n t_h(2a_d\omega_t+b_d\omega_{ed})+\alpha_n\right]\dot{\omega}_t+\alpha_n t_h(b_d\omega_t+2c_d\omega_{ed})\dot{\omega}_{ed}+\alpha_n t_h d_d j_d a_b$$
$$f_{d3}(\omega_t,\omega_{ed},a_b)=a_d\omega_t^2+b_d\omega_t\omega_{ed}+c_d\omega_{ed}^2+d_d a_b-\vartheta_d$$
$$f_{d4}(\omega_t,\omega_{ed})=e_d\omega_t^2+f_d\omega_t\omega_{ed}+g_d\omega_{ed}^2+h_d\omega_{ed}+i_d$$
$$f_{d5}(a_b)=j_d a_b$$
$$g_{d1}(\omega_t,a_b)=\alpha_n t_h d_d\left(k_d a_b^2\omega_t+l_d\right)$$
$$g_{d2}(\omega_t,a_b)=k_d a_b^2\omega_t+l_d$$
$$p_{d1}=-1$$
$$\Delta f_{d2}(\omega_t,\omega_{ed})=\left[\alpha_n t_h(2a_d\omega_t+b_d\omega_{ed})+\alpha_n\right]\Delta f_{dv1}+\alpha_n t_h(b_d\omega_t+2c_d\omega_{ed})\Delta f_{dv2}+\alpha_n t_h d_d\Delta f_{dv3}$$
$$\Delta g_{d1}(\omega_t,a_b)=\alpha_n t_h d_d\left(k_d a_b^2\omega_t+l_d\right)\Delta g_{dv3}$$
$$\Delta f_{d3}=\Delta f_{dv1},\Delta f_{d4}=\Delta f_{dv2},\Delta f_{d5}=\Delta f_{dv3},\Delta g_{d2}=\Delta g_{dv3}$$

where $a_d,b_d,c_d,d_d,e_d,f_d,g_d,h_d,i_d,j_d,k_d,l_d,\vartheta_d$ are constant coefficients, their specific values can be referred to ref. [16].

Appendix 5. The vehicle parameters are as follows:

$i_0=5.571$ - final reduction ratio;
$i_{g_1}=3.49$ - first position gear ratio of the automatic transmission;
$r_t=0.507$ - effective tire radius (m);
$I_e=3.189$ - rotational inertia of the engine flywheel (kg m²);
$\eta_k=0.98$ - total transmission efficiency;
$\eta_r=0.01$ - rolling resistance coefficient;
$M=10{,}000$ - vehicle nominal mass (kg);
$\varphi_s=0$ - nominal road grade (°);
$k_1=5.2\times10^{-3},k_2=-0.25,k_3=-1.1,k_4=145$ - engine fitting coefficients;
$t_1=-0.85,t_2=1.75,\tau_1=-7.19\times10^{-2}$, $\tau_2=3.97\times10^{-2}\tau_3=3.68\times10^{-2}$ - torque converter fitting coefficients for the forward transmit condition;
$\tau_{d1}=-2.1\times10^{-2},\tau_{d2}=6.76\times10^{-2},\tau_{d3}=-4.59\times10^{-2}$ - torque converter fitting coefficients for the reverse transmit condition;
$t_r=0.2$ - time constant of the dynamic response for the braking system;
$k_a=-4.3\times10^{-3},k_b=0.29$ - fitting coefficients of the heat fading efficiency model for the braking system;
η_1,η_2,σ_C - constant coefficients:

$$\eta_1=\left(0.232g_g\right)\left(\frac{0.377r\cdot60}{3.6i_g i_0\cdot2\pi}\right)^2,\ \eta_2=\frac{60}{6.283},\ \sigma_C=\left(\frac{M}{g_g\cdot18.8}\right)+\left(\frac{M\cdot1.55}{g_g^2\cdot18.8}\right);$$

$g_g=9.8\text{m}/\text{s}^2$.

8. Acknowledgment

We would like to thanks the data of vehicle supported by Tsinghua University, and the supports from the Young Scientists Fund of National Natural Science Foundation of China (51007003) and National Natural Science Foundation of China (51075010).

9. References

[1] Hedrick, J.K.; McMahon, D.H.; Swaroop, D. (1993). Vehicle Modeling and Control for Automated Highway Systems, *PATH Technical Report*, pp.1-73,UBC-ITS-PRR-93-24, Jan 01,1993

[2] Hedrick, J.K. (1998). Nonlinear Controller Design for Automated Vehicle Application, *Control '98. UKACC International Conference on (Conf. Publ. No. 455)*, pp.23-32,Swansea, UK, Sep 1-4,1998

[3] Rajamani, R.; Tan, H. S.; Law, B. K. and Zhang, W. B. (2000). Demonstration of Integrated Longitudinal and Lateral Control for the Operation of Automated Vehicles in Platoons. *IEEE Transactions on Control Systems Technology*, Vol.8, No.4, (July 2000), pp. 695-708, ISSN 1063-6536

[4] Yi, K.; Moon, I.; Kwon, Y. D. (2001). A vehicle-to-vehicle distance control algorithm for stop-and-go cruise control. *Intelligent Transportation Systems, 2001. Proceedings. 2001 IEEE*, pp. 478-482, ISBN: 0-7803-7194-1, Oakland, CA, USA, 2001

[5] Omae, M. (1999). Study on the vehicle platoon control system, *Doctor Dissertation*, Tokyo: University of Tokyo, 1999

[6] Fritz, A.; Schienlen, W. (1999). Automatic Cruise Control of a Mechatronically Steered Vehicle Convoy. *Vehicle System Dynamics*, Vol.32, (1999), pp. 331–344

[7] Schienlen, W.; Fritz, A. (1999). Nonlinear Cruise Control Concepts for Vehicle in Convoy. *Vehicle System Dynamics Supplement*, Vol.33,(1999), pp.256-269

[8] Naranjo, J.E.; Gonzalez, C.; Garcia, R.; de Pedro, T. (2006). ACC+Stop&go maneuvers with throttle and brake fuzzy control. *IEEE Transactions on Intelligent Transportation Systems*, Vol.7 ,No.4,(2006), pp. 213–225, ISSN: 1524-9050

[9] Daniele, C.; Schutter, B. De. (2008). Adaptive Cruise Control for a SMART Car: A Comparison Benchmark for MPC-PWA Control Methods. *IEEE Transactions on Control Systems Technology*, Vol.16, No.2, (March 2008), pp. 365-372, ISSN: 1063-6536

[10] Bifulco, G.N.; Simonelli, F.; Di Pace, R. (2008). Experiments toward a human-like adaptive cruise control, *Intelligent Vehicles Symposium*, pp. 919-924, ISBN 978-1-4244-2568-6, Eindhoven University of Technology, Eindhoven, The Netherlands, June 4-6, 2008

[11] Acarman, T.; Liu, Y. and Ozguner, U. (2006). Intelligent cruise control stop and go with and without communication, *Proceedings of the American Control Conference*, pp. 4356–4361, ISBN: 1-4244-0209-3, Minneapolis, Minnesota, USA, June 14-16, 2006

[12] Martinez, J. J. and Canudas-de-Wit, C. (2007). A Safe Longitudinal Control for Adaptive Cruise Control and Stop-and-Go Scenarios. *IEEE Transactions on Control Systems Technology*, Vol.15, No.2, (March 2007), pp.246-258, ISSN : 1063-6536

[13] Venhovens, P.; Naab, K.; Adiprasito, B. (2000). Stop and go cruise control, *Proceedings of Seoul 2000 FISITA World Automotive Congress*, pp.1-8, Seoul Korea, June 12-15, 2000

[14] Yamamura, Y.; Tabe, M.; Kanehira, M. (2001). Development of an adaptive cruise control system with stop and go capability,*SAE Technical Paper*, Detroit, MI, USA, March, 2001

[15] Bin,Y; Li, K. Q.; Ukawa, H.; Handa, M. (2006). Modeling and Control of Nonlinear Dynamic System for Heavy-Duty Trucks. *Proceedings of the Institution of Mechanical Engineers, Part D, Journal of Automobile Engineering,* Vol.220, No.10, Z(2006), pp. 1423-1435

[16] Bin, Y.; Li, K. Q.; Feng, N. L. (2008). Feedback Linearization Tracking Control of Vehicle Longitudinal Acceleration under Low-Speed Conditions. *Journal of Dynamic Systems, Measurement, and Control,* Vol.130, (2008), pp. 1-12

[17] Li, K. Q.; Bin, Y.; Ukawa, H.; Handa ,M. (2006). Study on Stop and Go Cruise Control of Heavy-Duty Vehicles. *Transaction of JSAE,* Vol.37, No.2, (2006), pp. 145-150, ISSN:0919-1364

[18] Bin,Y.; Li, K. Q,; Ukawa, H.; Handa, M. (2009). Nonlinear Disturbance Decoupling Control of Heavy-Duty Truck Stop and Go Cruise Systems. *Vehicle System Dynamics,* Vol.47, No.1, (January 2009), pp. 29-55

[19] Thomas D. Gillespie. (1992). Fundamentals of Vehicle Dynamics, Warrendale, *Society of Automotive Engineers, Inc,* 1992

[20] Swaroop, D.; Hetirick, J.K.; Chien, C.C.; Ioannou, P. (2004). A Comparison of Spacing and Headway Control Laws for Automatically Controlled Vehicles. *Vehicle System Dynamics,* Vol.23, No.8, (2004), pp. 597-625

[21] Bengtsson, J. (2001). Adaptive Cruise Control and Driver Modeling. *Research thesis, Department of automatic control, Lund Institute of Technology,* ISSN 0280–5316, Sweden

[22] Isidori, A.; Krener, A.; Gori-Giorgi, C.; Monaco, S. (1981). Nonlinear Decoupling via Feedback: A Differential Geometric Approach. *IEEE Transactions on Automatic Control,* Vol.26, No.2, (1981), pp. 331-345, ISSN: 0018-9286

[23] Nijmeijer, H. (1994). On dynamic state feedback in nonlinear control, *IEE Colloquium on Nonlinear Control,* pp.5/1~5/2, London, UK, May 24,1994

[24] Xia, X H. (1997). *Disturbance Decoupling Control*. Science Press, Beijing, 1997

[25] Morse, A.; Wonham, W. (1971). Status of noninteracting control. *IEEE Transactions on Automatic Control,* Vol.16, No.6, (Dec1971), pp. 568-581, ISSN: 0018-9286

[26] Conte, G.; Perdon, A. M. (1994). The disturbance decoupling problem by dynamic feedback for systems over a principal ideal domain, *Proceedings of the 33rd IEEE Conference on Decision and Control,* pp. 1276-1279, ISBN: 0-7803-1968-0, Lake Buena Vista, FL , USA, Dec 14-16 ,1994

[27] Xia, X H. (1993). Companion of Controlled invariant Distribution and Description of DDP Control Law, *Science in China,* pp. 130-136, Ser.A, 1993,

[28] Young, K.D. (1999). Variable structure systems, sliding mode and nonlinear control, *Springer,* London, 1999

[29] Hu, Y. M. (2003). *Variable Structure Control Theory and Application,* Science Press, Beijing

[30] Cheng, D. Z. (1988). *Geometric Theory of Nonlinear System*. Science Press, Beijing

[31] Byrnes, C.I.; Isidori, A.(1998). Output regulation for nonlinear systems: an overview, *Proceedings of the 37th IEEE Conference on Decision and Control,* pp.3069-3074, Tampa, FL, UNITED STATES, Dec 16-18, 1998

[32] Fliess, M.; Tevine, J.; Martin, P.; Rouchon, P.(1994). Nonlinear control and Lie-Backlund transformations: towards a new differential geometric standpoint, *1994*

Proceedings of the 33rd IEEE Conference on Decision and Control, pp.339-344, ISBN 0-7803-1968-0, Lake Buena Vista, FL , USA , Dec 14-16,1994
[33] Cheng, D. Z. (1987). Geometric Approach to Nonlinear Systems Part 1. Geometric Method and Geometric Preliminary, *Control Theory and Application*, Jan 1-9,1987
[34] Han, Z. Z.; Liu, J. H, et al.(1994). Characters of Nonlinear Control Systems (II). *Control and Decision*,Vol.9, No.5, (1994), pp. 91-97
[35] Wang, Y. F.; Xia, X. H.; Gao, W. B. (1994). Parameter Variations in Nonlinear Decoupling, *American Control Conference,* pp. 2700 - 2704 , ISBN: 0-7803-1783-1, June 29- July 1, 1994
[36] Isidori, A. (1985). *Nonlinear Control Systems,* Spronger Verlag,ISBN 0-387-15595-3, Berlin, Heidelberg
[37] Gao, W. B. (1988). *Nonlinear Control System Introduction*. Science Press, Beijing
[38] Slotine, J.; Li, W. (1991). *Applied Nonlinear Control,* Prentice-Hall Inc, ISBN 0-13-040890-5,Englewood Cliffs, New Jersey

Part 2

Control of Structures, Mechanical and Electro-Mechanical Systems

7

A Decentralized and Spatial Approach to the Robust Vibration Control of Structures

Alysson F. Mazoni, Alberto L. Serpa and Eurípedes G. de O. Nóbrega
University of Campinas - UNICAMP
Brazil

1. Introduction

Designing active controllers for minimizing mechanical vibration of structures is a challenging task which presents several levels of difficulties. Due to the continuous nature of the structures, they have an infinite number of degrees of freedom which leads to infinite vibration modes. This requires a model reduction and modal truncation considering the controller objectives, in order to achieve a viable numerical model which may allow the designed controller to perform satisfactorily within the frequency range of interest (Zhou & Doyle, 1997). But for real structures, even for truncated models, it may be expected a significant number of vibration modes to consider, conducting to mathematical and computational issues, besides the natural consequences of the reduction of the model order leading to unexpected behavior due to the controller feedback.

Considering the now vast literature in the vibration control area, there is no consensus regarding the most suitable control design method. Several techniques seem to give similar results, as shown in the works of Baz & Chen (2000); Bhattacharya et al. (2002); Hurlebaus et al. (2008). Linear matrix inequalities methods, due to powerful yet simple formulation and computational solution to implement the theory of robust control, present nowadays a slight predominance (Boyd et al., 1994; Zhou & Doyle, 1997). Several recent works that use this approach may be cited such as Barrault et al. (2007; 2008); Cheung & Wong (2009); Halim et al. (2008).

The $\mathcal{H}_\infty$ control technique emerged in the last decades as a robust control technique in the context of multiple-inputs and multiple-outputs (MIMO) feedback problems. The usual formulation involves the minimization of the $\mathcal{H}_\infty$ norm from the disturbances inputs to the performance outputs, corresponding to the minimization of the worst possible response. Vibration control of structures is a well reported application using this approach (Gawronski, 2004). Usually, performance outputs are selected based on the interest points distributed over the structure, and taken for the formulation of the objective function in the minimization problem.

However, the control problem, stated as the transfer matrix between the vibration actuator and sensor positions, has a known drawback. Because it does not clearly impose the desired behavior on the whole structure, it is not possible to guarantee the vibration level minimization beyond the sensor isolated position points. This approach may present acceptable reduction levels for simple structures, but more comprehensive methods are needed to achieve good results with real engineering structures, guaranteeing a vibration reduction through regions of the structure instead of isolated points.

The overall vibration energy distribution is then necessary to be considered, which renders the control problem always a non-collocated one. Additionally, real structures require in general a significant number of transducers, increasing the complexity of the system due to the number of transfer function combinations of inputs and outputs.

The spatial $\mathcal{H}_\infty$ control looks for an equivalent worst case output performance norm, in order to have a weighted performance over an specific region of the structure, instead of points. Addressing regions of interest instead of points, this methodology is attractive to the vibration control area. Particularly, for the common case of using a finite element model for the formulation of the spatial $\mathcal{H}_\infty$ control, it results in a simple formulation.

Some works have discussed the spatial $\mathcal{H}_\infty$ control, specially in the case of plate vibration (Halim, 2002; 2007; Halim et al., 2008). In these works, the spatial $\mathcal{H}_\infty$ design is solved trough a convenient algebraic manipulation which converts the spatial norm formulation to an equivalent ordinary $\mathcal{H}_\infty$ control problem.

Decentralized control is another promising approach recently studied for the vibration problem. Its basic architectural idea is to adopt several distributed controllers with lesser authority, instead of a big controller for the whole structure. Each controller accesses a subset of inputs and outputs, being responsible for a region of the structure. Decentralized control has been used for sound irradiation control of plates in Bianchi et al. (2004), with semi-active control in Casadei et al. (2010), using an optimal controller with static feedback in Jiang & Li (2010) and with decentralized velocity feedback in Zilletti et al. (2010). It is obviously useful for big structures in particular, where its constructive robustness represent an immediate advantage, since it may be implemented using independent microcontrollers and the system can easily accommodate actuator or sensor failures. Another advantage is the numerical simplicity of the controller algorithms, since each one deals with a smaller number of inputs and outputs. One problem is to decouple the controllers in order to avoid mutual undesired interference.

The purpose of this work is to compare an application of the spatial norm and the decentralized approaches to the vibration control of a plate including the $\mathcal{H}_\infty$ control technique, and adopting a linear matrix inequalities formulation.

The finite element method is used here to determine the vibration model of a plate. The plate is divided in Mindlin finite elements generating a discretized finite dimensional model that captures the vibration modes (Ferreira, 2008). The finite element method is suitable to determine the mass and stiffness matrices using interpolation functions of each finite element of the mesh, and the assemblage of these results for all the elements leads to a representation of the structure. This model is then used to generate the state-space model used to design the active controller.

This chapter is divided according to the following main topics: Structural modeling, where the main aspects of the dynamic equations, modal analysis fundamentals to describe the model, and model reduction are described; $\mathcal{H}_\infty$ control, where the optimization problem to minimize the $\mathcal{H}_\infty$ norm and aspects of multi-variable control are discussed; Spatial $\mathcal{H}_\infty$ control technique, where a more global and spatial vibration performance along the structure is considered as vibration reduction objective; Decentralized control, where controllers are designed in an independent form in order to reduce the design effort and also to increase the reliability in case of failures, is presented. Finally, the concluding remarks are presented.

The notation used in this work is: matrices are denoted by uppercase bold (**M**, **K**, **A** etc); vectors are denoted by lowercase bold (**x**, **z**, **y** etc); transposition of a matrix is denotes by the apostrophe (transpose of **C** is denoted by $\mathbf{C}'$ etc); time is denoted by variable t; frequency is denoted by ω.

2. Mechanical structures modeling

2.1 Dynamic equation

The movement equation of a generic structure can be written as

$$\mathbf{M}\ddot{\mathbf{q}}(t) + \mathbf{D}\dot{\mathbf{q}}(t) + \mathbf{K}\mathbf{q}(t) = \mathbf{B}_0\mathbf{f}(t), \tag{1}$$

where $\mathbf{q}(t)$ denotes the displacements, $\mathbf{M}$ is the mass matrix, $\mathbf{D}$ is the damping matrix, $\mathbf{K}$ is the stiffness matrix, $\mathbf{f}(t)$ is the vector of all external forces and $\mathbf{B}_0$ is a localization matrix for the external forces.

The main features of the dynamic response of a vibrating structure (Gawronski, 2004) are: 1) presence of resonance (amplification of response in specific frequencies); 2) the vibration models are decoupled (they can be excited independently); 3) the total response can be obtained by the summation of each mode contribution; 4) the impulse response consists of harmonic components, which are related to complex poles with small real parts; and 5) the system is controllable and observable.

2.2 Modal model

The structural model described in Eq. (1) can be represented also in modal coordinates (Ewins, 2000; Gawronski, 2004). The advantages of this kind of description are that the modal properties become evident such as the natural frequencies and damping factors for each structural mode.

The solution for the undamped free vibration is given by $\mathbf{q}(t) = \bar{\mathbf{q}}e^{j\omega t}$. Substituting this solution in the undamped free vibration movement equation it is obtained the eigen-problem given by

$$\mathbf{K}\bar{\mathbf{q}} = \omega^2\mathbf{M}\bar{\mathbf{q}}. \tag{2}$$

A structural system with n degrees of freedom presents n natural frequencies and n vibration modes. These natural frequencies and modes are determined through the solution of the eigen-problem related to the characteristic equation given by

$$det(\mathbf{K} - \omega^2\mathbf{M}) = 0 \tag{3}$$

The natural frequencies ω_i can be stored in a diagonal matrix $\mathbf{\Omega}$ given by

$$\mathbf{\Omega} = diag[\omega_1 \;\; \omega_2 \;\; \ldots \;\; \omega_n], \tag{4}$$

and the vibration modes $\boldsymbol{\phi}_i$ can be stored in the matrix $\mathbf{\Phi}$ according to

$$\mathbf{\Phi} = [\boldsymbol{\phi}_1 \;\; \boldsymbol{\phi}_2 \;\; \cdots \;\; \boldsymbol{\phi}_n]. \tag{5}$$

The model represented by the matrices $\mathbf{K}$ and $\mathbf{M}$ is the spatial model. The model denoted by the matrices $\mathbf{\Omega}$ and $\mathbf{\Phi}$ is the modal model.

A very important property is the orthogonality that allows the diagonalization of mass and stiffness matrices, i.e.,

$$\mathbf{\Phi}'\mathbf{M}\mathbf{\Phi} = diag[m_1 \;\; m_2 \;\; \ldots \;\; m_n] = diag[m_i], \tag{6}$$

$$\mathbf{\Phi}'\mathbf{M}\mathbf{\Phi} = diag[k_1 \;\; k_2 \;\; \ldots \;\; k_n] = diag[k_i]. \tag{7}$$

A specific situation which is mathematically convenient is the proportional damping where $\mathbf{D} = \alpha\mathbf{M} + \beta\mathbf{K}$. In this case it can be verified directly that the damping matrix is also diagonalized since $\mathbf{M}$ and $\mathbf{K}$ can be diagonalized, i.e.,

$$\mathbf{\Phi}'\mathbf{D}\mathbf{\Phi} = \alpha\mathbf{\Phi}'\mathbf{M}\mathbf{\Phi} + \beta\mathbf{\Phi}'\mathbf{K}\mathbf{\Phi} = diag[d_1 \ \ d_2 \ \ \ldots \ \ d_n] = diag[d_i] \tag{8}$$

Considering a coordinate transformation given by $\mathbf{p}(t) = \mathbf{\Phi}\mathbf{q}(t)$, and pre-multiplying the Eq. (1) by $\mathbf{\Phi}$, it is possible to obtain for proportional damping

$$\mathbf{\Phi}'\mathbf{M}\mathbf{\Phi}\ddot{\mathbf{p}}(t) + \mathbf{\Phi}'\mathbf{D}\mathbf{\Phi}\dot{\mathbf{p}}(t) + \mathbf{\Phi}'\mathbf{K}\mathbf{\Phi}\mathbf{p}(t) = \mathbf{\Phi}'\mathbf{B}_0\mathbf{f}(t), \tag{9}$$

which can be rewriten as

$$diag[m_i]\ddot{\mathbf{p}}(t) + diag[d_i]\dot{\mathbf{p}}(t) + diag[k_i]\mathbf{p}(t) = \mathbf{\Phi}'\mathbf{B}_0\mathbf{f}(t) \tag{10}$$

One can verify that Eq. (10) corresponds to a set of uncoupled second order differential equations similar to the movement equation of the one degree of freedom system. Each decoupled equation corresponds to a specific vibration mode of the system and can be written as

$$m_i\ddot{p}_i(t) + d_i\dot{p}_i(t) + k_ip_i(t) = \bar{f}_i(t), \tag{11}$$

or in a standard form of second order system as

$$\ddot{p}_i(t) + 2\xi\omega_i\dot{p}_i(t) + \omega_i^2p_i(t) = \gamma\omega_i^2\bar{f}_i(t). \tag{12}$$

It is possible to write for the i-mode the corresponding conjugate pair of poles

$$-\xi_i\omega_i \pm j\omega_i\sqrt{1-\xi_i^2} \tag{13}$$

where ω_i is the natural frequency and ξ_i is the non-dimensional damping factor, both related to the i vibration mode.

The modal model is a convenient way to include damping in models obtained by the finite element method, for example. The damping factor of each mode can be included independently. The proportional damping is not a mandatory hypothesis in this work since the control techniques can be applied to non-proportional damping also.

2.3 State-space model

The Eq. (1) can be rewriten as

$$\ddot{\mathbf{q}}(t) + \mathbf{M}^{-1}\mathbf{D}\dot{\mathbf{q}}(t) + \mathbf{M}^{-1}\mathbf{K}\mathbf{q}(t) = \mathbf{M}^{-1}\mathbf{B}_0\mathbf{f}(t). \tag{14}$$

Two kinds of external forces may be present in the active vibration control problem: the disturbance forces, denoted by $\mathbf{w}(t)$, and the control forces, denoted by $\mathbf{u}(t)$. Two kinds of outputs of the system can be defined: the measured outputs, denoted here by $\mathbf{y}(t)$, and the performance outputs, denoted by $\mathbf{z}(t)$.

Defining the state-space vector as $\mathbf{x}(t) = [\mathbf{q}(t) \ \ \dot{\mathbf{q}}(t)]'$, which corresponds to the displacements and velocities in this case, it is possible to write the state-space model in the form

$$\dot{\mathbf{x}}(t) = \mathbf{A}\mathbf{x}(t) + \mathbf{B}_1\mathbf{w}(t) + \mathbf{B}_2\mathbf{u}(t), \tag{15}$$

$$\mathbf{z}(t) = \mathbf{C}_1\mathbf{x}(t) + \mathbf{D}_{11}\mathbf{w}(t) + \mathbf{D}_{12}\mathbf{u}(t), \tag{16}$$

$$\mathbf{y}(t) = \mathbf{C}_2\mathbf{x}(t) + \mathbf{D}_{21}\mathbf{w}(t) + \mathbf{D}_{22}\mathbf{u}(t), \tag{17}$$

where the state-space matrix $\mathbf{A}$ is given by

$$\mathbf{A} = \begin{bmatrix} \mathbf{0} & \mathbf{I} \\ -\mathbf{M}^{-1}\mathbf{K} & -\mathbf{M}^{-1}\mathbf{K} \end{bmatrix}. \tag{18}$$

The matrices $\mathbf{B}_1$ and $\mathbf{B}_2$ are constructed with the structure $[\mathbf{0} \ \ \mathbf{M}^{-1}\mathbf{B}_0]'$ where the appropriate position matrix $\mathbf{B}_0$ is used for $\mathbf{w}(t)$ and for $\mathbf{u}(t)$. It is convenient to mention that the number of columns of $\mathbf{B}_1$ is the number of disturbances and the number of columns of $\mathbf{B}_2$ is the number of control forces.

The matrices $\mathbf{C}_1$, $\mathbf{D}_{11}$ and $\mathbf{D}_{12}$ are constructed to define the performance output in terms of the displacements, velocities and accelerations, or linear combinations of these values. The number of lines of $\mathbf{C}_1$ is the number of performances outputs to be monitored.

The matrices $\mathbf{C}_2$, $\mathbf{D}_{21}$ and $\mathbf{D}_{22}$ are constructed to specify the measured output also in terms of the displacements, velocities and accelerations. The number of lines of $\mathbf{C}_2$ is the number of measures.

This dynamic system can be represented in a compact form according to

$$\begin{bmatrix} \dot{\mathbf{x}}(t) \\ \mathbf{z}(t) \\ \mathbf{y}(t) \end{bmatrix} = \begin{bmatrix} \mathbf{A} & \mathbf{B}_1 & \mathbf{B}_2 \\ \mathbf{C}_1 & \mathbf{D}_{11} & \mathbf{D}_{12} \\ \mathbf{C}_2 & \mathbf{D}_{21} & \mathbf{D}_{22} \end{bmatrix} \begin{bmatrix} \mathbf{x}(t) \\ \mathbf{w}(t) \\ \mathbf{u}(t) \end{bmatrix}. \tag{19}$$

The transfer matrix of the system in Laplace domain variable s, relating each input to each output, can be writen as

$$\mathbf{P}(s) = \mathbf{C}(s\mathbf{I} - \mathbf{A})^{-1}\mathbf{B} + \mathbf{D}, \tag{20}$$

where

$$\mathbf{B} = [\mathbf{B}_1 \ \ \mathbf{B}_2], \quad \mathbf{C} = \begin{bmatrix} \mathbf{C}_1 \\ \mathbf{C}_2 \end{bmatrix}, \quad \mathbf{D} = \begin{bmatrix} \mathbf{D}_{11} & \mathbf{D}_{12} \\ \mathbf{D}_{21} & \mathbf{D}_{22} \end{bmatrix}. \tag{21}$$

2.4 State-space model in modal form

Considering that in modal coordinates the differential equations are decoupled for each mode, it is possible to reorganize the state-space in a modal form. Based on the standard form given by the Eq. (12), a state-space model for each mode can be generated.

One usual form of modal model (Gawronski, 2004) considers the states defined as

$$\mathbf{x}_i = \begin{bmatrix} \omega_i p_i \\ \xi_i \omega_i p_i + \dot{p}_i \end{bmatrix}. \tag{22}$$

For this case, the state-space matrix for the i-mode is given by

$$\mathbf{A}_i = \begin{bmatrix} -\xi_i\omega_i & \omega_i \\ -\omega_i & -\xi_i\omega_i \end{bmatrix}. \tag{23}$$

The state-space matrix will be a block diagonal matrix with the contribution of each mode in the form

$$\mathbf{A} = diag(\mathbf{A}_i). \tag{24}$$

This formulation is used in this work through the function `canon` with the option `modal` in the MATLAB® software.

2.5 Model reduction

A real structure is a continuous system with infinite degrees of freedom. It is necessary to have a finite dimensional representation for the system. This representation can be obtained by techniques such as finite elements or direct experimental identification. These two approaches lead to models that are finite dimensional but that can present a number of degrees of freedom yet considered excessive large. In this case, in order to have a feasible numerical treatment and feasible controller design, it is necessary to have a reduced order model.

The model reduction can be performed according to some techniques (Qu, 2004). The most usual and simple technique is the model truncation, where a number of modes is kept under a critical frequency value of interest. Upper frequency modes are simply discarded. This technique is adequate for the objectives of this work, and it is adopted here. Obviously the lost information can affect the dynamic representation of the structure and bring undesirable effects, such as spillover, implying the use of additional performance filtering to the model.

In most situations, the interest in the dynamic response of the structure is limited to a specific range of frequencies, and the model reduction can be performed considering this information. In the present case, the model reduction is conducted using the function `modreal` of the software MATLAB. This function performs the model reduction selecting the frequency ordered blocks of the modal model corresponding to the indicated frequency range, i.e., the selection is based on the blocks of the Eq. (24).

3. Structural and control models - plate vibration

It is considered in this work a finite element model of a plate. The MATLAB codes given in Ferreira (2008) were employed to obtain the mass and stiffness matrices considering the Mindlin plate formulation. The plate finite element has four nodes and three degrees of freedom in each node: rotations in axes x and y and displacement in axis z. The plate in this work was considered with all boundaries free. The finite element mesh is shown in Figure 1, and Table 1 shows the physical parameters used in the finite element model of the plate.

This finite element model presents 90 nodes with 3 degrees of freedom per node. This leads to a model of 270 degrees of freedom and 540 states. This model was reduced to a model with 24 states for control design purposes.

This model is used in this work to evaluate the spatial and decentralized $\mathcal{H}_\infty$ control techniques. There are, in this plate model, three convenient orientations for the transducers: horizontal, vertical and with an orientation of 45 degrees (representing identical actuation in the degrees of freedom in x and y directions of the same node). The placement of these sensors and actuators are indicated in Figure 1.

Height	1 m
Width	1 m
Thickness	2 mm
Density	2710 kg/m^3
Poisson Modulus	0.33
Young Modulus	70 GPa

Table 1. Physical properties of the plate

In order to have a more realistic dynamic system in the simulations, damping should be taken into account. In this case, it was included a modal damping of 3×10^{-6} to all vibration modes of the plate.

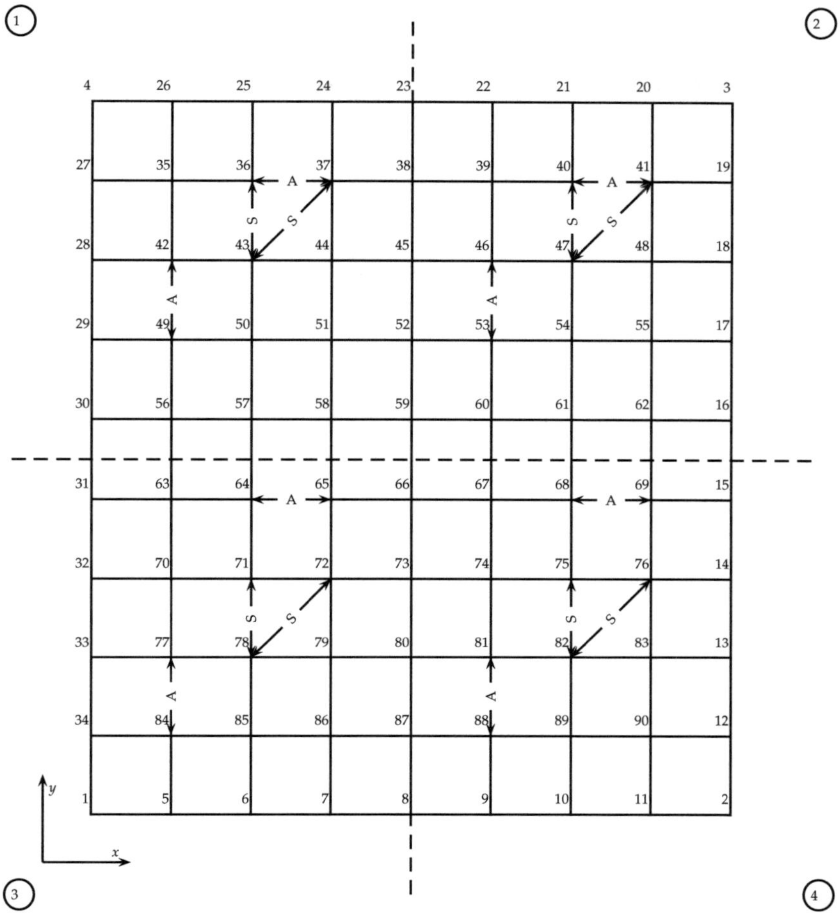

Fig. 1. Finite element mesh for the plate with four partitions - A denotes actuators and S denotes sensors

It is considered that the actuators and sensors are piezoelectric (PZT). The actuator receives a voltage and apply a pair of opposite moments in nearby nodes. The sensor generates a voltage proportional to its deformation, i.e., proportional to the difference between angles in nearby nodes. Figure 2 shows schematically the actuator and sensor representation used in this work.

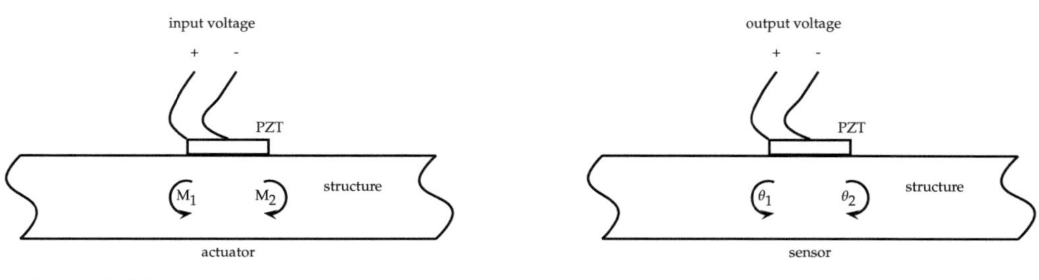

Fig. 2. PZT actuator and sensor relations to the respective degrees-of-freedom

	Actuators			Sensors		
	Number	DOF	Nodes	Number	DOF	Nodes
Disturbance	1 (w)	$\odot$	86			
Partition 1	2 (u_1)	$\leftrightarrow$	36 - 37	1 (z_1)	⤢	43 - 37
	3 (u_2)	$\updownarrow$	49 - 42	2 (y_1)	$\updownarrow$	43 - 36
Partition 2	4 (u_3)	$\leftrightarrow$	40 - 41	3 (z_2)	⤢	47 - 41
	5 (u_4)	$\updownarrow$	53 - 46	4 (y_2)	$\updownarrow$	47 - 40
Partition 3	6 (u_5)	$\leftrightarrow$	64 - 65	5 (z_3)	⤢	78 - 72
	7 (u_6)	$\updownarrow$	84 - 77	6 (y_3)	$\updownarrow$	78 - 71
Partition 4	8 (u_7)	$\leftrightarrow$	68 - 69	7 (z_4)	⤢	82 - 76
	9 (u_8)	$\updownarrow$	88 - 81	8 (y_4)	$\updownarrow$	82 - 75

Table 2. Definition and placement of actuators and sensors for the mesh in the Figure 1.

Table 2 shows actuators, sensors and nodes location for the mesh of Figure 1. The arrows indicate the respective degrees of freedom. The partition reveals which actuators and sensors are used in each local model for the case of the decentralized control. The disturbance is considered a force in the z direction applied in the node 86. Actuators numbered from 2 to 9 are chosen as control inputs. Sensors 2, 4, 6 and 8 are measuring outputs. The performance parameters are the sensors numbered as 1, 3, 5 and 7. The uncontrolled system was normalized to have an $\mathcal{H}_\infty$ norm equal to 1 (normalized plant).

4. $\mathcal{H}_\infty$ control formulation

The $\mathcal{H}_\infty$ control design method consists of designing a controller transfer function $\mathbf{K}(s)$ in a closed loop with a plant $\mathbf{P}(s)$ in order to minimize the $\mathcal{H}_\infty$ norm of the closed loop transfer function $\mathbf{T}(s)$ from the disturbance $\mathbf{w}$ to the performance $\mathbf{z}$ in the frequency domain ω. The loop is usually represented as in Figure 3.

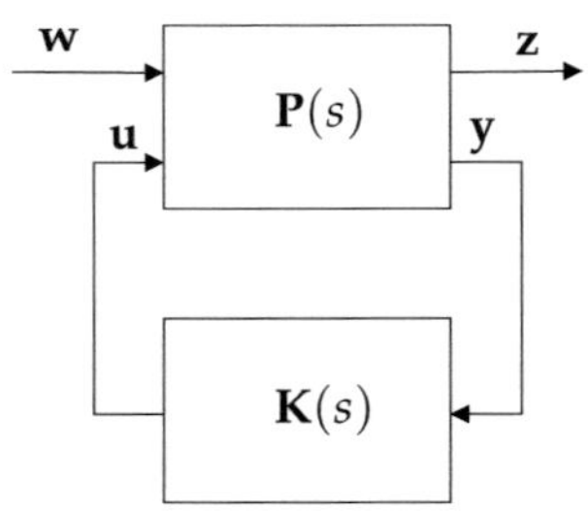

Fig. 3. $\mathcal{H}_\infty$ closed loop diagram

The $\mathcal{H}_\infty$ norm of a system, from the disturbance input $\mathbf{w}(t)$ to the performance output $\mathbf{z}(t)$ (Skelton et al., 1998), can be defined as

$$J_\infty = \|\mathbf{T}(s)\|_\infty = \frac{\int_0^\infty \mathbf{z}'(t)\mathbf{z}(t)\,dt}{\int_0^\infty \mathbf{w}'(t)\mathbf{w}(t)dt}. \tag{25}$$

The $\mathcal{H}_\infty$ norm can be calculated as

$$\|\mathbf{T}(s)\|_\infty = \sup_\omega \bar{\sigma}(\mathbf{T}(j\omega)),$$

where $\bar{\sigma}$ is the maximum singular value of the transfer function $\mathbf{T}(s)$ (Zhou & Doyle, 1997). This is a measure in the frequency domain of the worst response of $\mathbf{T}(s)$. If the worst response, in the sense of the higher amplitude, is achieved to an acceptable level, the performance is evidently guaranteed for all cases.

Weighting functions are used in this work to compel the performance output and control signals to follow the specified frequency distributions. In general, low-pass weighting functions $W_z(s)$ are used to balance the performance output levels and high-pass functions $W_u(s)$ are applied to the control forces. Commonly used filter functions (Zhou & Doyle, 1997) are:

$$W_p(s) = \left(\frac{s\sqrt[k]{M} + \omega_c}{s + \omega_c \sqrt[k]{\epsilon}} \right)^k, \qquad W_u(s) = \left(\frac{s + \omega_c \sqrt[k]{M}}{s\sqrt[k]{\epsilon} + \omega_c} \right)^k,$$

where ω_c is the cut frequency, k is the filter order, M is the gain at pass band and ϵ is the gain at rejection band.

Specifying the correct weighting functions for each problem is very important in the controller design process. They define the frequency regions where the disturbance signals response should be minimized and where the control signals should be effective, avoiding the excitation of neglected vibration modes in the model, which is fundamental to avoid the spillover effect (Balas, 1978).

An $\mathcal{H}_\infty$ controller design problem can be written as an optimization problem. The controller $\mathbf{K}(s)$ can be obtained by the minimization of the $\mathcal{H}_\infty$ norm of the closed-loop $\mathbf{T}(s)$, i.e.,

$$\begin{array}{rl} \min_{\mathbf{K}(s)} & \|\mathbf{T}(s)\|_\infty \\ \text{subjected to} & \mathbf{K}(s) \text{ stable} \\ & \mathbf{T}(s) \text{ stable.} \end{array}$$

This optimization problem can be considered a global design, since it involves all the inputs and outputs of the plant. Provided an acceptable level of vibration, a sub-optimal solution of this problem may be obtained solving the associated Riccati equations or by the solution of a linear matrix inequality problem (Boyd et al., 1994; Zhou & Doyle, 1997). The solution of this problem can be obtained using the MATLAB Robust Control Toolbox with the `hinfsyn` function for example.

5. Spatial $\mathcal{H}_\infty$ control

5.1 Spatial $\mathcal{H}_\infty$ norm

The $\mathcal{H}_\infty$ norm may be generalized considering a spatial distribution for the performance parameters. This can lead to a weighted response over the specified spatial region.

The spatial $\mathcal{H}_\infty$ norm (Skelton et al., 1998) for a dynamic system considering the disturbance input $\mathbf{w}(t)$ to the spatial performance output $\mathbf{z}(r,t)$ can be defined as

$$J_\infty = \frac{\int_0^\infty \int_{\mathcal{R}} \mathbf{z}'(t,r)\mathbf{Q}(r)\mathbf{z}(t,r)\, drdt}{\int_0^\infty \mathbf{w}'(t)\mathbf{w}(t)\, dt}, \tag{26}$$

where $\mathcal{R}$ denotes the spatial region and $\mathbf{Q}(r)$ is a spatial weighting function.

5.2 Spatial and non-spatial $\mathcal{H}_\infty$-control parallel

The spatial $\mathcal{H}_\infty$ norm allows to generalize the $\mathcal{H}_\infty$ control design problem. Consider a system in which the performance output $\mathbf{z}(r,t)$ depends both on space (r) and time (t), whilst the measured output $\mathbf{y}(t)$ depends only on time. The state space model may be described as:

$$\begin{aligned} \dot{\mathbf{x}}(t) &= \mathbf{A}\mathbf{x}(t) + \mathbf{B}_1\mathbf{w}(t) + \mathbf{B}_2\mathbf{u}(t) \\ \mathbf{z}(t,r) &= \mathbf{C}_1(r)\mathbf{x}(t) + \mathbf{D}_{11}(r)\mathbf{w}(t) + \mathbf{D}_{12}(r)\mathbf{u}(t) \\ \mathbf{y}(t) &= \mathbf{C}_2 x(t) + \mathbf{D}_{21}\mathbf{w}(t) + \mathbf{D}_{22}\mathbf{u}(t). \end{aligned} \quad (27)$$

It is possible to notice that r stands for a vector position and can represent two or three-dimensional problems. The definition of the spatial norm is a multiple integral depending on the problem dimensionality according to the Eq. (26).
The purpose of control design is to obtain a dynamic controller given by

$$\begin{aligned} \dot{\mathbf{x}}_k(t) &= \mathbf{A}_k\mathbf{x}_k(t) + \mathbf{B}_k y(t) \\ \mathbf{u}(t) &= \mathbf{C}_k\mathbf{x}_k(t) + \mathbf{D}_k\mathbf{y}(t), \end{aligned}$$

which reduces the particular norm of interest.
The spatial $\mathcal{H}_\infty$ problem is solved through the conversion to an equivalent punctual $\mathcal{H}_\infty$ with a modified performance output $\tilde{\mathbf{z}}(t)$, which is responsible for taking into account the desired vibration region. The ordinary $\mathcal{H}_\infty$ norm based on $\tilde{\mathbf{z}}(t)$ according to Equation (25) is

$$J_\infty = \frac{\int_0^\infty \tilde{\mathbf{z}}'(t)\tilde{\mathbf{z}}(t)\,dt}{\int_0^\infty \mathbf{w}'(t)\mathbf{w}(t)\,dt}. \quad (28)$$

Comparing equations (28) and (26) it is possible to establish the equivalence

$$\tilde{\mathbf{z}}'(t)\tilde{\mathbf{z}}(t) = \int_{\mathcal{R}} \mathbf{z}'(t,r)\mathbf{Q}(r)\mathbf{z}(r,t)\,dr. \quad (29)$$

This equivalence allows to convert the spatial $\mathcal{H}_\infty$ control design problem into the standard $\mathcal{H}_\infty$ problem with the modified performance output.
From Equation (27):

$$\mathbf{z}(t,r) = [\mathbf{C}_1(r)\ \ \mathbf{D}_{11}(r)\ \ \mathbf{D}_{12}(r)] \begin{bmatrix} \mathbf{x} \\ \mathbf{w} \\ \mathbf{u} \end{bmatrix},$$

and the equivalent punctual output

$$\tilde{\mathbf{z}}(t) = \mathbf{\Gamma} \begin{bmatrix} \mathbf{x} \\ \mathbf{w} \\ \mathbf{u} \end{bmatrix}. \quad (30)$$

Using Equation (29), it is possible to write

$$\begin{bmatrix} \mathbf{x} \\ \mathbf{w} \\ \mathbf{u} \end{bmatrix}' \mathbf{\Gamma}'\mathbf{\Gamma} \begin{bmatrix} \mathbf{x} \\ \mathbf{w} \\ \mathbf{u} \end{bmatrix} =$$

$$= \int_{\mathcal{R}} \begin{bmatrix} \mathbf{x} \\ \mathbf{w} \\ \mathbf{u} \end{bmatrix}' \begin{bmatrix} \mathbf{C}_1'(r) \\ \mathbf{D}_{11}'(r) \\ \mathbf{D}_{12}'(r) \end{bmatrix} \mathbf{Q}(r)\,[\mathbf{C}_1(r)\ \ \mathbf{D}_{11}(r)\ \ \mathbf{D}_{12}(r)] \begin{bmatrix} \mathbf{x} \\ \mathbf{w} \\ \mathbf{u} \end{bmatrix} dr,$$

$$\begin{bmatrix} \mathbf{x} \\ \mathbf{w} \\ \mathbf{u} \end{bmatrix}' \mathbf{\Gamma}'\mathbf{\Gamma} \begin{bmatrix} \mathbf{x} \\ \mathbf{w} \\ \mathbf{u} \end{bmatrix} =$$

$$= \begin{bmatrix} \mathbf{x} \\ \mathbf{w} \\ \mathbf{u} \end{bmatrix}' \int_{\mathcal{R}} \begin{bmatrix} \mathbf{C}_1'(r) \\ \mathbf{D}_{11}'(r) \\ \mathbf{D}_{12}'(r) \end{bmatrix} \mathbf{Q}(r) \left[\mathbf{C}_1(r) \;\; \mathbf{D}_{11}(r) \;\; \mathbf{D}_{12}(r)\right] dr \begin{bmatrix} \mathbf{x} \\ \mathbf{w} \\ \mathbf{u} \end{bmatrix}.$$

The equivalence results in the following equality

$$\mathbf{\Gamma}'\mathbf{\Gamma} =$$

$$= \int_{\mathcal{R}} \begin{bmatrix} \mathbf{C}_1'(r) \\ \mathbf{D}_{11}'(r) \\ \mathbf{D}_{12}'(r) \end{bmatrix} \mathbf{Q}(r) \left[\mathbf{C}_1(r) \;\; \mathbf{D}_{11}(r) \;\; \mathbf{D}_{12}(r)\right] dr. \tag{31}$$

By defining a spatial weighting function $\mathbf{Q}(r)$, the matrix $\mathbf{\Gamma}'\mathbf{\Gamma}$ can be found from Equation (31) and $\mathbf{\Gamma}$ may be determined. One should notice that $\mathbf{\Gamma}$ is the transformation that allows the punctual $\mathcal{H}_\infty$ problem to represent equivalently the spatial $\mathcal{H}_\infty$ problem.
Using Equation (30), the performance output $\tilde{\mathbf{z}}$ is defined as

$$\tilde{\mathbf{z}} = \mathbf{\Gamma} \begin{bmatrix} \mathbf{x} \\ \mathbf{w} \\ \mathbf{u} \end{bmatrix} = \left[\mathbf{\Pi} \;\; \mathbf{\Theta}_1 \;\; \mathbf{\Theta}_2\right] \begin{bmatrix} \mathbf{x} \\ \mathbf{w} \\ \mathbf{u} \end{bmatrix},$$

in which $\mathbf{\Pi}$, $\mathbf{\Theta}_1$ and $\mathbf{\Theta}_2$ are simultaneously defined as matrix partitions of $\mathbf{\Gamma}$ according to the signal dimensions.
So, the final plant model is written as

$$\begin{aligned} \dot{\mathbf{x}}(t) &= \mathbf{A}\mathbf{x}(t) + \mathbf{B}_1\mathbf{w}(t) + \mathbf{B}_2\mathbf{u}(t) \\ \tilde{\mathbf{z}}(t) &= \mathbf{\Pi}\mathbf{x}(t) + \mathbf{\Theta}_1\mathbf{w}(t) + \mathbf{\Theta}_2\mathbf{u}(t) \\ \mathbf{y}(t) &= \mathbf{C}_2\mathbf{x}(t) + \mathbf{D}_{21}\mathbf{w}(t) + \mathbf{D}_{22}\mathbf{u}(t). \end{aligned}$$

5.3 Calculation of Γ

Equation (31) defines $\mathbf{\Gamma}'\mathbf{\Gamma}$ as an integral of a square matrix of order $n + n_w + n_u$, where n is the number of plant states, n_w is the number of disturbances and n_u is the number of control signals. $\mathbf{\Gamma}$ has dimensions $p \times (n + n_w + n_u)$, where the number of lines p represents the number of performance outputs, i.e., the number of lines of $\tilde{\mathbf{z}}$. The number of elements of $\mathbf{\Gamma}$ is $p \times (n + n_w + n_u)$ and the number of elements of $\mathbf{\Gamma}'\mathbf{\Gamma}$ is $(n + n_w + n_u) \times (n + n_w + n_u)$. Since, $\mathbf{\Gamma}'\mathbf{\Gamma}$ is symmetric, the number of unknowns elements are $(n + n_w + n_u)(n + n_w + n_u + 1)/2$. A convenient choice is $p = n + n_w + n_u$, which amounts to a square matrix for $\mathbf{\Gamma}$, and in this case a Cholesky factorization can be applied in $\mathbf{\Gamma}'\mathbf{\Gamma}$ to obtain $\mathbf{\Gamma}$. Another possibility to determine Γ involves a specific situation related to finite element models as described in the next section.

5.4 Γ for the case of constant spatial weighting

Taking the spatial weighting function constant inside every element allows some simplifying results. In this case the spatial performance output can be discretized for the degrees of freedom that are the model states and the spatial performance output can be interpolated from the degrees of freedom. The integral of Equation (31) that defines $\mathbf{\Gamma}'\mathbf{\Gamma}$ may be approximated as

$$\Gamma'\Gamma \approx \sum_i f(r_i) \begin{bmatrix} \mathbf{C}_1'(r_i) \\ \mathbf{D}_{11}'(r_i) \\ \mathbf{D}_{12}'(r_i) \end{bmatrix} \mathbf{Q}(r_i) \left[\mathbf{C}_1(r_i) \;\; \mathbf{D}_{11}(r_i) \;\; \mathbf{D}_{12}(r_i) \right], \tag{32}$$

by supposing an integration method such as the gaussian quadrature (Bathe, 1995), where the values $f(r_i)$ represent the contribution to the specific degree of freedom. In this case, $f(r_i)$ can be considered the gauss weightings and r_i the respective integration points (in this case the degrees of freedom).

If the finite element mesh is homogeneous in terms of the element size, a simplification of a constant value of the integrand inside each element can be used leading to less calculations. The integral in Equation (31) becomes a summation according to

$$\Gamma'\Gamma = \sum_i \begin{bmatrix} \mathbf{C}_{1i}' \\ \mathbf{D}_{11i}' \\ \mathbf{D}_{12i}' \end{bmatrix} Q_i \left[\mathbf{C}_{1i} \;\; \mathbf{D}_{11i} \;\; \mathbf{D}_{12i} \right] A_i, \tag{33}$$

with A_i as an elementary length, area or volume, according to the dimension in the integral, and Q_i is the weighting function value related to point i in Equation (32). A simplification of notation, taking i to denote the corresponding r_i, was employed.

Defining $t_i = Q_i A_i$, it is possible to write Equation (33) as

$$\Gamma'\Gamma = \sum_i \left(\begin{bmatrix} \mathbf{C}_{1i}' \\ \mathbf{D}_{11i}' \\ \mathbf{D}_{12i}' \end{bmatrix} \sqrt{t_i} \right) \left(\sqrt{t_i} \left[\mathbf{C}_{1i} \;\; \mathbf{D}_{11i} \;\; \mathbf{D}_{12i} \right] \right).$$

This summation can be rewritten as

$$\Gamma'\Gamma = \left(\sum_i \sqrt{t_i} \begin{bmatrix} \mathbf{C}_{1i}' \\ \mathbf{D}_{11i}' \\ \mathbf{D}_{12i}' \end{bmatrix} \right) \left(\sum_i \sqrt{t_i} \left[\mathbf{C}_{1i} \;\; \mathbf{D}_{11i} \;\; \mathbf{D}_{12i} \right] \right) - \mathbf{X},$$

where

$$\mathbf{X} = \sum_i \sum_j^{i \neq j} t_i \begin{bmatrix} \mathbf{C}_{1i}' \\ \mathbf{D}_{11i}' \\ \mathbf{D}_{12i}' \end{bmatrix} \left[\mathbf{C}_{1j} \;\; \mathbf{D}_{11j} \;\; \mathbf{D}_{12j} \right].$$

Since finite element models are considered in this work, where the degrees of freedom are model states, the matrices $\left[\mathbf{C}_{1j} \;\; \mathbf{D}_{11j} \;\; \mathbf{D}_{12j} \right]$ are orthogonal. $\mathbf{C}_{1j}$ is a matrix of zeros with a one in the position j of the convenient degree of freedom. $\mathbf{D}_{11j}$ and $\mathbf{D}_{12j}$ are null since the displacement and velocities are the states. Acceleration outputs are not considered in this work. This yields $\mathbf{X} = \mathbf{0}$ and consequently

$$\Gamma'\Gamma = \left(\sum_i \sqrt{t_i} \begin{bmatrix} \mathbf{C}_{1i}' \\ \mathbf{D}_{11i}' \\ \mathbf{D}_{12i}' \end{bmatrix} \right) \left(\sum_i \sqrt{t_i} \left[\mathbf{C}_{1i} \;\; \mathbf{D}_{11i} \;\; \mathbf{D}_{12i} \right] \right)$$

where one can choose

$$\Gamma = \sum_i \sqrt{t_i} \left[\mathbf{C}_{1i} \;\; \mathbf{D}_{11i} \;\; \mathbf{D}_{12i} \right]. \tag{34}$$

In this way, a numerical definition of the output matrix $\mathbf{\Gamma}$ of the $\mathcal{H}_\infty$ spatial control is achieved for the particular case of finite element models, where the degrees of freedom are the states.

6. Decentralized $\mathcal{H}_\infty$ control

The decentralized control design problem can be obtained by imposing a block-diagonal structure to the controller. If the order of inputs and outputs in the transfer function respects physical proximity, a block diagonal structure for the controller can be obtained such as:

$$\mathbf{K}(s) = \begin{bmatrix} \mathbf{K}_1(s) & & & \\ & \mathbf{K}_2(s) & & \\ & & \ddots & \\ & & & \mathbf{K}_p(s) \end{bmatrix},$$

where $\mathbf{K}_i(s)$ are the local controllers.

It is difficult to formulate the decentralized control design with a problem structure that can be solved easily. When the optimization problem is formulated through linear matrix inequalities, the requirement to impose a particular structure in the decision variable $\mathbf{K}(s)$ represents a mathematical difficulty that can lead to a non-convex problem. This difficulty motivates the investigation of other approaches for the decentralized control.

One alternative is that the original plant can be divided in several local plants with their own inputs and outputs and with spatially close actuators and sensors. In this case, it is possible to design local controllers corresponding to each plant subdivision. The closed-loop can be generated by employing these controllers along with the original plant in all its input and output signals, i.e., it is possible to solve several optimization problems such as

$$\begin{array}{rl} \min_{\mathbf{K}_i(s)} & \|\mathbf{T}_i(s)\|_\infty \\ \text{subjected to} & \mathbf{K}_i(s) \text{ stable} \\ & \mathbf{T}_i(s) \text{ stable.} \end{array}$$

where the controllers $\mathbf{K}_i(s)$ are obtained. In this case, the closed-loop is a function of all controllers and of the global plant.

Through this approach no additional mathematical development is necessary, since the solution is taken as a combination of solutions of several simultaneous optimizations problems.

7. Simulated results

Using $\mathcal{H}_\infty$ control for both the centralized and decentralized designs the same configuration of actuators and sensors already described were adopted in order to permit to compare the results.

The control design is performed using the linear matrix inequalities formulation for the $\mathcal{H}_\infty$ controller design using the function `hinfsyn` of MATLAB 7.2 (default parameters).

The parameters of the weighting filters used in this work are shown in Table 3. The same filters were employed in all simulations of this work.

A simulation test is performed according to the presented configuration of inputs and outputs. A linear sine sweep of 10 s from 0 to 2 KHz is used as a disturbance signal in all cases.

	ω_c	k	M	ϵ
$W_z(s)$ - low-pass weight for performance	1500	1	0.1	0.001
$W_u(s)$ - high-pass weight for control force	2000	1	0.1	0.001

Table 3. Weighting filters parameters

7.1 Centralized control

The centralized control case is the ordinary punctual $\mathcal{H}_\infty$ applied to the finite element model of the plate described above. Control results for the centralized control are shown in time domain in Figure 4 and in frequency domain in Figure 5.

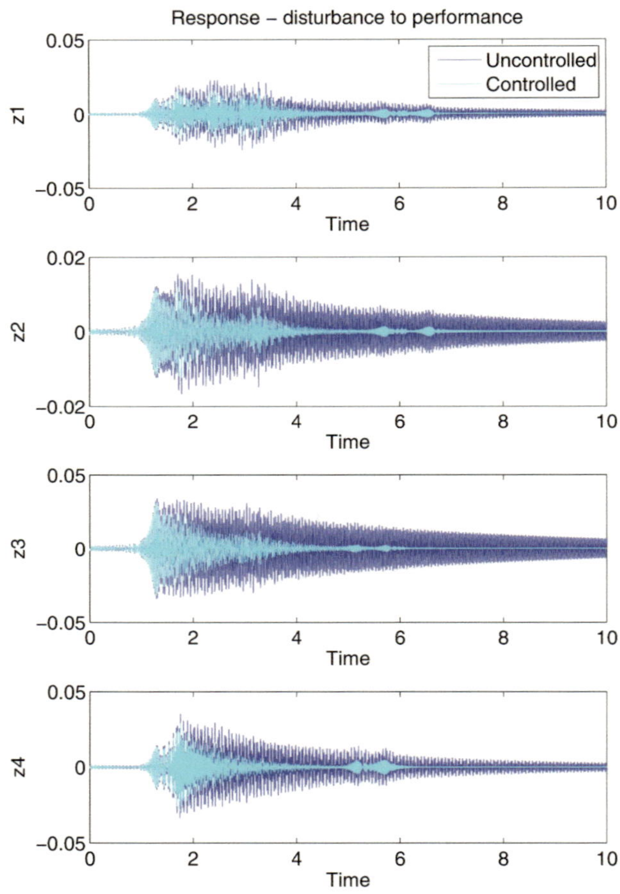

Fig. 4. Centralized control - control result from disturbance signal to spatial performance output

The time scale in Figure 4 is the duration of the sweep disturbance signal, and it may be interpreted as a frequency scale. It is possible to observe a good attenuation increasing as the

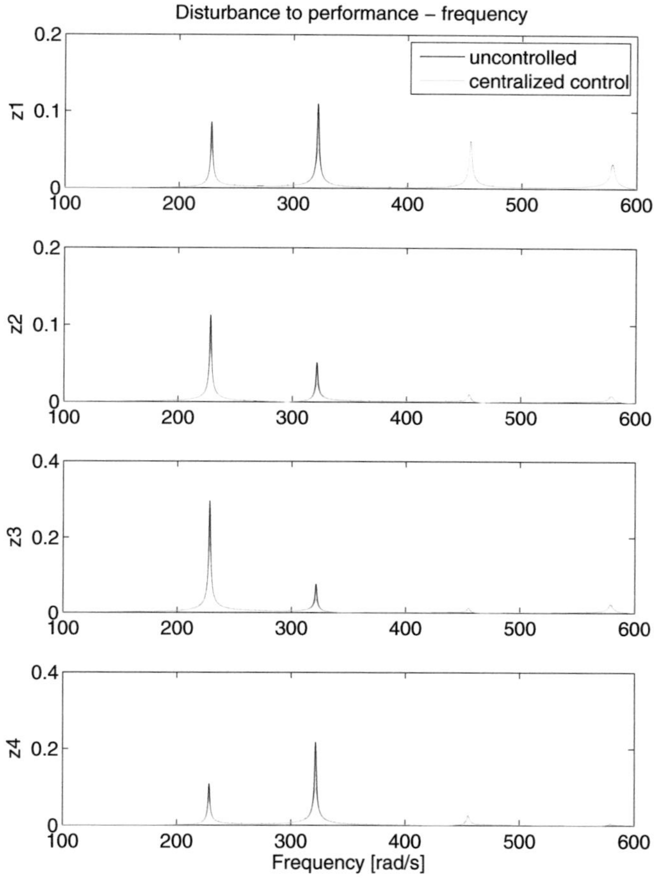

Fig. 5. Centralized control - control result from disturbance signal to spatial performance output in frequency domain

disturbance frequency increases. It is possible to notice also the presence of four predominant natural frequencies in the plate, but only in the attenuated response.
In the spectral response of Figure 5 it is possible to observe clearly the four natural frequencies, but only the first and the second peaks are attenuated, achieving a reduction of approximately 50% in the amplitude. The bigger reduction on the time response of Figure 4 for the highest frequencies is due to the low damping regularly found in these structures, and the respective transient response.

7.2 Decentralized control

Frequency and time domain results for the decentralized control in contrast with centralized control are shown in Figures 6 and 7.
In Figure 6 it is possible to see that the attenuation of the decentralized controller is practically the same, but just a bit less amplitude is present in the middle frequencies.

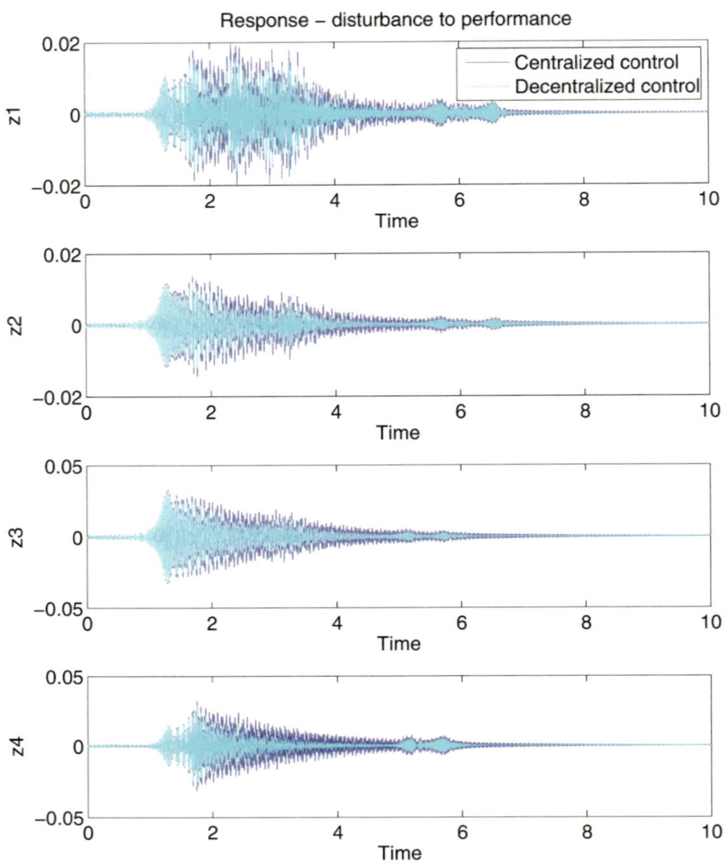

Fig. 6. Centralized and decentralized control - control result from disturbance signal to spatial performance output

In Figure 7 this slightly bigger attenuation is not present in any of the four peaks, with a complete superposition of the responses. Once again, the difference seen in Figure 6 is due to the low damping of the plate.

7.3 Spatial control

In the spatial control case the performance is the spatial output defined in the original design, which in this case is a constant and equal weighting of all the nodes except for the boundary of the plate.

The first four sensors are used as performance measurements with the same control loop design for the spatial performance defined for the whole plate, in order to compare it with $\mathcal{H}_\infty$ decentralized design. The control results obtained are shown in frequency and time response respectively in Figures 8 and 9.

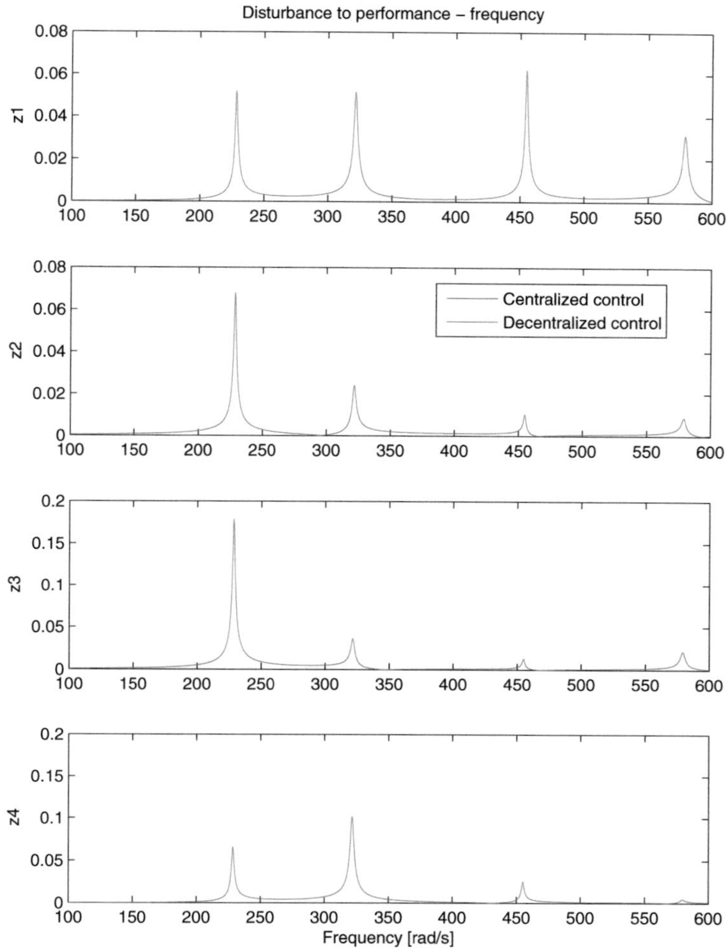

Fig. 7. Centralized and decentralized control - control result from disturbance signal to spatial performance output in frequency domain

In Figure 8 it is possible to notice that the spatial attenuation is bigger than in the decentralized controller results, in the low and middle frequency regions.
In Figure 9 the attenuation attained by the spatial controller is on the range around 10 and 20%, in comparison to the decentralized controller, which presented a similar result to the centralized controller on the order of 50%. This means that the spatial controller achieved indeed a good vibration attenuation result. But in the high frequency range, the two peaks once again were not attenuated.

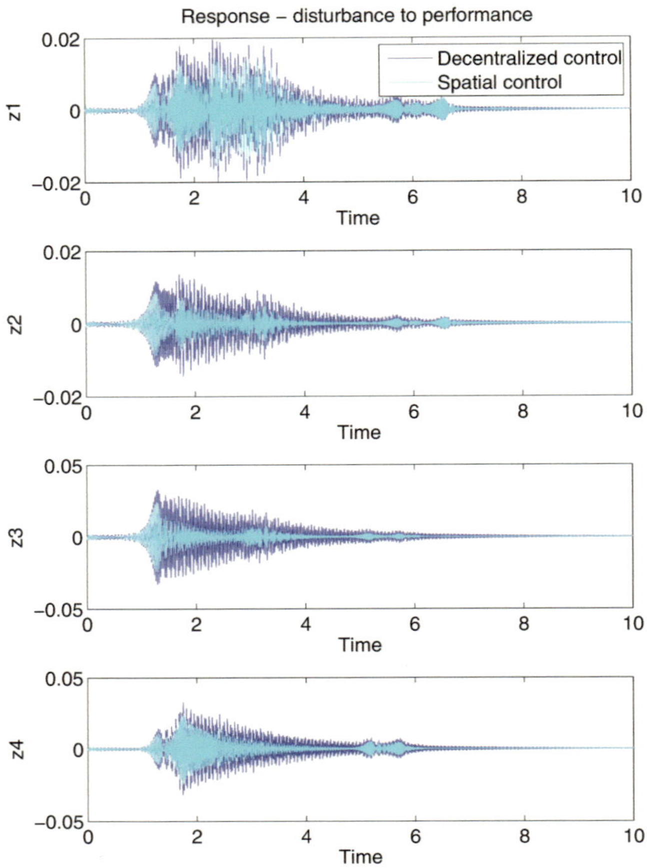

Fig. 8. Decentralized and spatial control - control results from disturbance signal to spatial performance outputs

7.4 Control signal comparison

The control signal levels for the spatial and decentralized controlled are presented in Figure 10 in the time domain. It is possible to have some insight into the behavior of the controllers results by analyzing these curves.

The closed loop frequency response, through the associated controllers efforts to attenuate the correspondent natural frequencies, is very clear on the curves of Figure 10. The spatial controller presented a bigger effort to attenuate the first and the second peaks, while the level of the decentralized controller does not reflect clearly the passage of the natural frequencies. That is the reason that made the spatial controller the better one. Other higher frequencies also presented controller effort, but the respective attenuation was not possible to be seen in the spectral responses in Figure 9. It would be interesting to further investigate if some other points of the region is presenting attenuation on these peaks.

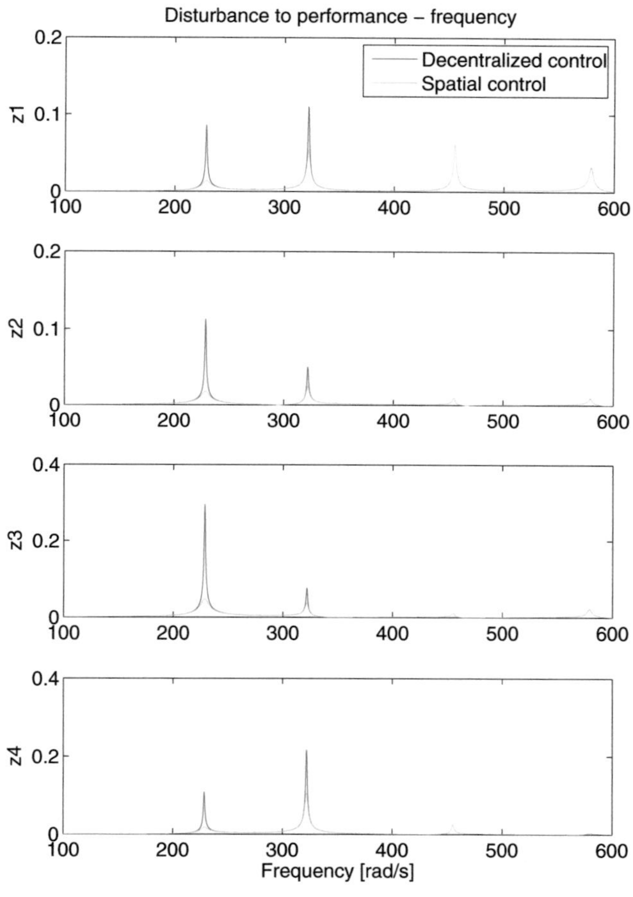

Fig. 9. Decentralized and spatial control - control results from disturbance signal to spatial performance outputs in frequency domain

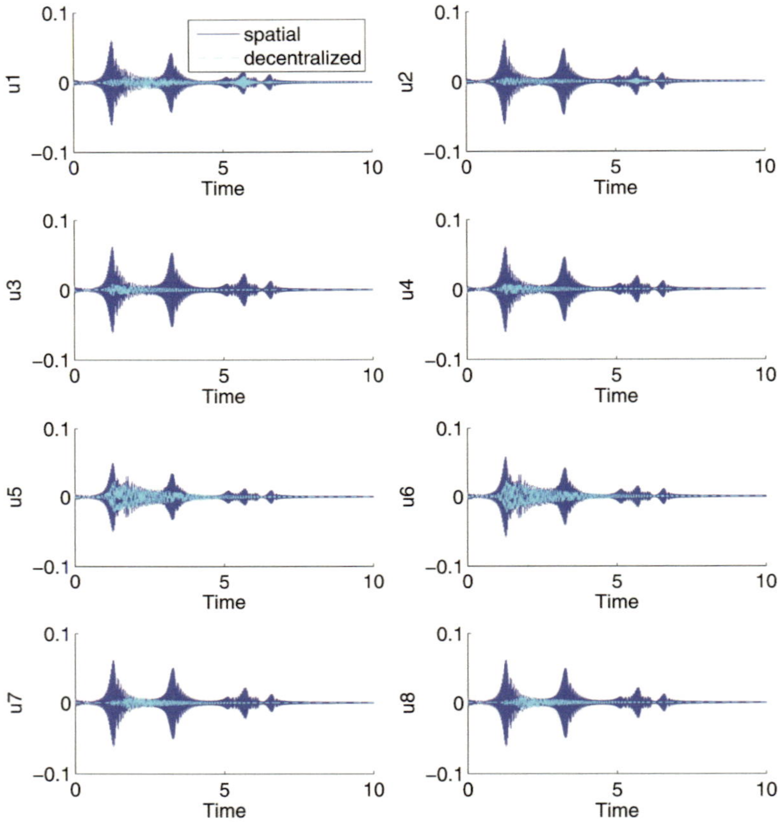

Fig. 10. Comparison of control signal in time domain between spatial and decentralized control

8. Concluding remarks

Two recently proposed $\mathcal{H}_\infty$ controller design methods dedicated to active structural vibration control were presented, and simulated results based on a finite element model of a plate were analyzed. The spatial norm based method aims to attenuate the vibration over entire regions of the structures, using the controller energy in a more effective way. The decentralized control method also tries to achieve a good energy distribution based on the application of the control effort through different controllers. A third controller, based on a standard $\mathcal{H}_\infty$ design for the complete plate, and using the same sensors and actuator, was evaluated also, to serve as a comparison base.

The decentralized control presented a similar behavior to the centralized one, but with a somewhat smaller control effort. Centralized control can demand more expensive equipment and is less robust in case of failures when compared to the decentralized approach. The results validate the option for a decentralized control as opposed to the regular centralized control.

The spatial control as compared to the decentralized control presented the better results in terms of attenuation. The analysis was based on the response on the same punctual performance points, instead of the complete region. But it is possible to affirm that a better attenuation on the complete region is present on the performance of this controller, based on the mathematical definition of the spatial norm.

A future investigation is related to the stability of the decentralized case, since each decentralized control can affect the others. In this work, this aspect was checked by the direct verification of the closed-loop stability, but only for the specific configuration of the four decentralized controllers considered here.

Also the choice of weighting function in the spatial control is an open problem, that heavily depends on the problem's practical requirements.

9. References

Balas, M. J. (1978). Feedback control of flexible systems, *IEEE Transactions on Automatic Control* 23: 673 - 679.

Barrault, G., Halim, D., Hansen, C. & Lenzi, A. (2007). Optimal truncated model for vibration control design within a specified bandwidth, *International Journal of Solids and Structures* 44(14-15): 4673 – 4689.

Barrault, G., Halim, D., Hansen, C. & Lenzi, A. (2008). High frequency spatial vibration control for complex structures, *Applied Acoustics* 69(11): 933 – 944.

Bathe, K.-J. (1995). *Finite Element Procedures (Part 1-2)*, Prentice Hall.

Baz, A. & Chen, T. (2000). Control of axi-symmetric vibrations of cylindrical shells using active constrained layer damping, *Thin-Walled Structures* 36(1): 1 – 20.

Bhattacharya, P., Suhail, H. & Sinha, P. K. (2002). Finite element analysis and distributed control of laminated composite shells using lqr/imsc approach, *Aerospace Science and Technology* 6(4): 273 – 281.

Bianchi, E., Gardonio, P. & Elliott, S. J. (2004). Smart panel with multiple decentralized units for the control of sound transmission. part iii: control system implementation, *Journal of Sound and Vibration* 274(1-2): 215 – 232.

Boyd, S., El Ghaoui, L., Feron, E. & Balakrishnan, V. (1994). *Linear Matrix Inequalities in System and Control Theory*, Vol. 15 of *Studies in Applied Mathematics*, SIAM.

Casadei, F., Ruzzene, M., Dozio, L. & Cunefare, K. A. (2010). Broadband vibration control through periodic arrays of resonant shunts: experimento investigation on plates, *Smart materials and structures* 19.

Cheung, Y. & Wong, W. (2009). H_∞ and H_2 optimizations of a dynamic vibration absorber for suppressing vibrations in plates, *Journal of Sound and Vibration* 320(1-2): 29 – 42.

Ewins, D. J. (2000). *Modal Testing: Theory, Practice and Application*, Research Studies Press, Ltd.

Ferreira, A. (2008). *MATLAB Codes for Finite Element Analysis: Solids and Structures*, Springer Publishing Company, Incorporated.

Gawronski, W. (2004). *Advanced Structural Dynamics and Active Control of Structures*, Springer-Verlag.

Halim, D. (2002). *Vibration analysis and control of smart structures*, PhD thesis, University of NewCastle – School of Electrical Engineering and Computer Science, New South Wales, Australia.

Halim, D. (2007). Structural vibration control with spatially varied disturbance input using a spatial method, *Mechanical Systems and Signal Processing* 21(6): 2496 – 2514.

Halim, D., Barrault, G. & Cazzolato, B. S. (2008). Active control experiments on a panel structure using a spatially weighted objective method with multiple sensors, *Journal of Sound and Vibration* 315(1-2): 1 – 21.

Hurlebaus, S., Stöbener, U. & Gaul, L. (2008). Vibration reduction of curved panels by active modal control, *Comput. Struct.* 86(3-5): 251–257.

Jiang, J. & Li, D. (2010). Decentralized guaranteed cost static output feedback vibration control for piezoelectric smart structures, *Smart Materials and Structures* 19(1): 015018.

Qu, Z.-Q. (2004). *Model Order Reduction Techniques with Applications in Finite Element Analysis*, Springer.

Skelton, R. E., Iwasaki, T. & Grigoriadis, K. M. (1998). *An Unified Algebraic Approach to Linear Control Design*, Taylor and Francis.

Zhou, K. & Doyle, J. C. (1997). *Essentials of Robust Control*, Prentice Hall.

Zilletti, M., Elliott, S. J. & Gardonio, P. (2010). Self-tuning control systems of decentralised velocity feedback, *Journal of Sound and Vibration* 329(14): 2738 – 2750.

8

Robust Control of Mechanical Systems

Joaquín Alvarez[1] and David Rosas[2]
[1]*Scientific Research and Advanced Studies Center of Ensenada (CICESE)*
[2]*Universidad Autónoma de Baja California*
Mexico

1. Introduction

Control of mechanical systems has been an important problem since several years ago. For free-motion systems, the dynamics is often modeled by ordinary differential equations arising from classical mechanics. Controllers based on feedback linearization, adaptive, and robust techniques have been proposed to control this class of systems (Brogliato et al., 1997; Slotine & Li, 1988; Spong & Vidyasagar, 1989).

Many control algorithms proposed for these systems are based on models where practical situations like parameter uncertainty, external disturbances, or friction force terms are not taken into account. In addition, a complete availability of the state variables is commonly assumed (Paden & Panja, 1988; Takegaki & Arimoto, 1981; Wen & Bayard, 1988). In practice, however, the position is usually the only available measurement. In consequence, the velocity, which may play an important role in the control strategy, must be calculated indirectly, often yielding an inaccurate estimation.

In (Makkar et al., 2007), a tracking controller that includes a new differentiable friction model with uncertain nonlinear terms is developed for Euler-Lagrange systems. The technique is based on a model and the availability of the full state. In (Patre et al., 2008), a similar idea is presented for systems perturbed by external disturbances. Moreover, some robust controllers have been proposed to cope with parameter uncertainty and external disturbances. H_∞ control has been a particularly important approach. In this technique, the control objective is expressed as a mathematical optimization problem where a ratio between some norms of output and perturbation signals is minimized (Isidori & Astolfi, 1992). It is used to synthesize controllers achieving robust performance of linear and nonlinear systems.

In general, the control techniques mentioned before yield good control performance. However, the mathematical operations needed to calculate the control signal are rather complex, possibly due to the compensation of gravitational, centrifugal, or Coriolis terms, or the need to solve a Hamilton-Jacobi-Isaacs equation. In addition, if an observer is included in the control system, the overall controller may become rather complex.

Another method exhibiting good robustness properties is the sliding mode technique (Perruquetti & Barbot, 2002; Utkin, 1992). In this method, a surface in the state space is made attractive and invariant using discontinuous terms in the control signal, forcing the system to converge to the desired equilibrium point placed on this surface, and making the controlled dynamics independent from the system parameters. These controllers display good performance for regulation and tracking objectives (Utkin et al., 1999; Weibing & Hung, 1993;

Yuzhuo & Flashner, 1998). Unfortunately, they often exhibit the chattering phenomenon, displaying high-frequency oscillations due to delays and hysteresis always present in practice. The high-frequency oscillations produce negative effects that may harm the control devices (Utkin et al., 1999). Nevertheless, possibly due to the good robust performance of sliding mode controllers, several solutions to alleviate or eliminate chattering have been developed for some classes of systems (Bartolini et al., 1998; Curk & Jezernik, 2001; Erbatur & Calli, 2007; Erbatur et al., 1999; Pushkin, 1999; Sellami et al., 2007; Xin et al., 2004; Wang & Yang, 2007).
In the previous works, it is also assumed that the full state vector is available. However, in practice it is common to deal with systems where only some states are measured due to technological or economical limitations, among other reasons. This problem can be solved using observers, which are models that, based on input-output measurements, estimate the state vector.
To solve the observation problem of uncertain systems, several approaches have been developed (Davila et al., 2006; Rosas et al., 2006; Yaz & Azemi, 1994), including sliding mode techniques (Aguilar & Maya, 2005; Utkin et al., 1999; Veluvolu et al., 2007). The sliding mode observers open the possibility to use the equivalent output injection to identify disturbances (Davila et al., 2006; Orlov, 2000; Rosas et al., 2006).
In this chapter, we describe a control structure designed for mechanical systems to solve regulation and tracking objectives (Rosas et al., 2010). The control technique used in this structure is combined with a discontinuous observer. It exhibits good performance with respect to parameter uncertainties and external disturbances. Because of the included observer, the structure needs only the generalized position and guarantees a good convergence to the reference with a very small error and a control signal that reduces significantly the chattering phenomenon. The observer estimates not only the state vector but, using the equivalent output injection method, it estimates also the plant perturbations produced by parameter uncertainties, non-modeled dynamics, and other external torques. This estimated perturbation is included in the controller to compensate the actual disturbances affecting the plant, improving the performance of the overall control system.
The robust control structure is designed in a modular way and can be easily programed. Moreover, it can be implemented, if needed, with analog devices from a basic electronic circuit having the same structure for a wide class of mechanical systems, making its analog implementation also very easy (Alvarez et al., 2009). Some numerical and experimental results are included, describing the application of the control structure to several mechanical systems.

2. Control objective

Let us consider a mechanical system with $n-$degree of freedom (DOF), modeled by

$$M(q)\ddot{q} + C(q,\dot{q})\dot{q} + G(q) + \Phi(q,\dot{q},\ddot{q})\theta + \gamma(t) = u = \tau_0 + \Delta_\tau. \qquad (1)$$

$q \in \mathbb{R}^n, \dot{q} = dq/dt, \ddot{q} = d^2q/dt^2$ denote the position, velocity, and acceleration, respectively; M and C are the inertia and Coriolis and centrifugal force matrices, G is the gravitational force, $\Phi\theta$ includes all the parameter uncertainties, and γ, which we suppose bounded by a constant σ, that is, $||\gamma(t)|| < \sigma$, denotes a external disturbance. τ_0 and Δ_τ are control inputs. Note that, under this formulation, the terms M, C, and G are well known. If not, it is known that they can be put in a form linear with respect to parameters and can be included in $\Phi\theta$ (Sciavicco & Siciliano, 2000).

We suppose that τ_0, which may depend on the whole state $(q, \dot{q})$, denotes a feedback controller designed to make the state $(q, \dot{q})$ follow a reference signal $(q_r, \dot{q}_r)$, with an error depending on the magnitude of the external disturbance γ and the uncertainty term $\Phi\theta$, but keeping the tracking error bounded. We denote this control as the "nominal control". We propose also to add the term Δ_τ, and design it such that it confers the following properties to the closed-loop system.

1. The overall control $u = \tau_0 + \Delta_\tau$ greatly reduces the steady-state error, provided by τ_0 only, under the presence of the uncertainty θ and the disturbance γ.
2. The controller uses only the position measurement.

Note that, for the nominal control, the steady state error is normally different to zero, usually large enough to be of practical value, and the performance of the closed-loop system may be poor. The role of the additional control term Δ_τ is precisely to improve the performance of the system driven by the nominal control.

The nominal control can be anyone that guarantees a bounded behavior of system (1). In this chapter we use a particular controller and show that, under some conditions, it preserves the boundedness of the state. In particular, suppose the control aim is to make the position q track a smooth signal q_r, and define the plant state as

$$e_1 = q - q_r, \quad e_2 = \dot{q} - \dot{q}_r. \tag{2}$$

Suppose also that the nominal control law is given by

$$\tau_0 = -M(\cdot)\left[K_p e_1 + K_v e_2 - \ddot{q}_r(t)\right] + C(\cdot)(e_2 + \dot{q}_r) + G(\cdot), \tag{3}$$

where K_p and K_v are $n \times n$-positive definite matrices. However, because the velocity is not measured, we need to use an approximation for the velocity error, which we denote as $\hat{e}_2 = \dot{\hat{q}} - \dot{q}_r$. This will be calculated by an observer, whose design is discussed in the next section. Suppose that the exact velocity error and the estimated one are related by $e_2 = \hat{e}_2 + \epsilon_2$. Then, if we use the estimated velocity error, the practical nominal control will be given by

$$\hat{\tau}_0 = -M(\cdot)(K_p e_1 + K_v \hat{e}_2 - \ddot{q}_r) + \hat{C}(\cdot)(\hat{e}_2 + \dot{q}_r) + G(\cdot). \tag{4}$$

Moreover, the approximated Coriolis matrix $\hat{C}$ can be given the form

$$\hat{C}(\cdot) = C(q, \dot{\hat{q}}) = C(\cdot, \hat{e}_2 + \dot{q}_r) = C(\cdot, e_2 + \dot{q}_r) - \Delta C(\cdot),$$

where $\Delta C = \mathcal{O}(\|\epsilon_2\|)$. Then the state space representation of system (1), with the control law (4), is given by

$$\begin{aligned} \dot{e}_1 &= e_2, \\ \dot{e}_2 &= -K_p e_1 - K_v e_2 + \xi(e, t) + \Delta u, \end{aligned} \tag{5}$$

where

$$\xi(\cdot) = -M^{-1}\left[(\hat{C} - MK_v)\epsilon_2 + \Delta C(e_2 + \dot{q}_r) + \Phi\theta + \gamma\right], \tag{6}$$

and $\Delta u = M^{-1}(\cdot)\Delta_\tau$ is a control adjustment to robustify the closed-loop system. When $\Delta u = 0$, a well established result is that, if

$$||\xi(e,t)|| < \rho_1||e|| + \rho_0, \;\; \rho_i > 0, \tag{7}$$

then there exist matrices K_p and K_v such that the state e of system (5) is bounded (Khalil, 2002). In fact, the bound on the state e can be made arbitrarily small by increasing the norm of matrices K_p and K_v.

The control objective can now be established as design a control input Δu that, depending only on the position, improves the performance of the control $\hat{\tau}_0$ by attenuating the effect of parameter uncertainty and disturbances, concentrated in ξ.

Note that disturbances acting on system (5) satisfy the matching condition (Khalil, 2002). Hence, it is theoretically possible to design a compensation term Δu to decouple the state e_1 from the disturbance ξ. The problem analyzed here is more complicated, however, because the velocity is not available.

In the next Section we solve the problem of velocity estimation using two observers that guarantee convergence to the states (e_1, e_2). Moreover, an additional property of these observers will allow us to have an estimation of the disturbance term ξ. This estimated perturbation will be used in the control Δu to compensate the actual disturbances affecting the plant.

3. Observation of the plant state

In this section we describe two techniques to estimate the plant state, yielding exponentially convergent observers.

3.1 A discontinuous observer

Discontinuous techniques for designing observers and controllers have been intensively developed recently, due to their robustness properties and, in some cases, finite-time convergence. In this subsection we describe a simple technique, just to show the observer performance.

The observer has been proposed in (Rosas et al., 2006). It guarantees exponential convergence to the plant state, even under the presence of some kind of uncertainties and disturbances.

Let us consider the system (5). The observer is described by

$$\begin{bmatrix} \dot{\hat{e}}_1 \\ \dot{\hat{e}}_2 \end{bmatrix} = \begin{bmatrix} \hat{e}_2 + C_2\epsilon_1 \\ -K_p e_1 - K_v \hat{e}_2 + \Delta u + C_1\epsilon_1 + C_0 \text{sign}(\epsilon_1) \end{bmatrix}, \tag{8}$$

where $\hat{e}_1 \in \mathbb{R}^n$ and $\hat{e}_2 \in \mathbb{R}^n$ are the states of the observer, $\epsilon_1 = e_1 - \hat{e}_1$. C_0, C_1, and C_2 are diagonal, positive-definite matrices defined by

$$C_i = \text{diag}\{c_{i1}, c_{i2}, \ldots, c_{in}\} \qquad \text{for } i = 0, 1, 2.$$

The signum vector function $\text{sign}(\cdot)$ is defined as

$$\text{sign}(v) = [\text{sign}(v_1), \text{sign}(v_2), \ldots, \text{sign}(v_n)]^T.$$

Then, the dynamics of the observation error $\epsilon = (\epsilon_1, \epsilon_2) = (e_1 - \hat{e}_1, e_2 - \hat{e}_2)$, are described by

$$\begin{bmatrix} \dot{\epsilon}_1 \\ \dot{\epsilon}_2 \end{bmatrix} = \begin{bmatrix} \epsilon_2 - C_2\epsilon_1 \\ -C_1\epsilon_1 - K_v\epsilon_2 - C_0\text{sign}(\epsilon_1) + \xi(e,t) \end{bmatrix}. \tag{9}$$

An important result is provided by (Rosas et al., 2006) for the case where $\rho_1 = 0$ (see equation (7)). Under this situation we can establish the conditions to have a convergence of the estimated state to the plant state.

Theorem 1. *(Rosas et al., 2006) If (7) is satisfied with $\rho_1 = 0$, then there exist matrices C_0, C_1, and C_2, such that system (9) has the origin as an exponentially stable equilibrium point. Therefore, $\lim_{t\to\infty} \hat{e}(t) = e(t)$.*

The proof of this theorem can be found in (Rosas et al., 2006). In fact, a change of variables given by $v_1 = \epsilon_1, v_2 = \epsilon_2 - C_2\epsilon_1$, allows us to express the dynamics of system (9) by

$$\begin{aligned} \dot{v}_1 &= v_2, \\ \dot{v}_2 &= -(C_1 + K_vC_2)v_1 - (C_2 + K_v)v_2 - C_0\text{sign}(v_1) + \xi(e,t), \end{aligned} \tag{10}$$

where v_1 and v_2 are vectors with the form

$$v_i = (v_{i1}, v_{i2}, \ldots, v_{in})^T; \;\; i = 1, 2.$$

Then system (10) can be expressed as a set of second-order systems given by

$$\begin{aligned} \dot{v}_{1i} &= v_{2i}, \\ \dot{v}_{2i} &= -\tilde{c}_{1i}v_{1i} - \tilde{c}_{2i}v_{2i} - c_{0i}\text{sign}(v_{1i}) + \xi_i(\cdot), \end{aligned} \tag{11}$$

where $\tilde{c}_{1i} = c_{1i} + k_{vi}c_{2i}$, $\tilde{c}_{2i} = c_{2i} + k_{vi}$, for $i = 1, \ldots, n$, and $|\xi_i| \leq \beta_i$, for some positive constants β_i. The conditions to have stability of the origin are given by

$$\tilde{c}_{1i} > 0, \tag{12}$$

$$\tilde{c}_{2i} > 0, \tag{13}$$

$$c_{0i} > 2\lambda_{\max}(P_i)\sqrt{\frac{\lambda_{\max}(P_i)}{\lambda_{\min}(P_i)}}\left(\frac{\tilde{c}_{1i}\beta_i}{\theta}\right), \tag{14}$$

for some $0 < \theta < 1$, where P_i is a 2×2 matrix that is the solution of the Lyapunov equation $A_i^T P_i + P_i A_i = -I$, and the matrix A_i is defined by

$$A_i = \begin{bmatrix} 0 & 1 \\ -\tilde{c}_{1i} & -\tilde{c}_{2i} \end{bmatrix}.$$

System (10) displays a second-order sliding mode (Perruquetti & Barbot, 2002; Rosas et al., 2010) determined by $v_1 = \dot{v}_1 = \ddot{v}_1 = 0$. To determine the behavior of the system on the sliding surface, the equivalent output injection method can be used (Utkin, 1992), hence

$$\ddot{v}_1 = -u_{eq} + \xi(e,t) = 0, \tag{15}$$

where u_{eq} is related to the discontinuous term $C_0\text{sign}(v_1)$ of equation (10). The equivalent output injection u_{eq} is then given by (Rosas et al., 2010; Utkin, 1992)

$$u_{eq} = \xi(e,t). \tag{16}$$

This means that the equivalent output injection corresponds to the perturbation term, which can be recovered by a filter process (Utkin, 1992). In fact, in this reference it is shown that the equivalent output injection coincides with the slow component of the discontinuous term in (10) when the state is in the discontinuity surface. Hence, it can be recovered using a low pass filter with a time constant small enough as compared with the slow component response, yet sufficiently large to filter out the high rate components.
For example, we can use a set of n second-order, low-pass Butterworth filter to estimate the term u_{eq}. These filters are described by the following normalized transfer function,

$$F_i(s) = \frac{\omega_{c_i}^2}{s^2 + 1.4142\omega_{c_i}s + \omega_{c_i}^2}, \qquad i = 1,\ldots,n, \tag{17}$$

where ω_{c_i} is the cut-off frequency of each filter. Here, the filter input is the discontinuous term of the observer, $c_{0_i}\text{sign}(v_{1i})$. By denoting the output of the filter set of as $x_f \in \mathbb{R}^n$, and choosing a set of constants ω_{c_i} that minimizes the phase-delay, it is possible to assume

$$\lim_{t\to\infty} x_f = \tilde{\xi}(\cdot) \approx \xi(\cdot), \tag{18}$$

where $\|\tilde{\xi}(\cdot) - \xi(\cdot)\| \le \tilde{\rho}$ for $\tilde{\rho} \ll \rho_0$.

3.2 An augmented, discontinuous observer

A way to circumvent the introduction of a filter is to use an augmented observer. To simplify the exposition, consider a 1-DOF whose tracking error equations have the form of system (5). An augmented observer is proposed to be

$$\begin{aligned}
\dot{\hat{e}}_1 &= w_1 + c_{21}(e_1 - \hat{e}_1),\\
\dot{w}_1 &= c_{11}(e_1 - \hat{e}_1) + c_{01}\text{sgn}(e_1 - \hat{e}_1),\\
\dot{\hat{e}}_2 &= w_2 + c_{22}(w_1 - \hat{e}_2) - K_p e_1 - K_v \hat{e}_2 + \Delta u,\\
\dot{w}_2 &= c_{12}(w_1 - \hat{e}_2) + c_{02}\text{sgn}(w_1 - \hat{e}_2).
\end{aligned} \tag{19}$$

If we denote the observation error as $\epsilon_1 = e_1 - \hat{e}_1$, $\epsilon_2 = e_2 - \hat{e}_2$, we arrive at

$$\begin{aligned}
\dot{\epsilon}_1 &= -c_{21}\epsilon_1 - w_1 + e_2,\\
\dot{w}_1 &= c_{11}\epsilon_1 + c_{01}\text{sgn}(\epsilon_1),\\
\dot{\epsilon}_2 &= -(K_v + c_{22})\epsilon_2 - w_2 - c_{22}(w_1 - e_2) + \xi,\\
\dot{w}_2 &= c_{12}(w_1 - e_2 + \epsilon_2) + c_{02}\text{sgn}(w_1 - e_2 + \epsilon_2).
\end{aligned} \tag{20}$$

A change of variables given by

$$\begin{aligned}
v_{11} &= \epsilon_1,\\
v_{12} &= -c_{21}\epsilon_1 - w_1 + e_2,
\end{aligned}$$

$$
\begin{aligned}
v_{21} &= w_1 - e_2 + \epsilon_2, \\
v_{22} &= \dot{v}_{21} = -c_{22}v_{21} - K_v\epsilon_2 + \dot{w}_1 - \dot{e}_2 - w_2 + \xi
\end{aligned}
$$

converts the system to

$$
\begin{aligned}
\dot{v}_{11} &= v_{12}, \\
\dot{v}_{12} &= -c_{11}v_{11} - c_{21}v_{12} - c_{01}\text{sgn}(v_{11}) + \dot{e}_2, \\
\dot{v}_{21} &= v_{22}, \\
\dot{v}_{22} &= -\tilde{c}_{12}v_{21} - c_{22}v_{22} - c_{02}\text{sgn}(v_{21}) + \tilde{\xi},
\end{aligned} \tag{21}
$$

where $\tilde{c}_{12} = c_{12} - K_v c_{22}$ and $\tilde{\xi}$ is a disturbance term that we suppose bounded. Under some similar conditions discussed in the previous section, particularly the boundedness of $\dot{e}_2$ and $\tilde{\xi}$, we can assure the existence of positive constants c_{ij} such that v_{ij} converges to zero, so $\hat{e}_1$ converges to e_1, w_1 and $\hat{e}_2$ to e_2, and w_2 converges to the disturbance ξ. This observer Hence we propose to use the redesigned control Δu, or Δ_τ, as (see equation (5))

$$
\Delta u = -w_2 \ \rightarrow -\xi, \quad \Delta_\tau = -M(\cdot)w_2
$$

to attenuate the effect of disturbance ξ in system (5) or in system (1), respectively.

4. The controller

As we mentioned previously, we propose to use the nominal controller (4) because the velocity is not available from a measurement. We can use any of the observers previously described, and replace the velocity e_2 by its estimation, $\hat{e}_2$. The total control is then given by

$$
\tau = \tau_0 + \Delta_\tau = -M(\cdot)\left[\nu + K_p e_1 + K_v\hat{e}_2 - \ddot{q}_r(t)\right] + C(\cdot)(\hat{e}_2 + \dot{q}_r) + G(\cdot), \tag{22}
$$

where ν is the redesigned control. This control adjustment is proposed to be $\nu = x_f$, where x_f is the output of filter (17), if the first observer is used (system (8)), or $\nu = w_2$, where w_2 is the last state of system (19), if the second observer is chosen.
The overall structure is shown in figure 1 when the first observer is used.
A similar structure is used for the second observer. An important remark is that the nominal control law (a PD-controller with compensation of nonlinearities in this case) can be chosen independently; the analysis can be performed in a similar way. However, this nominal controller must provide an adequate performance such that the state trajectories remain bounded.

5. Control of mechanical systems

To illustrate the performance of the proposed control structure we describe in this section its application to control some mechanical systems, a Mass-Spring-Damper (MSD), an industrial robot, and two coupled mechanical systems which we want them to work synchronized.

5.1 An MSD system

This example illustrates the application of the first observer (equation (8), Section 3.1).
Consider the MSD system shown in figure 2. Its dynamical model is given by equation (1),

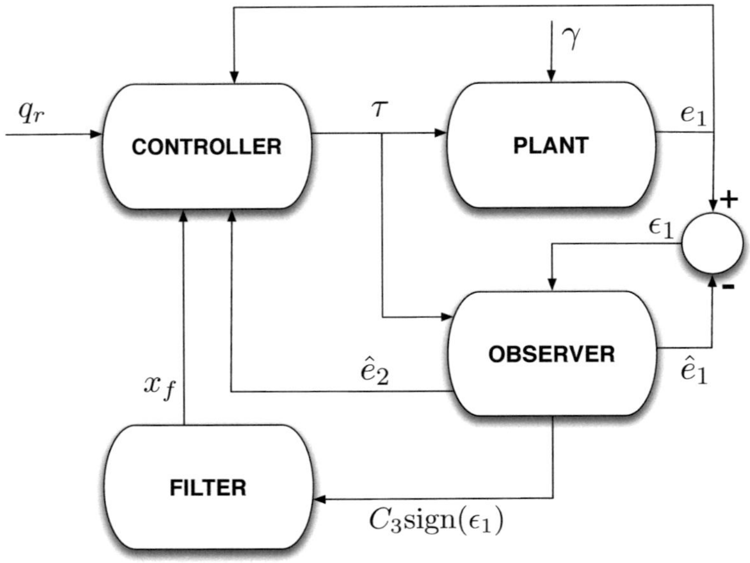

Fig. 1. The robust control structure.

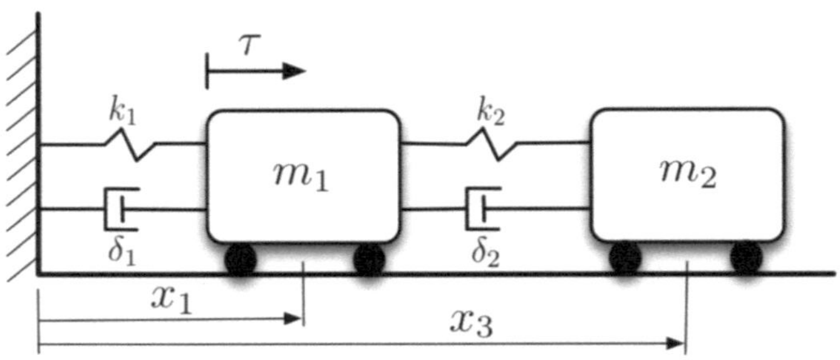

Fig. 2. Mass-spring-damper mechanical system.

with

$$M = \begin{pmatrix} m_1 & 0 \\ 0 & m_2 \end{pmatrix}, \quad C = \begin{pmatrix} \delta_1 + \delta_2 & -\delta_2 \\ -\delta_2 & \delta_2 \end{pmatrix}, \quad G = \begin{pmatrix} (k_1 + k_2)x_1 - k_2 x_3 \\ k_2(x_3 - x_1) \end{pmatrix}, \quad u = \begin{pmatrix} \tau \\ 0 \end{pmatrix},$$

where $x_1 = q_1$, $x_3 = q_2$. Consider that parameters k_i, δ_i, and m_i, for $i = 1, 2$, are known. Note also that the system is underactuated, and only one control input is driving the system at mass m_1. Therefore, we aim to control the position of mass 1 (x_1), and consider that the action of the second mass is a disturbance. Hence, the model of the controlled system is again given by equation (1), but now with $M = m_1$, $C = \delta_1$, $G = k_1 q$. If we denote $x_1 = q$, $x_2 = \dot{q}$, and $x = (x_1, x_2, x_3, x_4) = (x_1, \dot{x}_1, x_3, \dot{x}_3)$ (see figure 2), then

$$\Gamma(x, \dot{x}; \theta) = \Phi(x, \dot{x})\theta + \gamma = k_2(x_1 - x_3) + \delta_2(x_2 - x_4),$$

where x_3 and x_4 are the solutions of the system

$$\dot{x}_3 = x_4,$$
$$\dot{x}_4 = -\frac{k_2}{m_2}(x_3 - x_1) - \frac{\delta_2}{m_2}(x_4 - x_2),$$

groups the effect of uncertainty and disturbance terms $\Phi\theta + \gamma$ of equation (1).
Now denote as $e_1 = x_1 - q_r$, $\hat{e}_2 = \hat{x}_2 - \dot{q}_r$, then the nominal control input τ_0 is proposed as equation (3), that is,

$$\tau_0 = -m_1 \left[K_p e_1 + K_v \hat{e}_2 - \ddot{q}_r(t)\right] + k_1 x_1 + \delta_1 \hat{x}_2, \tag{23}$$

where K_p and K_v are positive constants. Because the velocity is not measured, in (23) we have used the estimation $\hat{x}_2 = \hat{e}_2 + \dot{q}_r$, delivered by the observer given by (8).
With an adequate selection of the constants K_p and K_v we can guarantee that the perturbation $\Gamma(\cdot)$ in (1) is bounded (see Section 2 and (Khalil, 2002)). Therefore, from equation (16), $u_{eq} = \Gamma(\cdot)$.
Using the filter (17), we can recover an estimation of the disturbance, denoted as x_f. Therefore, the redesigned control will be $\Delta_\tau = m_1 x_f$ which, added to (23), adjusts the nominal control input to attenuate the effect of the disturbance Γ.
A numerical simulation was performed with plant parameter values $k_1 = 10\left[k_g m/sec^2\right]$, $k_2 = 20\left[k_g m/sec^2\right]$, $\delta_1 = \delta_2 = 0.1\left[k_g m/sec\right]$, $m_1 = 1\,[kg]$, and $m_2 = 4\,[kg]$. The observer parameter values were set to $c_1 = 2$, $c_2 = 2$, and $c_0 = 3$, with controller gains $K_p = K_v = 10$, and filter frequencies $\omega_c = 500[rad/sec]$. In this simulation the nominal control τ_0 was applied from 0 to 15 sec. The additional control term Δ_τ is activated from 15 to 30 sec. The aim is to track the reference signal $q_r(t) = 0.25\sin(t)$.
Figure 3 shows the response of this controlled system.
Figures a) and b) show the convergence of the observer state to the plant state, in spite of disturbances produced by the mass m_2. Figure c) shows the disturbance identified by this observer. The response of the closed-loop system is presented in Figures d), e), and f). Here we see a tracking error when the additional control term Δ_τ is not present (from 0 to 15 seconds). However, when this term is incorporated to the control signal, at $t = 15$ sec, the tracking error tends to zero. It is important to note that, contrary to typical sliding mode controllers, the control input (Figure 3.f) does not contain high frequency components of large amplitude.

5.2 An industrial robot

This is an example of the application of the first observer (Section 3.1) to a real system.
In this section we show the application of the described technique to control the first two joints of a Selective Compliant Assembly Robot Arm (SCARA), shown in figure 4, used in the manufacturing industry, and manufactured by Sony®.
In this experiment we have an extreme situation because all parameters are unknown. The control algorithm was programed in a PC using the Matlab® software, and the control signals are applied to the robot via a data acquisition card for real-time PC-based applications, the DSpace® 1104. The desired trajectory, which was the same for both joints, is a sinusoidal signal given by $q_r(t) = \sin(t)$.

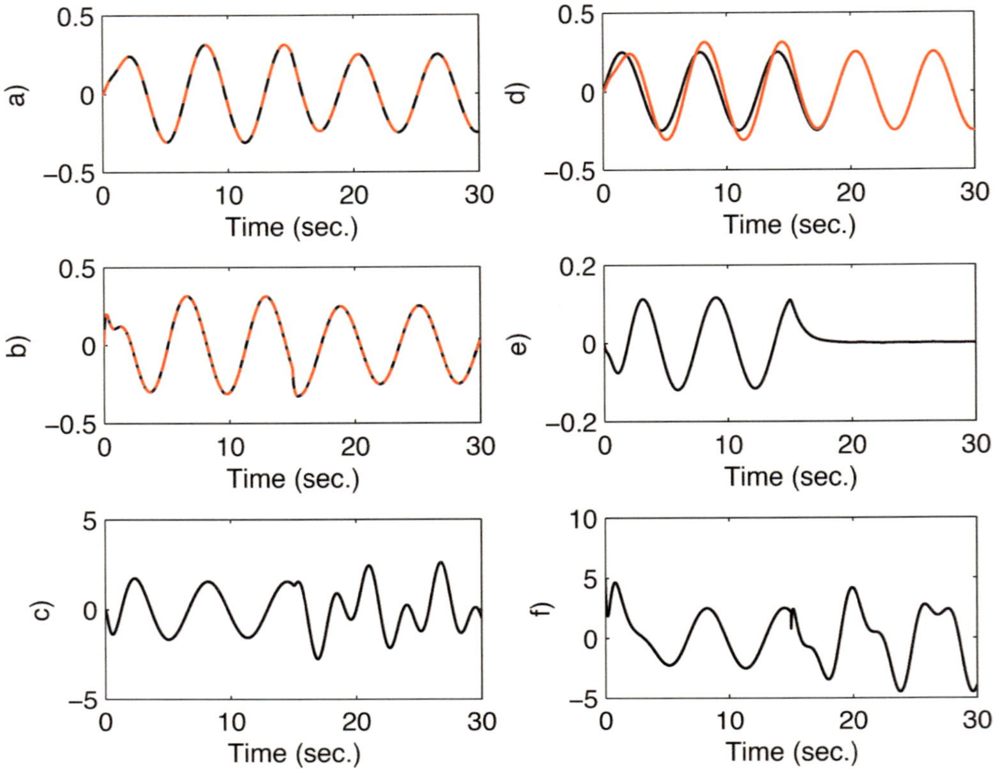

Fig. 3. Response of the closed-loop MSD system. a) x_1 (red) and $\hat{x}_1 = \hat{e}_1 + q_r$ (black); b) x_2 (red) and $\hat{x}_2 = \hat{e}_2 + \dot{q}_r$ (black), c) identified disturbance, x_f, d) reference q_r (black) and position x_1 (red); e) error $e_1 = x_1 - q_r$; f) control $\tau = \tau_0 + \Delta_\tau$.

In the design of the observer (8) the following matrices were selected,

$$C_0 = \begin{bmatrix} 300 & 0 \\ 0 & 300 \end{bmatrix}, \quad C_1 = C_2 = \begin{bmatrix} 25 & 0 \\ 0 & 25 \end{bmatrix}, \quad M^{-1} = \begin{bmatrix} 55.549 & 0 \\ 0 & 55.549 \end{bmatrix}.$$

A cut-off frequency $\omega_{ci} = 75$ rad/seg was selected for the filter(17). The control law is given by the controller (22), where

$$K_p = \begin{bmatrix} 668 & 0 \\ 0 & 391 \end{bmatrix}, K_v = \begin{bmatrix} 379 & 0 \\ 0 & 49 \end{bmatrix}.$$

Note that a nominal value of matrix M was used. Differences between nominal and the actual matrix $M(q)$ are supposed to be included in the perturbation term, as well as the Coriolis, centrifugal, and friction forces, external disturbances, parametric variations and coupling effects.

The perturbation terms $\xi_i(\cdot)$ for $i = 1, 2$ that correspond to perturbations present in the two joints are displayed in Figure 5.

Fig. 4. A SCARA industrial robot.

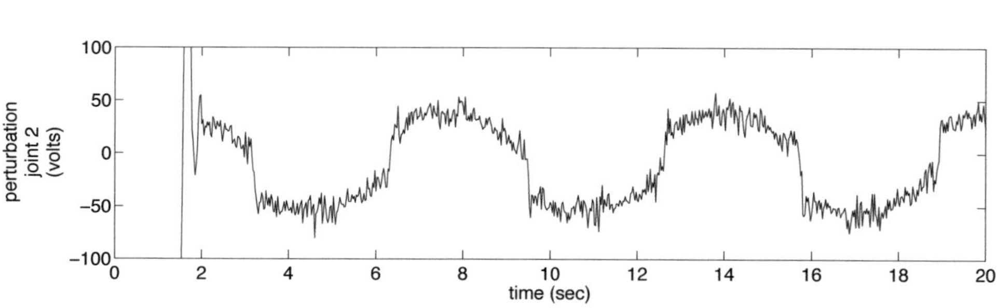

Fig. 5. Identified perturbation terms in the joints of an industrial robot. Up: joint 1 perturbation. Down: joint 2 perturbation.

To verify the observer performance, the observation errors $e_i = \theta_i - \hat{\theta}_i$, for $i = 1,2$, are displayed in Figure 6, showing small steady-state values.

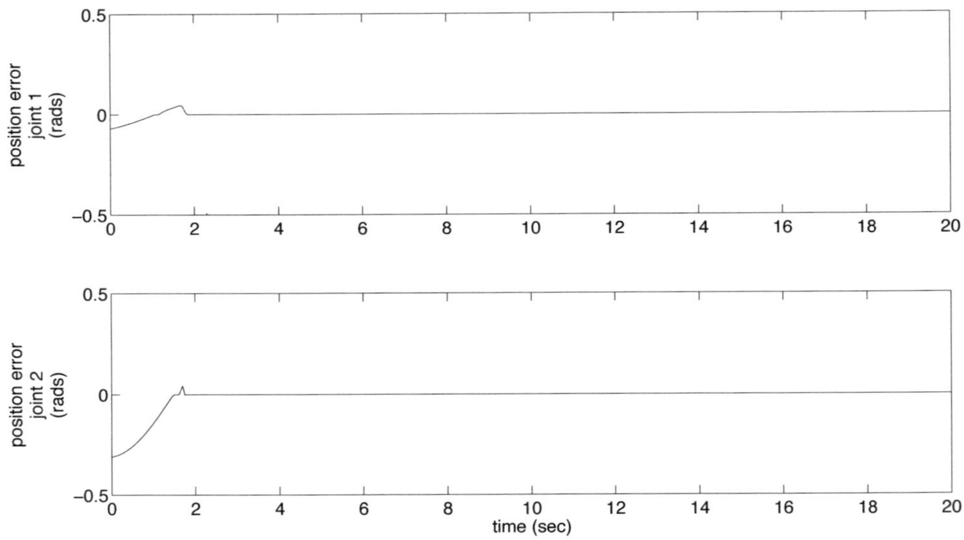

Fig. 6. Observation position errors of the industrial robot.

Figure 7 shows the system output and the reference. Control inputs for joints 1 and 2 are displayed in Figure 8.
Although these control inputs exhibit high frequency components with small amplitude, they do not produce harmful effects on the robot. Also, it is interesting to note that the control input levels remain in the dynamic range allowed by the robot driver, that is, between −12 V and +12 V.

5.3 Two synchronized mechanical systems

This example illustrates the practical performance of the proposed technique, using the augmented observer given by (19). It refers to a basic problem of synchronization.
Synchronization means correlated or corresponding-in-time behavior of two or more processes (Arkady et al., 2003). In some situations the synchronization is a natural phenomenon; in others, an interconnection system is needed to obtain a synchronized behavior or improve its transient characteristics. Hence, the synchronization becomes a control objective and the synchronization obtained in this way is called controlled synchronization (Blekhman et al., 1997). Some important works in this topic are given by (Dong & Mills, 2002; Rodriguez & Nijmeijer, 2004; Soon-Jo & Slotine, 2007).
In this subsection we present a simple application of the control technique to synchronize two mechanisms connected in the basic configuration, called master-slave (see figure 9).
The master system is the MSD described in Section 5.1, manufactured by the company ECP®, model 210, with only the first mass activated. The slave is a torsional system from the same company, with the first and third disks connected. The master sends its position x to the slave, and the synchronization objective is to make the slave track the master state, that is, the

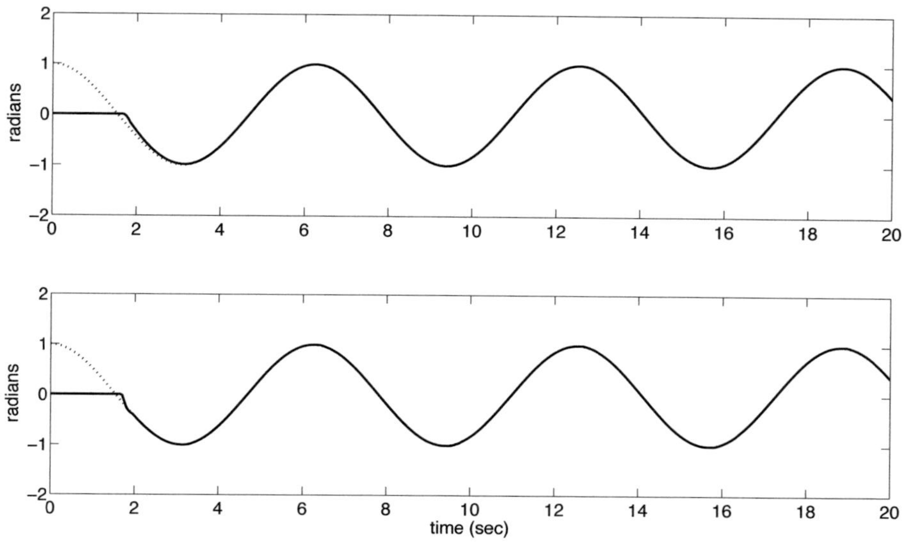

Fig. 7. Reference signal (dotted line) and position (solid line) for each joint of the industrial robot. Up: joint 1. Down: joint 2.

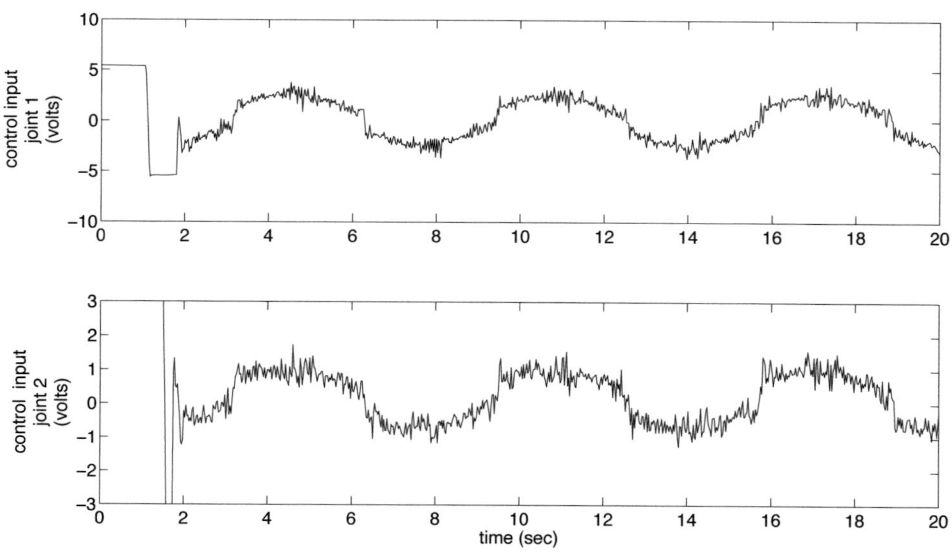

Fig. 8. Control input for each joint of the industrial robot.

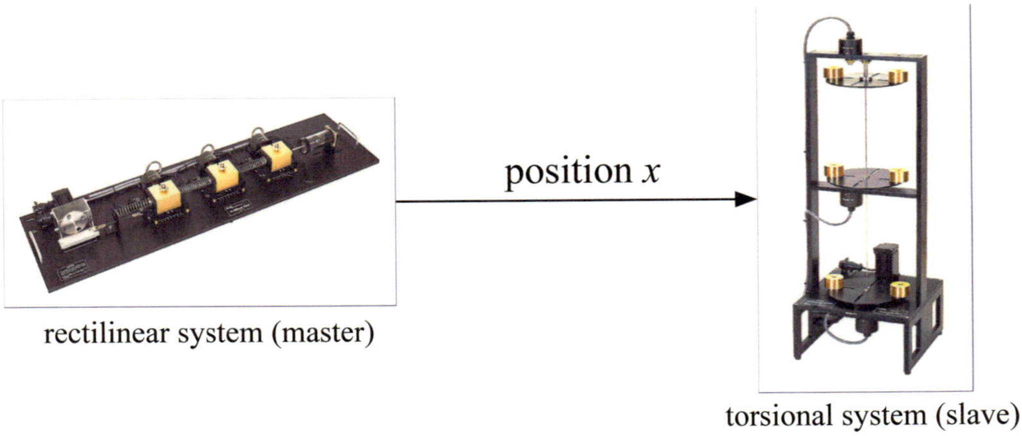

Fig. 9. Two synchronized mechanisms in a master/slave configuration. The master is the rectilinear system, model 210, from ECP®. The slave is the torsional system, model 205, from the same company.

angular position θ and velocity $\dot{\theta}$ of the torsional system must follow the position x and the velocity $\dot{x}$ of the master, respectively. The relation between the two states is 1cm of the master corresponds to 1rad of the slave.
The rectilinear system is modeled by

$$m\ddot{x} + c_m\dot{x} + k_m x + \gamma_m(t) = F(t),$$

where x is the position of the mass; m, c_m, and k_m are the mass, damping, and spring coefficients, respectively, and F is an external force driving the system. The torsional system is described as

$$J\ddot{\theta}_1 + c_t\dot{\theta}_1 + k_t(\theta_1 - \theta_2) + \gamma_t(t) = \tau_0 + \Delta_\tau,$$

where θ_1 and θ_2 are the angular positions of the first and third disks, respectively; J, c_t, and k_t are the inertia, damping, and spring coefficients of the first disk. γ_m and γ_t are external disturbances possibly affecting the systems. The force driving the MSD system is set as $F(t) = 1.5\sin(1.5\pi t)$. All positions are available, but the velocities are estimated with the second observer (19) (see Section 3.2).
The nominal values of the coefficients are given in Table 1.

System	Parameter	Value	Units
MSD	m	1.27	kg
	k_m	200	N/m
	c_m	2.1	N/m/sec
Torsional	J	0.0108	Kg-m^2
	c_t	0.007	N-m/rad/sec
	k_t	1.37	N-m/rad
Observer	$c_{11}, c_{12}, c_{21}, c_{22}$	500	
	c_{01}	50	
	c_{02}	100	

Table 1. Parameter values for the synchronization example.

If we define the synchronization error as

$$e_1 = x - \theta, \;\; e_2 = \dot{x} - \dot{\theta},$$

the control objective es to make $e = (e_1, e_2)$ converge to zero.
Let us consider the nominal control

$$\tau_0 = -J(k_p e_1 + k_v \hat{e}_2) + c_t \hat{\dot{\theta}}_1 + k_t \theta_1 - k_t \theta_2,$$

where $\hat{\dot{\theta}}_1$ and $\hat{e}_2$ are the estimated velocity and the estimated velocity error obtained from the observer. From the last equations it is possible to get the synchronization error dynamics as

$$\begin{aligned} \dot{e}_1 &= e_2 \\ \dot{e}_2 &= -k_p e_1 - k_v e_2 + \Delta u - \xi, \end{aligned}$$

where $\Delta u = J^{-1} \Delta_\tau$ and

$$\xi = (J c_t - k_v)\epsilon_2 + J^{-1}\gamma_t(t) - m^{-1}\left(c_m \dot{x} + k_m x - F(t) + \gamma_m(t)\right),$$

with $\epsilon_2 = e_2 - \hat{e}_2$.
We have then formulated this synchronization problem in the same framework allowing us to design a robust controller. Therefore, we can use one of the observers described previously, and use a redesign control $\Delta_\tau = J\hat{\xi}$.
We describe the results obtained from this controller to synchronize these devices. Figure 10 shows its performance, using the augmented observer (19).

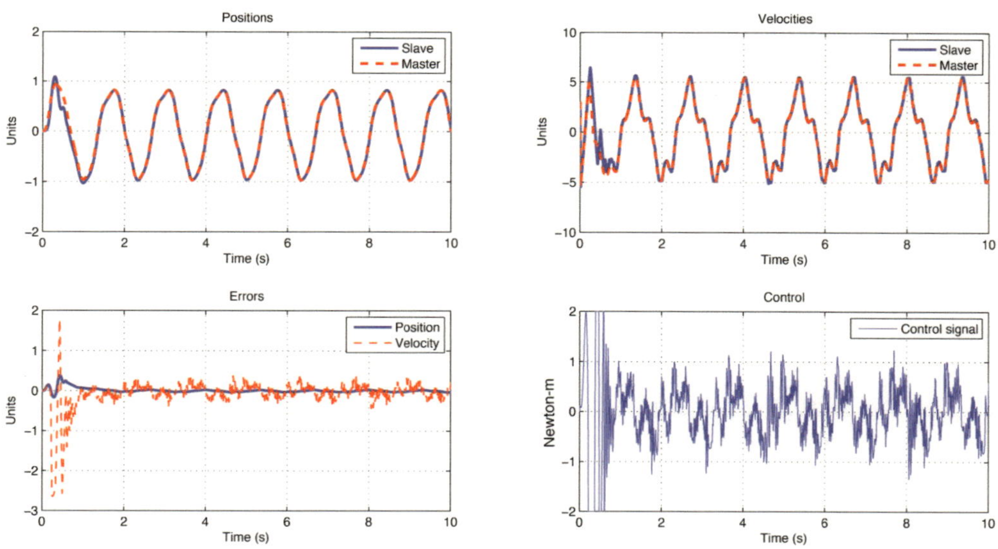

Fig. 10. Responses of the synchronized mechanisms (Figure 9). One unit corresponds to 1 cm (1 rad) for the position, or 1 cm/sec (1 rad/sec) for the velocity, of the master (slave) system.

This figure shows how the slave (torsional) system synchronizes with the master (rectilinear) system in about 1 sec. In 2 sec the synchronization error (position and velocity) is very small.

The control input designed for the slave is saturated at ± 2 N-m, and after 1 sec maintains its values between -1 and $+1$ N-m. This is accomplished even under the presence of the disturbance introduced by the third disk, which is not modeled.

6. Conclusions

A robust control structure for uncertain Lagrangian systems with partial measurement of the state has been presented. This control structure allows us to solve tracking and regulation problems and guarantees the convergence to a small neighborhood of the reference signal, in spite of nonvanishing disturbances affecting the plant.

This technique makes use of robust, discontinuous observers with a simple structure. An important property of these observers is its ability to estimate the disturbances acting on the plant, which can be conveniently incorporated in the control signal to increase the robustness of the controller and decrease the steady-state tracking error. The observer structure can even be built with conventional analog circuits, as it is described in (Alvarez et al., 2009). An adequate tuning of the observer parameters guarantees the convergence to the reference signal in an operation region large enough to cover practical situations.

The numerical simulations and the experimental results described in this chapter exhibited a good performance of the proposed technique, and the control signal showed values inside practical ranges.

An interesting and important problem that has been intensively studied recently is the synchronization of dynamical systems. Synchronization of mechanical systems is important as soon as two or more mechanical systems have to cooperate. The control technique described in this chapter has been applied to the simplest configuration, that is, the master/slave synchronization, exhibiting a good performance. This same control strategy, based on robust observers, can be also successfully applied to synchronize arrays of mechanical systems, connected in diverse configurations. A more detailed application can be found in (Alvarez et al., 2010).

7. Acknowledgement

We thank Jonatan Peña and David A. Hernandez for performing the experiments of Section 5.2 and 5.3.

8. References

Aguilar, R. & Maya, R. (2005), State estimation for nonlinear systems under model uncertainties: a class of sliding-mode observers, J. Process Contr. Vol. (15): 363-370.

Alvarez, J., Rosas, D. & Peña, J. (2009), Analog implementation of a robust control strategy for mechanical systems, IEEE Trans. Ind. Electronics, Vol. (56), No. 9: 3377-3385.

Alvarez, J., Rosas, D., Hernandez, D. & Alvarez, E. (2010), Robust synchronization of arrays of Lagrangian systems, Int. J. of Control, Automation, and Systems, Vol. (8), No. 5:1039-1047.

Arkady, P., Michael, R. & Kurths, J. (2003), Synchronization. A Universal Concept in Nonlinear Sciences, Cambridge Press: Cambridge.

Bartolini, G., Ferrara, A. & Usani, E. (1998), Chattering avoidance by second-order sliding mode control, IEEE Trans. Aut. Ctl. Vol. (43), No. 2: 241-246.

Blekhman, I. I., Fradkov, A. L., Nijmeijer, H. & A. Y. Pogromsky, (1997) On self-synchronization and controlled synchronization, Systems & Control Letters. (31): 299-305.

Brogliato, B., Niculescu, S. I. & Orhant, P. (1997), On the control of finite-dimensional mechanical systems with unilateral constraints, IEEE Trans. Aut. Ctl., Vol. (42), No. 2: 200-215.

Curk, B. & Jezernik, K. (2001), Sliding mode control with perturbation estimation: Application on DD robot mechanism, Robot. Vol. (19): 641-648.

Davila, J., Fridman, L. & Poznyak, A. (2006), Observation and identification of mechanical systems via second order sliding modes, Int. J. Control, Vol. (79), No. 10: 1251-1262.

Dong, S. & Mills, J. K. (2002), Adaptive synchronize control for coordination of multirobot assembly tasks, IEEE Trans. on Robotics and Automation, Vol. (18), No. 4: 498-510.

Erbatur, K., & Calli, B. (2007), Fuzzy boundary layer tuning as applied to the control of a direct drive robot, in Proc. IECON: 2858–2863.

Erbatur, K., Okyay, M. & Sabanovic, A. (1999), A study on robustness property of sliding-mode controllers: A novel design and experimental investigations, IEEE Trans. Ind. Electron., Vol. (46), no. 5: 1012–1018.

Isidori, A., &Astolfi, A. (1992), Disturbance attenuation and H U221e control via measurement feedback in nonlinear systems, IEEE Trans. Aut. Ctl., Vol. (37), No. 9: 1283-1293.

Khalil, H. (2002) Nonlinear Systems, New Jersey: Prentice Hall.

Makkar, C., Hu, G., Sawyer, W. G. & Dixon, W. E. (2007), Lyapunov-based tracking control in the presence of uncertain nonlinear parametrizable friction, IEEE Trans. Aut. Ctl., Vol (52), No. 10: 1988-1994.

Orlov, I. (2000), Sliding mode observer-based synthesis of state derivative-free model reference adaptive control of distributed parameter systems, J. Dyn. Syst-T ASME, Vol. (122), No. 4: 725-731.

Paden, B. & Panja, R. (1988). Globally asymptotically stable PD+ controller for robot manipulators, Int. J. Control Vol (47), No. 6: 1697–1712.

Patre, P. M., MacKunis, W., Makkar, C. & Dixon, W. E. (2008), Asymptotic tracking for systems with structured and unstructured uncertainties, IEEE Trans. Control Syst. Technol., Vol. (16), No. 2: 373-379.

Perruquetti, W. & Barbot, J. (Eds), (2002), Sliding Mode Control in Engineering, New York: Marcel Dekker.

Pushkin, K. (1999), Existence of solutions to a class of nonlinear convergent chattering free sliding mode control systems, IEEE Trans. Aut. Ctl., Vol. (44), No. 8: 1620-1624.

Rodriguez, A. & Nijmeijer, H. (2004), Mutual synchronization of robots via estimated state feedback: a cooperative approach, IEEE Trans. on Control Systems Technology, Vol. (12), No. 4: 542-554.

Rosas, D., Alvarez, J. & Fridman, L. (2006), Robust observation and identification of nDOF Lagrangian systems, Int. J. Robust Nonlin., Vol. (17): 842-861.

Rosas, D., Alvarez, J. & Peña, J. (2010), Control structure with disturbance identification for Lagrangian systems, Int. J. Non-Linear Mech., doi:10.1016/j.ijnonlinmec.2010.08.005.

Sciavicco, L. & Siciliano, B. (2000) Modelling and Control of Robots Manipulators, London: Springer-Verlag.

Sellami, A., Arzelier, D., M'hiri, R. & Zrida, J. (2007), A sliding mode control approach for systems subjected to a norm-bounded uncertainty, Int. J. Robust Nonlin., Vol. (17): 327-346.

Slotine, J. J. & Li, W. (1988). Adaptive manipulator control: A case study, IEEE Trans. Aut. Ctl., Vol. (33): 995-1003.

Soon-Jo, C & Slotine, E. (2007). Cooperative robot control and synchronization of Lagrangian systems, in Proc. 46th IEEE Conference on Decision and Control, New Orleans.

Spong, M. W. & Vidyasagar, M. (1989), Robot Dynamics and Control, New York: Wiley.

Takegaki, M. & Arimoto, S. (1981). A new feedback method for dynamic control manipulators, J. Dyn. Syst. Trans. ASME, Vol (103): 119–125.

Utkin, V. (1992), Sliding Modes in Control and Optimization, New York: Springer-Verlag.

Utkin, V., Guldner, J. & Shi, J. (1999), Sliding Mode Control in Electromechanical Systems, London, U.K.: Taylor & Francis.

Veluvolu, K., Soh, Y. & Cao, W. (2007), Robust observer with sliding mode estimation for nonlinear uncertain systems, IET Control Theory A., Vol. (5): 1533-1540.

Xin, X. J., Jun, P. Y. & Heng, L. T. (2004), Sliding mode control whith closed-loop filtering architecture for a class of nonlinear systems, IEEE T. Circuits-1, Vol. (51), No. 4: 168-173.

Wang, X. & Yang, G. (2007), Equivalent sliding mode control based on nonlinear observer for nonlinear non-minimum-phase systems, J. Dyn. Control. Syst., Vol. (13), No. 1: 25-36.

Weibing, G. & Hung, J. C., (1993), Variable structure control of nonlinear systems: A new approach, IEEE Trans. Ind. Electronics, Vol. (40), No. 1: 45-55.

Wen, J. & Bayard, D. (1988). New class of control law for robotic manipulators. Part 1: non-adaptive case, Int. J. Control, Vol (47), No. 5: 1361–1385.

Yaz, E. & Azemi, A. (1994), Robust/adaptive observers for systems having uncertain functions with unknown bounds, Proc. 1994 American Control Conference, Vol. (1): 73-74.

Yuzhuo, W. & Flashner, H., (1998), On sliding mode control of uncertain mechanical systems, in Proc. IEEE Aerosp. Conf., Aspen, CO, 85–92.

9

Robust Control of Electro-Hydraulic Actuator Systems Using the Adaptive Back-Stepping Control Scheme

Jong Shik Kim, Han Me Kim and Sung Hwan Park
School of Mechanical Engineering, Pusan National University
Republic of Korea

1. Introduction

Conventional hydraulic actuator (CHA) systems have been widely used as power units because they can generate very large power compared to their size. In general, a CHA system consists of an electric motor, a pump, a reservoir, various valves, hoses, which are used to transfer the working fluid and an actuator. CHA systems, however, have some problems such as environmental pollution caused by the leakage of the working fluid, maintenance load, heavy weight and limited installation space. These shortcomings can be overcome by compactly integrating the components of CHA systems and by applying a suitable control scheme for the electric motor. To overcome these shortcomings of CHA systems, electro-hydraulic actuator (EHA) systems have been developed, having merits such as smaller size, higher energy efficiency and faster response than existing CHA systems (Kokotovic, 1999). However, for the robust position control of EHA systems, system uncertainties such as the friction between the piston and cylinder and the pump leakage coefficient have to be considered.

To solve these system uncertainty problems of EHA systems and to achieve the robustness of EHA systems with system disturbance and bounded parameter uncertainties, Wang et. al. presented a sliding mode control and a variable structure filter based on the variable structure system (Wang, 2005). Perron et. al proposed a sliding mode control scheme for the robust position control of EHA systems showing volumetric capacity perturbation of the pump (Perron, 2005). However, these control methods have some chattering problem due to the variable structure control scheme. The chattering vibrates the system and may reduce the life cycle of the system. Jun et. al. presented a fuzzy logic self-tuning PID controller for regulating the BLDC motor of EHA systems which has nonlinear characteristics such as the saturation of the motor power and dead-zone due to the static friction (Jun, 2004). Chinniah et. al. used a robust extended Kalman filter, which can estimate the viscous friction and effective bulk modulus, to detect faults in EHA systems (Chinniah, 2006). Kaddissi et. al. applied a robust indirect adaptive back-stepping control (ABSC) scheme to EHA systems having perturbations of the viscous friction coefficient and the effective bulk modulus due to temperature variations (Kaddissi, 2006). However, in spite of the variation of the effective bulk modulus due to the temperature and pressure variations, Chinniah et. al. considered only the case of constant effective bulk modulus and Kaddissi et. al. used EHA systems that are not controlled by an electric motor but by a servo valve.

In this chapter, an ABSC scheme was proposed for EHA systems to obtain the desired tracking performance and the robustness to system uncertainties. Firstly, to realize a stable back-stepping control (BSC) system with a closed loop structure and to select new state variables, EHA system dynamics are represented with state equations and error equations. Defining the Lyapunov control functions, we can design a BSC system, which can guarantee exponential stability for the nominal system without system uncertainties. However, the BSC system cannot achieve robustness to system uncertainties. To overcome the drawback of the BSC system, an ABSC scheme for EHA position control systems with classical discrete disturbance observer was proposed. To evaluate the tracking performance and robustness of the proposed EHA position control system, both BSC and ABSC schemes were evaluated by computer simulation and experiment.

2. System modeling of EHA system

Figure 1 shows the simplified schematic diagram of an EHA system that consists of an electric servo motor, bi-directional gear pump and actuator. The servo motor rotates the gear pump, which, in turn, generates the flow rate. The pressure generated by the flow rate changes the position of the piston rod. The movement direction of the piston is related to the rotational direction of the servo motor. The chamber volumes of the actuator depend on the cross sectional area and the displacement of the piston as follows

$$\begin{cases} V_A(t) = V_{0A} + Ax(t) \\ V_B(t) = V_{0B} - Ax(t) \end{cases} \tag{1}$$

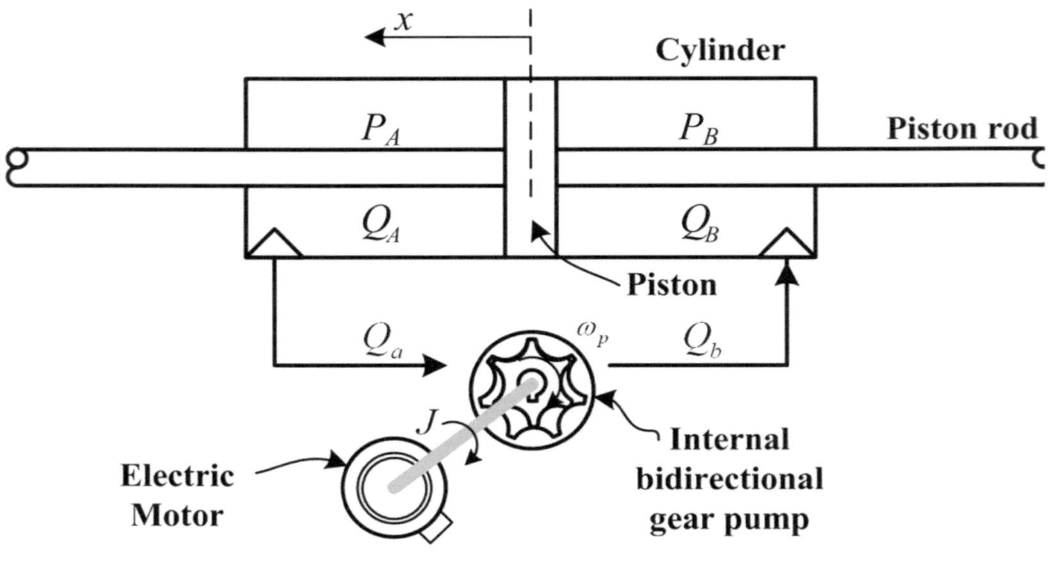

Fig. 1. Simplified schematic diagram of an EHA system

where V and V_0 are the chamber volume and the initial chamber volume, respectively, A and x are the pressure area of a double rod hydraulic cylinder and displacement of the piston, respectively, and subscripts 'A' and 'B' denote the chamber notations of the actuator. Considering the fluid compressibility and continuity principle for the actuator, the flow rate equations of both ports of the actuator can be represented as (Merritt, 1967)

$$\begin{cases} Q_A = A\dot{x} + \dfrac{V_{0A} + Ax}{\beta_e}\dot{P}_A + LP_A \\ Q_B = A\dot{x} - \dfrac{V_{0B} - Ax}{\beta_e}\dot{P}_B - LP_B \end{cases} \tag{2}$$

where Q is the flow rate in the actuator, β_e is the effective bulk modulus of the working fluid, and L and P are the actuator external leakage coefficient and the pressure in the chamber, respectively.

It is assumed that there is no fluid leakage of conduits because the conduits of EHA systems are very short and hard. Then, (2) can be expressed as

$$\begin{cases} Q_A = A\dot{x} + \dfrac{V_{0A} + Ax}{\beta_e}\dot{P}_A \\ Q_B = A\dot{x} - \dfrac{V_{0B} - Ax}{\beta_e}\dot{P}_B \end{cases} \tag{3}$$

The electric motor, directly connected to the hydraulic pump, changes the flow direction and adjusts the flow rate through the ports. In addition, the pressure generated by the continuous supply of flow in the actuator can produce a minute fluid leakage of the pump. Hence, the equations for the fluid leakage of the pump are expressed as

$$\begin{cases} Q_a = C_p\omega_p - L_f P_L \\ Q_b = -Q_a \end{cases} \tag{4}$$

where Q is the flow rate of the pump, whose subscripts a and b denote the ports of the pump, C_p is the volumetric capacity of the pump, ω_p is the rotational velocity of the electric motor, L_f is the leakage factor of the pump and the load pressure $P_L = P_A - P_B$. From (4), the inflow and outflow of the pump are expressed as functions of the rotational velocity ω_p.

In addition, the actuator dynamic equation of EHA systems is expressed as

$$(P_A - P_B)A = M\ddot{x} + F_f + F_{ex} \tag{5}$$

where M and x are the mass and displacement of the piston, respectively, F_f is the friction force between the cylinder and piston and F_{ex} is the external disturbance force.

In order to substitute (3) into (5), the derivative of (5) is expressed as

$$(\dot{P}_A - \dot{P}_B)A = M\dddot{x} + \dot{F}_f + \dot{F}_{ex} \tag{6}$$

In addition, it is assumed that the conduits connected between the actuator ports and the pump ports are very short. Then, the flow rates in (3) and (4) can be represented as $Q_A = Q_a$

and $Q_B = Q_b$. Substituting (1) through (4) into (6), therefore, the dynamic equation of EHA systems can be represented as

$$\ddot{x} = -\frac{1}{M}\left\{\beta_e A^2\left(\frac{1}{V_A}+\frac{1}{V_B}\right)\dot{x} + \dot{F}_f + \dot{F}_{ex}\right\} + \frac{\beta_e A}{M}\left(\frac{1}{V_A}+\frac{1}{V_B}\right)\left(L_f P_L - C_p \omega_p\right) \tag{7}$$

To represent the characteristics of the friction F_f between the piston and cylinder, the LuGre friction model is considered. The LuGre friction model is based on bristles analysis, which is represented with the average deflection force of bristles stiffness. The deflection displacement equation of bristles z, which is actually unmeasurable by experiment, is expressed as (Choi, 2004)(Lee, 2004)

$$\frac{dz}{dt} = \dot{x} - \frac{\sigma_0 |\dot{x}|}{g(\dot{x})} z \tag{8}$$

where σ_0 is the bristles stiffness coefficient, z is the unmeasurable internal state and the nonlinear function $g(\dot{x})$ depends on the material property, grade of lubrication and temperature; that is

$$g(\dot{x}) = F_c + F_s e^{-(\dot{x}/v_{sv})^2} \tag{9}$$

where F_c, F_s and v_{sv} represent the Coulomb friction force, static friction force, and Stribeck velocity between the cylinder and piston, respectively.

If the relative velocity of the contact materials increases gradually, the friction force decreases instantaneously and then it increases gradually again; this effect is called the Stribeck effect and the relative velocity is the Stribeck velocity. This phenomenon depends on the material property, grade of lubrication and temperature. The friction force F_f can be represented with the average deflection z and the velocity of the piston $\dot{x}$ as follows

$$F_f = \sigma_0 z + \sigma_1 \dot{z} + \mu \dot{x} \tag{10}$$

where σ_1 and μ represent the bristles damping and viscous friction coefficients, respectively.

3. Controller design for the EHA position control systems

The EHA position control system consists of the inner loop for the angular velocity control of the servo motor/pump and the outer loop for the position control of the piston. For the velocity control of the motor in the inner loop, Kokotovic et al. applied an adaptive control scheme so that the EHA position control systems can have robustness (Kokotovic, 1999). Habibi et. al. presented that if the inner loop dynamics are stable, the control gains of the PID velocity controller in the inner loop can have relatively large values and then the disturbance effect can be sufficiently rejected (Habibi, 1999). The velocity controller in the inner loop is very important because it regulates the electric motor. However, the case of (Kokotovic, 1999) is very complicated and the case of (Habibi, 1999), although it is theoretically possible, has a physical limitation that increases the control gains of the inner loop controller. Therefore, it is desirable to handle the controller in the outer loop rather than in the inner loop to improve the performance and robustness of EHA position control systems.

In this chapter, the BSC and ABSC schemes based on EHA system dynamics are considered as the position controller. Firstly, to design a BSC system, (7) is transformed to a general form (Slotine, 1999) as follows

$$\ddot{x} = f + bu \tag{11}$$

where

$$f = -\frac{1}{M}\left\{\beta_e A^2\left(\frac{1}{V_A}+\frac{1}{V_B}\right)\dot{x} + \dot{F}_f + \dot{F}_{ex} - \beta_e L_f A\left(\frac{1}{V_A}+\frac{1}{V_B}\right)P_L\right\},$$

$$b = -\frac{\beta_e A(V_A+V_B)C_p}{MV_AV_B},$$

$$u = \omega_p$$

Now, let (11) represent state equations as follows

$$\begin{cases}\dot{x}_1 = x_2\\ \dot{x}_2 = x_3\\ \dot{x}_3 = f + bu\end{cases} \tag{12}$$

And, in order to design the BSC system, new state variables are defined as follows

$$z_1 = x_1 - x_d \tag{13}$$

$$z_2 = x_2 - \alpha_1(z_1) \tag{14}$$

$$z_3 = x_3 - \alpha_2(z_1, z_2) \tag{15}$$

where x_d is the desired position input, and α_1 and α_2 are the functions for new state variables, which can be obtained through the following BSC design procedure.

Step 1.

From (13), the state equation for z_1 can be described as

$$\dot{z}_1 = z_2 + \alpha_1(z_1) - \dot{x}_d \tag{16}$$

$\alpha_1(z_1)$ is the virtual control which should be selected to guarantee the stability of the control system through the Lyapunov control function(LCF) which is defined as

$$V_1(z_1) = \frac{1}{2}z_1^2 \tag{17}$$

Then,

$$\dot{V}_1(z_1) = z_1\dot{z}_1 = z_1[\alpha_1(z_1) - \dot{x}_d] + z_1z_2 \tag{18}$$

From (18), if $\alpha_1(z_1) = -k_1z_1 + \dot{x}_d$, (16) can be exponentially stable when $t \to \infty$. And $k_1(>0)$ is a design parameter.

Step 2.
From (14), the state equation for z_2 can be described as

$$\dot{z}_2 = z_3 + \alpha_2(z_1, z_2) - \dot{\alpha}_1(z_1) \tag{19}$$

where

$$\dot{\alpha}_1(z_1) = \frac{\partial \alpha_1}{\partial z_1}\dot{z}_1 + \frac{\partial \alpha_1}{\partial \dot{x}_d}\ddot{x}_d = -k_1\dot{z}_1 + \ddot{x}_d = -k_1(z_2 - k_1 z_1) + \ddot{x}_d$$

Since (19) includes the information of (16), the second LCF for obtaining the virtual control to guarantee the stability of the control system can be selected as

$$V_2(z_1, z_2) = V_1(z_1) + \frac{1}{2}z_2^2 \tag{20}$$

Then,

$$\dot{V}_2(z_1, z_2) = \dot{V}_1(z_1) + \dot{z}_2 z_2 = -k_1 z_1^2 + z_2 z_3 + z_2[z_1 + \alpha_2(z_1, z_2) - \dot{\alpha}_1(z_1)] \tag{21}$$

If the virtual control α_2 in the last term of (21) is defined as

$$\alpha_2(z_1, z_2) = -k_2 z_2 - z_1 + \dot{\alpha}_1(z_1)$$

where $k_2(>0)$ is a design parameter, then another expression of α_2 can be rearranged as

$$\alpha_2(z_1, z_2) = -(k_1 + k_2)z_2 - (1 - k_1^2)z_1 + \ddot{x}_d \tag{22}$$

Therefore

$$\dot{V}_2 = z_2 z_3 - k_1 z_1^2 - k_2 z_2^2 \tag{23}$$

Step 3.
From (15), the state equation for z_3 is described as

$$\dot{z}_3 = \dot{x}_3 - \dot{\alpha}_2(z_1, z_2) = f + bu - \dot{\alpha}_2(z_1, z_2) \tag{24}$$

where

$$\dot{\alpha}_2(z_1, z_2) = \frac{\partial \alpha_2}{\partial z_1}\dot{z}_1 + \frac{\partial \alpha_2}{\partial z_2}\dot{z}_2 + \frac{\partial \alpha_2}{\partial \dot{\alpha}_1}\ddot{\alpha}_1 = -\dot{z}_1 - k_2\dot{z}_2 + \ddot{\alpha}_1 \tag{25}$$

and

$$\ddot{\alpha}_1 = \frac{\partial \dot{\alpha}_1}{\partial z_1}\dot{z}_1 + \frac{\partial \dot{\alpha}_1}{\partial z_2}\dot{z}_2 + \frac{\partial \dot{\alpha}_1}{\partial \ddot{x}_d}\dddot{x}_d = k_1^2\dot{z}_1 - k_1\dot{z}_2 + \dddot{x}_d \tag{26}$$

Substituting (16), (19), and (26) into (25), (25) can be rearranged as

$$\dot{\alpha}_2(z_1, z_2) = (k_1^2 - 1)\dot{z}_1 - (k_1 + k_2)\dot{z}_2 + \dddot{x}_d \tag{27}$$

Since (24) uses the information of z_1 and z_2 due to the property of the design procedure of the back-stepping control, the third LCF for (15) can be defined as

$$V_3(z_1,z_2,z_3) = V_2(z_1,z_2) + \frac{1}{2}z_3^2 \tag{28}$$

Then,

$$\dot{V}_3(z_1,z_2,z_3) = \dot{V}_2 + z_3\dot{z}_3 = -k_1 z_1^2 - k_2 z_2^2 + z_3(z_2 + f + bu - \dot{\alpha}_2) \tag{29}$$

If the last term of (29) for satisfying the system stability is defined as

$$-k_3 z_3 = z_2 + f + bu - \dot{\alpha}_2 \tag{30}$$

then the BSC law can be selected as

$$u = \frac{1}{b}(\dot{\alpha}_2 - k_3 z_3 - z_2 - f) \tag{31}$$

From (31), if the information of f is assumed to be known, the negative semi-definite of $\dot{V}_3$ can be obtained by substituting (31) into (29) as

$$\dot{V}_3(z_1,z_2,z_3) = \dot{V}_2 + z_3\dot{z}_3 = -k_1 z_1^2 - k_2 z_2^2 - k_3 z_3^2 \le 0 \tag{32}$$

From (32), it is found that EHA position control systems using the BSC law of (31) can guarantee exponential stability.

If system uncertainties can be exactly known, the BSC law of (31) can achieve the desired tracking performance and the robustness to the system uncertainties of EHA systems. However, the BSC law of (31) will cause a tracking error and does not achieve the robustness to the system uncertainties because the value of f cannot be exactly known. To improve the tracking performance and the robustness to the system uncertainties, the value of f, in which system uncertainties are included , should be estimated.

Therefore, in this chapter, an ABSC scheme is proposed, which is the BSC scheme with the estimator of f. In order to design the ABSC system, the BSC law of (31) is modified as

$$u = \frac{1}{b}(\dot{\alpha}_2 - k_3 z_3 - z_2 - \hat{f}) \tag{33}$$

where $\hat{f}$ is the estimator of the system uncertainties.

Substituting (33) into (12), (12) is modified as

$$\begin{cases} \dot{x}_1 = x_2 \\ \dot{x}_2 = x_3 \\ \dot{x}_3 = \tilde{f} + \dot{\alpha}_2 - k_3 z_3 - z_2 \end{cases} \tag{34}$$

where $\tilde{f} = f - \hat{f}$.

From (13), (14), and (15), these equations are the error equations for the velocity, acceleration and jerk of the piston, which include $\alpha_1(z_1)$ and $\alpha_2(z_1,z_2)$ that guarantee the exponential stability of EHA position control systems. Substituting these equations into (34), the error dynamics can be represented as

$$\begin{cases} \dot{z}_1 = z_2 - k_1 z_1 \\ \dot{z}_2 = z_3 - k_2 z_2 - z_1 \\ \dot{z}_3 = \tilde{f} - k_3 z_3 - z_2 \end{cases} \tag{35}$$

From (35), the LCF is defined as

$$V_4 = \frac{1}{2}z_1^2 + \frac{1}{2}z_2^2 + \frac{1}{2}z_3^2 + \frac{1}{2\gamma}\tilde{f}^2 \tag{36}$$

where γ is a positive constant.
The derivative of (36) can be described as

$$\dot{V}_4 = z_1\dot{z}_1 + z_2\dot{z}_2 + z_3\dot{z}_3 + \frac{1}{\gamma}\tilde{f}\dot{\tilde{f}} = -k_1 z_1^2 - k_2 z_2^2 - k_3 z_3^2 + \tilde{f}(z_3 - \frac{1}{\gamma}\dot{\hat{f}}) \le 0 \tag{37}$$

From (37), an estimation rule to guarantee the system stability can be obtained as

$$\dot{\hat{f}} = \dot{f} - \gamma z_3 \tag{38}$$

Equation (38) uses the information of z_3, which depends on the information of z_1 and z_2. Therefore, (38) closely relates to α_1 and α_2, which guarantee the stability of BSC systems. However, (38) cannot be used to the estimation rule because the value of f is unknown. On the other hand, if f is assumed as a lumped uncertainty, system uncertainty f can be estimated by $\hat{f} = -\gamma z_3$. However, since the value of f for the EHA system is changed according to the operating condition, it cannot be assumed as the lumped uncertainty. Therefore, to obtain the value of f, the classical discrete disturbance observer scheme was used. Assuming that the sampling rate of the control loop is very fast, the classical discrete disturbance observer expressed by the difference equation is induced from (31) as follows

$$f(k-1) = bu(k) - \dot{\alpha}_2(k) + k_3 z_3(k) + z_2(k) \tag{39}$$

To analyze the stability of the proposed control scheme, (38) is substituted into (37). Then,

$$\dot{V}_4 = -k_1 z_1^2 - k_2 z_2^2 - k_3 z_3^2 = -\mathbf{z}^{\mathrm{T}}\mathbf{K}\mathbf{z} < 0 \tag{40}$$

where $\mathbf{K}$ is the diagonal matrix of k_1, k_2 and k_3, $\mathbf{z}=[z_1\ z_2\ z_3]^{\mathrm{T}}$, and $\dot{V}_4 = 0$ if $\mathbf{z} = 0$.
If $\mathbf{z}$ is bounded, (40) can be defined as

$$\Phi(t) = \mathbf{z}^{\mathrm{T}}\mathbf{K}\mathbf{z} \ge 0 \tag{41}$$

Integrating (41) from 0 to t, the following result can be obtained

$$\int_0^t \Phi(\tau)d\tau = V_4(\mathbf{z}(0),\tilde{f}(0)) - V_4(t) \tag{42}$$

Applying Barbalat's Lemma(Krstic, 1995) to (42), we can obtain that $\Phi(t) \to 0$ as $t \to \infty$. Therefore,

$$\lim_{t\to\infty} \int_0^t \Phi(\tau)d\tau \le V_4(\mathbf{z}(0),\tilde{f}(0)) < \infty \tag{43}$$

4. Computer simulation

In order to evaluate the validity of the proposed control scheme for EHA position control systems, a sinusoidal reference input was considered as follows

$$x_d = \sin(0.25\pi t) + \sin(0.05\pi t) \text{ [cm]} \tag{44}$$

This sinusoidal reference input is suitable for evaluating the tracking performance and the robustness of EHA position control systems because it reflects the various changes in the magnitudes of the velocity and position of the piston. Table 1 shows the system parameters of the EHA system which are used to computer simulation. Figure 2 shows the block diagram of the EHA position control system.

Notation	Description	Unit
V_0	Initial volume of the chamber	3.712×10^{-4} m^3
A	Pressure area of the piston	5.58×10^{-3} m^2
M	Piston mass	5 kg
β_e	Effective bulk modulus	1.7×10^{3} MPa
L_f	Leakage factor of the pump	3.16×10^{-16} m^3/Pa
σ_0	Bristles stiffness coefficient	5.77×10^{6} N/m
σ_1	Bristles damping coefficient	2.28×10^{4} N/m/s
$\omega_{p\max}$	Maximum speed of the motor	178 rad/s
C_p	Volumetric capacity of the pump	1.591×10^{-6} m^3/rad
μ_0	Coulomb friction coefficient	370 N
μ_1	Static friction	217 N
μ_2	Viscous friction coefficient	2318 N/m/s
v_{sv}	Stribeck velocity	0.032 m/s

Table 1. System parameters of the EHA

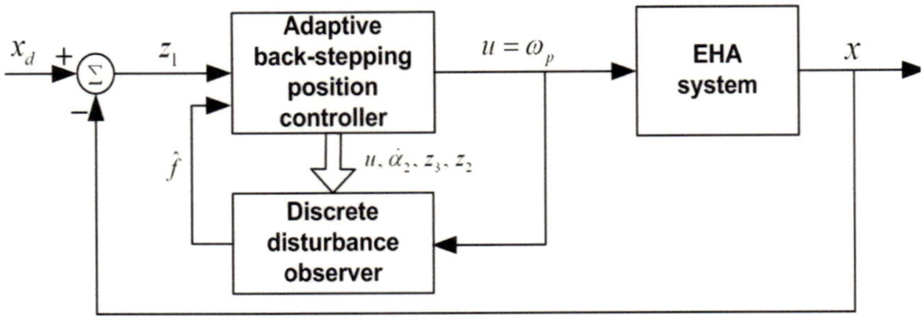

Fig. 2. Block diagram of the EHA position control system

Figure 3 shows the tracking errors of the BSC and ABSC systems for the sinusoidal reference. This result shows that the ABSC system has better tracking performance than the BSC system and has error repeatability precision of higher reliability than the BSC system. In addition, in both position and control schemes relatively large tracking errors occur at the nearly zero velocity regions. This is due to the effect of dynamic friction characteristics, which produce an instantaneous large force at the nearly zero velocity regions. For the transient response region of the initial operation of EHA position control systems, the ABSC system with the estimator for system uncertainties yields approximately 40% improvement compared with the BSC system without the estimator because the f in (31) including system uncertainties is estimated by (43), as shown in Fig. 4. Figure 4 shows the estimated value $\hat{f}$ for the system uncertainties of EHA systems obtained by the proposed adaptive rule. The estimated value plat a role in the consideration of nonlinearity and uncertainties included in EHA systems.

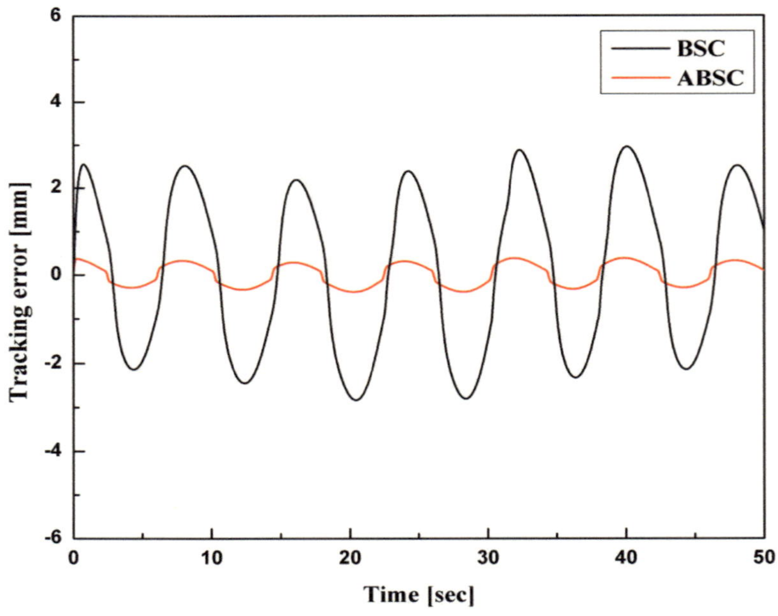

Fig. 3. Tracking errors of the BSC and ABSC systems for the sinusoidal reference input

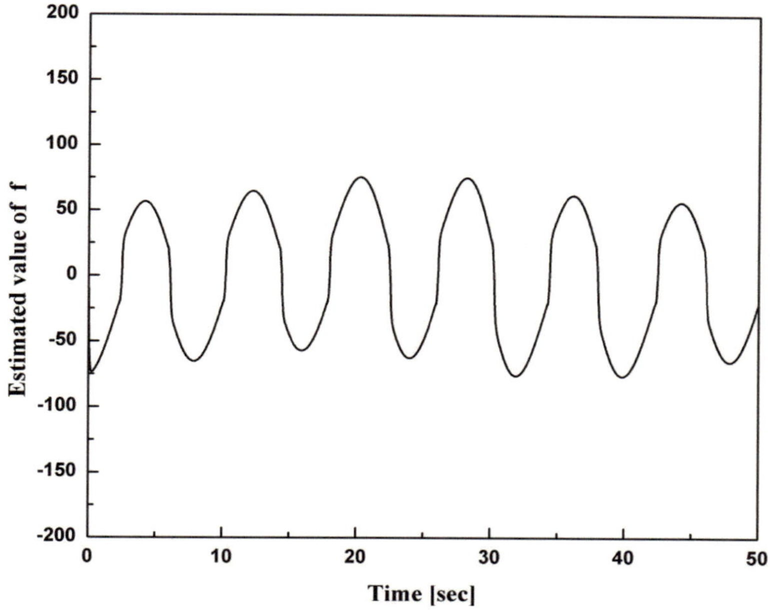

Fig. 4. Estimated value for the system uncertainties of the ABSC system for the sinusoidal reference input

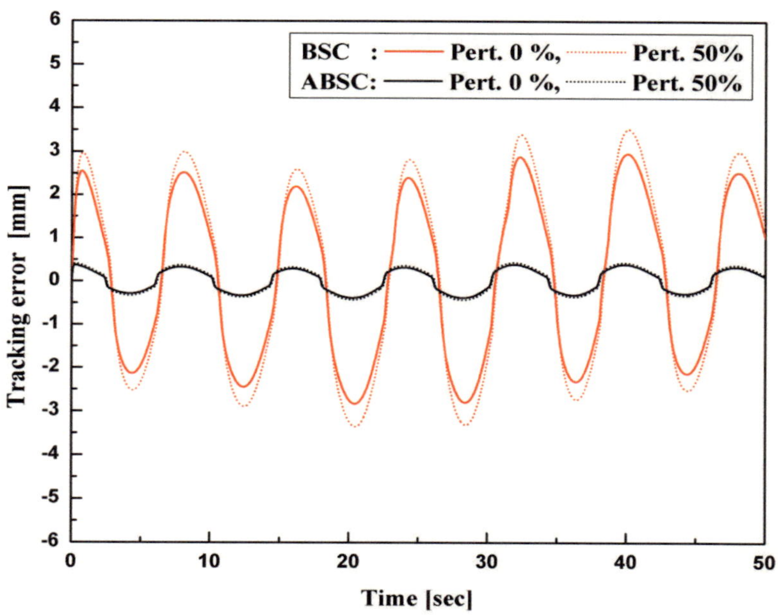

Fig. 5. Tracking errors of the BSC and ABSC systems with the perturbation of the system parameters

Figure 5 shows the tracking errors of the BSC and ABSC systems having perturbations of the system parameters such as the Coulomb friction, viscous friction and pump leakage coefficient in the EHA system for the sinusoidal reference input. It was assumed that the system parameters have a perturbation of 50%. From Fig. 5, it was found that the perturbations of the system parameters of the EHA system are closely related with the tracking performance of the EHA system. Table 2 shows the tracking RMS errors of the BSC and ABSC systems according to the perturbation of the system parameters. The variations of the tracking RMS errors due to the 50% perturbation of the system parameters for the BSC and ABSC systems are 17.6% and 3.02%, respectively. These results show that the proposed position control scheme has the desired robustness to system uncertainties such as the perturbation of the viscous friction, Coulomb friction and pump leakage coefficient.

Control scheme	Perturbation ratio	RMS value
BSC	0%	1.878 mm
	50%	2.209 mm
ABSC	0%	0.265 mm
	50%	0.273 mm

Table 2. Tracking RMS errors of the BSC and ABSC systems according to the perturbations of the system parameters

5. Experimental results and discussion

Figure 6 shows the experimental setup of the EHA system. To evaluate the effectiveness of the proposed control system, the PCM-3350(AMD Geode processor, 300MHz) was used. The control algorithms were programmed by Turbo-C++ language on MS-DOS, in order to directly handle the PCM-3718 as a data acquisition board. The PCM-3718 is a fully multifunctional card with PC/104 interface. In addition, to measure the position of the piston, an LVDT(linear variable differential transformer) sensor was used. The sampling rate was set to 1 kHz.

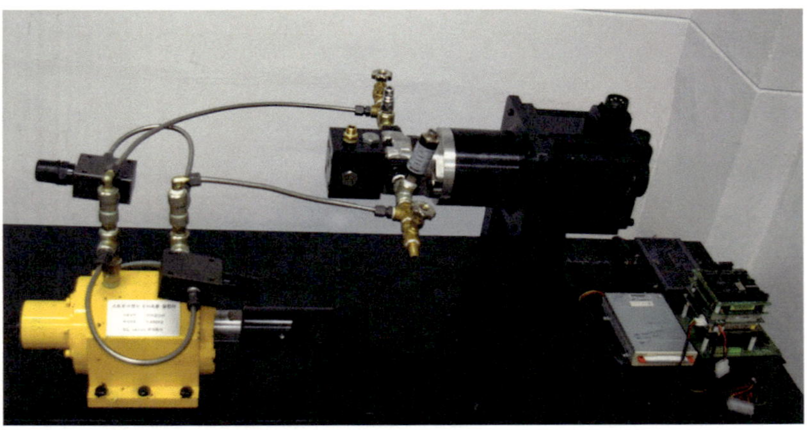

Fig. 6. Experimental setup of the EHA system

Figure 7 shows the tracking errors of the BSC and ABSC systems for the sinusoidal reference input, which was used in the computer simulation. The tracking error of the BSC system is relatively large when the direction of the piston is changed because the BSC system cannot compensate the friction of the EHA system. In addition, the tracking error of the BSC varies according to the direction of the piston because of the system uncertainties of the EHA system. However, the ABSC system has better tracking performance than the BSC system because the ABSC system can effectively compensate the system uncertainties as well as the nonlinear friction effects by using the estimated value $\hat{f}$, which is shown in Fig. 8.
Figure 9 shows the speed of the motor as the control input for the sinusoidal reference input. Figure 10 shows the tracking errors of the BSC and ABSC systems for the square wave type reference input. The characteristics of the transient responses of the BSC and the ABSC systems are almost same. In the BSC system, however, steady-state error occurs relatively large in the backward direction. This shows that the BSC system cannot compensate the system uncertainties of the EHA system. But we can show that the ABSC system can effectively compensate the system uncertainties regardless of the piston direction. Figure 11 shows the estimated value $\hat{f}$ for the system uncertainties of the ABSC system for the square wave type reference input. The estimated value $\hat{f}$ for the system uncertainties makes the desired tracking performance and robustness to the EHA system with system uncertainties. Figure 12 shows the speed of the motor as the control input for the square wave type reference input.

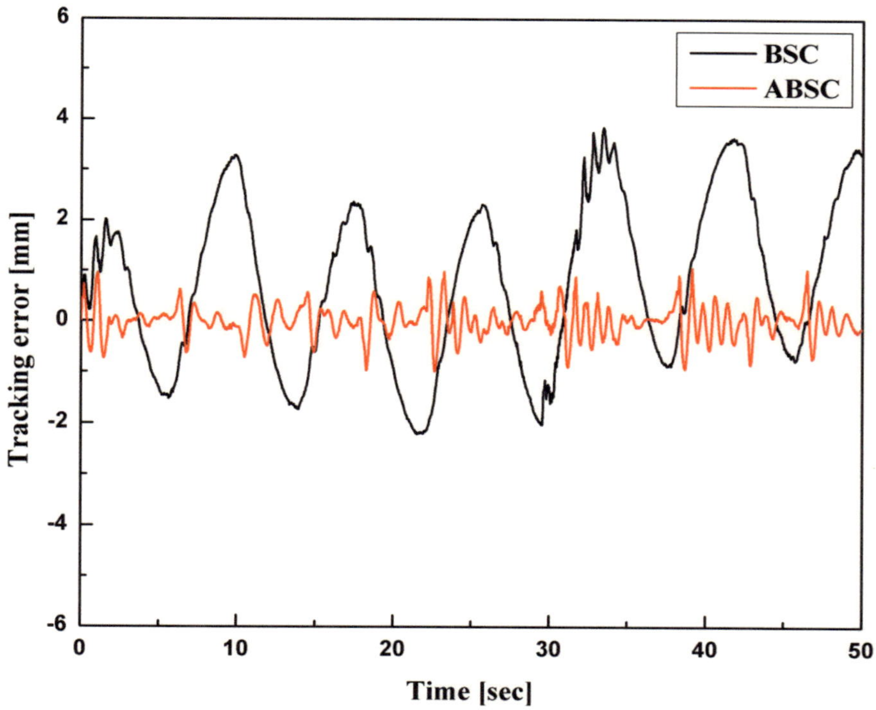

Fig. 7. Tracking errors of the BSC and ABSC systems for the sinusoidal reference input

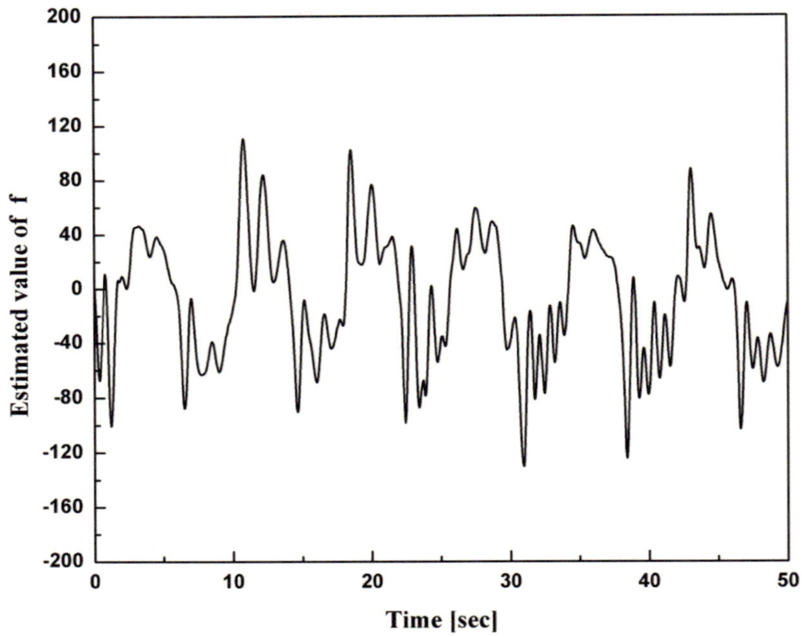

Fig. 8. Estimated value for the system uncertainties of the ABSC system for the sinusoidal reference input

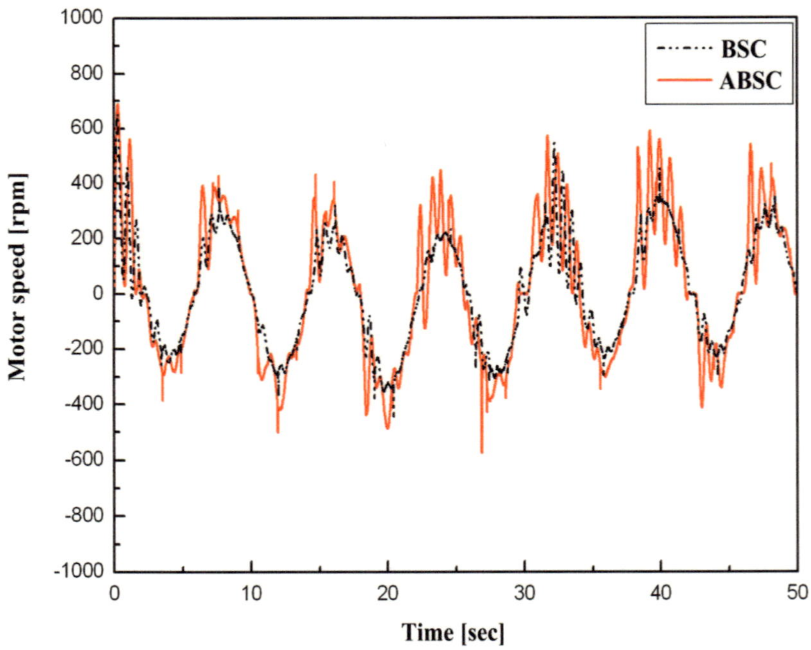

Fig. 9. Speed of the motor as the control input for the sinusoidal reference input

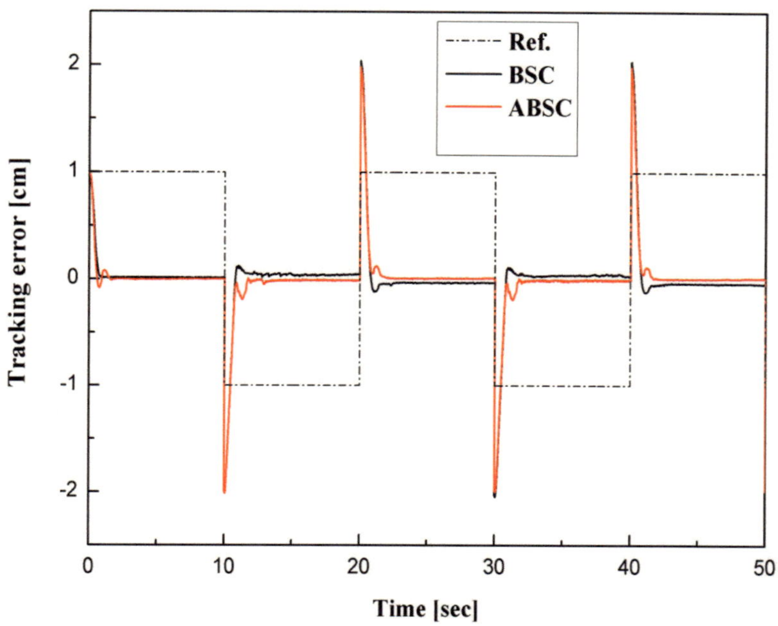

Fig. 10. Tracking errors of the BSC and ABSC systems for the square wave type reference input

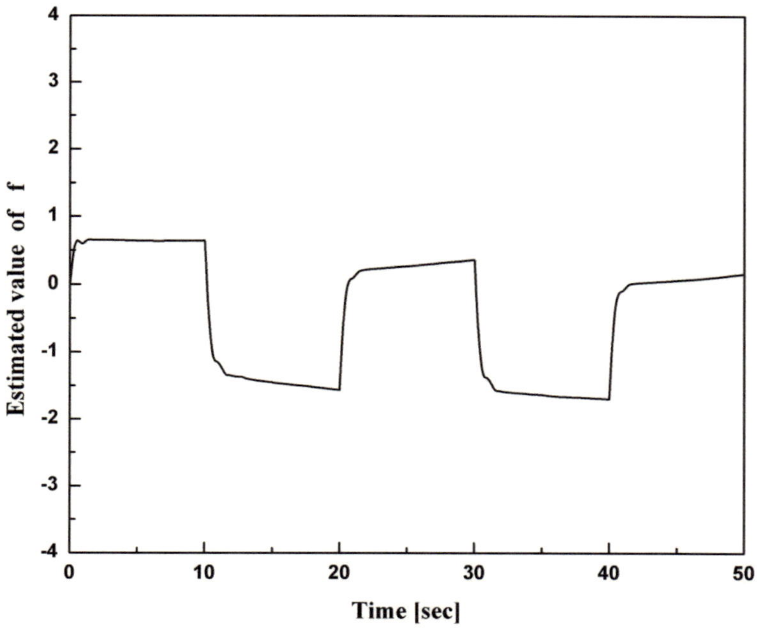

Fig. 11. Estimated value for the system uncertainties of the ABSC system for the square wave type reference input

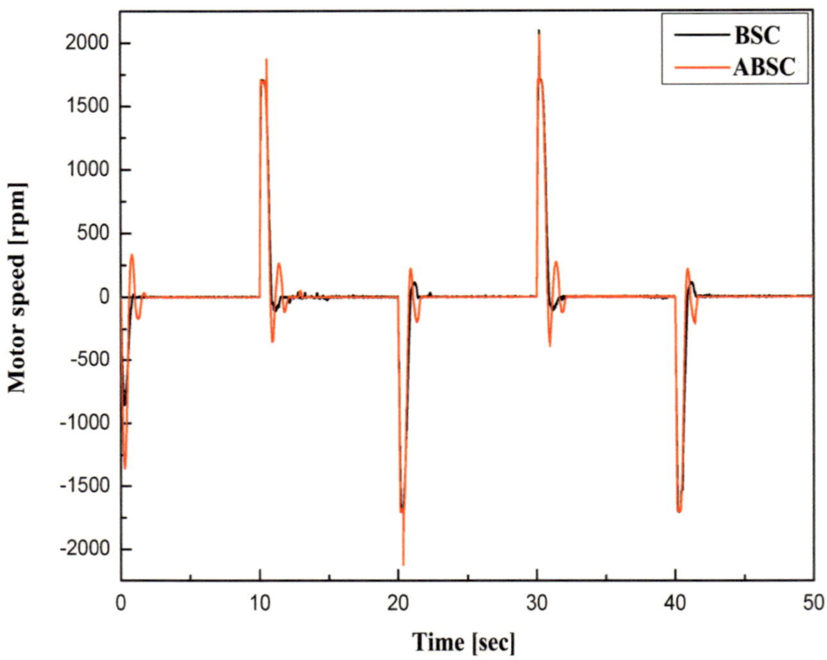

Fig. 12. Speed of the motor as the control input for the square wave type reference input

Table 3 shows the tracking RMS errors of the BSC and ABSC systems for the sinusoidal reference input and the square wave type reference input at steady-state. From Table 3, it was found that using the ABSC system instead of the BSC system yields about 5 times improvement in the tracking performance of the EHA position control system.

Control system	Sinusoidal reference input	Square wave type reference input at steady state
BSC	1.762 mm	0.395 mm
ABSC	0.309 mm	0.114 mm

Table 3. Tracking RMS errors of the BSC and ABSC systems

6. Conclusion

A robust position control of EHA systems was proposed by using the ABSC scheme, which has robustness to system uncertainties such as the perturbation of viscous friction, Coulomb friction and pump leakage coefficient. Firstly, a stable BSC system based on the EHA system dynamics was derived. However, the BSC scheme had a drawback: it could not consider system uncertainties. To overcome the drawback of the BSC scheme, the ABSC scheme was proposed having error equations for the velocity, acceleration and jerk of the piston, respectively, which were induced by the BSC scheme. To evaluate the performance and robustness of the proposed EHA position control system, BSC and ABSC schemes were implemented in a computer simulation and experiment. It was found that the ABSC scheme can yield the desired tracking performance and the robustness to system uncertainties.

7. References

Y. Chinniah, R. Burton and S. Habibi (2006), Failure monitoring in a high performance hydrostatic actuation system using the extended kalman filter, *Int. J. Mechatronics 16(10)* , pp. 643-653.

J. J. Choi, J. S. Kim and S. I. Han (2004), Pre-sliding friction control using the sliding mode controller with hysteresis friction compensator, *KSME Int'l J. 18(10)*, pp. 1755-1762.

S. Habibi and A. Goldenberg (2000), Design of a new high-performance electro-hydraulic actuator, IEEE Trans. *Mechatronics 5(2)*, pp. 158-164.

L. Jun, F. Yongling, Z. Guiying, G. Bo and M. Jiming (2004), Research on fast response and high accuracy control of an airborne brushless DC motor, *Proc. 2004 IEEE Int. Conf. Robotics and Biomimetics*, Shenyang, China, pp. 807-810.

C. Kaddissi, J. P. Kenne and M. Saad (2006), Indirect adaptive control of an electro-hydraulic servo system based on nonlinear backstepping, *IEEE Int. Symposium Ind. Electron*, Montreal, Quebec, Canada, pp. 3147-3153.

V. V. Kokotovic, J. Grabowski, V. Amin and J. Lee (1999), Electro hydraulic power steering system, *Int. Congress & Exposition*, Detroit, Michigan, USA, pp. 1-4.

M. Krstic, I. Kanellakopoulos and P. Kokotovic (1995), Nonlinear and Adaptive Control Design, *Wiley Interscience*, New York, USA.

K. J. Lee, H. M. Kim and J. S. Kim (2004), Design of a chattering-free sliding mode controller with a friction compensator for motion control of a ball-screw system, *IMechE J. of Systems and Control Engineering*, 218, pp. 369-380.

H. E. Merritt (1967), Hydrostatic Control Systems, Wiley, New York, USA.

M. Perron, J. de Lafontaine and Y. Desjardins (2005), Sliding-mode control of a servomotor-pump in a position control application, *IEEE Conf. Electrical and Computer Eng*, Saskatoon, Canada, pp. 1287-1291.

J. J. Slotine and W. Li (1991), Applied Nonlinear Control, *Pearson Education*, New Jersey, USA.

S. Wang, R. Burton and S. Habibi (2005), Sliding mode controller and filter applied to a model of an electro-hydrostatic actuator system, *ASME Int. Mechanical Engineering Congress & Exposition*, Orlando, Florida, USA, pp. 1-10.

10

Discussion on Robust Control Applied to Active Magnetic Bearing Rotor System

Rafal P. Jastrzebski[1], Alexander Smirnov[1], Olli Pyrhönen[1] and Adam K. Piłat[2]

[1]*Dept. of Electrical Engineering, LUT Energy, Lappeenranta University of Technology*
[2]*Dept. of Automatics, AGH University of Science and Technology, Krakow*
[1]*Finland*
[2]*Poland*

1. Introduction

Since the 1980s, a stream of papers has appeared on system uncertainties and robust control. The robust control relies on $\mathcal{H}_\infty$ control and μ synthesis rather than previously favored linear-quadratic Gaussian control. However, highly mathematical techniques have been difficult to apply without dedicated tools. The new methods have been consolidated in the practical applications with the appearance of software toolboxes, such as Robust Control Toolbox from Matlab. This chapter focuses on the application of this toolbox to the active magnetic bearing (AMB) suspension system for high-speed rotors.

AMBs are employed in high-speed rotating machines such as turbo compressors, flywheels, machine tools, molecular pumps, and others (Schweitzer & Maslen, 2009). The support of rotors using an active magnetic field instead of mechanical forces of the fluid film, contact rolling element, or ball bearings enables high-speed operation and lower friction losses. Other major advantages of AMBs include no lubrication, long life, programmable stiffness and damping, built-in monitoring and diagnostics, and availability of automatic balancing. However, AMB rotor system forms an open-loop unstable, multiple-input multiple-output (MIMO) coupled plant with uncertain dynamics that can change over time and that can vary significantly at different rotational speeds. In practical systems, the sensors are not collocated with the actuators, and therefore, the plant cannot always be easily decoupled. Additionally, the control systems face a plethora of external disturbances.

The major drawback of an AMB technology is a difficulty in designing a high-performance reliable control and its implementation. For such systems, the μ and $\mathcal{H}_\infty$ control approaches offer useful tools for designing a robust control (Moser, 1993; Zhou et al., 1996).

The high-performance and high-precision control for the nominal plant without uncertainties can be realized by using model-based, high-order controllers. In the case of control synthesis, which is based on the uncertain plant model, there is a tradeoff between the nominal performance (time- and frequency-domain specifications) and the robustness. The modeled uncertainties cannot be too conservative or otherwise obtaining practical controllers might be not feasible (Sawicki & Maslen, 2008). Moreover, too complex uncertainty models lead to increased numerical complexity in the control synthesis. The models applied for the control

synthesis of AMBs can vary from a point mass (Oliveira et al., 2006) to very complex MIMO plants (Li, Lin, Allaire & Luo, 2006).
The literature presents different weighting or interconnection design schemes. Each of the schemes has its contradictive objectives and tradeoffs. For the point mass levitated systems, the load uncertainty is typically applied. As an example, Li, Lin, Allaire & Luo (2006) present an **S/T/KS** scheme, where the **S**, **T**, **K**, and **G** are the sensitivity, complementary sensitivity, controller, and plant transfer functions. The corresponding weights are tuned using engineering judgment and manual trial and error simulations. Losch (2002) splits the available design schemes to signal-based and the loop-shaping schemes. The signal-based schemes are considered to be more complex and conservative. The loop-shaping schemes, for example, discussed by Losch (2002) include **KS/SG/T** for the control of the rigid rotor and **KS/SG/T/S** for the control of the flexible rotor. Another loop-shaping procedure is developed by Glover & McFarlane (1989). It applies robust stabilization of normalized coprime factorization of the plant using two weights: pre- and post-compensators. Skogestad & Postlethwaite (2005) give a general recommendation on the selection of these weights.
This chapter reviews different weighting schemes for building the robust control of AMB systems. The presentation starts with the point mass levitation and then undertakes non-gyroscopic and gyroscopic coupled AMB rotor systems. The aim of the robust control is to stabilize the rotor suspension independently to the assumed uncertainties. The robustness must be satisfied in the full range of the operating frequencies and for the selected range of the state variables. The work studies how to select the optimal control weighting functions for selected schemes based on genetic algorithms and experimental data obtained from the test rig. The Linear Parameter-Varying (LPV) technique is applied to suppress the influence of the variable rotational speed on the plant dynamics, thus reducing the uncertainty set. The real-time controller operating conditions are considered. The nonlinear simulations of the synthesized controllers and the accurate plant models in Simulink are compared with experimental results.

2. Suspension of the point mass

2.1 Introduction

The main component of the AMB system is an electromagnet that is used for the levitation purposes to keep the ferromagnetic object (e.g. rotor) levitated. The electromagnetic force value is controlled by the coil current steered by the external regulator. The introduction to the robust control is described by the example of Active Magnetic Suspension (AMS), which is also referred as Active Magnetic Levitation (AML). The robust approach can be applied to the uncertainty of the electromagnetic actuator and the levitated object mass. The controller synthesis and experiments are devoted to the MLS2EM (InTeCo, 2008) system (see Fig. 1) that extends the standard single electromagnet AML and represents one axis of the typical four horse-shoe AMB configuration.

2.2 Why robust control is required

In the classical state-feedback control approach for locally linearized AML model (Pilat, 2002) the mass uncertainty affects the control quality and object position. For the designed state-feedback controller with different closed-loop properties the 90 % mass perturbation has been introduced and presented with Bode diagrams in Fig. 2. One can find the influence of the mass change on the phase and amplitude depending on the designed controller properties.

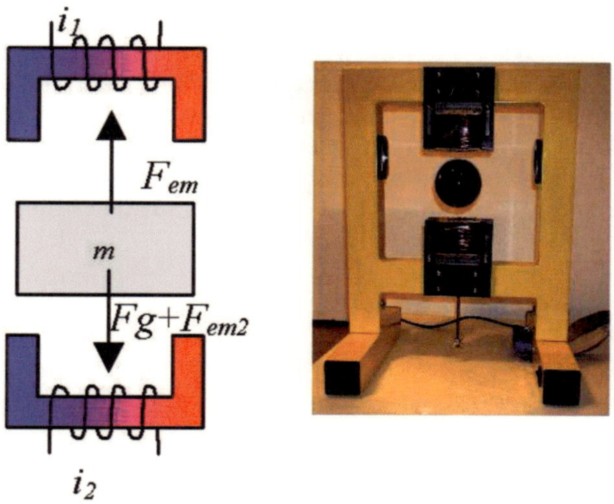

Fig. 1. Dual electromagnet Active Magnetic Levitation System - concept and test-rig.

The closed-loop characteristics remain unchanged due to the fixed and non-robust structure of the controller.

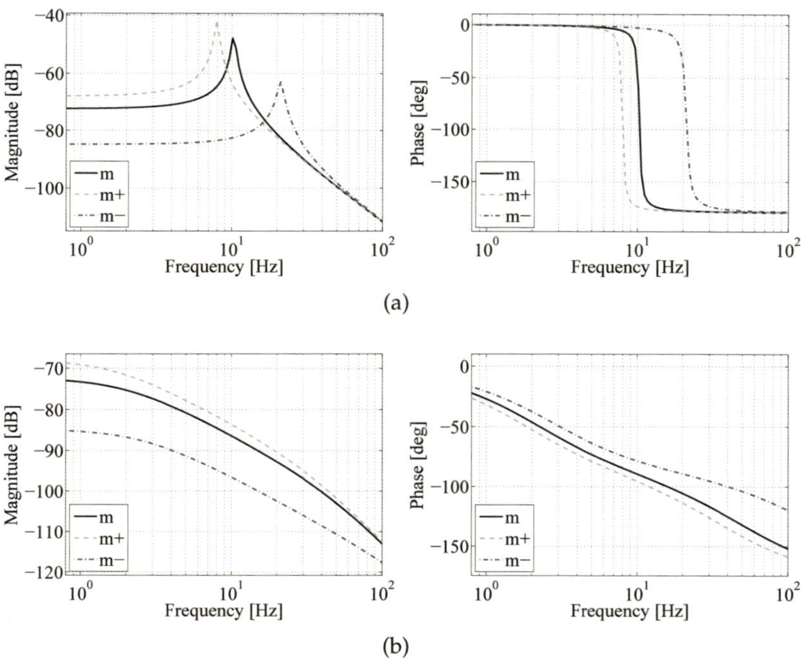

Fig. 2. Influence on the mass perturbation for the state feedback controller: a) for k = 250 $\mathrm{N\,m^{-1}}$, c = 0.2 $\mathrm{N\,s\,m^{-1}}$, b) for for k = 250 $\mathrm{N\,m^{-1}}$, c = 20 $\mathrm{N\,s\,m^{-1}}$.

The robust controller can be realized in the intelligent form by the application of the Fuzzy-Logic approach (Pilat & Turnau, 2005), where the controller is pre-tuned and optimized at the modelling and simulation stage, or by the application of an on-line adopted neural network (Pilat & Turnau, 2009), where the weights and biases are updated while the real-time control is pending. Another approach is based on the linear control theory and parameter uncertainty. Some applications to the magnetic levitation and bearing systems can be found in (Fujita et al., 1995; Gosiewski & Mystkowski, 2008; Mystkowski & Gosiewski, 2009). The following section will present a robust controller design to stabilize the levitated object independently to its mass uncertainty. More detailed, simulation results and comparison to the state feedback controller can be found in (Pilat, 2010).

2.3 AML modelling and control

2.3.1 Nonlinear and linear AML model

The open loop structurally unstable model of the current driven single electromagnet AML (Pilat, 2009) is given by Equation (1).

$$\ddot{x_1} = -K_{\text{em}} \frac{(i_0 + i)^2}{m(x_{10} + x_1)^2} + g, \tag{1}$$

where: x_1 - object displacement with respect to the x_{10} [m], x_{10} - nominal object distance from the electromagnet surface [m] (x_1>0), x_2 - object velocity [m s^{-1}], m - object mass [kg], g - gravity acceleration [m s^{-2}], K_{em} - actuator constant describing its construction [N m^2 A^{-2}], i - coil current [A], i_0 nominal coil current for the object distance x_{10}. This research will use the laboratory setup (Fig. 1b) characterized by the following parameter values: m = 0.056 kg, K_{em} = 5.594·10^{-5} N m^2 A^{-2}. By analyzing the nonlinear model one can observe that the variable mass affects the system dynamics so that heavier objects require an increase in the coil current when the actuator construction remains the same. It means that the controller should react to the variable load using the robustness property. The steady-state coil current depends on the nominal object distance and the levitated object mass and the actuator design 2.

$$i_0(x_{10}, m) = x_{10}\sqrt{mgK_{\text{em}}^{-1}}. \tag{2}$$

One can notice that the mass variation with respect to the nominal object mass is a source of demand for the coil current change. This should be satisfied automatically by the controller. To perform the controller synthesis for a chosen object position a linear model is required. The nonlinear model is linearized in the steady-state point $\boldsymbol{x}_0 = [0\ 0]^T$ resulting in the linear model in the form $\dot{\boldsymbol{x}} = \mathbf{A}\boldsymbol{x} + \mathbf{B}\boldsymbol{u}$, where:

$$\mathbf{A} = \begin{bmatrix} 0 & 1 \\ m^{-1}\alpha_0 & 0 \end{bmatrix}, \quad \mathbf{B} = \begin{bmatrix} 0 \\ m^{-1}\beta_0 \end{bmatrix} \tag{3}$$

with: $\alpha_0 = 2K_{\text{em}} i_0^2 x_{10}^{-3}$ kg s^{-2}, $\beta_0 = -2K_{\text{em}} i_0 x_{10}^{-2}$ kg m A^{-1} s^{-2}.

2.3.2 Robust controller design

The $\mathcal{H}_2$, $\mathcal{H}_\infty$ and μ-synthesis theory allows to perform an analysis and synthesis of the robust control systems (Battachatyya et al., 1995; Gu et al., 2005a; Kwakernaak, 1993; 2002) in the case of model-system uncertainties and perturbations. In the AML, the exact physical value of the levitated object mass is not known, but can be measured before an experiment. When applying the AML in real applications the mass value can vary. It can be assumed that the mass value is known with a certain, known interval. Thus, we can represent the mass as follows:

$$m = \overline{m}(1 + p_{\mathrm{m}}\delta_{\mathrm{m}}), \tag{4}$$

where $\overline{m}$ is the nominal value of m, and p_{m} and δ_{m} represent the relative perturbation on the object mass. The $\delta_{\mathrm{m}} \in [-1, 1]$ allows to perturb the mass vs. nominal value with a given ratio $p_{\mathrm{m}} \in [0, 1]$ corresponding to the percentage uncertainty. Let G_{ML0} denote the open-loop dynamics of the AMS taking into account the uncertainty of the levitated object mass. Thus, the AMS dynamics is given in the following form:

$$\begin{bmatrix} \dot{x}_1 \\ \dot{x}_2 \\ y_{\mathrm{m}} \\ y \end{bmatrix} = \mathbf{G}_{\mathrm{ML0}} \begin{bmatrix} x_1 \\ x_2 \\ u_{\mathrm{m}} \\ u \end{bmatrix}, \tag{5}$$

where:

$$\mathbf{G}_{\mathrm{ML0}} = \begin{bmatrix} \mathbf{A} & \mathbf{B}_1 & \mathbf{B}_2 \\ \mathbf{C}_1 & \mathbf{D}_1 & \mathbf{0} \\ \mathbf{C}_2 & \mathbf{D}_2 & \mathbf{0} \end{bmatrix}, \mathbf{A} = \begin{bmatrix} 0 & 1 \\ \overline{m}^{-1}\alpha_0 & 0 \end{bmatrix}, \mathbf{B}_1 = \begin{bmatrix} 0 \\ -p_{\mathrm{m}} \end{bmatrix}, \mathbf{B}_2 = \begin{bmatrix} 0 \\ \overline{m}^{-1}\beta_0 \end{bmatrix}, \tag{6}$$

$$\mathbf{C}_1 = \begin{bmatrix} \overline{m}^{-1}\alpha_0 & 0 \end{bmatrix}, \mathbf{C}_2 = \begin{bmatrix} 1 & 0 \end{bmatrix}, \mathbf{D}_1 = -p_{\mathrm{m}}, \mathbf{D}_2 = \overline{m}^{-1}\beta_0.$$

Note that the G_{ML0} depends only on the nominal AML parameters and the possible perturbation of a nominal object mass. The objective is to design the robust feedback controller $K(s)$ applied in the form:

$$u(s) = K(s)y(s). \tag{7}$$

The stability (8) of the nominal plant model as well as closed-loop robust stability (9) must be fulfilled.

$$\left\| \begin{bmatrix} W_p(1 + \mathbf{G}_{\mathrm{ML0}}K)^{-1} \\ W_u K(1 + \mathbf{G}_{\mathrm{ML0}}K)^{-1} \end{bmatrix} \right\|_\infty < 1 \tag{8}$$

$$\left\| \begin{bmatrix} W_p(1 + F_u(\mathbf{G}_{\mathrm{ML0}}, \delta_{\mathrm{m}})K)^{-1} \\ W_u K(1 + F_u(\mathbf{G}_{\mathrm{ML0}}, \delta_{\mathrm{m}})K)^{-1} \end{bmatrix} \right\|_\infty < 1 \tag{9}$$

The closed loop system with the designed controller, mass uncertainty and added weighting functions is presented in Fig. 3. The performance criterion is to have transfer functions from d to e_p and e_u small in the sense of $|| \cdot ||_\infty$ for all possible mass uncertainties. The weighting functions are used to reflect the relative significance of the performance requirement over different frequency ranges.

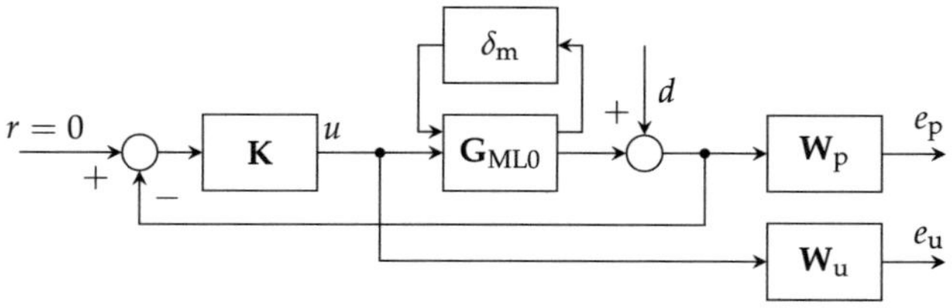

Fig. 3. AML closed loop system with an uncertain mass.

Thus, a key point in the controller design is to develop the sensitivity function to satisfy the required closed-loop performance over a specified frequency range. There are many possible approaches to propose the weighting functions, for example they can be chosen as follows:

$$W_p(s) = \frac{w_{n0}}{w_{d1}s + w_{d0}}. \tag{10}$$

The control weighting function $W_u(s)$ is chosen as a scalar value of 10^{-3}. By adjusting the values of w_{n0}, w_{d1}, and w_{d0} the performance of the robust controller could be tuned up.
The robust μ-synthesis based on the D-K iteration procedure involving a set of optimizations produces the controller in a continuous form. The resulting controller order can be high and depend on the mass perturbation, formulation of the weighting function, and the number of iterations executed to find the optimal controller. The obtained 3rd order controller has the following parameters: a_2=-1.473·10^6, a_1=-8.457·10^7, a_0 = -8.552·10^8, b_2 = 1.648·10^3, b_1 = 3.014·10^5, b_0 = 3.012·10^4 and it is given by equation (11).

$$K(s) = \frac{a_2 s^2 + a_1 s + a_0}{s^3 + b_2 s^2 + b_1 s + b_0}. \tag{11}$$

2.3.3 Real-time experiments

The realization of the AML controller is carried out using the MATLAB/Simulink and additional toolboxes. When steered from the PC-based platform, the I/O board is installed in the PC and RTW/RTWT toolboxes are applied to provide a real-time simulation in the Windows environment. When the control unit is based on the dSPACE controller and the Control Desk toolbox, the real-time controller is executed on the target embedded platform.
The dual electromagnet AML system driven by a frequency-based current hardware feedback controller was used to test the performance of the robust controller. The MLS2EM system was steered from the PC with a FastDAQ custom I/O board (Pilat & Piatek, 2008) from MATLAB/Simulink via RTWT at a sampling frequency of F_S = 4 kHz. The extra force generated in the programmable way and produced by the lower electromagnet was attracting the levitated object and therefore simulating mass uncertainty. To show the performance of the robust controller, the experimental data has been filtered to remove the high frequencies from the measured signals.
In the case of a step-type load representing a narrow mass change of 15 % the object is brought down to the desired level in 100 ms. The maximal overshoot versus desired object position is equal to 317 µm while for the triangular load corresponding to the low-frequency mass change of 33% is equal to 237 µm.

2.3.4 Conclusions to AML robust control design

The analytical robust control approach requires a good model of the system at the operating point. The parameter uncertainty does not cancel the structural nonlinearities, but is satisfactory for the required control performance. In some cases, the obtained high-order controller structure could not be realized by the hardware resources. In this case, the order reduction under special attendance of the controller quality is required.

3. Modelling of the AMB rotor systems

The second case study plant is a laboratory test stand with an AMB-supported custom rotor. The machine was originally a solid rotor induction motor for general industrial high-speed applications with the rated speed 12000 rpm. The original machine was produced by Rotatek Finland Oy. The AMB setup consist of two radial actuators and one axial actuator. The control

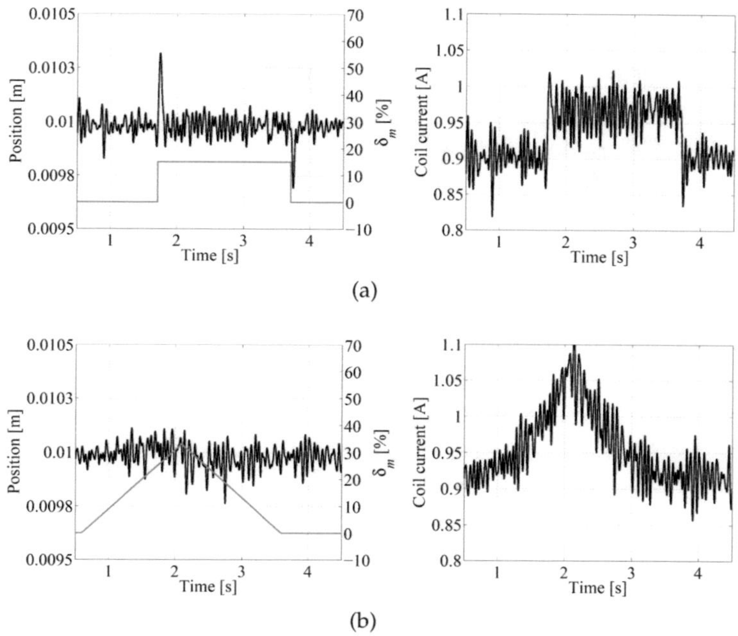

Fig. 4. Real-time experimens: a) narrow load change, b) slow load change.

layout comprises the inner current control loop and the outer position control loop. This section focuses on the radial suspension.

The studied AMB system is non-symmetric and non-collocated. The rotor is of a long rotor type without a significant gyroscopic effect. The machine is subcritical, that is, the maximum rotational speed is below the first flexible bending mode. From the radial position control point of view, the measured outputs are rotor displacements in two axes in two sensor planes and the applied control signals are four control currents of two radial eight-pole magnetic bearings. The system parameters are presented in Table 1.

Current stiffness and position stiffness	$k_i = 268\,\mathrm{N\,A^{-1}}$ and $k_x = 992\,\mathrm{N\,mm^{-1}}$
Rotor mass	46.2 kg
Rotor transverse moment of inertia	$4.8\,\mathrm{kg\,m^2}$
Rotor polar moment of inertia	$0.041\,\mathrm{kg\,m^2}$
Damping ratio of 1-3 flexible modes	0.0041, 0.0022, 0.0043
DC link voltage	150 V
Bias current and maximum currents	2.5 A and 10 A
Equivalent coil inductance and resistance	$L = 0.042$ H and $R = 0.43\,\Omega$
Equivalent average modulation delay	$\tau_{\mathrm{PWM}} = 25\,\mu\mathrm{s}$
Nominal magnetic air-gap lengths	0.6 mm

Table 1. Key AMB system parameters and their nominal values.

The technical details of the plant are given by Jastrzebski (2007), Jastrzebski & Pöllänen (2009) and Jastrzebski et al. (2010). The plant model comprises the actuator model and the rotor model.

3.1 Modelling of an AMB radial actuator

For each input-output channel, a complete nominal actuator model consists of a 2nd-order system with a pulse width modulation (PWM) delay and a motion-induced back electromotive force. The magnetic force relation for a single axis in each actuation plane in the close vicinity of the operating point is assumed to be

$$f = k_i i_\mathrm{c} + k_x x, \tag{12}$$

where k_i and k_x denote the current stiffness and the position stiffness. i_c and x are the control current and the position at the location of the bearings, respectively. Each of the inner current control loops is modeled as

$$\begin{bmatrix} \dot{i}_\mathrm{c} \\ \dot{u} \end{bmatrix} = \mathbf{A}_\mathrm{a} \cdot \begin{bmatrix} i_\mathrm{c} \\ u \end{bmatrix} + \mathbf{B}_\mathrm{ar}\dot{x} + \mathbf{B}_\mathrm{a} i_\mathrm{c,ref}, \quad i_\mathrm{c} = \mathbf{C}_\mathrm{a} \begin{bmatrix} i_\mathrm{c} \\ u \end{bmatrix}, \tag{13}$$

$$\mathbf{A}_\mathrm{a} = \begin{bmatrix} -\frac{R}{L} & \frac{1}{L} \\ -\frac{G_\mathrm{p}}{\tau_\mathrm{PWM}} & -\frac{1}{\tau_\mathrm{PWM}} \end{bmatrix}, \quad \mathbf{B}_\mathrm{ar} = \begin{bmatrix} -\frac{k_i}{L} \\ 0 \end{bmatrix}, \quad \mathbf{B}_\mathrm{a} = \begin{bmatrix} 0 \\ \frac{G_\mathrm{p}+G_\mathrm{ff}}{\tau_\mathrm{PWM}} \end{bmatrix}, \quad \mathbf{C}_\mathrm{a} = \begin{bmatrix} 1 \\ 0 \end{bmatrix}. \tag{14}$$

$i_\mathrm{c,ref}$ is the reference control current provided by the position control loop. G_p and G_ff are the proportional and feed-forward gains of the inner controllers.

3.2 Modeling of a mechanical subsystem

The rotor is modeled using a finite element method (FEM) custom code (Jastrzebski, 2007). The FEM model has 32 nodes, which corresponds to 128 degrees of freedom. The FEM code model is tuned to better correlate with the results of an experimental modal analysis of the free-free rotor and the results of frequency responses of the AMB levitated rotor (Jastrzebski et al., 2010). It is sufficient to retain only few lowest frequency modes. We apply the reduced unsupported rotor model for the controller synthesis. The model retains three flexible bending modes calculated at standstill in each plane (in the xz and yz planes). The equation of motion for the rotor spinning with the rotational speed Ω in the modal coordinates is

$$\mathbf{M}^\mathrm{m}\ddot{\boldsymbol{\eta}}^\mathrm{m} + (\mathbf{D}^\mathrm{m} + \Omega\mathbf{G}^\mathrm{m})\,\dot{\boldsymbol{\eta}}^\mathrm{m} + \mathbf{K}^\mathrm{m}\boldsymbol{\eta}^\mathrm{m} = \boldsymbol{f}^\mathrm{m}. \tag{15}$$

The matrices of the mechanical system description $\mathbf{M}^\mathrm{m}$, $\mathbf{K}^\mathrm{m}$, $\mathbf{G}^\mathrm{m}$ and $\mathbf{D}^\mathrm{m}$ are the diagonal mass matrix, the diagonal stiffness matrix, the skew-symmetric gyroscopic matrix, and the damping matrix, respectively. $\boldsymbol{f}^\mathrm{m}$ and $\boldsymbol{\eta}^\mathrm{m}$ are the vector of the modal forces acting on the rotor and the vector of modal coordinates. In fact, the first four coordinates correspond to the rigid rotor modes in the center of gravity coordinates. This reduced rotor model has in total ten coordinates.

In order to include the bearing stiffness matrices $\mathbf{K}_i$ and $\mathbf{K}_x$ in the rotor model, a transformation $\mathbf{C}_\mathrm{f}$ from the position of actuators to the center of mass is applied. Additionally, the model has to provide rotor displacements in the position of sensors and velocities at the location of bearings. Thus, another transformation matrixes are necessary $\mathbf{C}_\mathrm{s}$ and $\mathbf{C}_\mathrm{b}$.

Finally, after removing the superscript 'm' for modal, the state space equations of the rotor-bearing system has the following form

$$\begin{aligned}
\mathbf{A}_{\mathrm{r}} &= \begin{bmatrix} \mathbf{0} & \mathbf{I} \\ -\mathbf{M}^{-1}\left(\mathbf{K}-\mathbf{C}_{\mathrm{f}}^{T}\mathbf{K}_{x}\mathbf{C}_{\mathrm{f}}\right) & -\mathbf{M}^{-1}\left(\mathbf{D}+\Omega\mathbf{G}\right) \end{bmatrix}, \\
\mathbf{B}_{\mathrm{r}} &= \begin{bmatrix} \mathbf{0} \\ -\mathbf{M}^{-1}\mathbf{C}_{\mathrm{f}}^{T}\mathbf{K}_{i} \end{bmatrix}, \\
\mathbf{C}_{\mathrm{rs}} &= \begin{bmatrix} \mathbf{C}_{\mathrm{s}} \; \mathbf{0} \end{bmatrix}, \\
\mathbf{C}_{\mathrm{rb}} &= \begin{bmatrix} \mathbf{0} \; \mathbf{C}_{\mathrm{b}}. \end{bmatrix}
\end{aligned} \tag{16}$$

3.3 Complete nominal plant model

The resulting equation for the full system, which combines the rotor and actuator has the following form

$$\mathbf{A} = \begin{bmatrix} \mathbf{A}_{\mathrm{a}} & \mathbf{B}_{\mathrm{ar}}\mathbf{C}_{\mathrm{rb}} \\ \mathbf{B}_{\mathrm{r}}\mathbf{C}_{\mathrm{a}} & \mathbf{A}_{\mathrm{r}} \end{bmatrix}, \quad \mathbf{B} = \begin{bmatrix} \mathbf{B}_{\mathrm{a}} \\ \mathbf{0} \end{bmatrix}, \quad \mathbf{C} = \begin{bmatrix} \mathbf{0} \; \mathbf{C}_{\mathrm{r}} \end{bmatrix}. \tag{17}$$

The open-loop transfer function of the plant in the Laplace domain using the state variable form can be written as

$$\mathbf{y} = \mathbf{G}\left(s\right)\boldsymbol{u} = \mathbf{C}\left(s\mathbf{I}-\mathbf{A}\right)^{-1}\mathbf{B}\boldsymbol{u}. \tag{18}$$

$\mathbf{G}(s)$ is a transfer function matrix of the plant. $\boldsymbol{u}$ and $\boldsymbol{y}$ are the vectors of the control currents and the measured displacements, respectively. $\mathbf{A}$, $\mathbf{B}$, and $\mathbf{C}$ are the state matrix, the input matrix, and the output matrix in the state-space representation, respectively. The combined actuator and rotor models form a coupled plant, which has 28 states. The coupling between the transversal and tilting rotor movements is caused by the radial actuators. The coupling between the xz and yz planes appears as a result of the gyroscopic coupling (Jastrzebski, 2007). In order to decrease the condition number of the plant, the MIMO coupled plant model applies a similarity transformation leading to a normalized per-unit (pu) system. Such a per-unit plant is less prone to numerical problems when designing a controller.

3.4 Modeling of uncertainties

We divide uncertainties into dynamic perturbations and disturbance signals (Gu et al., 2005b). The disturbances originate from the inverter, motor and the load transmitted forces, as well as sensor and actuator noise. These disturbances are difficult to measure and model but some rough estimations can be applied. The dynamic perturbations comprise unstructured uncertainties, when the perturbations are only considered by upper and lower bounds, and the structured uncertainties when the perturbations appear in particular parameters (Skogestad & Postlethwaite, 2005). The former ones are unmodelled dynamics of the base and truncated high-frequency modes of the rotor. The letter ones are neglected nonlinearities of the actuators and sensors and other variations of nominal system parameters.
The structured uncertainties in the actuator include:

- variable current stiffness and position stiffness because of modeling inaccuracies, actuator nonlinearities, and changes of the operational point (± 10 %)
- variation of electrical parameters of the actuator (± 10 %)

The effects of hysteresis and time delays (of the modulation, digital control, and sensors) can be neglected for the applied system components.

The structured uncertainties when considering mechanical models and position sensors are:

- variable mass resulting from external low-frequency loads depending on applications, e.g., in compressors and pumps (± 10 %)
- variable sensor gain and offset (± 5 %)
- uncertain sensor locations (± 1 %)
- variation of the rotational speed
- uncertainties in the modal mass and damping matrices (± 2 %)

The uncertainty in the sensor locations emulates uncertainty of the mode shapes of the bearings, which are more difficult to implement. The most notable variations occur because of changes in the operational speed and shifting of the operational point. The shifting occurs for the nonzero reference position, in the presence of sensing errors, rotor runout, and external forces.

All or some of the dynamic perturbations can be lumped into a single perturbation block Δ. Δ is referred to as an unstructured uncertainty and it is complex whereas the parametric uncertainties are assumed to be real.

Differences between the measured frequency responses of the test-rig and the nominal plant model that are significant and otherwise not covered by the structured uncertainties (Jastrzebski et al., 2010) are modeled as an unstructured uncertainty. In particular, the structural resonance of the base of the machine at about 1130 rad/s (180 Hz), is modeled using an output multiplicative uncertainty $\mathbf{\Delta} = \mathbf{\Delta}_\text{o}$.

The uncertain plant with a multiplicative output uncertainty in each input-output channel is

$$\mathbf{G}_\text{p} = (\mathbf{I} + w_\text{o}\mathbf{\Delta}_\text{o})\,\mathbf{G},\ \|\mathbf{\Delta}_\text{o}\|_\infty \leq 1. \tag{19}$$

$w_\text{o}(s)$ is an uncertainty scalar weight with appropriately selected coefficients a_i such as

$$|w_\text{o}(j\omega)| \geq l_\text{o}(\omega)\,\forall\omega,\ l_\text{o}(\omega) = \max_{\bar{\sigma}}\left(\left(\mathbf{G}_\text{p} - \mathbf{G}\right)\mathbf{G}^{-1}(j\omega)\right), \tag{20}$$

$$w_\text{o}(j\omega) = \frac{a_0 s^2 + a_1 s + a_2 s^2}{s^2 + a_3 s + a_4 s^2}. \tag{21}$$

$\|.\|_\infty$ and $\bar{\sigma}$ denote the $\mathcal{H}_\infty$ norm and the maximum singular value (Skogestad & Postlethwaite, 2005), respectively. When the uncertain parts are separated from the dynamics, the system can be presented in a well-known upper linear fractional transformation (LFT) $F\,(\mathbf{M}, \mathbf{\Delta})$, where $\mathbf{M}$ represents a standard interconnection of the system with uncertainties taken out. Now, $\mathbf{\Delta}$ consists of both the unstructured and parametric uncertainties. The uncertain block $\mathbf{\Delta}$ is a diagonal matrix.

The analysis and design are more difficult when the structural uncertainties are real numbers. Unfortunately, the use of the lumped full model also results in a pessimistic analysis and a conservative design (Gu et al., 2005b). Therefore, a proper selection of modeled uncertainties is not straightforward.

After analyzing the suspension of the point mass, which can be treated equivalently to the axial suspension of the rotor, we focus on the radial suspension.

4. $\mathcal{H}_\infty$ control of the AMB rotor system with insignificant gyroscopic effect

Different weighting schemes are applicable to form a cost function subject to the $\mathcal{H}_\infty$ norm in (sub)optimization problems. Perhaps the most commonly applied schemes are the $\mathbf{S}/\mathbf{KS}$ and $\mathbf{S}/\mathbf{T}$ schemes. The $\mathbf{S}/\mathbf{KS}$ scheme can achieve nominal performance in tracking or disturbance rejection and robust stability against the additive perturbations. The weighted mixed sensitivity $\mathbf{S}/\mathbf{T}$ scheme can achieve nominal performances and robust stability against multiplicative perturbations. However, in the aforementioned schemes there is a danger of pole-zero cancellation between the nominal model and the controller (Sefton & Glover, 1990). They are also limited by the condition of the number of the right-half plane poles. In the perturbed system the number should be the same as in the nominal one (Lunz, 1989). In response to these limitations, another weighting scheme features robust stabilization against normalized coprime factor perturbation of the nominal plant. The $\mathcal{H}_\infty$ loop-shaping design relaxes the right-half plane restrictions and produces no pole-zero cancellation. The solution is obtained directly without the need for iterations.

4.1 Loop-shaping Glover-McFarlane control of an AMB rotor system

A loop-shaping $\mathcal{H}_\infty$ design procedure was introduced by Glover & McFarlane (1989). Later, it was extended to the two-degrees of freedom problem by Limebeer et al. (1993). The approach gained its popularity as it does not require a γ-iteration and provides a result by solving two Riccati equations.

To achieve a controller based on a loop-shaping technique, two weights should be selected. This is a pre-compensator $\mathbf{W}_1$ and a post-compensator $\mathbf{W}_2$. They alter the open-loop transfer function of the plant $\mathbf{G}$ to the desired shape $\mathbf{G}_\mathrm{s}$. Selection of weights depends on the performance and robustness criteria. Additionally, weights can be selected based on the presented multiplicative uncertainties. Structured uncertainty is not supported directly by the method. In a case where there are many sources of uncertainties, other methods prove to be conservative or too difficult to apply. Hence, multiplicative uncertainty approximated on the representative set of plants is a useful solution.

After selecting weights and multiplying the nominal plant from left and right, the system is stabilized with an $\mathcal{H}_\infty$ controller $\mathbf{K}_\mathrm{s}$ see Fig. 5(a). The final controller $\mathbf{K}$ is obtained as

$$\mathbf{K} = \mathbf{W}_1\mathbf{K}_\mathrm{s}\mathbf{W}_2. \tag{22}$$

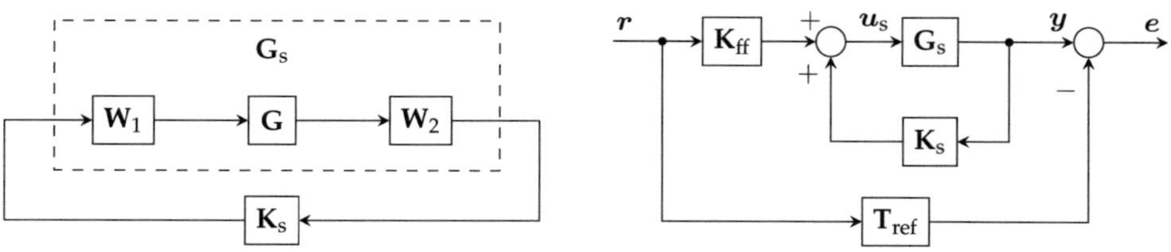

(a) Glover-McFarlane controller structure (b) Two degrees-of-freedom design problem

Fig. 5. Loop-shaping controllers

As a next step, a feedforward part is added. For that, a reference transfer function $\mathbf{T}_\mathrm{ref}$ should be chosen. The feedforward controller $\mathbf{K}_\mathrm{ff}$ is obtained by minimizing the following problem

$$\|(\mathbf{I} - \mathbf{G}_\mathrm{s}\mathbf{K}_\mathrm{s})^{-1}\mathbf{G}_\mathrm{s}\mathbf{K}_\mathrm{ff} - \mathbf{T}_\mathrm{ref}\|_\infty \leq \gamma \tag{23}$$

The described method is based mainly on the weight selection. The pre-compensator is usually a low-pass filter. Additional features can be included in the weight such as a notch filter, which is particularly useful in the described application to suppress the flexible modes. The post-compensator is used to emphasize one output over the other. As in the AMB system, all the outputs are equivalent, and this weight is a constant diagonal matrix. The last weight $\mathbf{T}_{\text{ref}}$ should describe the desired transfer function of a closed-loop plant. Thus, a first-order transfer function with a steady state gain equals one, and a crossover frequency is chosen to correspond to the desired bandwidth of the system.

This procedure for an AMB system was applied by Fujita et al. (1993), where the authors give a review how to choose weights based on the multiplicative uncertainties in the system. The applied model of the system was relatively simple accounting only for rigid modes. The loop-shaping method was combined with a μ-synthesis procedure by Lanzon & Tsiotras (2005) to guarantee performance specifications and tolerate structured uncertainties. The resulting controller was successfully applied to the AMB system.

In this work, for the controller synthesis, the procedure suggested by (Skogestad & Postlethwaite, 2005, ch. 9.4.2) is applied. The process is the same as described earlier with an addition of calculation gains to ensure the desired steady-state response. The reference function $\mathbf{T}_{\text{ref}}$ is chosen as

$$\mathbf{T}_{\text{ref}} = \mathbf{I}_{4\times4}\frac{1}{\frac{1}{\omega_{\text{bw}}}s+1}, \tag{24}$$

where $\omega_{\text{bw}} = 215$ rad/sec is the bandwidth of the closed-loop system.

The pre-compensator transfer function is chosen to be a low-pass filter with a DC gain of 78.1 dB. Additionally, the weight includes a notch filter as a second-order transfer function. The damping frequency is $1.664 \cdot 10^3$ rad s^{-1} and the damping ration is 0.08. The final transfer function is as follows

$$\mathbf{W}_1 = \mathbf{I}_{4\times4}\frac{s+110}{s+0.01365} \cdot \frac{\left(\frac{s}{1.664\cdot10^3}\right)^2 + \frac{0.3\cdot0.08}{1.664\cdot10^3}s+1}{\left(\frac{s}{1.664\cdot10^3}\right)^2 + \frac{0.3}{1.664\cdot10^3}s+1}. \tag{25}$$

Using the weight (25) presented in Fig. 6(a) and the reference function (24), a two-degrees of freedom controller is synthesised. The order of the controller is reduced to fit the real-time implementation using a Schur method (Safonov & Chiang, 1988). The evaluation of the controller with a μ-analysis is presented in Fig. 6(b).

The plot shows that the controller is capable of handling the modeled structured uncertainty. All the values are below one. As it was expected, the highest values are spotted near the first flexible mode. However, the use of a notch filter helps to alleviate the problem. The experimental evaluation of the controller is discussed in section 4.6.

4.2 $\mathcal{H}_\infty$ signal-based control of an AMB rotor system

For the simple weighting schemes in the frequency domain, for example, such as the $\mathbf{S}/\mathbf{KS}$ mixed sensitivity problem, which provides good tracking and limits control energy, we are not able to include more complex specifications. The more versatile schemes are signal-based $\mathcal{H}_\infty$ approaches. However, the more complex the resulting lower LFT becomes, the more difficult the selection of the multiple design weights and the more complex and difficult the numerical solution of the minimization problem are. We modify the Skogestad & Postlethwaite (2005) scheme by lifting up the restrictions on all of the states but instead adding application-specific

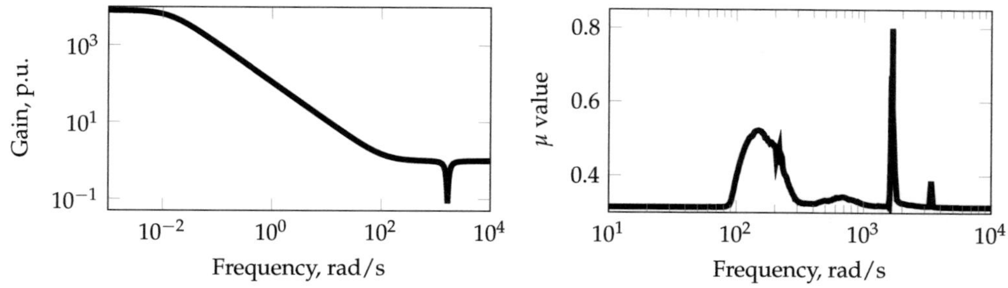

(a) Pre-compensator shape in the frequency domain

(b) μ-analysis

Fig. 6. $\mathcal{H}_\infty$ loop-shaping design

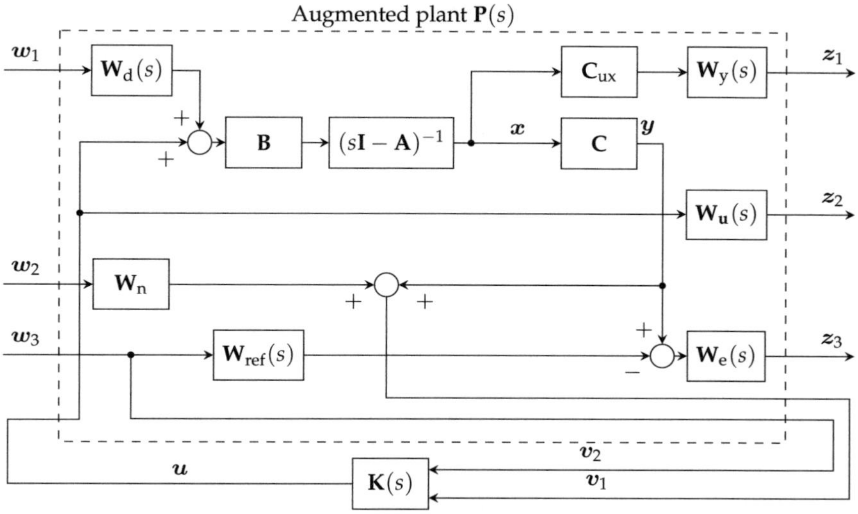

Fig. 7. 2DOF $\mathcal{H}_\infty$ signal-based design problem.

voltage limitation and displacement limitations at critical locations other than sensor planes. Now, a magnetic bearing system can be described using a general control configuration (Fig. 7). The stabilizing controller can be found by minimizing the $\mathcal{H}_\infty$ norm from the exogenous inputs $\boldsymbol{w} = [\boldsymbol{w}_1, \boldsymbol{w}_2, \boldsymbol{w}_3]^\mathrm{T}$ to the exogenous outputs $\boldsymbol{z} = [\boldsymbol{z}_1, \boldsymbol{z}_2, \boldsymbol{z}_3]^\mathrm{T}$. The inputs to the controller $\boldsymbol{v}_1$ and $\boldsymbol{v}_2$ are the vector of the position reference signals and the vector of the distorted signals received from the displacement sensors, respectively. The augmented plant input vectors $\boldsymbol{w}_1$, $\boldsymbol{w}_2$ and $\boldsymbol{w}_3$ are the vector of the input distortion signals, the vector of the output distortion signals, and the vector of the reference signals, respectively. The augmented plant output vectors $\boldsymbol{z}_1$, $\boldsymbol{z}_2$, and $\boldsymbol{z}_3$ are the vector of the voltages and displacements obtained using the output matrix $\mathbf{C}_{ux}$, weighted system input signals, and the vector of the weighted position error signals, respectively.

The signal limitations result in the unitary weight for the normalized plant special outputs $\mathbf{W}_y = \mathbf{I}$. The sensor noise spectrum is approximated by the first-order high-pass filter $\mathbf{W}_\mathrm{n}$ with the dc gain and the high-frequency gain equal to 0.2 % and 5 % of the measuring range. The crossover frequency is 250 Hz.

The weights $\mathbf{W}_d(s)$, $\mathbf{W}_u(s)$, and $\mathbf{W}_e(s)$ are defined as first-order transfer functions multiplied by a 4×4 unitary matrices. The coefficients of these functions are treated as design parameters. Applying the $\mathcal{H}_\infty$ control problem, the admissible controller $\mathbf{K}$ is found, if one exists, so that for $\gamma > 0$

$$\|\mathbf{F}_l(\mathbf{P},\mathbf{K})\|_\infty = \max_\omega \bar{\sigma}\left(\mathbf{F}_l\left(\mathbf{P},\mathbf{K}\right)(j\omega)\right) \leq \gamma. \tag{26}$$

4.3 Control design specification

For all of the schemes, the appropriate choice of the weighting functions, which provide guaranteed stability, robustness margin, and tracking nominal performance, is required. The structure and order of the weights should deliver enough flexibility but without too much added complexity in the optimization. The order of the weights together with the order of the nominal plant decide if and how easily the satisfactory solution can be obtained using necessary numerical procedures.

When defining the control design specification the major points are:

- closed-loop stability
- limitations of actuators, i.e., maximum coil current and limited DC link voltage and as a result, limited force slue rate
- no steady-state error
- sufficient input disturbance rejection and noise rejection of the sensors (output)
- robust stability and robust performance
- minimization of the output sensitivity peak

Additionally, the minor objectives are: desired step responses and the closed-loop bandwidth within the desired range.

The most of the listed objectives are easily tested. However, in order to test robust stability, the structural singular value μ has to be computed (Gu et al., 2005b; Skogestad & Postlethwaite, 2005; Zhou, 1998)

$$\mu_\Delta^{-1}(\mathbf{M}) := \min_{\mathbf{\Delta}\in\mathbf{\Delta}_s} \left\{\bar{\sigma}(\mathbf{\Delta}) : \det\left(\mathbf{I} - \mathbf{M}\mathbf{\Delta}\right) = 0\right\}. \tag{27}$$

$\mathbf{M}$ is the interconnected closed-loop system transfer function matrix and $\mathbf{\Delta} \in \mathbf{\Delta}_s$ represents uncertainties. $\mathbf{M}(s)$ is formed with respect to the uncertainty set $\mathbf{\Delta}_s$

$$\mathbf{\Delta}_s = \left\{diag\left[\delta_1 I_{1r}, \cdots, \delta_s I_{rs}, \Delta_1, \cdots, \Delta_f\right] : \delta_i \in C, \Delta_j \in C^{m_j \times m_j}\right\}, \tag{28}$$

where $\sum_{i=1}^{s} r_i + \sum_{j=1}^{f} m_j = n$ with $\mathbf{M} \in C^{n\times n}$. n equals to the dimension of the block $\mathbf{\Delta}$. s and f are the dimensions of the scalar and full uncertainty blocks, respectively.

The structural singular value of $\mathbf{M}(s)$ is a measure of the robust stability of the uncertain system

$$\mu_\Delta\left(\mathbf{M}(s)\right) := \sup_{\omega\in R} \mu_\Delta\left(\mathbf{M}(j\omega)\right). \tag{29}$$

For normalized uncertainties, the system in a standard configuration is robustly stable if $\mathbf{M}(s)$ is stable and $\mu_\Delta\left(\mathbf{M}(s)\right) < 1$. The robust performance requires that the closed-loop control system performs satisfactorily even in the presence of the defined plant uncertainties. The robust performance problem can be solved by generalizing to the robust stabilization problem with the uncertainty block replaced by $\tilde{\mathbf{\Delta}} \in \tilde{\mathbf{\Delta}}_{\mathrm{s}} := \mathrm{diag}\left\{\mathbf{\Delta}, \mathbf{\Delta}_{\mathrm{p}}\right\}$, where the uncertainties are normalized and the fictious performance uncertainty block is bounded by the norm $\left\|\mathbf{\Delta}_{\mathrm{p}}\right\|_{\infty} \leq 1$. $\mathbf{\Delta}_{\mathrm{p}}$ is unstructured with appropriate dimensions defined by the exogenous inputs and error outputs of the system $\mathbf{M}$ to represent system performance specifications.

4.4 Weighting functions as design parameters

An elegant solution to alleviate the weight selection procedure is the μ-synthesis approach. For the system with a specified uncertainty set, the algorithm gives the weights that result in a robustly stable controller obtained by an $\mathcal{H}_\infty$ synthesis. The main drawback is that there is no analytical solution for the problem. The procedure is iterative and computationally expensive. It results in a controller of a very high order. What is more, the performance requirements are again specified as initial weights, and the designer should also choose a specific weighting scheme. Last but not least, the resulting order of the controller depends on the complexity of the applied weighting scheme, plant order, and applied uncertainties. Detailed interconnections lead to controllers, which are difficult to implement and are not transparent.

For complex systems, such as the flexible AMB rotor system, finding appropriate performance weights by trial and error is very time consuming. To find the weights that produce a design meeting the multiple requirements, we could use the optimization based on the method of inequalities (Whidborne et al., 1994) or the linear matrix inequalities (Scherer et al., 1997). Another option for such a multibjective design is to apply a basic genetic algorithm (GA) (Jastrzebski et al., 2010).

For a signal-based weighting scheme, the coefficients of the weights $\mathbf{W}_{\mathrm{d}}(s)$, $\mathbf{W}_u(s)$, and $\mathbf{W}_e(s)$ are limited by its minimum and maximum selected values because of numerical reasons. Also for numerical reasons and to prevent unwanted pole-zero cancellation (Gu et al., 2005b) the stable and minimum-phase weights are applied. For continuous-time weights in the Laplace domain both the zeros and poles of a minimum phase weight must be strictly inside the left-half s-plane.

4.5 Genetic algorithm approach to multiobjective synthesis

The $\mathcal{H}_\infty$ optimization and the μ-synthesis result in the complex controllers of the higher order than the plant. In an effort to obtain a lower-order controller, we could reduce the plant model by truncating the high-frequency modes beyond the actuator bandwidth prior to the controller synthesis. In the iterative design, the resulting lower-order controller could be tested in each iteration together with the higher-order non-reduced plant against the multi-objective performance function. An alternative procedure to obtain a lowest-order controller is to use the detailed plant model for the synthesis and to apply the controller-order reduction afterwards. Both approaches replace a direct design of the low-order controller.

For the signal-based weighting scheme, some of the signal weights are kept constant while the others, which are the free parameters in the optimizations, are varied in order to reach the optimal design in the multi-objective control design problem. The basic genetic algorithm search is improved by limiting the feasible solution space. This improves the numerical conditioning, and the weights without physical relevance are excluded from the solutions (Jastrzebski et al., 2010).

The design objectives are normalized by the desired limiting values and are proportional to the square of the following performance indices:

- Output sensitivity peak $M_S = \|\mathbf{S}_\text{o}\|_\infty$ and the closed-loop bandwidth (the frequency where $\bar{\sigma}(\mathbf{S}_\text{o})$ first crosses 0.7 from below.
- High controller gain (a small maximum singular value of the sensitivity) at low frequencies $\bar{\sigma}\left(\mathbf{S}_\text{o}(\omega)\right)$, where $\omega \to 0$.
- Input disturbance attenuation $M_{T\text{i}} = \left\| \mathbf{K}_\text{fb}\mathbf{G}\mathbf{S}_\text{i} \; \mathbf{G}\mathbf{S}_\text{i} \right\|_\infty$ and output disturbance attenuation $M_{T\text{o}} = \left\| \mathbf{G}\mathbf{K}_\text{fb}\mathbf{S}_\text{o} \; \mathbf{K}_\text{fb}\mathbf{S}_\text{o} \right\|_\infty$.
- Value of γ.

The norms $M_{T\text{i}}$ and $M_{T\text{o}}$ minimize the usage of control signals and the plant output signals in the presence of the input and output distortion signals, respectively. The sensitivity functions are defined as

$$\mathbf{S}_\text{o} = \left(\mathbf{I} + \mathbf{G}\mathbf{K}\right)^{-1}, \quad \mathbf{S}_\text{i} = \left(\mathbf{I} + \mathbf{K}\mathbf{G}\right)^{-1}. \tag{30}$$

After applying GA and obtaining a final $\mathcal{H}_\infty$ controller the closed-loop uncertain system is

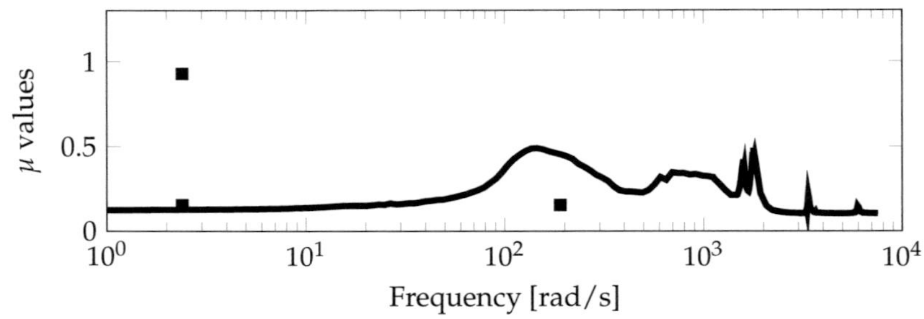

Fig. 8. μ analysis for robust stability.

tested for robust stability (Fig. 8) and performance. In order to limit design conservativeness in $\mathcal{H}_\infty$ control the uncertainties in the plant model, which is applied for design synthesis, are limited to the uncertain speed. In the case of the structured uncertainty, the computation of the structural singular value μ has to be applied and the μ synthesis remains an open problem. When applying the weights obtained using the GA to the μ synthesis, the μ synthesis cannot considerably improve the initial γ value.

4.6 Control validation

The controller achieved in section 4.1 is applied for radial AMBs in the test rig. First, the frequency responses are compared. The output sensitivity function for the B-end of the rotor is measured. The results are presented in Fig. 9. The theoretical values coincide the values obtained from the prototype. One peak that is not presented in the theoretical model corresponds to the natural frequency of the foundation. The foundations were not taken into account in the described synthesis method. Additionally, the values vary between 100 and 300 rad s^{-1}. This can be explained by the water-bed effect. The surface below one and above one closed by the curve must be equal. Thus, these lower values compensate the higher frequency peak.

The next experiment was carried out in the time domain. Two type of step responses are measured; the reference step response and an input disturbance step response. The

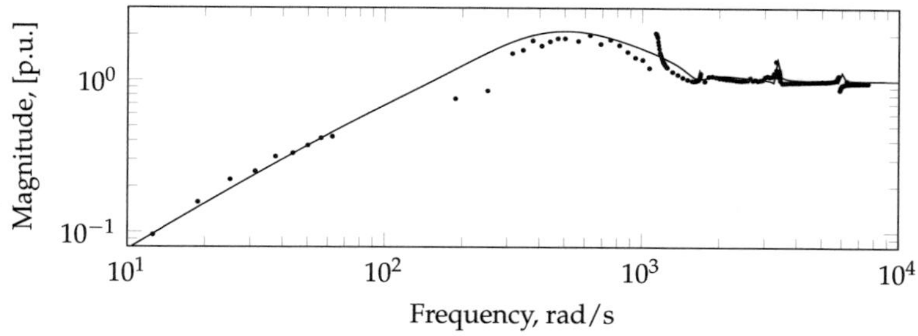

Fig. 9. Output sensitivity function.

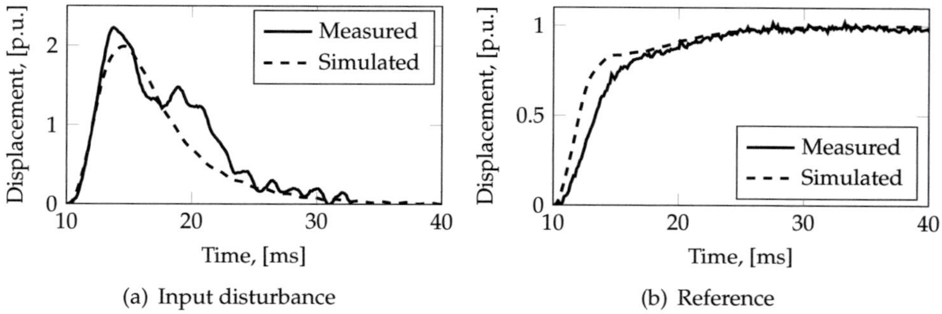

Fig. 10. Step responses.

results of the measurements are presented in Fig. 10. The reference response shows a good correspondence in the settling time. The disagreement with rise time can be explained by the presented nonlinearities of the system or position of the rotor away from the operational point. Disturbance response shows slightly higher maximum amplitude. Additional oscillations come from the first flexible mode, which is difficult to suppress by the feedback.

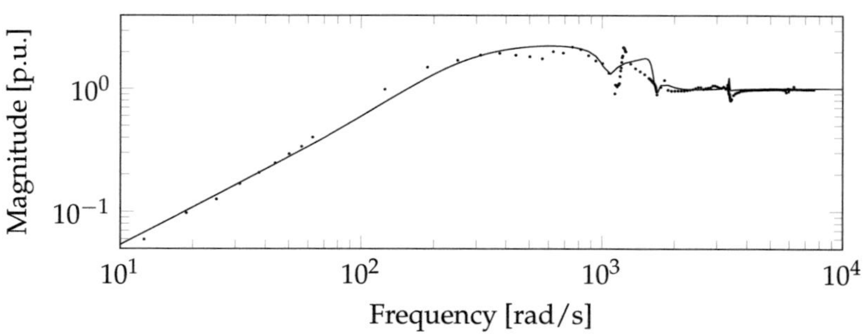

Fig. 11. Output sensitivity function of the H_∞ control.

Similarly, the step responses of the signal-based $\mathcal{H}_\infty$ control (from section 4.2) are presented in Fig. 12. The measured positions for the step responses are filtered by the controller with a relatively low bandwidth of about 110 rad/s. The measured output sensitivity function (Fig. 11) does not differ significantly from the analytical result. The peak unaccounted in the analytically computed values is caused by the structural mode of the base. For the rotational speed in the range from 0 to 6000 rpm the output sensitivity peek varies from 2.618 to 2.625 pu.

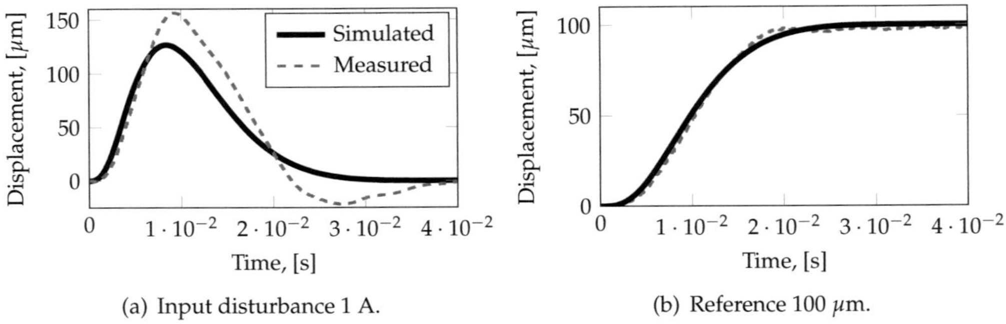

Fig. 12. Step responses of the $\mathcal{H}_\infty$ signal-based design problem.

5. LPV method applied to a gyroscopic AMB rotor system

The rotating speed of the rotor in AMBs is a source of significant uncertainty. What is more important is that it affects the frequency of the flexible modes; these are known to be challenging to suppress with the feedback control (Li, Lin & Allaire, 2006). Especially, the problem is significant for highly gyroscopic systems. The system is considered as a gyroscopic one if a polar moment of inertia is greater than the diametral one $I_p > I_d$ or the rotational speed is significant (Schweitzer & Maslen, 2009). A good illustration to the problem is a Campbell diagram presented in Fig. 13.

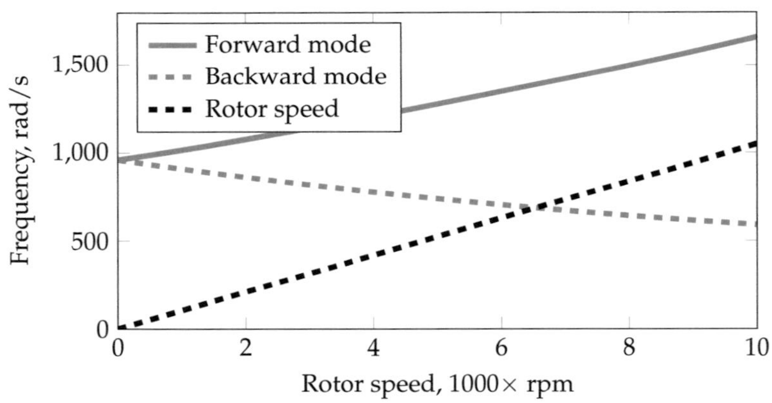

Fig. 13. Campbell diagram.

The significant splitting of the first flexible mode causes the corresponding peak to shift considerably in the frequency response. That affects the controller significantly as it must stabilize the system all the way from the start and up to the nominal speed. The most challenging part of the acceleration curve is around the points where the rotor crosses the flexible mode. In Fig. 13 this happens around 6500 rpm at that point the structural mode gets in resonance with the rotor speed resulting in a significant oscillation magnitude. To overcome this problem, a set of different controllers are synthesized for different rotational speeds. In that case a problem during the switch appears as different controllers have different levels of signals for the same operational point. The problem gets even more significant in MIMO systems. This drawback is treated with bumpless switch techniques (Li (2007); Turner

& Walker (2000)). Another approach to avoid bumps is to interpolate the controllers along with a changing parameter, which is the rotor speed in this case. The last method provides some restrictions, such that interpolated controllers should have the same order and structure. The interpolated controllers are called gain-scheduled controllers (Leith & Leithead, 2000). In particular, the implementation of the AMBs can be presented as a system where the rotor speed is a linear varying parameter. Thus, it is referred to as a category of LPV systems, for which a special LPV gain-scheduling methods can be applied. These methods are free of the above-mentioned drawbacks and provide a unified approach for controller synthesis.
An idea to systematically utilize the rotor speed for the controller adjustment in AMBs was proposed by different authors in different ways. One of the first examples is presented in the work of Matsumura et al. (1996). The authors synthesize a robust loops-shaping controller, which is able to reject sinusoidal disturbance with the rotor rotational frequency. It is carried out by adopting additional boundary constraints for an $\mathcal{H}_\infty$ problem. The model used by the authors contained only rigid modes and the rotor under consideration was without unbalance. Lu et al. (2008) applied an LPV technique for an AMB system. Their model also contained only rigid modes, and additionally, a special technique was used to identify uncertainty and provide weighting functions for the controller synthesis. The authors presented the controller in a set of parameter-dependent LMIs via a Lyapunov function. The basic controller was a general $\mathcal{H}_\infty$ problem with weighting functions.
The problem of LPV controllers was also investigated in the work of Li (2007), where the author compared an LFT approach with a Lyapunov function approach and additionally, with a "frozen" $\mathcal{H}_\infty$ controller. The model used was highly accurate, including not only higher-frequency modes but also structural resonances resulting in a nominal model with as many as 48 states. The author provided the comparison based only on the theoretical γ values of an $\mathcal{H}_\infty$ controllers.
Here, the system model in an LPV form is presented. Based on the Lyapunov function approach to an LPV gain-scheduling a controller synthesis procedure is described. The achieved controllers are compared with the optimal $\mathcal{H}_\infty$ controllers based on the maximum singular values. Additionally, simulations with a non-linear model and unbalance presented in the rotor are discussed. An LPV model of an AMB system is obtained by the same linearization around the operational point as an ordinary AMB model. The system is assumed to be in the operational point in the center of magnetic forces. Only small deviations from that point are considered as a rotor displacement. The rotor movement is translated to the center of mass resulting in a system with five degrees of freedom. The system is decoupled between the z and x, y axes, providing four states for the radial case. The states are displacements in the x and y and rotations around the x and y axes. They are respectively denoted as $\boldsymbol{q}_c^T = [x\ y\ \alpha\ \beta]$. It can be seen that the model (16) and the following (17) has the form of an LPV system

$$\begin{aligned} \dot{\boldsymbol{x}} &= \mathbf{A}(\Omega)\boldsymbol{x} + \mathbf{B}\boldsymbol{u}, \\ \boldsymbol{y} &= \mathbf{C}\boldsymbol{x}. \end{aligned} \tag{31}$$

In general, there are two approaches to the controller synthesis for the LPV plants. The first approach is based on a Lyapunov function. A quadratic Lyapunov function was used to achieve a set of parameter dependent LMIs by Becker & Packard (1994). Later, less conservative results were obtained by incorporating boundaries on parameters variation rates by Wu et al. (1996). As LMIs in this approach are parameter dependent, in general, there are an infinite number of them to solve. Usually, it is suggested to grid the space of varying parameters and achieve a solution for a limited number of points. The method is proposed for a small number of varying parameters. In an AMB controller synthesis one parameter can be considered as varying. However, the order of the system is rather high and griding with more than five points results in an unrealistically long synthesis time (Li, 2007). Another drawback is that an implementation requires a matrix inversion in real time that is quite challenging with the desired sampling rate on available microcontrollers.

The second one is based on a small gain theorem. The plant is considered as a linear time invariant (LTI) system, which is closed by the feedback loop with a varying parameter. Thus, the full system can be presented as a lower fractional transformation (LFT) of an LTI part and a parameter as presented in Fig. 14(a). Solutions for continuous- and discrete-time cases in a form of LMIs were presented by Apkarian & Gahinet (1995). The solution is conservative compared with the first approach as the realness of parameters is not used. Helmersson (1995) provided an additional research to reduce conservatism by introducing a rate of variation of parameters, which, however, leads to infinite dimensional solvability conditions. A more detailed overview of the research in that field is given by (Leith & Leithead, 2000).

In this work the first approach is taken and in particular, its extension to an affine system proposed by Apkarian et al. (1995). It can be seen from (16) that the system provides a convex set on the speed parameter Ω. Thus, the problem of infinite LMIs is avoided by solving them only at vertexes of the varied parameter. An additional benefit is that the controller implementation is simple and does not require matrix inversion. However, the system considered by Apkarian et al. (1995) neglects the varying state, input, and output transformations, and provides sufficient results under the assumption of a slow-varying parameter as discussed by Leith & Leithead (2000). The slow-varying parameter assumption is valid for an AMB system as the speed variation is relatively slow compared with state variations.

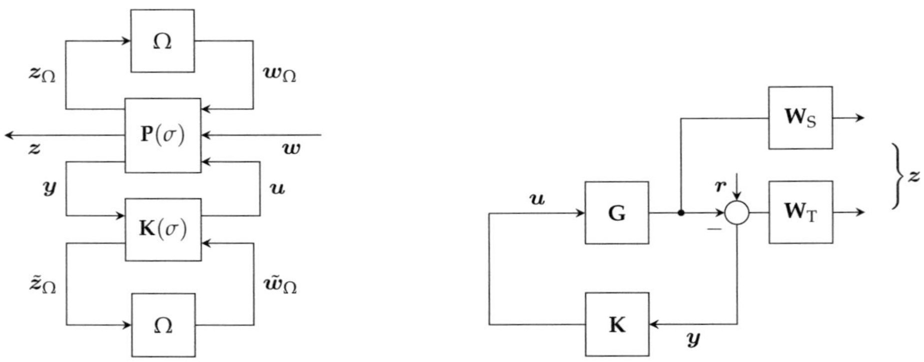

(a) LPV controller structure (b) Mixed sensitivity control problem

Fig. 14. Controller structures.

The general linear $\mathcal{H}_\infty$ problem is to find a controller $\mathbf{K}$ such that it minimizes (26). The generalized plant $\mathbf{P}$ is obtained from the original plant by providing additional exogenous inputs $\boldsymbol{w}$ and outputs $\boldsymbol{z}$ with specified weights to tune the desired system. The particular scheme for the mixed sensitivity problem is presented in Fig. 14(b). The following weights were used to shape the plant to the desired objectives

$$\mathbf{W}_\mathrm{S} = \mathbf{I}_{4\times 4} \cdot 0.5\frac{s+144}{s+0.144}, \quad \mathbf{W}_\mathbf{T} = \mathbf{I}_{4\times 4} \cdot 0.5\frac{s+0.01}{s+10}. \tag{32}$$

The general rule is that the weight for the sensitivity function is a low-pass filter and for the complementary sensitivity one is a high-pass filter.
Having the generalized plant **P** in the form

$$\mathbf{P}(s) = \mathbf{D} + \mathbf{C}(s\mathbf{I} - \mathbf{A})^{-1}\mathbf{B} \tag{33}$$

we follow the authors of (Apkarian et al., 1995) and solve the following LMIs

$$\begin{pmatrix} \mathcal{N}_\mathrm{R} & 0 \\ \hline 0 & \mathbf{I} \end{pmatrix}^T \begin{pmatrix} \mathbf{A}_i\mathbf{R} + \mathbf{R}\mathbf{A}_i^T & \mathbf{R}\mathbf{C}_{1i}^T & \mathbf{B}_{1i} \\ \mathbf{C}_{1i}\mathbf{R} & -\gamma\mathbf{I} & \mathbf{D}_{11i} \\ \hline \mathbf{B}_{1i}^T & \mathbf{D}_{11i}^T & -\gamma\mathbf{I} \end{pmatrix} \begin{pmatrix} \mathcal{N}_\mathrm{R} & 0 \\ \hline 0 & \mathbf{I} \end{pmatrix} < 0, \quad i = 1, \ldots, r, \tag{34}$$

$$\begin{pmatrix} \mathcal{N}_\mathrm{S} & 0 \\ \hline 0 & \mathbf{I} \end{pmatrix}^T \begin{pmatrix} \mathbf{A}_i\mathbf{S} + \mathbf{S}\mathbf{A}_i^T & \mathbf{S}\mathbf{B}_{1i} & \mathbf{C}_{1i}^T \\ \mathbf{B}_{1i}^T\mathbf{S} & -\gamma\mathbf{I} & \mathbf{D}_{11i}^T \\ \hline \mathbf{C}_{1i} & \mathbf{D}_{11i} & -\gamma\mathbf{I} \end{pmatrix} \begin{pmatrix} \mathcal{N}_\mathrm{S} & 0 \\ \hline 0 & \mathbf{I} \end{pmatrix} < 0, \quad i = 1, \ldots, r, \tag{35}$$

$$\begin{pmatrix} \mathbf{R} & \mathbf{I} \\ \mathbf{I} & \mathbf{S} \end{pmatrix} \geq 0, \tag{36}$$

where the bases of the null spaces of $(\mathbf{B}_2^T, \mathbf{D}_{12}^T)$ and $(\mathbf{C}_2, \mathbf{D}_{21})$ are denoted $\mathcal{N}_\mathrm{R}$ and $\mathcal{N}_\mathrm{S}$, respectively. Next, the unique solution $\mathbf{X}_\mathrm{cl}$ of the matrix equation $\mathbf{\Pi}_2 = \mathbf{X}_\mathrm{cl}\mathbf{\Pi}_1$ should be computed, where

$$\mathbf{\Pi}_2 = \begin{pmatrix} \mathbf{S} & \mathbf{I} \\ \mathbf{N}^T & \mathbf{0} \end{pmatrix}, \quad \mathbf{\Pi}_1 = \begin{pmatrix} \mathbf{I} & \mathbf{R} \\ \mathbf{0} & \mathbf{M}^T \end{pmatrix} \tag{37}$$

and the matrices **M** and **N** are such that

$$\mathbf{M}\mathbf{N}^T = \mathbf{I} - \mathbf{R}\mathbf{S}. \tag{38}$$

The controllers for each vertex can be found by solving the following LMI

$$\begin{pmatrix} \mathbf{A}_{\mathrm{K}i}^T\mathbf{X}_\mathrm{cl} + \mathbf{X}_\mathrm{cl}\mathbf{A}_{\mathrm{K}i} & \mathbf{X}_\mathrm{cl}\mathbf{B}_{\mathrm{K}i} & \mathbf{C}_{\mathrm{K}i}^T \\ \mathbf{B}_{\mathrm{K}i}^T\mathbf{X}_\mathrm{cl} & -\gamma\mathbf{I} & \mathbf{D}_{\mathrm{K}i}^T \\ \mathbf{C}_{\mathrm{K}i} & \mathbf{D}_{\mathrm{K}i} & -\gamma\mathbf{I} \end{pmatrix} < 0. \tag{39}$$

Having the state-space matrices for each vertex $\mathbf{A}_{\mathrm{K}i}$, $\mathbf{B}_{\mathrm{K}i}, \mathbf{C}_{\mathrm{K}i}, \mathbf{D}_{\mathrm{K}i}$ the controller for the particular point is obtained as

$$\begin{bmatrix} \mathbf{A}_\mathrm{K}(\Omega) & \mathbf{B}_\mathrm{K}(\Omega) \\ \mathbf{C}_\mathrm{K}(\Omega) & \mathbf{D}_\mathrm{K}(\Omega) \end{bmatrix} = \sum_{i=1}^{r} \alpha_i \begin{bmatrix} \mathbf{A}_{\mathrm{K}i} & \mathbf{B}_{\mathrm{K}i} \\ \mathbf{C}_{\mathrm{K}i} & \mathbf{D}_{\mathrm{K}i} \end{bmatrix}, \tag{40}$$

where α_i is such that

$$\Omega = \left\{ \sum_{i=1}^{r} \alpha_i\omega_i : \alpha_i \geq 0, \sum_{i=1}^{r} \alpha_i = 1 \right\}. \tag{41}$$

The r denotes the total number of vertices and ω_i the particular vertex. The parameter Ω is measured in real time and the controllers are updated on each step. For the controller testing, we use the system with the Campbell diagram (see Fig. 13) presented above. The polar moment of inertia $I_\mathrm{p} = 10.6\ \mathrm{kg\,m^2}$ is greater than the diametral moment $I_\mathrm{d} = 0.59\ \mathrm{kg\,m^2}$. However, the speed varies from 0 to 10 000 rpm, and thus, the system has significant gyroscopic effect. The controller is synthesized using the above-mentioned weighting functions (32). As the main objective is the stability of the system, a maximum singular value of sensitivity functions is evaluated. A good starting point for a comparison is an output sensitivity function $\mathbf{S}_o$ (Li, Lin & Allaire, 2006). Additionally, an output complementary sensitivity function $\mathbf{T}_o$ is used for the evaluation.

$$\mathbf{T}_o = \mathbf{GK}(\mathbf{I} + \mathbf{GK})^{-1}. \tag{42}$$

The evaluation of the controller is carried out in each point of a variation parameter. It means that the closed loop transfer functions (30) and (42) are calculated at each rotor speed and their peak value is found. The peak values of the MIMO system are defined as maximum singular values in the same fashion as in (26). The achieved results are presented in Fig. 15.

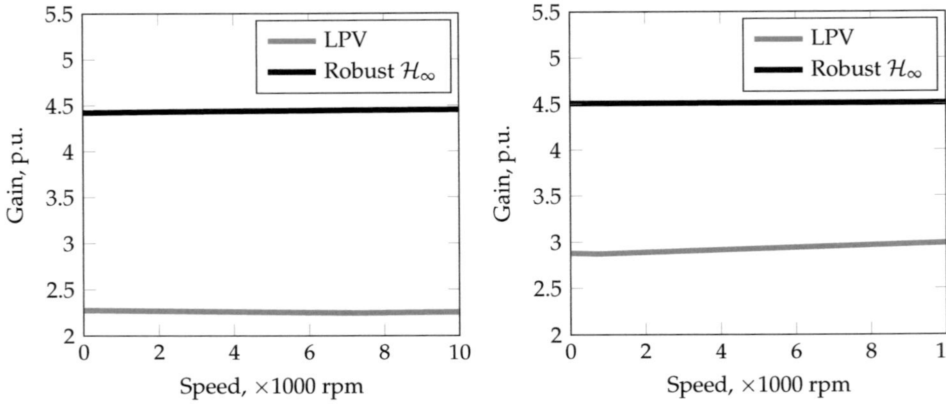

(a) Output sensitivity (b) Output complementary sensitivity

Fig. 15. Maximum values of sensitivity functions.

It is seen that an LPV controller provides a stable system with low values of sensitivity functions. For comparison a robust $\mathcal{H}_\infty$ controller is synthesized using the same weights and the same structure. In this synthesis, speed is treated as a structured uncertainty. In Fig. 15 it is seen that the robust controller has higher peak values and additionally, the values do not change over the parameter variation. The LPV controller peak values are smaller and have some deviation. In general, Fig. 15 shows that the LPV controller provides a greater stability margin.

The previous assessment was based on a theoretical model and provides the basic insight into the stability margins. For a deeper evaluation, simulations with a non-linear model are carried out. The rotor for the simulations is considered to have an imbalance of 0.01 kg, and a system with three flexible modes is used. The force-current relations are non-linear; they include the actuator delay and are based on look-up tables from the switch-reluctance network model.

A typical case of the rotor acceleration is simulated. The speed increases linearly from zero up to the maximum value. The beginning and ending phases of acceleration are smoothed to avoid unrealistic sharp edges. The results are evaluated for an LPV and a robust $\mathcal{H}_\infty$ controllers. The displacements at the A-end in the x direction are presented in Fig. 16.

The LPV controller shows worse performance for the transient response. The magnitude of oscillations is significantly higher around two times. The oscillations take place around the

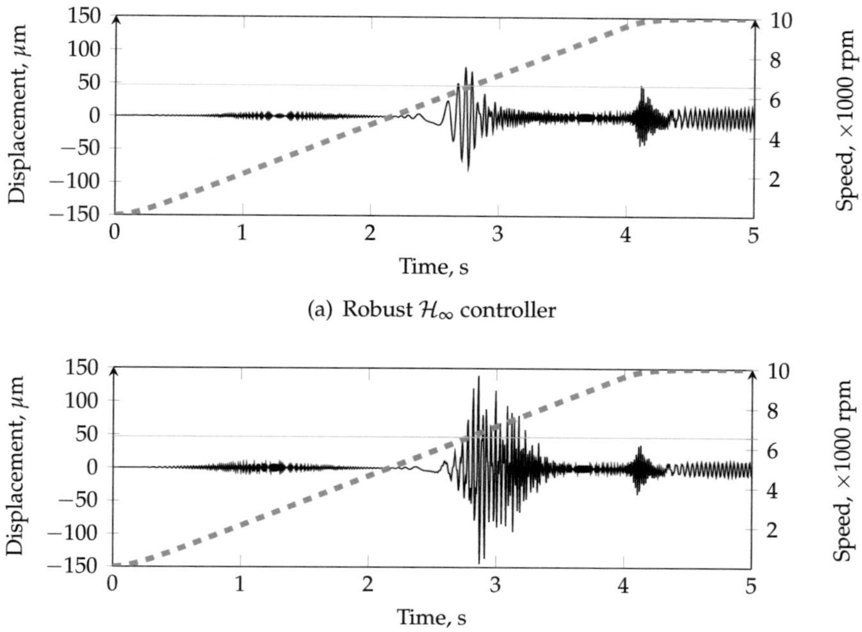

(a) Robust $\mathcal{H}_\infty$ controller

(b) LPV controller

Fig. 16. Rotor acceleration responses.

point of 6500 rpm where the system crosses the first flexible mode. The second point where the system experiences oscillations is close to the maximum speed and it can be explained by the deceleration of the rotor. The LPV controller has a lower magnitude of oscillations around this point; the difference is 35 %. Such a behavior can be explained by an adaptive nature of an LPV controller. In each step, the gains are modified according to the rotational speed. During the acceleration process, the system does not have enough time to adapt. This results in a higher amplitude of oscillations. During the later deceleration phase, the coefficients do not change that fast and performance is better. The speed of the parameter variation is a significant problem for the LPV controllers, and usually the main point of conservatism in that approach (Leith & Leithead, 2000).

The second simulation experiment in the steady state proves that LPV controller provides a better performance. In this experiment, a step disturbance to the x channel of the rotor A-end is applied at the maximum rotational speed. The simulation results are presented in Fig. 17. The magnitude of the disturbance response for an LPV controller is about three times smaller than that of a robust controller. Additionally, the LPV controller does not have coupling between different ends, so the disturbance does not propagate through the system.

6. Real-time operating conditions

The AMB-based system requires hard-real time controllers. In the case of a robust control strategy, the control law is of higher complexity than other solutions. Therefore, the implementation of the control law must fulfill the requirements of the target control system such as finite precision of the arithmetic and number format and available computational

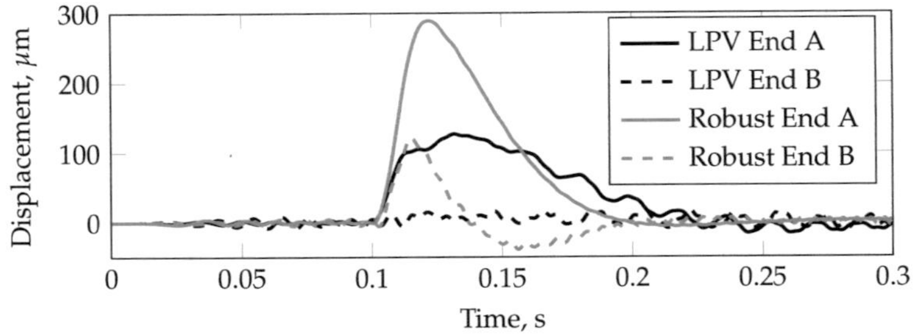

Fig. 17. Step disturbance response for controllers in the x direction.

power. The digital control realization requires a digital controller that matches the continuous form in the operating frequency range. The controllers for the radial suspension of the AMB rotor system are tested using a dSpace DS1005-09 digital control board and a DS4003 Digital Input/Output system board as a regulation platform. The Simulink and Real-time Workshop software are applied for automatic program code generation. The selected sampling rate is 10 kHz. The resolution of the applied ADCs is 16 bits. The control setup limits the maximum number of states of the implemented controllers to 28 states.

7. Conclusions

The chapter discusses options and feasible control solutions when building uncertain AMB rotor models and when designing a robust control for the AMB rotor systems. The review of the AMB systems is presented. The recommendations for difficult weight selection in different weighting schemes are given. Design-specific problems and trade-offs for each controller are discussed. It is shown that the operating conditions of the selected real-time controllers satisfy the control quality requirements. The resulting order of the controller depends on the complexity of the applied weighting scheme, plant order, and applied uncertainties. The detailed interconnections lead to controllers, which are difficult to implement and are not transparent. However, the too simple weighting schemes cannot provide sufficient design flexibility with respect to the multi-objective specification. For the systems with considerably gyroscopic rotors and high rotational speeds, the LPV method provides a significantly better solution than nonadaptive robust control methods.

8. Acknowledgement

This chapter was partially founded by AGH Research Grant no 11.11.120.768

9. References

Apkarian, P. & Gahinet, P. (1995). A convex characterization of gain-scheduled H_∞ controllers, *Automatic Control, IEEE Transactions on* 40(5): 853–864.

Apkarian, P., Gahinet, P. & Becker, G. (1995). Self-scheduled H_∞ control of linear parameter-varying systems: a design example, *Automatica* 31(9): 1251–1261.

Battachatyya, S. P., Chapellat, H. & Keel, L. H. (1995). *Robust Control The Parametric Approach*, Prentice Hall.

Becker, G. & Packard, A. (1994). Robust performance of linear parametrically varying systems using parametrically-dependent linear feedback, *Systems & Control Letters* 23(3): 205–215.

Fujita, M., Hatake, K. & Matsumura, F. (1993). Loop shaping based robust control of a magnetic bearing, *Control Systems Magazine, IEEE* 13(4): 57–65.

Fujita, M., Namerikawa, T., Matsamura, F. & Uchida, K. (1995). mi-synthesis of an electromagnetic suspension system, *IEEE Transactions on Automatic Control* 40: 530–536.

Glover, K. & McFarlane, D. (1989). Robust stabilization of normalized coprime factor plant descriptions with H_∞-bounded uncertainty, *Automatic Control, IEEE Transactions on* 34(8): 821–830.

Gosiewski, Z. & Mystkowski, A. (2008). Robust control of active magnetic suspension: Analytical and experimental results, *Mechanical Systems and Signal Processing* 22: 1297–1303.

Gu, D., Petkov, P. & Konstantinov, M. (2005a). *Robust Control Design with MATLAB*, Springer.

Gu, D.-W., Petkov, P. & Konstantinov, M. (2005b). *Robust Control Design with MATLAB*, Springer, Leipzig, Germany.

Helmersson, A. (1995). *Methods for robust gains scheduling*, PhD thesis, Linkoping University.

InTeCo (2008). *MLS2EM, Magnetic Levitation User's Guide*, InTeCo, Poland.

Jastrzebski, R. (2007). *Design and Implementation of FPGA-based LQ Control of Active Magnetic Bearings*, PhD thesis, LUT, Finland.

Jastrzebski, R., Hynynen, K. & Smirnov, A. (2010). *H*-infinity control of active magnetic suspension, *Mechanical Systems and Signal Processing* 24(4): 995–1006.

Jastrzebski, R. & Pöllänen, R. (2009). Centralized optimal position control for active magnetic bearings - comparison with decentralized control, *Electrical Engineering* 91(2): 101–114.

Kwakernaak, H. (1993). Robust control and hinf-optimization tutorial paper, *Automatica* 29: 253–273.

Kwakernaak, H. (2002). H2-optimization theory and applications to robust control design, *Annual Reviews in Control* 26: 45–56.

Lanzon, A. & Tsiotras, P. (2005). A combined application of H infin; loop shaping and mu;-synthesis to control high-speed flywheels, *Control Systems Technology, IEEE Transactions on* 13(5): 766–777.

Leith, D. J. & Leithead, W. E. (2000). Survey of gain-scheduling analysis and design, *International Journal of Control* 73(11): 1001–1025.

Li, G. (2007). *Robust stabilization of rotor-active magnetic bearing systems*, PhD thesis, University of Virginia.

Li, G., Lin, Z. & Allaire, P. (2006). Uncertainty classification of rotor-amb systems, *Proc. of* 11th *International Symposium on Magnetic Bearings*.

Li, G., Lin, Z., Allaire, P. & Luo, J. (2006). Modeling of a high speed rotor test rig with active magnetic bearings, *Journal of Vibration and Acoustics* 128: 269–281.

Limebeer, D. J. N., Kasenally, E. M. & Perkins, J. D. (1993). On the design of robust two degree of freedom controllers, *Automatica* 29(1): 157–168.

Losch, F. (2002). *Identification and Automated Controller Design for Active Magnetic Bearing Systems*, Swiss Federal Institute of Technology, ETH Zurich.

Lu, B., Choi, H., Buckner, G. D. & Tammi, K. (2008). Linear parameter-varying techniques for control of a magnetic bearing system, *Control Engineering Practice* 16(10): 1161–1172.

Lunz, J. (1989). *Robust Multivariable Feedback Control*, Prentice Hall, London.

Matsumura, F., Namerikawa, T., Hagiwara, K. & Fujita, M. (1996). Application of gain scheduled H_∞ robust controllers to a magnetic bearing, *Control Systems Technology, IEEE Transactions on* 4(5): 484–493.

Moser, A. (1993). Designing controllers for flexible structures with H-infinity/μ-synthesis, *IEEE Control Systems* pp. 79–89.

Mystkowski, A. & Gosiewski, Z. (2009). Uncertainty modeling in robust control of active magnetic suspension, *Solid State Phenomena* 144: 22–26.

Oliveira, V., Tognetti, E. & Siqueira, D. (2006). Robust controllers enhanced with design and implementation processes, *IEEE Trans. on Education* 49(3): 370–382.

Pilat, A. (2002). *Control of Magnetic Levitation Systems*, PhD thesis, AGH University of Science and Technology.

Pilat, A. (2009). Stiffness and damping analysis for pole placement method applied to active magnetic suspension (in polish), *Automatyka* 13: 43–54.

Pilat, A. (2010). mi-synthesis of robust controller for active magnetic levitation system, *MSM 2010 : Mechatronic Systems and Materials : 6th international conference : 5-8 July, Opole, Poland*.

Pilat, A. & Piatek, P. (2008). Multichannel control and measurement board with parallel data processing (in polish), *in* L. Trybus & S. Samolej (eds), *Recent advances in control and automation*, Academic Publishing House EXIT, pp. 00–00.

Pilat, A. & Turnau, A. (2005). Self-organizing fuzzy controller for magnetic levitation system, *Computer Methods and Systems, Krakï£¡w, Poland*, pp. 101–106.

Pilat, A. & Turnau, A. (2009). Neural adapted controller learned on-line in real-time, *14 International Conference on Methods and Models in Automation and Robotics, 19-21 August, Miedzyzdroje, Poland*.

Safonov, M. G. & Chiang, R. Y. (1988). A Schur Method for Balanced Model Reduction, *American Control Conference, 1988*, pp. 1036–1040.

Sawicki, J. & Maslen, E. (2008). Toward automated amb controller tuning: Progress in identification and synthesis, *Proc. of* 11th *International Symposium on Magnetic Bearings*, pp. 68–74.

Scherer, C., Gahinet, P. & Chilali, M. (1997). Multiobjective output-feedback control via LMI optimization, *IEEE Transations on Automatic Control* 42(7): 896–911.

Schweitzer, G. & Maslen, E. (2009). *Magnetic Bearings: Theory, Design, and Application to Rotating Machinery*, Springer, New York.

Sefton, J. & Glover, K. (1990). Pole/zero cancellations in the general [infinity] problem with reference to a two block design, *Systems and Control Letters* 14(4): 295–306.

Skogestad, S. & Postlethwaite, I. (2005). *Multivariable Feedback Control Analysis and Design*, 2 edn, John Wiley & Sons Ltd., England.

Turner, M. C. & Walker, D. J. (2000). Linear quadratic bumpless transfer, *Automatica* 36(8): 1089–1101.

Whidborne, J., Postlethwaite, I. & Gu, D.-W. (1994). Robust controller design using H_∞ loop-shaping and the method of inequalities, *IEEE Transations on Control Systems Technology* 2(2): 455–461.

Wu, F., Yang, X. H., Packard, A. & Becker, G. (1996). Induced L2-norm control for LPV systems with bounded parameter variation rates, *International Journal of Robust and Nonlinear Control* 6(9-10): 983–998.

Zhou, K. (1998). *Essentials of Robust Control*, Prentice-Hall, Upper Saddle River, NJ.

Zhou, K., Doyle, J. & Glover, K. (1996). *Robust and Optimal Control*, Prentice-Hall, Englewood Cliffs, NJ.

Part 3

Distillation Process Control and Food Industry Applications

11

Reactive Distillation: Control Structure and Process Design for Robustness

V. Pavan Kumar Malladi[1] and Nitin Kaistha[2]

[1]*Department of Chemical Engineering, National Institute of Technology Calicut, Kozhikode,*

[2]*Department of Chemical Engineering, Indian Institute of Technology Kanpur, Kanpur, India*

1. Introduction

Reactive Distillation (RD) is the combination of reaction and distillation in a single vessel (Backhaus, 1921). Over the past two decades, it has emerged as a promising alternative to conventional "reaction followed by separation" processes (Towler & Frey, 2002). The technology is attractive when the reactant-product component relative volatilities allow recycle of reactants into the reactive zone via rectification/stripping and sufficiently high reaction rates can be achieved at tray bubble temperature. For equilibrium limited reactions, the continuous removal of products drives the reaction to near completion (Taylor & Krishna, 2000). The reaction can also significantly simplify the separation task by reacting away azeotropes (Huss et al., 2003). The Eastman methyl acetate RD process that replaced a reactor plus nine column conventional process with a single column is a classic commercial success story (Agreda et al., 1990). The capital and energy costs of the RD process are reported to be a fifth of the conventional process (Siirola, 1995).

Not withstanding the potentially significant economic advantages of RD technology, the process integration results in reduced number of valves for regulating both reaction and separation with high non-linearity due to the reaction-separation interaction (Engell & Fernholtz, 2003). Multiple steady states have been reported for several RD systems (Jacobs & Krishna, 1993; Ciric & Miao 1994; Mohl et al., 1999). The existence of multiple steady states in an RD column can significantly compromise column controllability and the design of a robust control system that effectively rejects large disturbances is a principal consideration in the successful implementation of the technology (Sneesby et al., 1997).

In this Chapter, through case studies on a generic double feed two-reactant two-product ideal RD system (Luyben, 2000) and the methyl acetate RD system (Al-Arfaj & Luyben, 2002), the implications of the non-linear effects, specifically input and output multiplicity, on open and closed loop column operation is studied. Specifically, steady state transitions under open and closed loop operation are demonstrated for the two example systems. Input multiplicity, in particular, is shown to significantly compromise control system robustness with the possibility of "wrong" control action or a steady state transition under closed loop operation for sufficiently large disturbances.

Temperature inferential control system design is considered here due to its practicality in an industrial setting. The design of an effective (robust) temperature inferential control system requires that the input-output pairings be carefully chosen to avoid multiplicity in the vicinity of the nominal steady state. A quantitative measure is developed to quantify the severity of the multiplicity in the steady-state input output relations. In cases where an appropriate tray temperature location with mild non-linearity cannot be found, it may be possible to "design" a measurement that combines different tray temperatures for a well-behaved input-output relation and consequently robust closed loop control performance. Sometimes temperature inferential control (including temperature combinations) may not be effective and one or more composition measurements may be necessary for acceptable closed loop control performance. In extreme cases, the RD column design itself may require alteration for a controllable column. RD column design modification, specifically the balance between fractionation and reaction capacity, for reduced non-linearity and better controllability is demonstrated for the ideal RD system. The Chapter comprehensively treats the role of non-linear effects in RD control and its mitigation via appropriate selection/design of the measurement and appropriate process design.

2. Steady state multiplicity and its control implications

Proper regulation of an RD column requires a control system that maintains the product purities and reaction conversion in the presence of large disturbances such as a throughput change or changes in the feed composition etc. This is usually accomplished by adjusting the column inputs (e.g. boil-up or reflux or a column feed) to maintain appropriate output variables (e.g. a tray temperature or composition) so that the purities and reaction conversion are maintained close to their nominal values regardless of disturbances. The steady state variation in an output variable to a change in the control input is referred to as its open loop steady state input-output (IO) relation. Due to high non-linearity in RD systems, the IO relation may not be well behaved exhibiting gain sign reversal with consequent steady state multiplicity.

From the control point of view, the multiplicity can be classified into two types, namely, input multiplicity and output multiplicity as shown in Figure 1. In case of output multiplicity, multiple output values are possible at a given input value (Figure 1(a)). Input multiplicity is implied when multiple input values result in the same output value (Figure 1(b)).

To understand the implications of input/output multiplicity on control, let us consider a SISO system. Let the open loop IO relation exhibit output multiplicity with the nominal operating point denoted by '*'(Figure 1(a)). Under open loop operation, a large step decrease in the control input from u_0 to u_1 would cause the output to decrease from y_0 to y_1. Upon increasing the input back to u_0, the output would reach a different value y_0' on the lower solution branch. For large changes in the control input (or alternatively large disturbances), the SISO system may exhibit a steady state transition under open loop operation. For RD systems, this transition may correspond to a transition from the high conversion steady state to a low conversion steady state. The transition can be easily prevented by installing a feedback controller with its setpoint as y_0. Since the output values at the three possible steady states corresponding to u_0 are distinct, it is theoretically possible to drive the system to the desired steady state with the appropriate setpoint (Kienle & Marquardt, 2003). Note

that for the IO relation in Figure 1(a), the feedback controller would be reverse acting for y_0/y_0' and direct acting for y_0'' as the nominal steady state.
The implications of input multiplicity in an IO relation are much more severe. To understand the same, consider a SISO system with the IO relation in Figure 1(b) and the point marked '*' as the nominal steady state. Assume a feedback PI controller that manipulates u to maintain y at y_0. Around the nominal steady state, the controller is direct acting. Let us consider three initial steady states marked a, b and c on the IO relation, from where the controller must drive the output to its nominal steady state. At a, the initial error (y^{SP}-y) is positive and the controller would decrease u to bring y to the desired steady state. At b, the error is again positive and the system gets driven to the desired steady state with the controller reducing u. At c, due to the y^{SP} crossover in the IO relation, the error signal is negative and the direct acting controller would increase u, which is the *wrong* control action. Since the IO relation turns back, the system would settle down at the steady state marked '**'. For large disturbances, a SISO system with input multiplicity can succumb to wrong control action with the control input saturating or a steady state transition if the IO relation exhibits another branch with the same slope sign as the nominal steady state. Input multiplicity or more specifically, multiple crossovers of y^{SP} in the IO relationship thus severely compromise control system robustness.

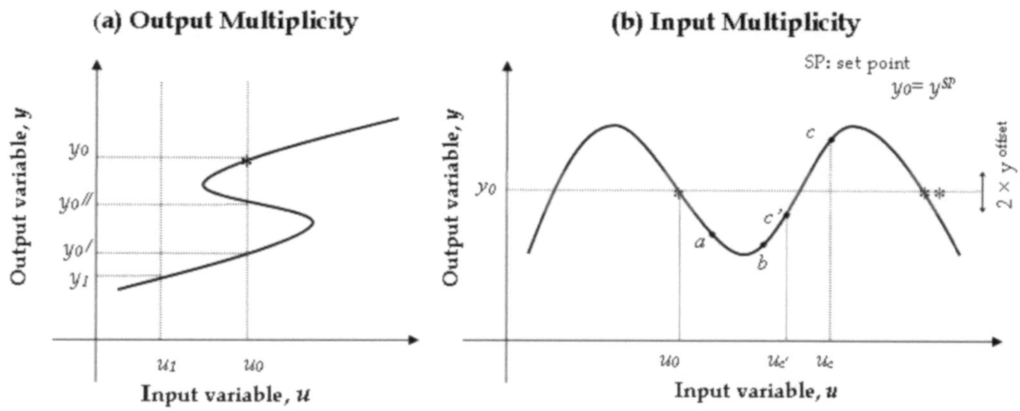

Fig. 1. Steady state multiplicity, (a) Output multiplicity, (b) Input multiplicity

The suitability of an input-output (IO) pairing for RD column regulation can be assessed by the steady state IO relation. Candidate output variables should exhibit good sensitivity (local slope in IO relation at nominal operating point) for adequate muscle to the control system where a small change in the input drives the deviating output back to its setpoint. Of these candidate sensitive (high open loop gain) outputs, those exhibiting output multiplicity may be acceptable for control while those exhibiting input multiplicity may compromise control system robustness due to the possibility of wrong control action. The design of a robust control system for an RD column then requires further evaluation of the IO relations of the sensitive (high gain) output variables to select the one(s) that are monotonic for large changes in the input around the nominal steady state and avoid multiple y^{SP} crossovers. If such a variable is not found, the variable with a y^{SP} crossover point (input multiplicity), that is the furthest from the nominal operating point should be selected. It may also be possible to combine different outputs to design one that avoids crossover (input multiplicity). The magnitude $|u_0 - u_c|$, where u_c is the input value at the nearest y^{SP} crossover can be used as a criterion to screen out candidate outputs. For robustness, Kumar & Kaistha (2008) define the rangeability, r, of an IO relation as

$$r = |u_0 - u_c'|$$

where u_c' is obtained for $y = y^{SP} - y^{offset}$ as shown in Figure 1(b). The offset from the actual crossover point ensures robustness to disturbances such as a bias in the measurement. In extreme cases, where a suitable output variable is not found that can effectively reject large disturbances, the RD column design may require alteration for improving controllability. Each of these aspects is demonstrated in the following example case studies on a hypothetical two-reactant two-product ideal RD column and an industrial scale methyl acetate RD column.

3. RD control case studies

To demonstrate the impact of steady state multiplicity on RD control, two double feed two-reactant two-product RD columns with stoichiometric feeds (neat operation) are considered in this work. The first one is an ideal RD column with the equilibrium reaction $A + B \leftrightarrow C + D$. The component relative volatilities are in the order $\alpha_C > \alpha_A > \alpha_B > \alpha_D$ so that the reactants are intermediate boiling. The RD column consists of a reactive section with rectifying and stripping trays respectively above and below it. Light fresh A is fed immediately below and heavy fresh B is fed immediately above the reactive zone. Product C is recovered as the distillate while product D is recovered as the bottoms. The rectifying and stripping trays recycle the reactants escaping the reactive zone and prevent their exit in the product streams. This hypothetical ideal RD column was originally proposed by Luyben (2000) as a test-bed for studying various control structures (Al-Arfaj & Luyben, 2000).

In terms of its design configuration, the methyl acetate column is similar to the ideal RD column with light methanol being fed immediately below and heavy acetic acid being fed immediately above the reactive section. The esterification reaction $CH_3COOH + CH_3OH \leftrightarrow CH_3COOCH_3 + H_2O$ occurs in the reactive zone with nearly pure methyl acetate recovered as the distillate and nearly pure water recovered as the bottoms.

Figure 2 shows a schematic of the two RD columns. The ideal RD column is designed to process 12.6 mol s^{-1} of stoichiometric fresh feeds to produce 95% pure C as the distillate product and 95% pure D as the bottoms product. Alternative column designs with 7 rectifying, 6 reactive and 7 stripping trays or 5 rectifying, 10 reactive and 5 stripping trays are considered in this work. For brevity, these designs are referred to as 7/6/7 and 5/10/5 respectively. The methyl acetate RD column is designed to produce 95% pure methyl acetate distillate. The 7/18/10 design configuration reported by Singh et al. (2005) is studied here. Both the columns are operated neat with stoichiometric feeds. The reaction and vapor liquid equilibrium model parameters for the two systems are provided in Table 1.

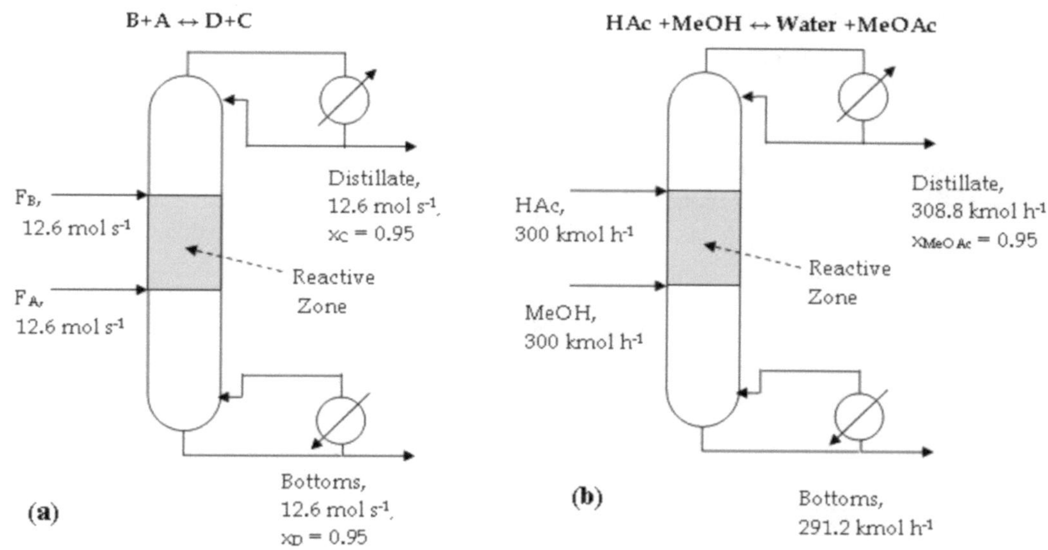

Fig. 2. Schematics of example RD columns. (a) Ideal, (b) Methyl acetate

	Ideal RD column	**Methyl acetate RD column**
Reaction	B+A ↔ D+C	Acetic acid + Methanol ↔ Water + Methyl Acetate
Relative volatility	$\alpha_C : \alpha_A : \alpha_B : \alpha_D$ $= 8:4:2:1$	Extended Antonie Equations are used for the estimation of saturation vapour pressure, temperature dependent
Liquid phase activity	Ideal	Wilson
Vapour phase	Ideal	Ideal with Marek Method (Marek, 1995) (Vapour dimerization of Acetic acid)
Reaction kinetics	$r_C(mol/mol\ s^{-1})$ $= k_f x_A x_B - k_b x_C x_D$	$r_{MeOAc} = \frac{M_C k_f \left[a_{HAc} a_{MeOH} - a_{MeOAc} a_{H_2O} / K_{eq} \right]}{\left[1 + K_{HAc} a_{HAc} + K_{MeOH} a_{MeOH} + K_{MeOAC} a_{MeOAC} + K_{H_2O} a_{H_2O} \right]^2}$
	$k_f = 2.4260 \times 10^{16} e^{15098.1/T}$ $k_b = 2.11768 \times 10^{-6} e^{5032.47/T}$	$K_{HAc} = 3.18; K_{MeOH} = 4.95; K_{H_2O} = 10.5$ $K_{MeOAc} = 0.82$ $k_f(kmol/kg_{cat}/h) = 69.42 \times 10^9 e^{-52275.93/(RT)}$ $K_{eq} = 2.32 e^{782.98/T}$
Heat of reaction	- 41840 kJ/kmol Temperature independent	- 33566.80 kJ/kmol at 330 K Temperature dependent

Table 1. VLE and reaction parameters of the example RD systems

3.1 Output multiplicity effects

To demonstrate the impact of output multiplicity on column operation, the 7/6/7 design with 1 kmol reaction holdup per reactive tray is considered for the ideal RD system. For 95% pure distillate and 95% pure bottoms, the reflux ratio and vapor boilup is found to be 2.6149 and 28.32 mol s^{-1}, respectively. For the methyl acetate RD column, the 7/18/10 design is considered. At the nominal design, the reflux ratio and reboiler duty is 1.875 and 4.6021 MW respectively for 95% methyl acetate distillate and 96.33% water bottoms.

3.1.1 Ideal RD column

The variation in the bottoms *D* purity with respect to the vapor boilup at constant reflux rate in the 7/6/7 ideal RD column design is shown in Figure 3(a). Both input and output multiplicity are present in the relation with respect to the nominal steady state. Output multiplicity is observed with three distinct purities for the product *D* other than the basecase

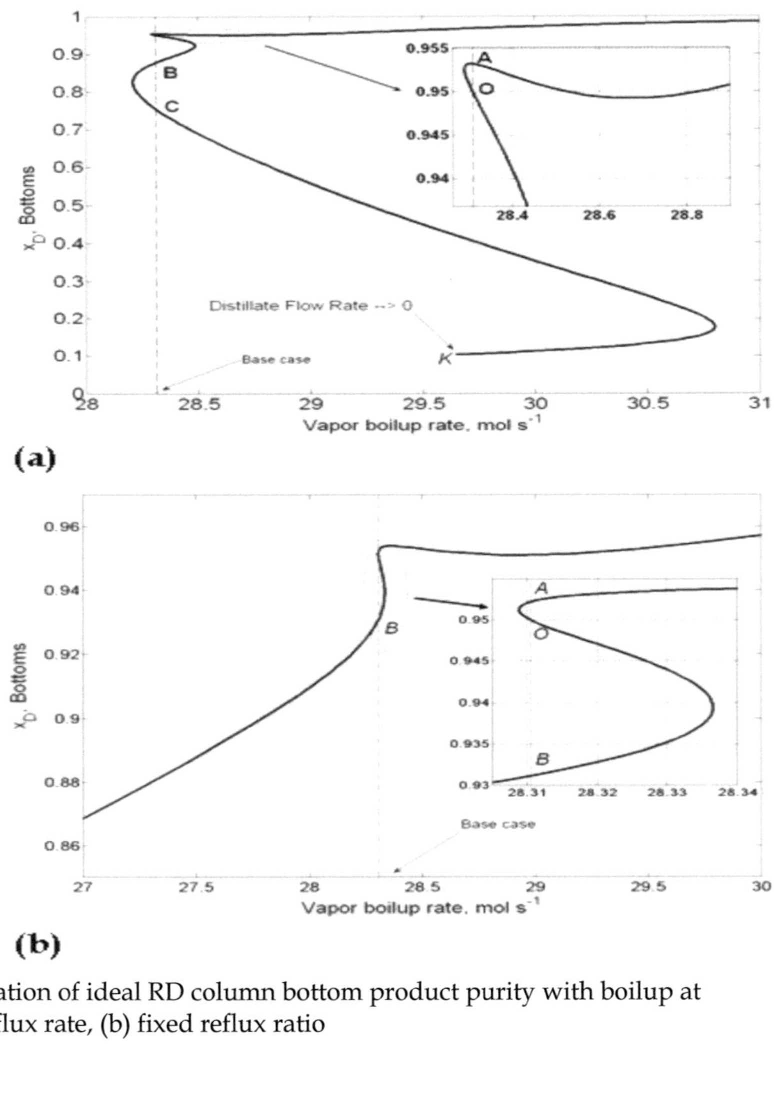

Fig. 3. Variation of ideal RD column bottom product purity with boilup at (a) fixed reflux rate, (b) fixed reflux ratio

design purity of 95%. At point *K* on the solution diagram, the distillate flow rate almost reaches 0 beyond which a steady solution is not found.

Figure 3(b) shows that IO relation of bottoms purity with vapor boilup at constant reflux to distillate ratio, a common operating policy implemented on distillation columns. Output multiplicity at the nominal steady state is evident in the Figure. Notice that a feasible steady state solution now exists for boilups below its nominal value, unlike for column operation at fixed reflux rate. From the column operation standpoint, maintaining reflux in ratio with the distillate is therefore a more pragmatic option as a feasible steady state exists for large changes in the vapor boilup in either direction.

To understand the implication of the observed steady state solution diagrams on column operation, the dynamic column response to a ±5% pulse change of one hour duration in the vapor boilup is obtained at a fixed reflux rate or at a fixed reflux ratio. The reflux drum and bottom sump levels are maintained using respectively the distillate and the bottoms flow (P controller with gain 2). The dynamic response is plotted in Figure 4. At constant reflux rate (Figure 4(a)), for the -5% boilup step change, the distillate rate quickly goes down to zero corresponding to no feasible solution in the solution diagram. For the +5% pulse change, the distillate rate settles at a slightly higher value of 12.623 mol s^{-1} (nominal value: 12.6 mol s^{-1}) implying an open loop steady state transition. This new steady state corresponds to Point *B* in the bifurcation diagram in Figure 3(a). For the -5% pulse, the distillate valve shuts down due to the absence of a feasible steady state solution for a large reduction in the boilup.

At fixed reflux ratio, a stable response is obtained for the ±5% pulse in boilup (Figure 4(b)). The column however ends up transitioning to different steady states for a +5% and a -5% pulse change, respectively. This is in line with the bifurcation diagram in Figure 3(b) with the column transitioning to a high conversion steady state (*A*) or a low conversion steady state (*B*) solution under open loop column operation.

Given the possibility of an open loop steady state transition due to output multiplicity, a PI controller is implemented that adjusts the reflux rate/reflux ratio to hold the distillate purity at 95%. The loop is tuned using the ATV method (Astrom & Hagglund, 1984) with Tyreus-Luyben settings (Tyreus & Luyben, 1992). At constant reflux rate, a boilup pulse change of -5% is handled with the column returning to its nominal steady state. In addition, a -5% step change is also handled with a stable response implying the existence of a steady state solution (feasibility) at low boilups with the distillate purity held constant. This is in contrast to the no feasible solution at reduced boilups for column operation at constant reflux rate. With the composition control loop on automatic, an unstable response is however observed for a large -20% step change which is likely due to the absence of a feasible steady state for low boilups at constant distillate composition. With the composition control loop, a +5% pulse change in the vapor boilup does not result in a steady state transition unlike for column operation at constant reflux and the column returns to its nominal steady state.

The implementation of a feedback loop controlling distillate purity by adjusting the reflux ratio results in the column returning to its nominal steady state for a ±5% pulse change in the boilup. The open loop steady state transition observed for the same pulse disturbance at constant reflux ratio is thus prevented. In addition, a -20% step change in the boilup results in a stable response with the column settling at a new steady state implying feasibility. These dynamic results serve to highlight that the implementation of feedback control serves to mitigate the non-linear effects of output multiplicity so that an open loop steady transition is prevented (Dorn et al., 1998). Feedback control also ensures feasible operation over a larger disturbance range.

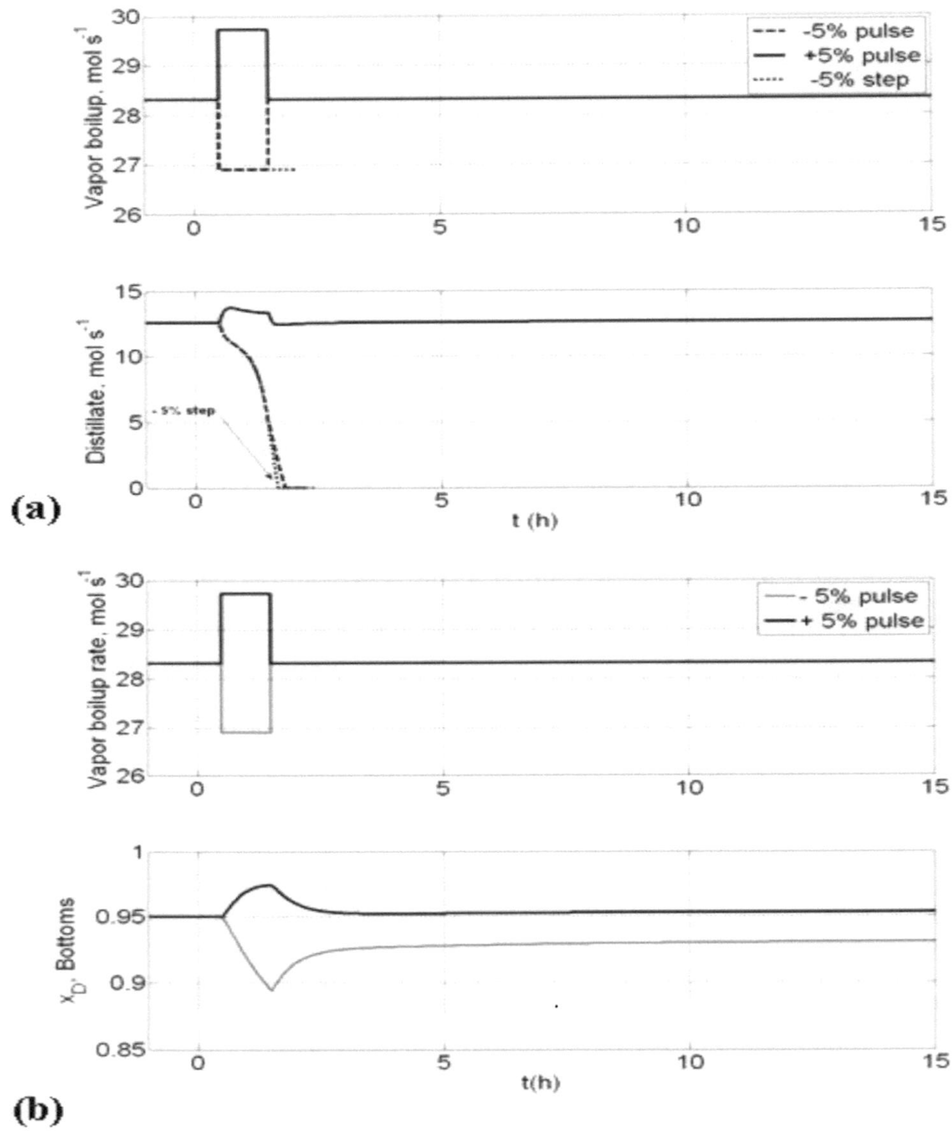

Fig. 4. Open loop dynamics of ideal RD column (7/6/7 design), (a) fixed reflux rate, (b) fixed reflux ratio

3.1.2 Methyl acetate RD column

The 7/18/10 methyl acetate RD column design is studied (Singh et al., 2005). The steady state variation of reaction conversion with respect to reboiler duty at a fixed reflux ratio and a fixed reflux rate is shown in Figure 5. At fixed reflux ratio, the nominal steady state is unique with a 97.77% conversion while two additional low conversion steady states (conversion: 72.95% and 59.66%) are observed at fixed reflux rate. The column dynamic response to a 5 hour duration -3% pulse in the reboiler duty at alternatively, a fixed reflux rate, a fixed reflux ratio or controlling a reactive tray temperature using reflux rate is shown in Figure 6. The liquid levels in the reflux and reboiler drums are controlled using the distillate and bottoms, respectively (P controller with gain 2). Whereas the column returns to its nominal steady state for a fixed reflux ratio or for reactive tray temperature control using reflux, a steady state transition to a low conversion steady state is observed at a fixed reflux rate. This transition is attributed to the output multiplicity at constant reflux rate in Figure 5. Maintaining the reflux in ratio with the distillate is thus a simple means of avoiding output multiplicity and the associated open loop column operation issues (Kumar & Kaistha, 2008).

3.2 Input multiplicity and its implications on controlled variable selection

As discussed, the existence of input multiplicity in an IO pairing can severely compromise control system robustness due to the possibility of wrong control action. In this section, we demonstrate wrong control action in the ideal and methyl acetate RD systems. We also demonstrate the systematic use of steady state IO relations to choose CVs (controlled variables) that are better behaved (more robust) in terms of their multiplicity behavior and the consequent improvement in control system robustness for the two example RD systems.

3.2.1 Ideal RD column

The 5/10/5 design with 1 kmol reaction holdup per reactive tray is considered here. For 95% distillate and bottoms purities, the reflux ratio and vapor boilup are respectively 2.6915 and 29.27 mol s^{-1} respectively. As with the 7/6/7 design, maintaining reflux in ratio with the distillate mitigates nonlinear effects and is therefore implemented. The simplest policy of operating the column at fixed reflux ratio is first considered.

At a fixed reflux ratio, there are three available inputs for control, namely the fresh A feed (F_A), the fresh B feed (F_B) and the vapor boilup (V_S). Of these, one of the inputs must be used

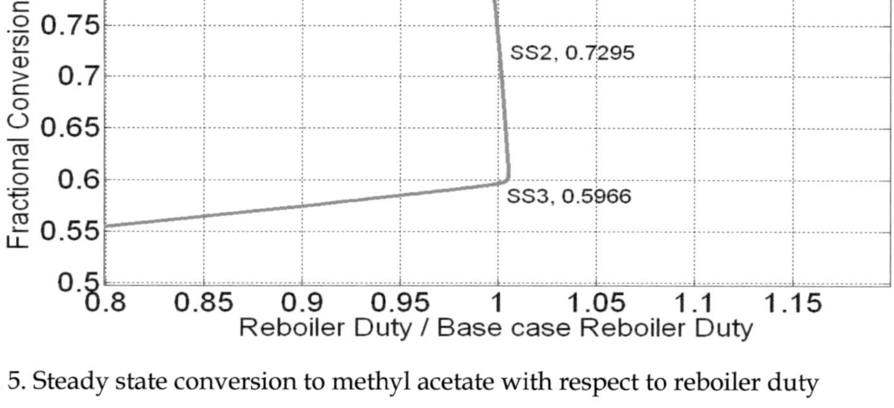

Fig. 5. Steady state conversion to methyl acetate with respect to reboiler duty

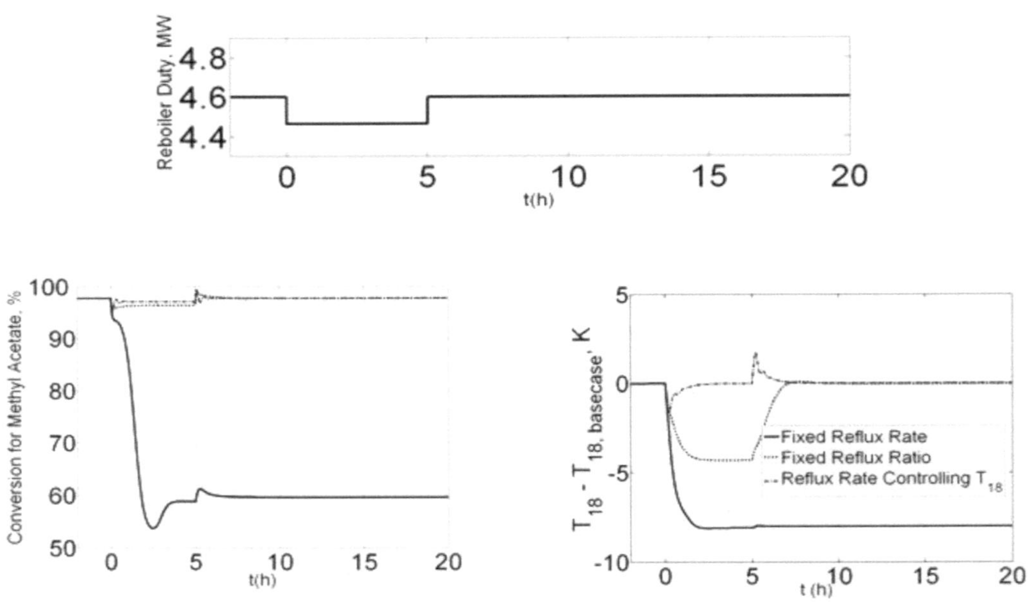

Fig. 6. Dynamic response of methyl acetate RD column for a pulse change in reboiler duty

to set the production rate (throughput) with the remaining two inputs available for column regulation. F_B is chosen as the throughput manipulator as the dynamic response of the tray temperatures (potential controlled outputs) to F_B is sluggish compared to V_S or F_A due to the associated large liquid hydraulic lags. V_S and F_A would thus be more effective manipulation handles for column regulation. From sensitivity analysis, a stripping tray temperature is the most sensitive to a change in F_A. Accordingly, F_A is paired with the sensitive stripping tray temperature (T_2, bottom-up tray numbering). V_S is then used as the manipulation handle for controlling a non-stripping (reactive or rectifying) tray temperature. Sensitivity analysis shows T_{18} to be the most sensitive rectifying tray temperature with T_{12} being the most sensitive reactive tray temperature, which is however lower than T_{18}. We therefore consider two alternative pairings namely T_{18}-V_S or T_{12}-V_S. A schematic of the two-temperature control structure is shown in Figure 7. The Niederlinski Index and Relative Gain Array of the two alternative control loop pairings are also given in the Figure and are found to be acceptable. These local metrics suggest T_{18} to be the better controlled variable.

The steady state input-output relations between the manipulated and controlled variables are now evaluated for multiplicity. The variation of three tray temperatures (T_2, T_{18} and T_{12}) with respect to all three inputs (F_B, F_A and V_S) is plotted in Figure 8. For easy comparison, the difference in the temperature from its nominal value is plotted with respect to percentage change in the inputs around the nominal steady state. Input-output relations are nearly monotonic with respect to V_S with an increase in V_S causing the tray temperature to increase. Although gain sign reversal is seen in T_{12} and T_2 for large negative change in V_S, the IO relations remain away from a crossover. On the other hand, crossover is seen with respect to F_B. In the T_{18}–F_B IO relation, crossover is observed at -22.5% and -30.8% and +22.7% change in F_B. With respect to F_A, directionality in response is observed with no change in T_{12} or T_{18} for an increase in F_A but a visible change for a decrease in F_A. The response of T_2 (controlled using F_A) is better behaved with gain sign reversal for a decrease in F_A. But the IO relation remains away from crossover for a ±35% input change.

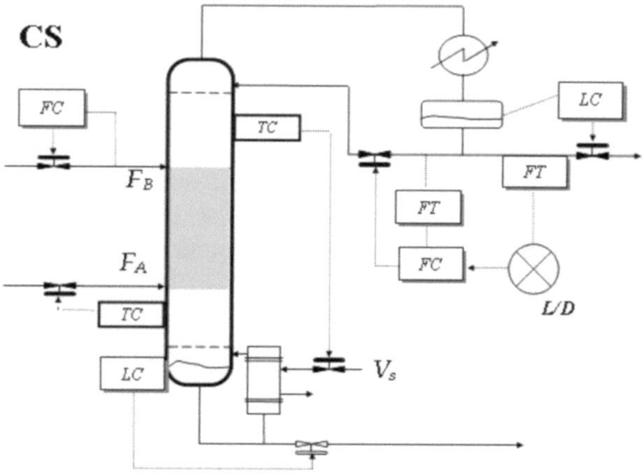

CS pairing	NI	RGA
$V_S - T_{18}$; $F_A - T_2$	0.8862	$\Lambda = \begin{matrix} & V_S & F_A \\ T_{18} & 1.128 & -0.128 \\ T_2 & -0.128 & 1.128 \end{matrix}$
$V_S - T_{12}$; $F_A - T_2$	0.7345	$\Lambda = \begin{matrix} & V_S & F_A \\ T_{12} & 1.361 & -0.361 \\ T_2 & -0.361 & 1.361 \end{matrix}$

Fig. 7. Two temperature control structure with *Niederlinski Index (NI)* and *Relative Gain Array (RGA)* of control loop pairings

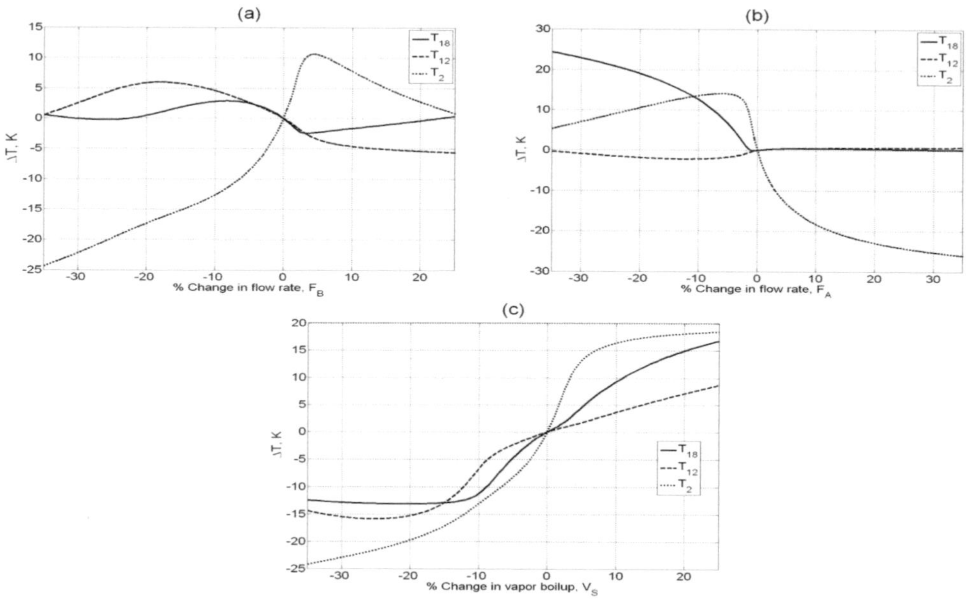

Fig. 8. Open loop variation of ideal RD column tray temperatures with inputs (F_B, F_A and V_S)

The open loop IO relation that a control loop 'sees' can be significantly different depending on whether the other loop is on manual (its input is fixed) or automatic (its output is fixed). To evaluate the same, open loop IO relations for the T_{18}-V_S pairing and T_2-F_A pairing are obtained with the output for the other loop (T_2 or T_{18}) maintained at its setpoint (nominal value). Similarly the T_{12}-V_S (T_2 fixed) and T_2-F_A (T_{12} fixed) IO relations are also obtained.

These are shown in Figure 9. The nominal steady state is marked *O* and the corresponding crossover points are marked *A, B* etc. A non-nominal steady state on a solution branch is stable if the local slope in the IO relation has the same sign as for the nominal steady state O, else it is unstable. Accordingly, the stable solution branch is shown as a continuous curve while the unstable solution branch is shown as a dashed curve.

For the T_{18}-V_S and T_2-F_A pairing, the input multiplicity steady states *A* and *B* are unstable with respect to controller action (reverse or direct) as the local slope sign of at least one of the IO relations is opposite the nominal slope sign. Steady state *C* on the other hand is stable. Disturbances that push the column towards *A* i.e., cause a large decrease in F_A/V_S, can result in wrong control action with saturation of a control input. On the other hand, disturbances that cause large increases in F_A/V_S can result in a closed loop steady state transition to steady state solution *C*. For the T_{12}-V_S and T_2-F_A pairing, both the input multiplicity steady states *A′* and *B′* are unstable with respect to controller action so that wrong control action with consequent valve saturation is expected for large changes in F_A/V_S in either direction (increase or decrease).

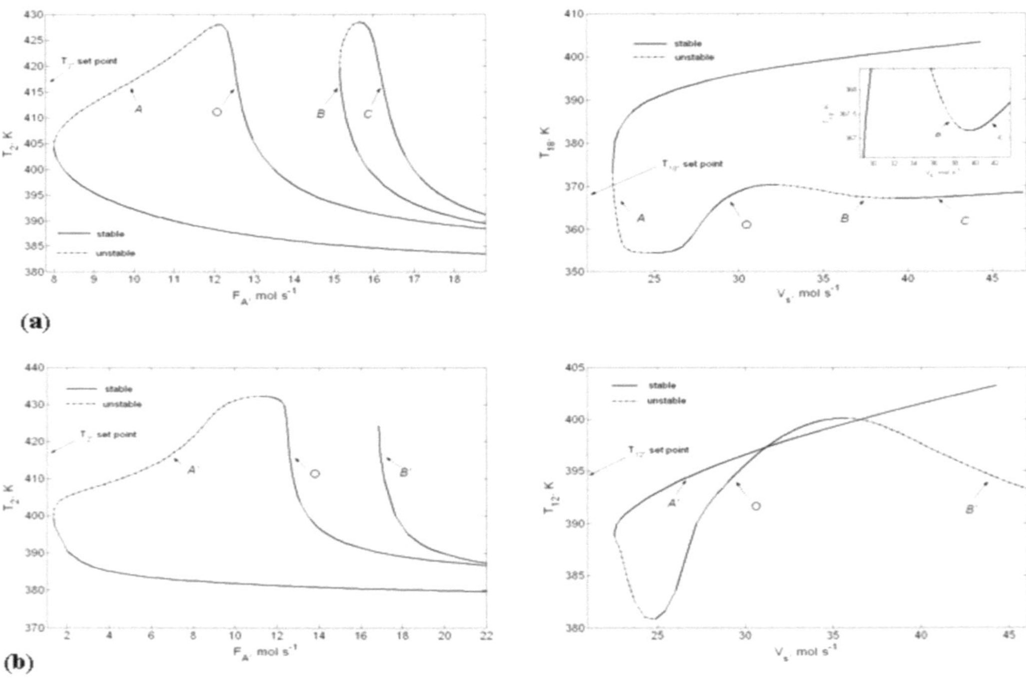

Fig. 9. Ideal RD column IO relations,
(a) T_2-F_A (fixed T_{18}) & T_{18}-V_S (fixed T_2) (b) T_2-F_A (fixed T_{12}) & T_{12}-V_S (fixed T_2)

Which pairing (T_{18}-V_S/T_2-F_A versus T_{12}-V_S/T_2-F_A) would handle larger disturbances without succumbing to wrong control action depends on the degree of tightness of control of the outputs. Usually tightest tray temperature control is usually possible with boilup as the manipulation handle. T_{18}/T_{12} is therefore likely to be controlled tightly without significant deviations from its nominal setpoint. Larger deviations in T_2 (controlled using F_A) can result in wrong control action due to input multiplicity corresponding to higher F_A feed into the column (Figure 8 and Figure 9). In the T_2-F_A IO relation (Figure 9), notice that a crossover in T_2 occurs earlier when T_{18} is held constant compared to when T_{12} is held constant. Accordingly, one would expect controlling T_{12} to handle larger disturbances without wrong control action.

Using T_{18}/T_{12} and T_2 as controlled variables to manipulate V_S and F_A respectively, two different series of step changes are given to the throughput manipulator F_B to demonstrate the impact of input multiplicity under closed loop operation. The temperature controllers are tuned individually using the relay feedback test. The T_{18}-V_S loop must be detuned by a factor of 5 from its Tyreus Luyben settings to avoid a highly oscillatory response while not detuning is necessary when the T_{12}-V_S loop is implemented. In the first (second) series of step changes, the F_B flow rate value is decreased (increased) to 15% (20%) and then 30% (40%) below its basecase value at time 0 and 15 hr respectively, and then restored back to its nominal value of 12.6 mol s^{-1} at 30 hour. The closed dynamic results for these step changes when T_{18} is controlled are shown in Figure 10(a).

For the first series of step changes, stable closed loop responses are obtained for the changes made at 0 and 15 hr (Figure 10(a)). Tight control of the product purities with less than 1% deviations is achieved suggesting that two-point temperature inferential control provides effective column regulation holding the reaction and separation close to the nominal steady state. Upon restoration of the F_B flow rate to its nominal value at 30 hrs with a large 30% step increase, the F_A and V_S valves are completely closed. A sudden large increment of F_B

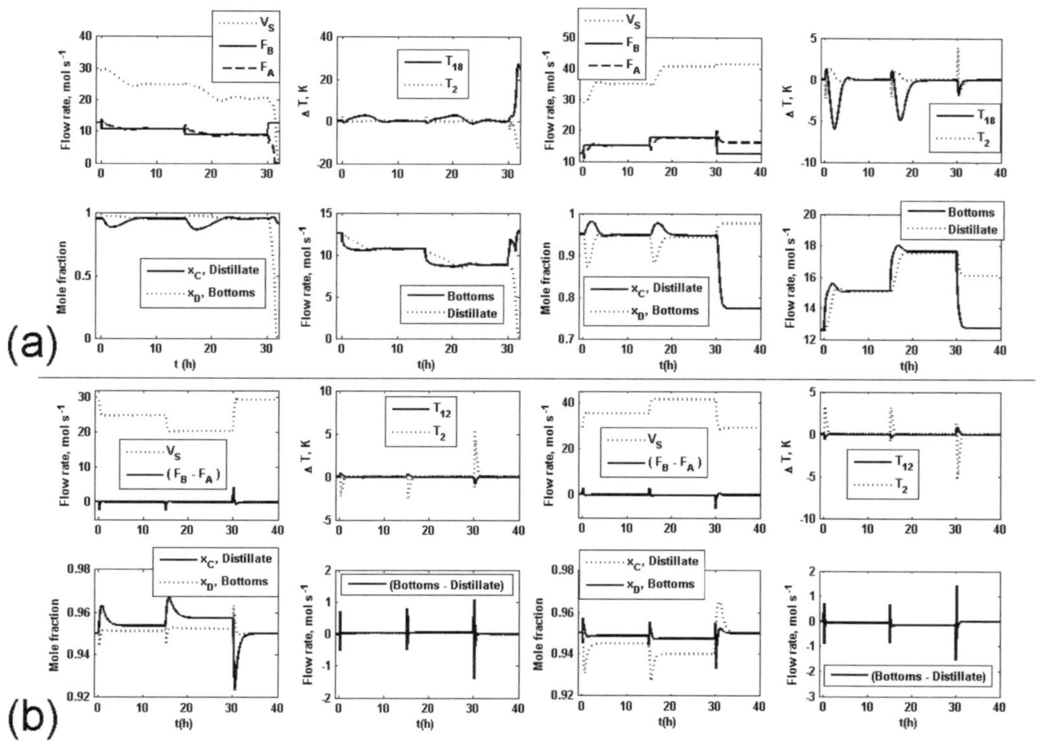

Fig. 10. The closed loop dynamics of ideal RD column for the two different series of step changes in F_B when (a) T_{18} (b) T_{12} is controlled variable

flow rate from 8.82 to 12.6 mol s^{-1} brings the column operation in the vicinity of point A in Figure 9 (relatively low F_A) with the consequent wrong control action causing a valve shutdown.
For the second series of step changes (+20%, +20% and -40%), a stable and well behaved response is observed for the two +20% step changes with acceptably small product purity deviations. However, for the -40% step change to bring F_B back to its nominal value, the column drifts to new steady state, i.e., settles at steady state C in Figure 9. The large F_B flow value decrease 17.64 to 12.6 mol s^{-1} at 30 hr, results in excess A input which causes a steady state transition to the stable steady state C in Figure 9. The same series of step changes in F_B (-15%, -15%, +30% and +20%, +20%, -40%) is effectively handled with no valve saturation or steady state transition due to wrong control action when T_{12} is used as the controlled variable manipulating V_S instead of T_{18}. The closed loop dynamic response is shown in Figure 10(b). The small steady state product purity deviations for the large throughput changes again highlight two-point temperature inferential control as an effective means of column regulation.
These results clearly demonstrate that proper choice of the controlled output variable can significantly improve the robustness of the control system in rejecting large disturbances. The results also highlight that the conventional wisdom of choosing controlled variables using local steady state metrics such as open loop gain or Niederlinski Index/relative gain may lead to the wrong conclusions. In the current example, the open loop sensitivity and relative gain for the T_{18}-V_S pairing are better than for the T_{12}-V_S pairing. A more comprehensive bifurcation analysis however reveals T_{12} to be the more robust CV. Such a comprehensive steady state analysis is strongly recommended for designing robust control systems for highly non-linear RD systems.

3.2.2 Methyl acetate RD column

In this RD column, column trays are numbered from top to bottom with the condenser as tray 0. As seen earlier, column operation at fixed reflux ratio avoids output multiplicity. Accordingly, the simple constant reflux ratio policy is implemented leaving the remaining three inputs, namely acetic acid feed (F_{HAc}), methanol feed (F_{MeOH}) and reboiler duty (Q_r) for column regulation. Sensitivity analysis shows that the temperature of tray 18 in the reactive section is very sensitive with respect to F_{HAc} and Q_r. In the stripping section, temperature of tray 34 is sensitive to all three inputs. Based on these sensitivities, two decentralized temperature inferential control structures, labelled CS1 and CS2, are synthesized, which are schematically depicted in Figure 11. In CS1, Q_r is the throughput manipulator, F_{HAc} controls a reactive tray and F_{MeOH} controls a stripping tray. This control structure was originally proposed by Roat et al. (1986). In CS2, F_{HAc} is the throughput manipulator with a reactive tray temperaure controlled using Q_r and a stripping tray temperature controlled using F_{MeOH}.
Further analysis is now conducted to check for multiplicity in the IO relations. As shown in Figure 12(a), all reactive tray temperatures (including the most sensitive T_{18}) exhibit input multiplicity with respect to changes in F_{HAc} and Q_r. To quantify the severity of input multiplicity, the rangeability (with a 3K offset) of the reactive tray temperatures with respect to Q_r and F_{HAc} are reported in Table 2. Even as reactive tray temperature, T_{18}, is the most sensitive to F_{HAc} and Q_r as evidenced from the slope at the nominal steady state in Figure 12(a), its rangeability is lower compared to reactive tray temperature T_{20}. To eliminate a

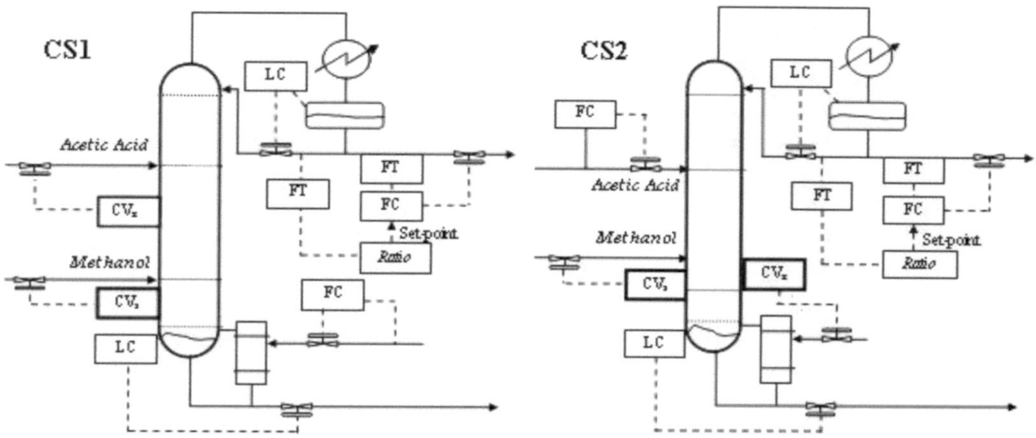

Fig. 11. Schemetics of two temperature control structures used for the methyl acetate RD column

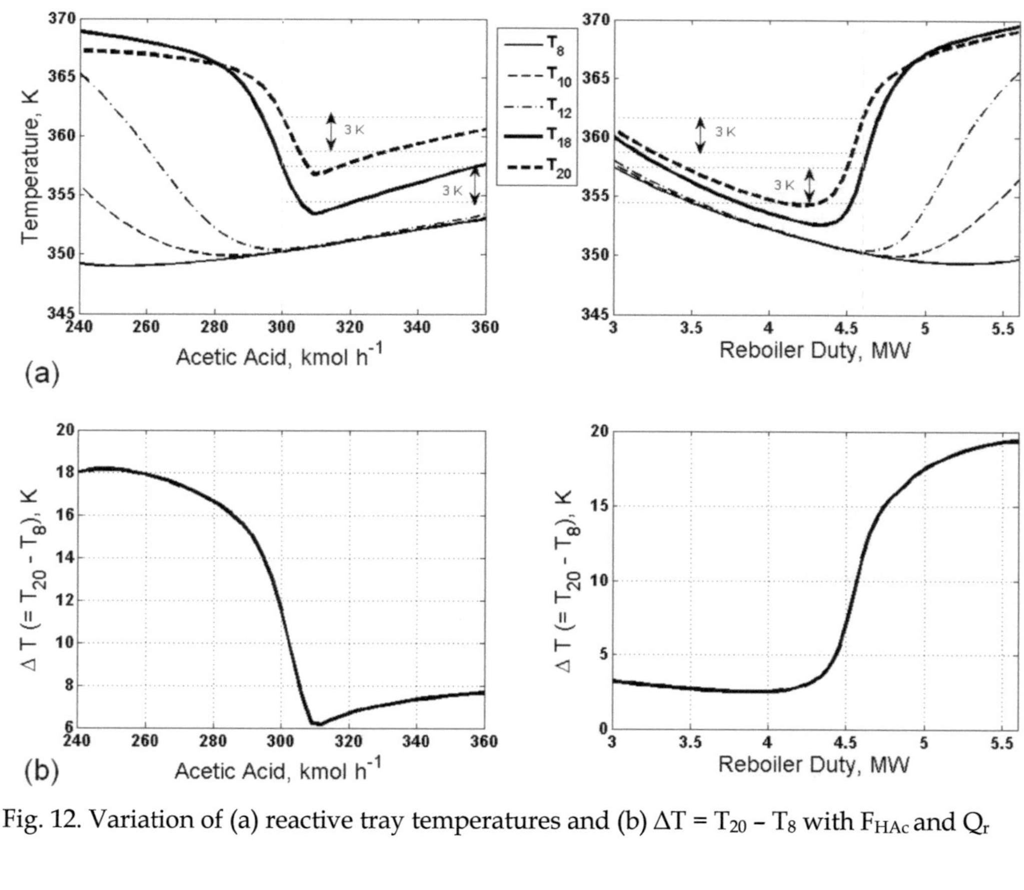

Fig. 12. Variation of (a) reactive tray temperatures and (b) $\Delta T = T_{20} - T_8$ with F_{HAc} and Q_r

crossover in the IO relations for high rangeability, we also consider a combination of tray temperatures. The difference between two reactive tray temperatures ($\Delta T = T_{20} - T_8$) was found to avoid input multiplicity with respect to F_{HAc} and Q_r with the corresponding IO relations in Figure 12(b).

In the T_{34}-F_{MeOH} IO relation, a crossover does not occur (data not shown) so that this pairing is fixed in both CS1 and CS2. For the reactive tray temperature control loop, there are three candidate controlled outputs in both CS1 and CS2, namely, T_{18}, T_{20} and ΔT ($T_{20} - T_8$). Superscripts '*a*', '*b*' and '*c*' are appended to the control structure label (CS1 or CS2) corresponding to T_{18}, T_{20} and ΔT, respectively, as the controlled reactive zone measurement. Note that T_{18} exhibits the highest sensitivity but low rangeability, T_{20} exhibits reasonable sensitivity with higher rangeability while ΔT exhibits the best rangeability with reasonable sensitivity. The three variants of each control structure are tested using rigorous dynamic simulations for the maximum throughput change handled in the worst-case direction. From the IO relations in Figure 12(a), for CS1, a step decrease in Q_r is the worst-case direction due to input multiplicity at reduced Q_r while for CS2, a step increase in F_{HAc} is the worst-case direction due to input multiplicity at increased F_{HAc}.

Tray Number	HAc		Reboiler Duty		MeOH	
	Decrease	Increase	Decrease	Increase	Decrease	Increase
16	>20	0	0	>20	>20	0
17	>20	0	6.9	>20	>20	0
18	>20	8.8	17.32	>20	>20	0
19	>20	10.9	24.63	>20	>20	0
20	>20	11.1	25.19	>20	>20	0
21	19.2	7	31.14	>20	>20	0

Values are in % change about their basecase values
A 3K offset is used in calculating rangeability

Table 2. Rangeability of reactive tray temperatures

The PI temperature loops are systematically tuned (Kumar & Kaistha, 2008). The two level controllers are P only with a gain of 2. The column pressure is assumed fixed, which is reasonable as in practice tight pressure control is achieved by manipulating the condenser duty. Also instantaneous flow control is assumed which is again reasonable in that tray temperature dynamics are significantly slower than flow dynamics.

Table 3 reports the maximum throughput step change handled by the different variants of the two control structures. $CS1^a$ and $CS1^b$ fail for a 20% and 30% throughput decrease respectively while $CS1^c$ effectively handles 40% (larger changes not tested). The throughput increase for which $CS2^a$ and $CS2^b$ fail are respectively 25% and 40% while $CS2^c$ works even for a 50% throughput increase (larger increase not tested). The trend in both CS1 and CS2 is in direct agreement with the increasing rangeability of the controlled outputs T_{18} ($CS1/2^a$), T_{20} ($CS1/2^b$) and ΔT ($CS1/2^c$). The result confirms the direct relationship between control system robustness and input multiplicity with rangeability being a useful metric for selecting 'robust' controlled variables. The result also shows that a well designed controlled variable such as ΔT with high rangeability and acceptable sensitivity results in a robust control system that effectively rejects large disturbances.

CS	CS1[a]		CS1[b]		CS1[c]		CS2[a]		CS2[b]		CS2[c]	
	-15%	40%	-25%	40%	-40%	40%	-40%	20%	-40%	35%	-40%	40%

Table 3. Maximum throughput change in either direction handled by the control structures

For the sake of brevity, the dynamic response to throughput change for CS1 and CS2 is not shown and may be found in Kumar & Kaistha (2008). These dynamic results show that controlling ΔT better prevents the breakthrough of heavy acetic acid from the reactive zone. In fact, the cause of input multiplicity in the IO relations is heavy acetic acid moving down and breaking through the reactive zone. This breakthrough would occur if the F_{HAc} is sufficiently increased above F_{MeOH} or if Q_r is sufficiently reduced, which results in the input multiplicity in the IO relations in Figure 12(a). For successful regulation of the RD column, such accumulation or breakthrough of acetic acid must be prevented and the same is effectively achieved by controlling ΔT.

In this example, an appropriate temperature based measurement could be designed that does not exhibit output multiplicity for robust column control. If such a temperature-based measurement is not evident for an RD system, controlling an appropriate tray composition may be considered. Even as online composition measurements are expensive, the additional expense would be justified in order to make the practical implementation of RD technology feasible.

4. RD design for controllability

The two case studies on control of RD columns clearly demonstrate that the existence of steady state multiplicity can result in hard-to-fathom nonlinear dynamic phenomena such as an open loop or a closed loop steady state transition, which can be particularly confusing for operators. In extreme cases where the non-linear effects cannot be sufficiently mitigated by appropriate choice/design of the controlled variable (including composition control), it may be necessary to alter the design of the column to mitigate the non-linearity for better controllability.

How to alter the column design to mitigate the non-linear effects? Several researchers have attempted to address this question for the ideal RD system with often contradictory claims (Huang et al., 2006; Kumar & Kaistha, 2008a, 2008b). To us, it appears that design modifications that help prevent escape of reactants from the reactive zone improve the controllability. To that end, for RD systems with exothermic reactions, the extension of the reactive zone into the stripping section with catalyst redistribution helps prevent the breakthrough of the heavy reactant from the reactive zone. Alternatively, the lower feed tray location may be moved up into the reactive zone. Reduced energy consumption has been demonstrated using a catalyst redistribution and lower feed tray location alteration. With respect to the original 5/10/5 ideal RD column design, controllability improves with catalyst redistribution only but deteriorates significantly when the lower feed tray location is moved up. A combination of the two provides acceptable controllability with significant energy savings. The extension of reactive zone into the rectifying section or upper feed tray alteration does not help improve controllability or energy consumption as the exothermic reaction causes the light reactant to escape up the top. For an endothermic reaction however, such a strategy may have merit (Huang et al., 2006).

For the methyl acetate column studied earlier, input multiplicity caused the control system to succumb to wrong control action for large throughput changes. Redistributing the catalyst onto the adjacent eight stripping trays results in significantly improved

controllability and energy savings (Kumar & Kaistha, 2008b). Figure 13 plots the variation in the methyl acetate purity with reboiler duty at a fixed reflux rate for this alternative design. Notice that unlike the original 7/18/10 design with conventional feed tray locations, the revised design does not exhibit output multiplicity with respect to the nominal steady state (compare with Figure 5). The non-linearity is thus mitigated in this alternative design with expectedly improved control performance. Thus for example, where CS1 for the original design with the most sensitive reactive and stripping tray temperatures as the controlled outputs succumbs to wrong control action for a -20% step change in the reboiler duty, the corresponding change is easily handled in the revised design (Kumar & Kaistha, 2008b).
The IO relation of product purity (top or bottom) with respect to a column input can be a useful tool to screen out poor designs exhibiting output/input multiplicity with respect to the nominal steady state. To demonstrate this for the ideal RD system, we consider the 7/6/7 design which is the most difficult to control using temperature inferential control (Luyben, 2000). The catalyst hold up on each tray is kept fixed at 1 kmol. Keeping the distillate rate equal to the fresh feed rate, the reflux ratio can be adjusted for reaction conversions of 90%, 95% or 98.5% with corresponding product purities of 90%, 95% and 98.5%. As shown in Figure 14, for a column pressure of 9 bars, the distillate and bottoms purity IO relations exhibit input and output multiplicity with respect to the nominal steady state for high conversions (and purities) of 95% and 98.5%. The multiplicity disappears for 90% conversion suggesting that high conversion RD columns are likely to exhibit multiplicity and therefore susceptible to consequent non-linear dynamic phenomena.

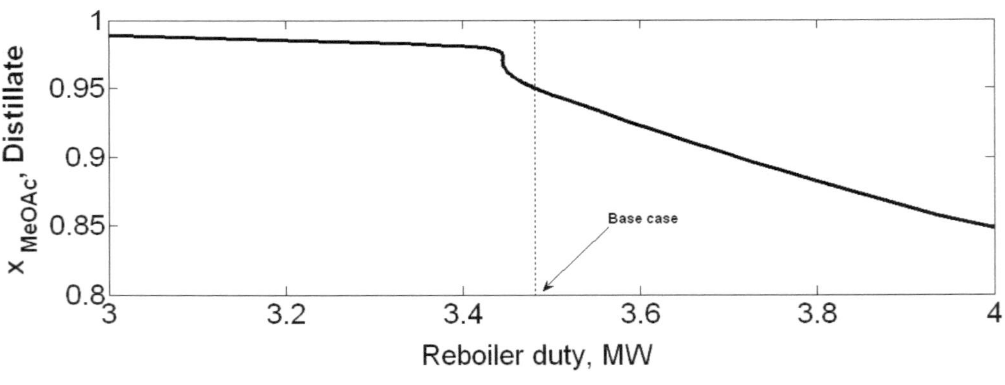

Fig. 13. Steady state variation of methyl acetate purity with respect to reboiler duty

We now consider column re-design for the highest considered conversion (and purity) of 98.5%. Holding the number of stripping trays equal to the number of rectifying trays, the number of reactive trays is increased and the IO relation of the distillate purity with respect to vapor boilup at constant reflux ratio is obtained. Similarly, holding the number of reactive trays constant, the number of stripping trays (equal to rectifying trays) is altered and the distillate purity-boilup IO relation is generated. Table 4 reports whether input or output multiplicity is observed in the different designs. From the Table, observe that simply reducing the number of rectifying (and stripping) trays from 7 to 4 causes the IO relation to be well behaved with no input/output multiplicity. The boilup is however too high and the design is uneconomical. No multiplicity is also observed for column designs with higher number of reactive trays and not too many fractionation trays, specifically, in the 4/9/4 and 7/12/7 designs. Of these, the latter consumes much less energy with a 30.17% lower boilup than the former. This design thus appears to be a good one both from the process economics and controllability perspectives.

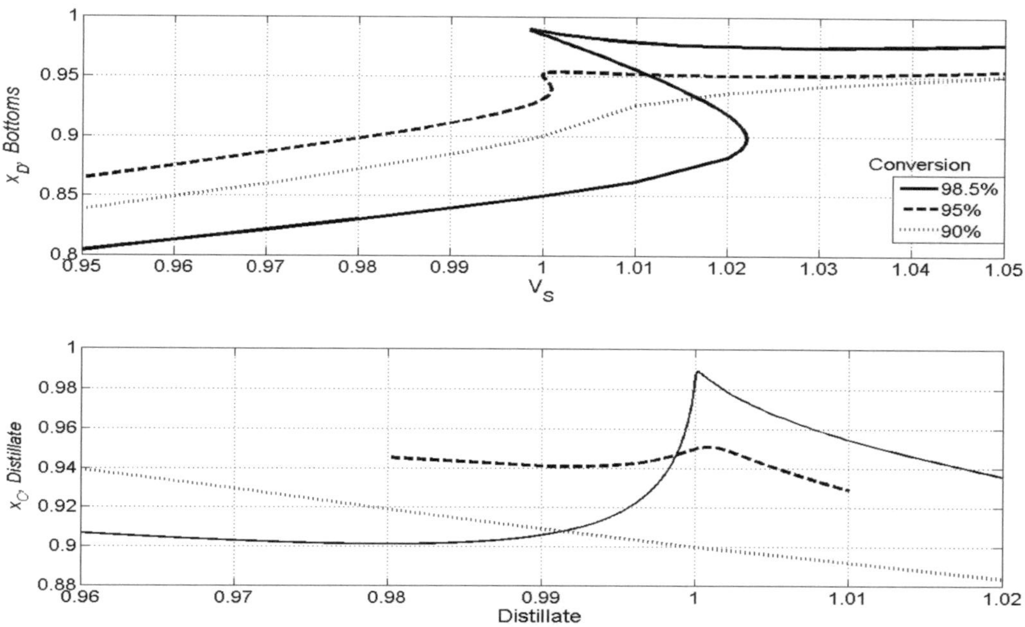

Fig. 14. Variation of $x_C{}^D$ & $x_D{}^B$ with vapour boilup and distillate for ideal RD 7/6/7 design

Design	Input multiplicity	Output multiplicity	Reflux ratio	Vapor boilup, mol s^{-1}
4/6/4	No	No	12.4400	151.4698
7/6/7	Yes	Yes	3.2841	36.1073
10/6/10	Yes	Yes	2.8155	30.2030
13/6/13	Yes	Yes	2.7311	29.1397
4/9/4	No	No	4.0037	45.1734
7/9/7	Yes	Yes	2.8312	30.4013
10/9/10	Yes	Yes	2.7868	29.8415
13/9/13	Yes	Yes	2.7774	29.7223
4/12/4	No	No	3.0407	33.0401
7/12/7	No	No	2.9055	31.3368
10/12/10	No	Yes	2.9007	31.2766
13/12/13	No	Yes	2.8996	31.2621
16/12/16	No	Yes	2.8989	31.2543

Table 4. Nature of the IO relation of bottom product purity versus vapour boilup (Ideal RD)

The multiplicity trends in the Table also suggest that excess fractionation capacity causes output multiplicity to 'appear' in the IO relations (compare e.g. 4/9/4 design with 7/9/7 design). The process design must therefore seek the appropriate balance between reaction capacity and fractionation capacity for well behaved IO relations (Bisowarno et al., 2004). For an economical design, sufficient reaction capacity must be provided.

We have dynamically tested both the 4/9/4 and 7/12/7 designs of 95% conversion ($x_{C,\ Distillate} = x_{D,\ Bottoms} = 0.95$) using two-point temperature inferential control structures similar to the ones studied earlier. Large throughput changes (up to 40%) in either direction are handled without wrong control action suggesting that these designs are inherently more controllable. This simple example demonstrates the power of steady state bifurcation analysis in arriving at economical RD column designs with good controllability.

5. Conclusions

To conclude, we hope that this Chapter has convinced the reader that a systematic evaluation of steady state multiplicity in RD systems is fundamental for designing robust control systems that effectively reject large disturbances. Specifically, the possibility of an open loop steady state transition due to output multiplicity and wrong control action under closed loop operation due to input multiplicity has been demonstrated for the example systems studied here. To improve the robustness of the control system, the controlled variables should be selected with care for a larger operating window around the nominal steady state without a crossover in the IO relation. In conjunction with local linear tools such as open loop gain, Niederlinski Index and relative gain, the proposed rangeability metric is a useful tool for selecting 'robust' controlled variables and rejecting poor choices that may potentially succumb to non-linear dynamic phenomena. The steady state IO relations (bifurcation analysis) can also help in arriving at an inherently more controllable and economical RD process design.

6. References

Agreda, V.H., Partin, L.R., Heise, W.H. (1990). High-purity methyl acetate via reactive distillation, *Chemical Engineering Progress, 86(2)*, pp.40 – 46.

Al-Arfaj, M.A., Luyben, W.L. (2000). Comparison of alternative control structures for an ideal two-product reactive distillation column, *Industrial and Engineering Chemitsry Research*, 39(9), pp.3298-3307.

Al-Arfaj, M.A., Luyben, W.L. (2002). Comparative control study of ideal and methy acetate reactive distillation, *Chemical Engineering Science, 57*, pp.5039-5050.

Astrom, K. J., Hagglund. (1984). Automatic tuning of simple regulators with specifications on phase and amplitude margins, *Automatica*, 20, pp.645-651.

Backhaus, A. A. (1921). Continuous Processes for the Manufacture of Esters, US patent 1400849.

Bisowarno, B., Tian, Y.-C., Tadé, M. O. (2004). Interaction of separation and reactive stages on ETBE reactive distillation columns, *AIChE Journal*, 50(3), pp.646-653.

Ciric, A.R., Miao, P. (1994). Steady state multiplicities in an ethylene glycol reactive distillation column, *Industrial and Engineering Chemistry Research, 33*, pp.2738-2748.

Dorn, C., Guttinger, T.E., Wells, G.J., Morari, M., Kienle, A., Klein, E., Gilles, E.-D. (1998). Stabilization of an unstable distillation column, *Industrial and Engineering Chemistry Research*, 37(2), pp.506-515.

Engell, S., Fernholz, G. (2003). Control of a reactive separation process, *Chemical Engineering and Processing, 42,* pp.201-210.

Huang, K., Nakaiwa, M., Tsutsumi, A. (2006). Towards further internal heat integration in design of reactive distillation columns. - Part 2: The process dynamics and operation, *Chemical Engineering Science*, 61(16), pp.5377-5392.

Huss, R. S., Chen, F., Malone, M. F., Doherty, M. F. (2003). Reactive distillation for methyl acetate production, *Computers and Chemical Engineering, 27,* pp.1855-1866.

Jacobs, R., Krishna, R. (1993). Multiple solutions in reactive distillation for methyl-tert-butyl ether synthesis, *Industrial and Engineering Chemistry Research, 32*, pp.1706-1709.

Kienle, A., Marquardt, W. (2003). Nonlinear dynamics and control of Reactive distillation processes. In K. Sundmacher and A. Kienle, Eds.; *Reactive distillation – status and future directions*. Wiley-VCH:Weinheim, 241-281. ISBN 978-3527305797

Kumar, M.V.P., Kaistha, N. (2008). Role of multiplicity in reactive distillation control system design, *Journal of Process Control*, 18 (7-8), pp.692-706.

Kumar, M.V.P., Kaistha, N. (2008a). Internal heat integration and controllability of double feed reactive distillation columns, 1. Effect of feed tray location, *Industrial and Engineering Chemistry Research*, 47(19), pp.7294-7303.

Kumar, M.V.P., Kaistha, N. (2008b). Internal heat integration and controllability of double feed reactive distillation columns, 1. Effect of catalyst redistribution, *Industrial and Engineering Chemistry Research*, 47(19), pp.7304-7311.

Luyben, W.L. (2000). Economic and dynamic impact of use of excess reactant in reactive distillation systems, *Industrial and Engineering Chemistry Research, 39*, pp.2935-2946.

Mohl, K., Kienle, A., Gilles, E., Rapmund, P., Sundmacher, K., Hoffmann, U. (1999) Steady-state multiplicities in reactive distillation columns for the production of fuel ethers MTBE and TAME: theoretical analysis and experimental verification, *Chemical Engineering Science, 54(8)*, pp.1029-1043.

Roat, S., Downs, J., Vogel, E., Doss, J., 1986. Integration of rigorous dynamic modeling and control system synthesis for distillation columns. In Morari, M., McAvoy, T.J., (Eds), Chemical Process Control; CPC III, Elsevier: Amesterdam, The Netherlands.

Siirola, J. J. (1995). An industrial perspective on process synthesis, *A.I.Ch.E. Symposium Series, 91 (304)*, pp.222-233.

Singh, B.P., Singh, R., Kumar, M.V.P., Kaistha, N. (2005). Steady state analysis of reactive distillation using homotopy continuation method, *Chemical Engineering Research and Design*, 83A, 959-968.

Sneesby, M.G., Tade, M.O., Datta, R., Smith, T.N. (1997). ETBE synthesis via reactive distillation. 2. Dynamic simulation and control aspects, *Industrial and Engineering Chemistry Research, 36*, pp.1870-1881.

Taylor, R., Krishna, R. (2000). Modeling of reactive distillation, *Chemical Engineering Science, 55,* 5183–5229.

Towler, G.P., Frey, S.J., (2001). Reactive Distillation, Kulpratipanja, S., Ed. In *Reactive Separation Processes*, Taylor & Francis. ISBN 978-1560328254

Tyreus, B.D., Luyben, W.L., 1992. Tuning PI controllers for integrator/deadtime processes, *Industrial and Engineering Chemistry Research*, 31, pp.2625-2628.

12

Robust Multivariable Control of Ill-Conditioned Plants – A Case Study for High-Purity Distillation

Kiyanoosh Razzaghi and Farhad Shahraki
Department of Chemical Engineering,
University of Sistan and Baluchestan, Zahedan
Iran

1. Introduction

Distillation is one of the most important unit operations in the chemical industry. Among various distillation operations, control of high-purity column poses difficult control due to a number of characteristics of these systems, including strong directionality, ill-conditioning and strongly nonlinear behavior. At the same time, the potential benefits that can be obtained through tight and economic control of the product compositions are very large. This is due to different reasons including the large energy consumption required by the columns and the market requirements which are becoming stricter and stricter. Because of these obvious features of high-purity distillation, this type of column has been studied extensively.

Control systems for chemical processes are typically designed using an approximate, linear, time-invariant model of the plant. The actual plant may differ from the nominal model due to many sources of uncertainty, such as nonlinearity, the selection of low-order models to represent a plant with inherently high-order dynamics, inaccurate identification of model parameters due to poor measurements or incomplete knowledge, and uncertainty in the manipulative variables. Considering the differences between the actual plant and nominal model, it is necessary to insure that the control system will be stable and meet some predetermined performance criteria when applied to the actual plant. The identification and control of distillation columns have been subjects of frequent study due to the ill-conditioned nature of the distillation process. An ill-conditioned plant is very close to singular, and unless care is taken, very small errors can make the model useless. In distillation, this means that a model may have features that are in conflict with physical knowledge (Luyben, 1987; Jacobsen & Skogestad, 1994; Böling & Häggblom, 1996). In addition, ill-conditioned dynamics of high-purity distillation columns leads to high sensitivity to uncertainties in the manipulated variables (Skogestad & Morari, 1988). This effect causes even small errors in the manipulated variables show significant deterioration of the product quality, a fact which explains why open-loop control of high-purity distillation columns is hardly ever satisfactory. The model of a high-purity distillation process has a steady-state gain matrix with a high condition number. The gain matrix is almost singular and its determinant may be affected by quite small model errors, and if determinant of the gain matrix of the model and that of the plant have different signs, no controller with integral action exists that can stabilize both the model and plant (Grosdidier et al., 1985).

Many control design techniques have been applied to the high-purity distillation columns (e.g. Georgiou et al., 1988; Skogestad and Lundström, 1990; Sriniwas et al., 1995; Christen et al., 1997; Shin et al., 2000; Razzaghi & Shahraki, 2005, 2007; Biswas et al., 2009). Some possible improvements for linear multivariable predictive control of high-purity distillation columns are proposed by Trentacapilli et al. (1997) and a simple way of inserting a local model that contains part of the process nonlinearity into the controller is described also. In addition, a reliable model of the column is generally considered as a prerequisite for the design of efficient two-product control by multivariable methods. Another important aspect of distillation control design is the choice of a good configuration. In fact, poor control performance can result from the improper choice of manipulated/controlled variable pairing (Hurowitz et al., 2003). Some authors have been considered control configuration selection (Shinskey, 1984; Skogestad and Morari, 1987a; Finco et al., 1989; Stichlmair, 1995; Heath et al., 2000; Hurowitz et al., 2003; Luyben, 2005; Hori & Skogestad, 2007; Razzaghi & Shahraki, 2009), but there is no general agreement among these authors in choosing the best control configuration, however, a complete review in this field is performed by Skogestad et al. (1990). The main works for selection of manipulated/controlled variable pairings have focused upon using controllability measures, such as relative gain array (Bristol, 1966) and structured singular value μ (Doyle, 1982). The relative gain array (RGA) provides a steady-state measure of coupling in multivariable systems and can be used to evaluate the steady-state coupling of configurations. RGA is still the most commonly used tool for control structure selection for single-loop controllers. Shinskey (1984) used the relative gain array to choose configuration which is applied widely in industry. Several authors such as Skogestad et al. (1990) and Kariwala et al. (2006) have demonstrated practical applications of the RGA that it depends on the plant model only, that it is scaling independent and that all possible configurations can be evaluated base on the a single matrix. The structured singular value (SSV) approach provides necessary and sufficient conditions for robust stability and performance for the situation in which uncertainty occurs simultaneously and independently in various parts of the overall control system (e.g. input and output uncertainty) but the perturbation matrix is still norm-bounded. One of the most difficult steps in analysing the robust stability and performance of any control system is the specification of an estimate of the uncertainty associated with the nominal process model. This is a critical step because an overestimation of the model inaccuracy will lead to extensively poor control performance and an underestimation may lead to instability (McDonald et al., 1988). Several papers discuss ways in which model inaccuracy can be described and methods that can be used for assessing robust stability. The most common multivariable approaches that use singular values (Doyle and Stein, 1981; Arkun et al., 1984) and structured singular values assume that the actual plant can be described by a norm-bounded perturbation matrix in the frequency domain. In chemical process control, nonlinearity is one of the most significant sources of model inaccuracy. We usually have some knowledge about the structure of model inaccuracy due to nonlinearity, however, and this knowledge should be exploited in our robustness studies. In formulating the SSV problem, use of physically-based uncertainty description is important. Simplified models that predict gain and time constant changes as the process is perturbed over the expected operating regime can be used to characterise the uncertainty (McDonald et al., 1988).

The objective in this chapter is to show that acceptable closed-loop performance can be achieved for an ill-conditioned high-purity distillation column by use of the structured singular value μ. The distillation column model used in this case study is a high-purity column, referred to as "column at operating point A" by Skogestad and Morari (1988). Table 1 summarizes the steady-state data of the model in detail. The following simplifying assumptions are also made for the column: (1) binary separation, (2) constant relative volatility, and (3) constant molar flows. To include the effect of neglected flow dynamics, we will add uncertainty when designing and analysing controller.

Column data	
Relative volatility	$\alpha = 1.5$
Number of theoretical trays	$N_T = 40$
Feed tray (1 = reboiler)	$N_F = 21$
Feed composition	$z_F = 0.50$
Operating data	
Distillate composition	$y_D = 0.99$
Bottom composition	$x_B = 0.01$
Distillate to feed ratio	$D/F = 0.500$
Reflux to feed ratio	$L/F = 2.706$

Table 1. Steady-state data for distillation column.

2. Process description

A simple two time-constant dynamic model presented by Skogestad and Morari (1988) is chosen as the basis for the controller design. The model is derived assuming the flow and composition dynamics to be decoupled, and then the two separate models for the composition and flow dynamics are simply combined. The nominal model of the column is given by

$$\begin{aligned} dy_D &= \frac{87.8}{1+194s} dL + \left(\frac{1.4}{1+15s} - \frac{87.8}{1+194s} \right) dV, \\ dx_B &= \frac{108.2}{1+194s} g_L(s)\, dL + \left(-\frac{1.4}{1+15s} - \frac{108.2}{1+194s} \right) dV. \end{aligned} \tag{1}$$

$g_L(s)$ expresses the liquid flow dynamics:

$$g_L(s) = \frac{1}{[1+(2.46/n)s]^n} \tag{2}$$

where n is the number of trays in the column (N_T - 1). Fig. 1 shows a schematic of a binary distillation column that uses reflux and vapor boilup as manipulated inputs for the control of top and bottom compositions, respectively. This is denoted as the LV-configuration (structure). This structure is commonly used in industry for one-point composition control.

However, severe interactions often make two-point control difficult with this configuration. Although the closed-loop system may be extremely sensitive to input uncertainty when the LV-configuration is used, while it is shown that it is possible to obtain good control behavior (i.e. good performance) with the LV-configuration when model uncertainty and possible changes in the operating point are included (Skogestad and Lundström, 1990). The simultaneous control of overhead and bottoms composition in a binary distillation column using reflux and steam flow as the manipulated variables often proves to be particularly difficult because of the coupling inherent in the process. The result of this coupling, which cause the two control loops to interact, leads to a deterioration in the control performance of both composition control loops compared to their performance if the objective were control of only one composition. Since high-purity distillation columns can be very sensitive to uncertainties in the manipulated variables, it is important for successful implementation that a controller guarantees its performance in the presence of uncertainties. This particular design task is frequently solved by modeling a multiplicative uncertainty for a nominal plant model and subsequently calculating the controller using μ-synthesis (Doyle, 1982).

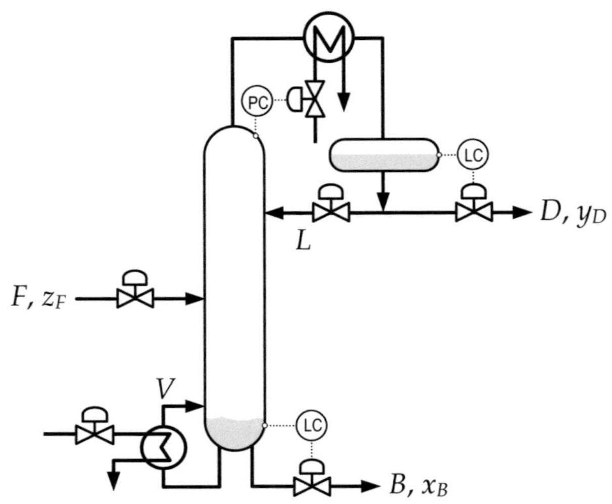

Fig. 1. Schematic of a binary distillation column using the LV-configuration. L and V: manipulated inputs; x_B and y_D: controlled outputs.

2.1 General control problem formulation

Fig. 2 shows general control problem formulation, where G is the generalized plant and C is the generalized controller. The controller design problem is divided into the analysis and synthesis phases. The controller C is synthesized such that some measure, in fact a norm, of the transfer function from w to z is minimized, e.g. the H_∞-norm. Then the controller design problem is to find a controller C (that generates a signal u considering the information from v to mitigate the effects of w on z) minimizing the closed-loop norm from w to z. For the analysis phase, the scheme in Fig. 2 is to be modified to group the generalized plant G and the resulting synthesized controller C in order to test the closed-loop performance achieved with C. To get meaningful controller synthesis problems, weights on the exogenous inputs w and outputs z are incorporated. The weighting matrices are usually frequency dependent and typically selected such that the weighted signals are of magnitude one, i.e. the norm from w to z should be less than one.

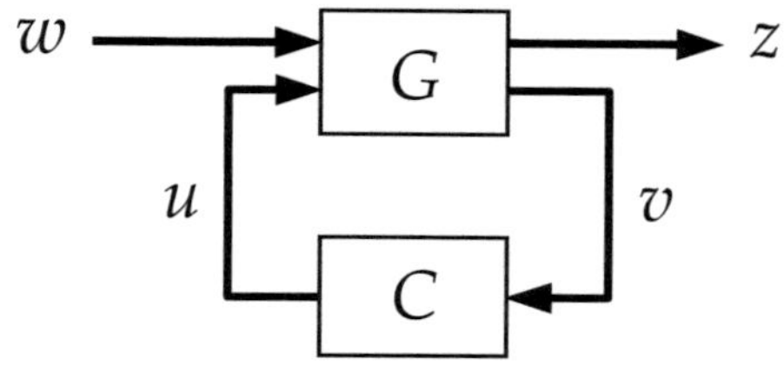

Fig. 2. General control problem formulation with no model uncertainty.

Once the stabilizing controller C is synthesized, it rests to analyze the closed-loop performance that it provides. In this phase, the controller for the configuration in Fig. 2 is incorporated into the generalized plant G to form the system N, as it is shown in Fig. 3. The expression for N is given by

$$N = G_{11} + G_{12}C(\mathbf{I} - G_{22})^{-1}G_{21} \equiv F_l(G,C) \tag{3}$$

where $F_l(G, C)$ denotes the lower Linear Fractional Transformation (LFT) of G and C. In order to obtain a good design for C, a precise knowledge of the plant is required. The dynamics of interest are modeled but this model may be inaccurate and may not reflect the changes suffered by the plant with time. To deal with this problem, the concept of model uncertainty comes out. The plant G is assumed to be unknown but belonging to a class of models, $\mathbf{P}$, built around a nominal model G_o. The set of models $\mathbf{P}$ is characterized by a matrix Δ, which can be either a full matrix or a block diagonal matrix that includes all possible perturbations representing uncertainty to the system. The general control configuration in Fig. 2 may be extended to include model uncertainty as it is shown in Fig. 4.

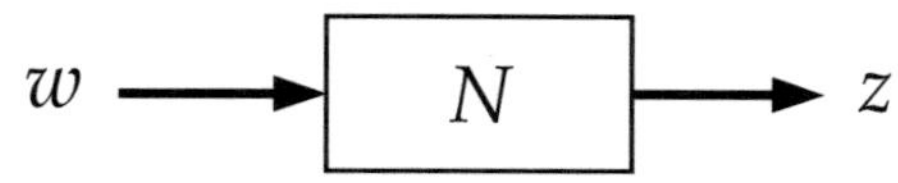

Fig. 3. General block diagram for analysis with no model uncertainty.

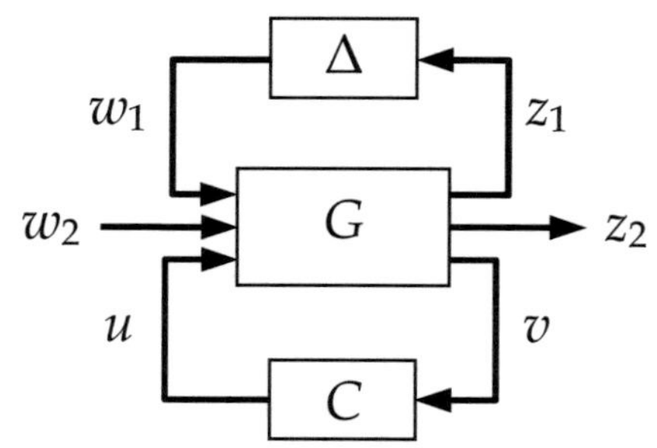

Fig. 4. General control problem formulation including model uncertainty.

The block diagram in Fig. 4 is used to synthesize the controller C. To transform it for analysis, the lower loop around G is closed by the controller C and it is incorporated into the generalized plant G to form the system N as it is shown in Fig. 5. The same lower LFT is obtained as in Eq. (3) where no uncertainty was considered.

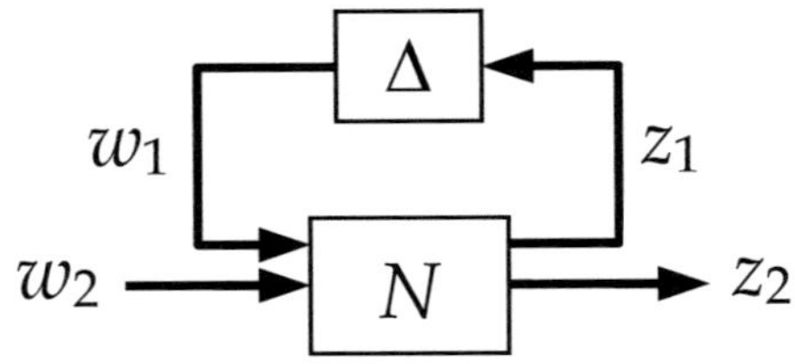

Fig. 5. General block diagram for analysis including model uncertainty.

To evaluate the relation between $w = [w_1\ w_2]^T$ and $z = [z_1\ z_2]^T$ for a given controller C in the uncertain system, the upper loop around N is closed with the perturbation matrix Δ. This results in the following upper LFT:

$$F_u(N,\Delta) \equiv N_{22} + N_{21}\Delta(\mathbf{I} - N_{11})^{-1}N_{12}\,. \tag{4}$$

To represent any control problem with uncertainty by the general control configuration in Fig. 4, it is necessary to represent each source of uncertainty by a single perturbation block Δ_i, normalized such that $\bar{\sigma}(\Delta_i) \le 1$. The individual uncertainties Δ_i are combined into one large block diagonal matrix Δ,

$$\Delta = \mathrm{diag}\{\Delta_1, \Delta_2, \ldots, \Delta_m\}\,, \tag{5}$$

satisfying

$$\bar{\sigma}(\Delta) \le 1\,. \tag{6}$$

Structured uncertainty representation considers the individual uncertainty present on each input channel and combines them into one large diagonal block. This representation avoids the norm-physical coupling at the input of the plant that appears with the full perturbation matrix Δ in an unstructured uncertainty description. Consequently, the resulting set of plants is not so large as with an unstructured uncertainty description and the resulting robustness analysis is not so conservative (Balas et al., 1993).

2.2 Robust performance and robust stability

For obtaining good set point tracking, it is obvious that some performance specifications must be satisfied in spite of unmeasured disturbances and model-plant mismatch, i.e. uncertainty. The performance specification should be satisfied for the worst-case combination of disturbances and model-plant mismatch (robust performance). In order to achieve robust performance, some specifications have to be satisfied. The following terminologies are used:

1. *Nominal Stability*—The closed-loop system has Nominal Stability (NS) if the controller C internally stabilizes the nominal model G_o, i.e. the four transfer matrices N_{11}, N_{12}, N_{21} and N_{22} in the closed-loop transfer matrix N are stable.
2. *Nominal Performance*—The closed-loop system has Nominal Performance (NP) if the performance objectives are satisfied for the nominal model G_o, i.e. $\| N_{22} \|_\infty < 1$.

3. *Robust Stability*—The closed-loop system has Robust Stability (RS) if the controller C internally stabilizes every plant $G \in \mathbf{P}$, i.e. in Fig. 5, $F_u(N,\Delta)$ is stable and $\|\Delta\|_\infty \leq 1$.
4. *Robust Performance*—The closed-loop feedback system has Robust Performance (RP) if the performance objectives are satisfied for $G \in P$, i.e. in Fig. 5, $\|F_u(N,\Delta)\|_\infty < 1$ and $\|\Delta\|_\infty \leq 1$.

The structured singular value is used as a robust performance index. To use this index one must define performance using the H_∞ framework. The H_∞-norm of a transfer function $G(s)$ is the peak value of the maximum singular value over all frequencies

$$\|G(s)\|_\infty \equiv \sup_\omega \bar{\sigma}(G(j\omega)) . \tag{7}$$

Uncertainties are modeled by the perturbations and uncertainty weights included in G. These weights are chosen such that $\|\Delta\|_\infty \leq 1$ generates the family of all possible plants to be considered (Fig. 4). Δ may contain both real and complex perturbations, but in this case study only complex perturbations are used. The performance is specified by weights in G which normalized w_2 and z_2 such that a closed-loop H_∞-norm from w_2 to z_2 of less than one (for worst-case Δ) means that the control objectives are achieved. Fig. 6 is used for robustness analysis where N is a function of G and C, and Δ_P ($\|\Delta_P\|_\infty \leq 1$) is a fictitious "performance perturbation" connecting z_2 to w_2.

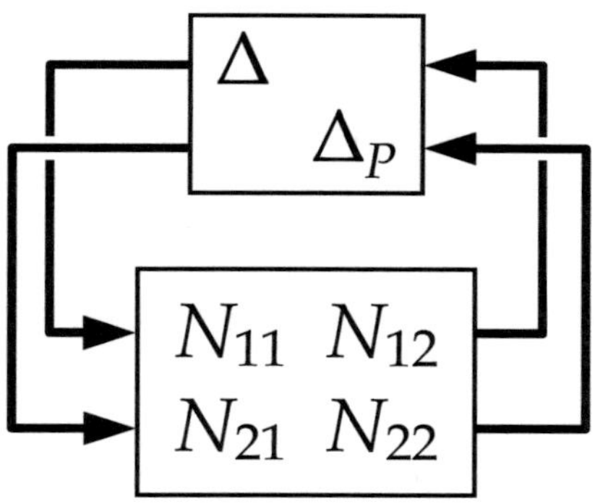

Fig. 6. General block diagram for robustness analysis.

Provided that the closed-loop system is nominally stable, the condition for robust performance (RP) is

$$\text{RP} \Leftrightarrow \mu_{\text{RP}} = \sup_\omega \mu_\Delta(N(j\omega)) < 1 , \tag{8}$$

where $\Delta = \text{diag}\{\Delta, \Delta_P\}$. μ is computed frequency-by-frequency through upper and lower bounds. Here we only consider the upper bound which is derived by the computation of non-negative scaling matrices D_l and D_r defined within a set $\mathbf{D}$ that commutes with the structure Δ:

$$\mu_\Delta(N) \leq \inf_{D \in \mathbf{D}} \bar{\sigma}(D_l N D_r^{-1}) , \tag{9}$$

where $\mathbf{D} = \left\{ D \middle| D\Delta = \Delta D \right\}$. A detailed discussion on the specification of such a set **D** of scaling matrices can be found in Packard and Doyle (1993).

2.3 Design procedure

The design procedure of a control system usually involves a mathematical model of the dynamic process, the plant model or nominal model. Consequently, many aspects of the real plant behavior cannot be captured in an accurate way with the plant model leading to uncertainties. Such plant-model mismatching should be characterized by means of disturbances signals and/or plant parameter variations, often characterized by probabilistic models, or unmodelled dynamics, commonly characterized in the frequency domain.

The modern approach to characterizing closed-loop performance objectives is to measure the size of certain closed-loop transfer function matrices using various matrix norms. Matrix norms provide a measure of how large output signals can get for certain classes of input signals. Optimizing these types of performance objectives, over the set of stabilizing controllers is the main thrust of recent optimal control theory, such as L_1, H_2, H_∞ and optimal control (Balas et al., 1993). Usually, high performance specifications are given in terms of the plant model. For this reason, model uncertainties characterization should be incorporated to the design procedure in order to provide a reliable control system capable to deal with the real process and to assure the fulfillment of the performance requirements. The term *robustness* is used to denote the ability of a control system to cope with the uncertain scheme. It is well known that there is an intrinsic conflict between performance and robustness in the standard feedback framework (Doyle and Stein, 1981; Chen, 1995). The system response to commands is an open-loop property while robustness properties are associated with the feedback. Therefore, one must make a trade-off between achievable performance and robustness. In this way, a high performance controller designed for a nominal model may have very little robustness against the model uncertainties and the external disturbances. For this reason, worst-case robust control design techniques such as μ-synthesis, have gained popularity in the last thirty years.

3. Modeling of the uncertain system

Analyzing the effect of uncertain models on achievable closed-loop performance and designing controller to provide optimal worst-case performance in the face of the plant uncertainty are the main features that must be considered in robust control of an uncertain system. Skogestad et al. (1988) recommended a general guideline for modeling of uncertain systems. According to this, three types of uncertainty can be identified:

1. Uncertainty of the manipulated variables which is referred to input uncertainty.
2. Uncertainty because of the process nonlinearity, and
3. Unmodelled high-frequency dynamics and uncertainty of the measured variables which is referred to output uncertainty.

Fig. 7(a) shows a block diagram of a distillation column with related inputs (u, d) and outputs (y, y_m). In Fig. 7(b), we have added two additional blocks to Fig. 7(a). One is the controller C, which computes the appropriate input u based on the information about the process y_m. The other block, Δ, represents the model uncertainty. $\hat{G}$ and G are models only, and the actual plant is different depending on Δ. Based on the measurements y_m, the objective of the controller C is to generate inputs u that keep the outputs y as close as possible to their set points in spite of disturbances d and model uncertainty Δ. The controller C is often non-square, as there are usually more measurements than manipulated variables. For the design of the controller C, information about the expected model uncertainty should be taken into account. Usually, there are two main ways for adding uncertainty to a constructed model: additive and multiplicative uncertainty. Fig. 7(c) represents additive uncertainty. In this case, the perturbed plant gain G_p will be $G + \Delta$ where Δ is unstructured uncertainty. Fig. 7(d) represents multiplicative uncertainty where the perturbed plant is equal to $G\,(\mathbf{I} + \Delta)$.

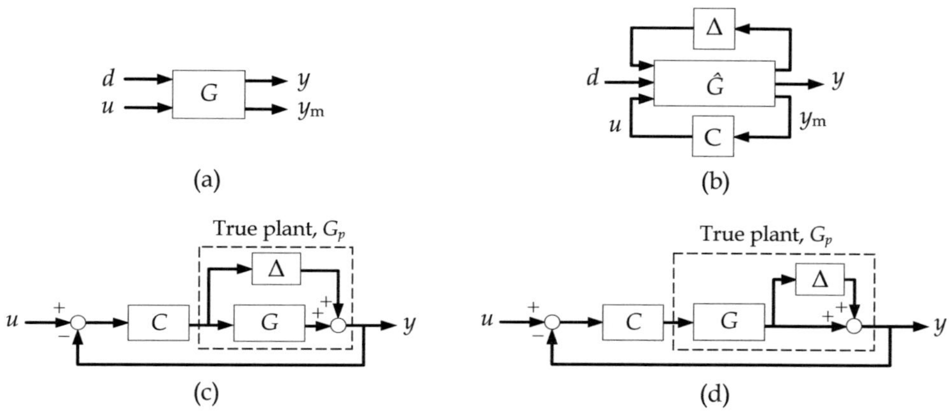

Fig. 7. (a) Schematic representation of distillation column; (b) general structure for studying any linear control problem; (c) additive unstructured uncertainty, $G_p = G + \Delta$; (d) multiplicative unstructured uncertainty, $G_p = G\,(\mathbf{I} + \Delta)$.

Here we will consider only input and output uncertainties:

Input uncertainty – Input uncertainty always occurs in practice and generally limits the achievable closed-loop performance (Skogestad et al., 1988). Ill-conditioned plants can be very sensitive to errors in the manipulated variables. The bounds for the relative errors of the column inputs u are modeled in the frequency domain by a multiplicative uncertainty with two frequency-dependent error bounds w_u. These two bounds are combined in the diagonal matrix $W_u = w_u\mathbf{I}$. In this case

$$\tilde{u}(j\omega) = \left[\mathbf{I} + \Delta_u(j\omega)W_u(j\omega)\right]u(j\omega) \quad \text{with} \quad \left\|\Delta_u(j\omega)\right\|_\infty \le 1 . \tag{10}$$

The value of the bound W_u is almost very small for low frequencies (we know the model very well there) and increases substantially as we go to high frequencies where parasitic parameters come into play and unmodelled structural flexibility is common. If all flow measurements are carefully calibrated, an error bound of 10% for the low frequency range is reasonable (Christen et al., 1997). This error bound is not common among the researchers (e.g. Skogestad and Lundström, 1990, used an error bound of 20% at steady state). Higher errors must be assumed in the higher frequency range. Because of uncertain or neglected high-frequency dynamics or time delays, the input error exceeds 100%. The following weight is used as input uncertainty weight

$$w_u(s) = 0.1\frac{1+10s}{1+s}. \tag{11}$$

The weight is shown graphically as a function of frequency in Fig. 8.

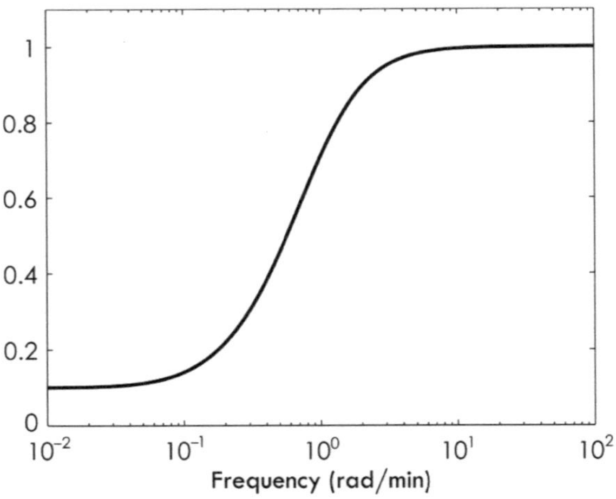

Fig. 8. Input uncertainty weight $|w_u(j\omega)|$ as a function of frequency.

Output uncertainty – Due to the nonlinear vapor/liquid equilibrium, the gains of the individual transfer functions between the two manipulated inputs and controlled outputs may change in opposite directions (gain directionality). This behavior can be described with independent multiplicative uncertainties for the two outputs of the model and a diagonal weighting matrix $W_y = w_y\mathbf{I}$. In mathematical form we can write

$$\tilde{y}(j\omega) = \left[\mathbf{I} + \Delta_y(j\omega)W_y(j\omega)\right]y(j\omega) \quad \text{with} \quad \left\|\Delta_y(j\omega)\right\|_\infty \le 1. \tag{12}$$

For the low-frequency range, an uncertainty of 10% is assumed for the description of uncertainties in the measured outputs. The uncertainty weight is

$$w_y(s) = 0.1\frac{1+180\,s}{1+2.5\,s}, \tag{13}$$

which has large gains in the high-frequency range that takes the effect of unmodelled dynamics into account.

Performance – The performance weight used in this study is the same in Skogestad and Morari (1988). The weight is defined as

$$w_P(s) = 0.5\frac{1+10\,s}{10\,s}. \tag{14}$$

3.1 Controller

Skogestad and Lundström (1990) proposed two different approaches to tune controllers. The first approach is to fix the performance specification and minimize μ_{RP} by adjusting the

controller tunings. The performance requirement is satisfied if μ_{RP} is less than one, and lower μ_{RP} values represent a better design. The second approach is to fix the uncertainty and find what performance can be achieved. In this approach, we adjust the time constant in the performance weight to make the optimal μ_{RP} values equal to one. The latter approach has two disadvantages: (1) it introduces an outer loop in the μ calculations, and (2) it may be impossible to achieve μ_{RP} equal to one by adjusting the time constant in the performance weight. Here the first approach is used for tuning the controller because of the mentioned disadvantages of the second approach.
A diagonal PID controller based on internal model control (IMC) (Rivera et al., 1986) is used to investigate the process. Optimal setting for single-loop PID controller is found by minimizing μ_{RP}. Furthermore, a μ-optimal controller is designed since it gives a good indication of the best possible performance of a linear controller.

3.2 Analysis of controller

Comparison of controller is based mainly on computing μ for robust performance. The main advantage of using the μ-analysis is that it provides a well-defined basis for comparison. μ-analysis is a worst-case analysis. It minimizes the H∞-norm with respect to the structured uncertainty matrix Δ. A worst-case analysis is particularly useful for ill-conditioned systems in the cross-over frequency range (Gjøsæter and Foss, 1997). This is due to the fact that such systems may provide large difference between nominal and robust performance.
The value of μ_{RP} is indicative of the worst-case response. If $\mu_{RP} > 1$, then the "worst-case" does not satisfy our performance objective, and if $\mu_{RP} < 1$ then the "worst-case" is better than required by our performance objective. Similarly, if $\mu_{NP} < 1$ then the performance objective is satisfied for the nominal case. However, this may not mean very much if the system is sensitive to uncertainty and μ_{RP} is significantly larger than one. It is shown that this is the case, for example, if an inverse-based controller is used for the distillation column (Skogestad and Morari, 1988). Controller was obtained by minimizing $\sup_\omega \mu_{RP}$ for the model using the input and output uncertainties and performance weight. The plots for RP for the μ-optimal controller are of particular interest since they indicate the best achievable performance for the plant. μ provides a much easier way of comparing and analyzing the effect of various combinations of controllers, uncertainty and disturbances than the traditional simulation approach. One of the main advantages with the μ-analysis as opposed to simulations is that one does not have to search for the worst-case, i.e. μ finds it automatically (Skogestad and Lundström, 1990).

3.3 Synthesis of controller

The structured singular value provides a systematic way to test for both robust stability and robust performance with a given controller C. In addition to this analysis tool, the structured singular value can be used to synthesize the controller C. The robust performance condition implies robust stability, since

$$\sup_\omega \mu_\Delta(N) \geq \sup_\omega \mu_\Delta(G) . \tag{15}$$

Therefore, a controller designed to guarantee robust performance will also guarantee robust stability. Provided that the interconnection matrix N is a function of the controller C, the μ-optimal controller can be found by

$$\text{minimize}\ \{\sup_{\omega} \mu_{\Delta}(N)\} \tag{16}$$

At the present time, there is no direct method to find the controller C by minimizing (16), however, combination of μ-analysis and H_∞-synthesis which is called μ-synthesis or DK-iteration (Zhou et al., 1996) is a special method that attempts to minimize the upper bound of μ. Thus, the objective function (16) is transformed into

$$\min_{C}\left(\inf_{D_l, D_r \in \mathbf{D}} \sup_{\omega} \bar{\sigma}(D_l N D_r^{-1})\right) \tag{17}$$

The DK-iteration approach involves to alternatively minimize

$$\sup_{\omega} \bar{\sigma}(D_l N D_r^{-1}) \tag{18}$$

for either C or D_l and D_r while holding the other constant. For fixed D_l and D_r, the controller is solved via H_∞ optimization; for fixed C, a convex optimization problem is solved at each frequency. The magnitude of each element of $D_l(j\omega)$ and $D_r(j\omega)$ is fitted with a stable and minimum phase transfer function and wrapped back into the nominal interconnection structure. The procedure is carried out until $\sup_{\omega} \bar{\sigma}(D_l N D_r^{-1}) < 1$. Although convergence in each step is assured, joint convergence is not guaranteed. However, DK-iteration works well in most cases (Balas et al., 1993; Packard and Doyle, 1993). The optimal solutions in each step are of supreme importance to success with the DK-iteration. Moreover, when C is fixed, the fitting procedure plays an important role in the overall approach. Low order transfer function fits are preferable since the order of the H_∞ problem in the following step is reduced yielding controllers of low order dimension. Nevertheless, the method is characterized by giving controllers of very high order that must be reduced applying model reduction techniques (Glover, 1984).

3.4 Simulation

Simulations are carried out with the nonlinear model of the column and using single-loop controller, which generally is insensitive to steady-state input errors (Skogestad and Morari, 1988). In addition, input and output uncertainties are included to get a realistic evaluation of the controller. Simulations are for both cases with and without uncertainty.

4. Model analysis

4.1 RGA-analysis of the model

Let $\times$ denote element-by-element multiplication. The RGA of the matrix G (Bristol, 1966) is defined as

$$\Lambda(G) = G \times (G^{-1})^T . \tag{19}$$

For 2×2 systems

$$\text{RGA} = \begin{pmatrix} \lambda_{11} & \lambda_{12} \\ \lambda_{21} & \lambda_{22} \end{pmatrix} = \begin{pmatrix} \lambda_{11} & 1-\lambda_{11} \\ 1-\lambda_{11} & \lambda_{11} \end{pmatrix} \ \text{and}\ \ \lambda_{11} = \frac{1}{1-\left(g_{12}g_{21}/g_{11}g_{22}\right)}, \tag{20}$$

where g_{ij}s are open-loop gain from the jth input to the ith output of the process. The RGA has been considered as an important MIMO system information for feedback control. Controllers with large RGA elements should generally be avoided, because otherwise the closed-loop system is very sensitive to input uncertainty (Skogestad and Morari, 1987b). Fig. 9 shows the magnitude of the diagonal element of the RGA (λ_{11}). As seen in the figure, the plant is ill-conditioned at low frequencies, while at higher frequencies, the value of the RGA-element drops. This says that only based on the RGA plot, making a decision on the ill-conditionedness of the control problem may be misleading. On the other hand, the bandwidth area is located in a frequency range where the RGA elements are small or at lower frequencies where the RGA elements are large.

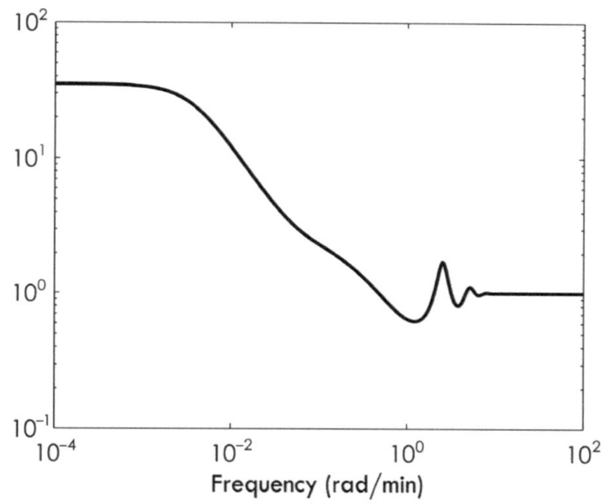

Fig. 9. Plot of $|\lambda_{11}|$ as a function of frequency.

4.2 Ill-conditionedness and process gain directionality

The common definition of an ill-conditioned plant is that it has a model with a large condition number (γ). The condition number is defined as the ratio between the largest and smallest singular values ($\bar{\sigma}/\underline{\sigma}$) of a process model. However, the condition number depends on the scaling of the process model. This problem arises from the scaling dependency of the Singular Value Decomposition (SVD). To eliminate the effect of scaling, the minimized condition number ($\gamma_{\min}$) is defined as the smallest possible condition number that can be achieved by varying the scaling. Close relationship between $\gamma_{\min}$ and RGA is proposed by Grosdidier et al. (1985). For 2 × 2 systems

$$\gamma_{\min}(G) = \|\Lambda(G)\|_1 + \sqrt{\|\Lambda(G)\|_1^2 - 1}\,, \tag{21}$$

where the 1-norm of the RGA is defined as

$$\|\Lambda\|_1 = \max_j \sum_{i=1}^{m} |\lambda_{ij}|\,. \tag{22}$$

According to the above relationship, a 2 × 2 system with small RGA elements always has a small $\gamma_{\min}$. In particular, if $0 \leq \lambda_{11} \leq 1$ the minimized condition number is always equal to one. A process model with a large span in the possible gain of the model is said to show high directionality and a process model with the smallest singular value equal to the largest singular value is said to show no directionality. Waller et al. (1994) suggest redefined definition of process directionality. The definition divides the concept of process directionality into two parts. The minimized condition number is connected to stability aspects, whereas the condition number of a process model scaled according to the weight of the variables is connected to performance aspects. Fig. 10 shows the largest and smallest singular values and condition number of the process model as a function of frequency.

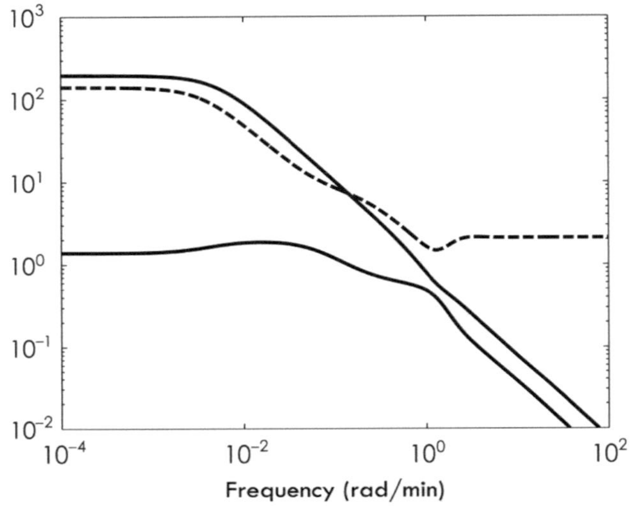

Fig. 10. Singular values (—) and condition number (----) of the distillation column.

The condition number of the process is about 10 times lower at high frequencies than at low frequencies (steady state). Fig. 11(a) represents the values of γ and $\gamma_{\min}$ as a function of frequency. Values of γ and $\gamma_{\min}$ match each other from low to intermediate frequencies, but $\gamma_{\min}$ approaches one at high frequencies. For 2 × 2 systems (Grosdidier et al., 1985):

$$\left\|\Lambda\right\|_1 - \frac{1}{\gamma_{\min}(G)} \leq \gamma_{\min}(G) \leq \left\|\Lambda\right\|_1 . \quad (23)$$

Consequently, for 2 × 2 systems the difference between these quantities is at most one and $\left\|\Lambda\right\|_1$ approaches $\gamma_{\min}$ as $\gamma_{\min} \to \infty$. Since $\left\|\Lambda\right\|_1$ is much easy to compute than $\gamma_{\min}$, it is the preferred quantity to use. In Fig. 11(b), $\gamma_{\min}$ and $\left\|\Lambda\right\|_1$ are plotted as a function of frequency. The value of $\gamma_{\min}$ at low frequencies is approximately twice $\left\|\Lambda\right\|_1$. At high frequencies, both $\gamma_{\min}$ and $\left\|\Lambda\right\|_1$ approach one (after $\omega = 20$ rad/min). This is in agreement with the result obtained from λ_{11}-vs-frequency plot (Fig. 9). Since $\gamma_{\min}$ is independent of scaling, therefore it is better to use $\gamma_{\min}$ instead of γ, which is scale dependent.

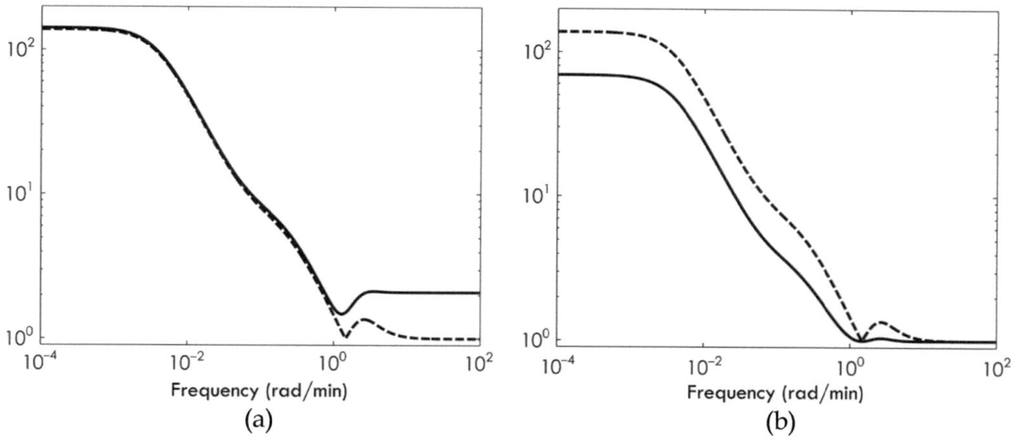

Fig. 11. (a) Plots of γ and γ_{min} as a function of frequency (— γ and ---- γ_{min}); (b) plots of $||\Lambda||_1$ and γ_{min} as a function of frequency (— $||\Lambda||_1$ and ---- γ_{min}).

4.3 Synthesis of the controller

The plots of the singular values of the sensitivity functions $S = (\mathbf{I} + GC)^{-1}$ demonstrate good disturbance rejection properties, which indicate the closed-loop system is insensitive to uncertainties in inputs and outputs (Fig. 12(a)). The tracking properties of this controller are also adequate, which is illustrated by plots of the complementary sensitivity function, $T = \mathbf{I} - S$ (Fig. 12(b)). Up to the mid-frequency range, the singular values are close to one and the maximum of the upper singular values is slightly greater than one.

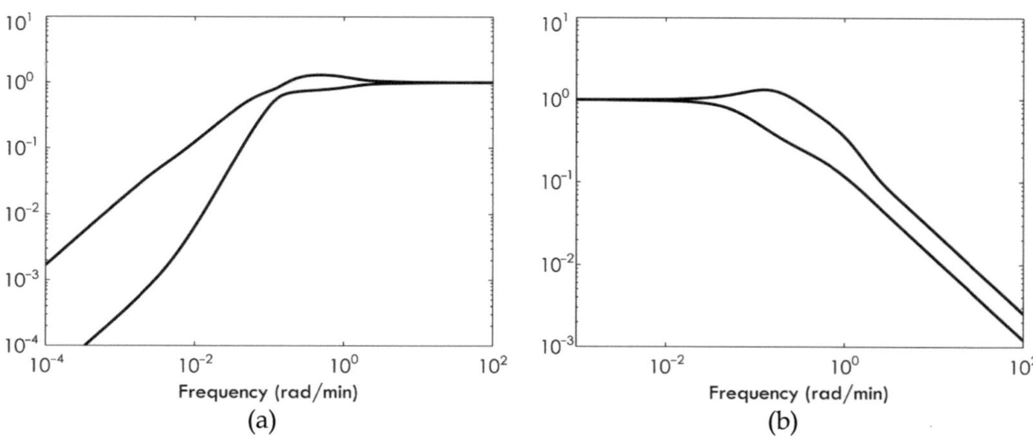

Fig. 12. Singular values of the closed-loop system. (a) Sensitivity function; (b) complementary sensitivity function.

4.4 PID-tuning of the controller

Table 2 summarizes the PID controller setting that is used for the column. Fig. 13 shows μ-plots of the controller. From a maximum peak-value point of view, it is seen that both robust and nominal performance plots are less than one which satisfy the criterion. The plots approach 0.5 as frequency approaches infinity.

Type of controller	k	τ_I (min)	τ_D (min)
PID Controller			
Top composition control loop	0.37	5.16	0.58
Bottom composition control loop	0.20	3.70	1.18
μ-Optimal Controller			
Top composition control loop	0.26	3.43	1.33
Bottom composition control loop	0.31	4.71	0.67

Table 2. Tuning parameters for PID and μ-optimal controllers.

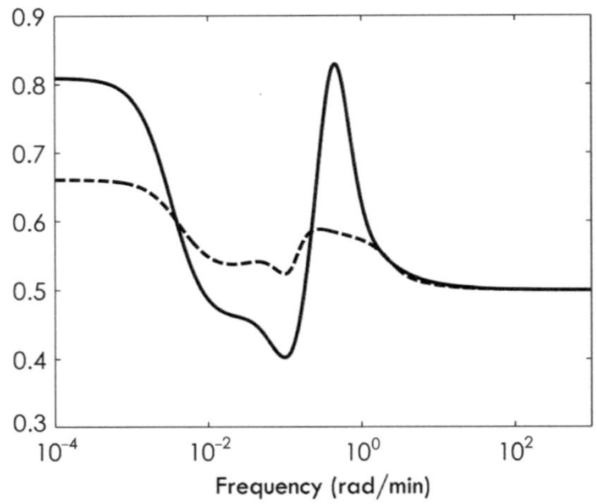

Fig. 13. μ plots for PID controller: — robust performance; ---- nominal performance.

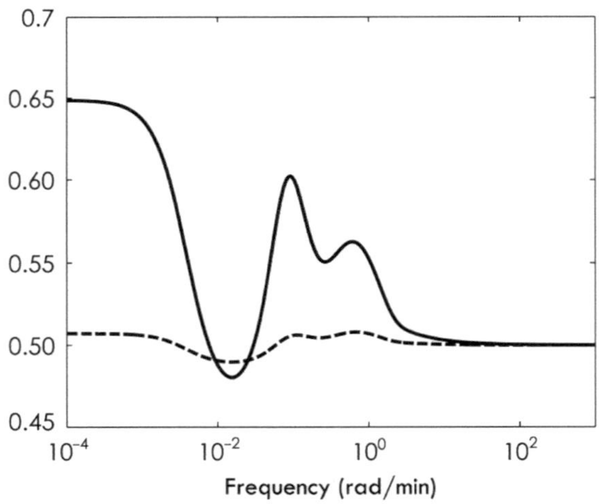

Fig. 14. μ-plots for the μ-optimal controller: — robust performance; ---- nominal performance.

4.5 Comparison with μ-optimal controller

Nominal and robust performance plots of the μ-optimal controller is shown in Fig. 14. Comparison of nominal performance of the controllers shows that for the μ-optimal controller, the plot is nearly flat over a large frequency range which indicates that an optimal controller is achieved. Comparing robust performance of the controllers indicates that obtaining robust performance with the LV-configuration is also possible. This is also in agreement with the results presented by Skogestad and Lundström (1990).

5. Simulations

Simulations of a set-point change in y_D using the PID- and μ-optimal controllers are shown in Figs. 15 and 16, respectively. As it is seen, the introduced uncertainties do not seriously affect the performance of the μ-optimal controller, while for the PID-controller, the effect of uncertainties is more rather the μ-optimal controller. It should be noted that the reference signal is filtered by a prefilter with a time constant of 5 min. Fig. 16 also shows that the PID controller has a slow return to steady state. This is due to the high μ_{NP} value at lower frequencies compared with the μ-optimal controller (Figs. 13 and 14). In Table 3, numerical values of μ for nominal and robust performance are presented.

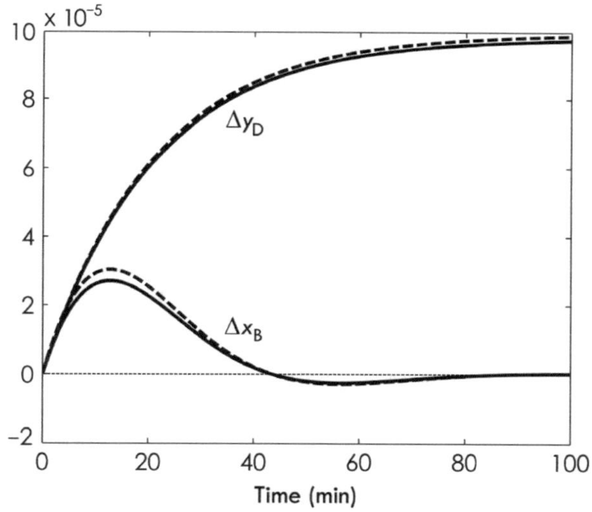

Fig. 15. Closed-loop response to small set-point change in y_D (μ-optimal controller): — no uncertainty; ---- 10% uncertainty on input and output.

Controller	Nominal Performance	Robust Performance
PID	0.661	0.830
μ-optimal (both input and output uncertainties)	0.506	0.648
μ-optimal (only input uncertainty)	0.611	0.721

Table 3. μ values of the controllers.

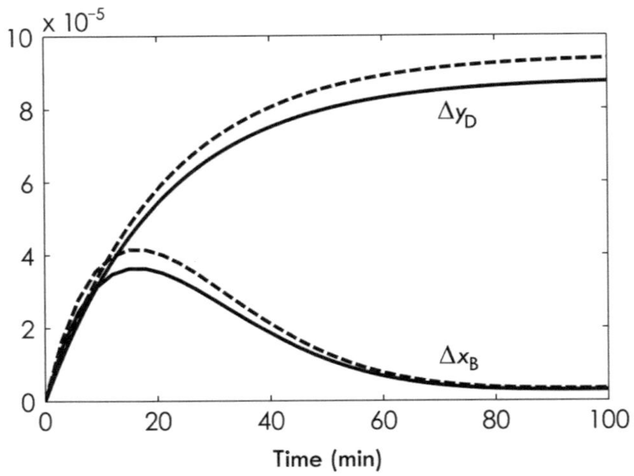

Fig. 16. Closed-loop response to small set-point change in y_D (PID controller): — no uncertainty; ---- 10% uncertainty on input and output.

Fig. 17 shows the closed-loop response of the μ-optimal controller to a 20% increase in feed flow rate. In Fig. 18, the closed-loop response for both controllers is shown simultaneously. As the figure shows, the PID controller needs considerably more times to reach steady state than the μ-optimal controller (see next page for the figures).

5.1 Effect of output uncertainty

Fig. 19 shows the effect of output uncertainty on closed-loop response of the μ-optimal controller. For the case that both input and output uncertainties are considered, the response is faster than for the case that only input uncertainty is considered, however, this difference is not so large. The reason for this again returns to the μ_{NP} values at low frequencies. The μ-values of nominal performance for the case including both input and output uncertainties is close to the case where only input uncertainty included (Table 3)

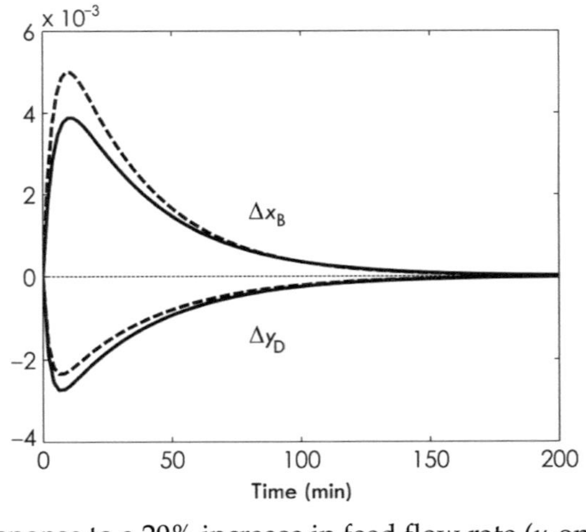

Fig. 17. Closed-loop response to a 20% increase in feed flow rate (μ-optimal controller): — no uncertainty; ---- 10% uncertainty on inputs and outputs.

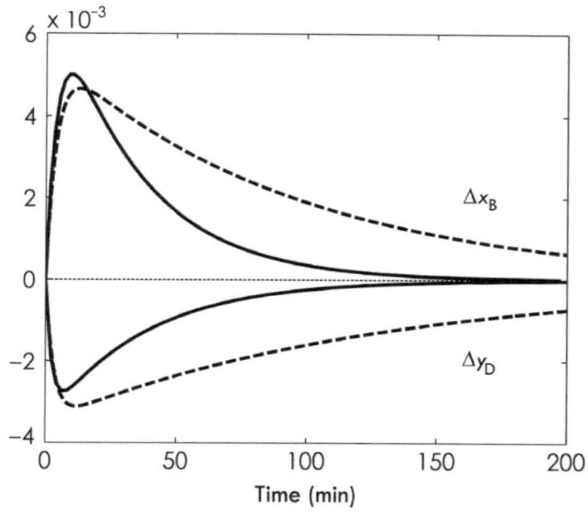

Fig. 18. Closed-loop response to a 20% increase in feed flow rate (including input and output uncertainties): — μ-optimal controller; ---- PID controller.

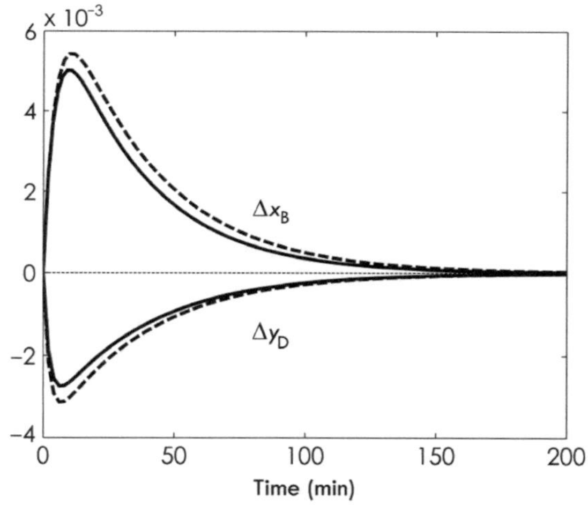

Fig. 19. Closed-loop response to a 20% increase in feed flow rate (μ-optimal controller): — both input and output uncertainty; ---- input uncertainty only.

6. Discussion

The structured singular value (μ) is used to investigate the robust performance and robust stability of the PID controller. The control problem formulation used in this study is using weighted input and output uncertainties. Although other sources of uncertainty could be included, however, these two are the most severe uncertainties that may be considered. The inclusion of both input and output uncertainty prevents the control system from becoming sensitive to the uncertainties, as may happen with inverting controllers.

The solution of the problem leads to the inequality of Eq. (8). The numerical solution of this design task is difficult. At present, there is no direct method to synthesize a μ-optimal

controller, however, combination of μ-analysis and H_∞-synthesis which called μ-synthesis or DK-iteration, often yields good results. This algorithm has two drawbacks. Firstly, the algorithm cannot guarantee convergence, and secondly, the algorithm requires a scaling of the plant in each iteration step, which increases the order of the plant.
The μ analysis advantageously avoids dealing explicitly with the bad condition of the plant. With the μ-approach, the upper bound for the bandwidth of the control system is provided by the uncertainty model, whereas the lower bound is a matter of optimization. μ-synthesis is ideally suited to deal with complex uncertainty models which takes into account such aspects as various operating points. A difficulty that one may encountered in synthesis of controller is high computation time, because the μ approach requires scaling in each iteration. If, however, loop-shaping ideas are used to form the augmented plant, H_∞-synthesis may be used to advantage. In this case, the results are as good as with the μ-synthesis, but are obtained with less numerical efforts (Christen et al., 1997).
In this case study, the LV-configuration is used. The use of this configuration for columns with high condition number may be doubtful, but under special considerations, this configuration may yield acceptable performance. It is shown (Skogestad and Lundström, 1990) that it is possible to achieve good control behavior using the LV-configuration for two-point composition control provided measurement delays are not too large (typically less than 1–2 min). In addition, severe interactions and poor control often reported with the LV-configuration may be almost eliminated if the loops are tuned sufficiently tight. However, this does not imply that the LV-configuration is the best structure to use. Shinskey (1984) showed that the use of the $(L/D)(V/B)$-configuration is probably better in most cases, and in particular for columns with large reflux.

7. Concluding remarks

Based on a structured uncertainty model, which describes the column dynamics within the entire operating range, a decentralized PID controller is calculated using the μ-synthesis technique. The controller was found to be robust with respect to model-plant mismatch, provided the RGA values of the column transfer function are not too large in the cross-over frequency range. The response of the system is improved by using a μ-optimal controller.
In spite of high condition number of the process, nominal and robust performance is achieved by insertion of input and output uncertainties in the control system and using the structured singular value to synthesis the controller. Good set-point tracking and disturbance rejection of the controller is observed by simulations that carried out for the closed-loop system. It was also shown that good control performance can be obtained by using the LV-configuration which is difficult to implement for two-point control. The obtained results also verify the findings of Skogestad and Lundström (1990).

Symbols

B	Bottom product
C	Controller
D	Distillate, scaling matrix
$\mathbf{D}$	Set of scaling matrices
F	Feed flow rate, Linear Fractional Transformation (LFT)
g_L	Liquid flow dynamics
G	Plant transfer function

$\mathbf{I}$	Identity matrix
L	Reflux
n	Number of trays in the column
N	Lower LFT
N_T	Number of theoretical trays in the column
$\mathbf{P}$	Set of all possible plants
S	Sensitivity function
T	Complementary sensitivity function
u	Uncertain input
$\tilde{u}$	Weighted input
V	Vapor boilup
w	Scalar weight, input signal
W	Diagonal matrix weight
x_B	Bottom composition
y	Output
$\tilde{y}$	Weighted output
y_D	Distillate composition
z	Output signal
z_F	Feed composition
$\|\cdot\|_1$	1-norm
$\|\cdot\|_\infty$	∞-norm

Greek letters

α	Relative volatility
γ	Condition number
Δ	Perturbation matrix
Λ	Relative gain array
λ_{ij}	i, j element of the RGA
μ	Structured singular value (SSV)
σ	Singular value
τ	Time constant
ω	Frequency (rad/min)

Subscripts

D	Derivative
I	Integral
l	Lower, left
min	Minimized
NP	Nominal performance
o	Nominal
P	Performance
r	Right
RP	Robust performance
u	Input, upper
y	Output

8. References

Arkun, Y., Manousiouthakis, B. & Palazoglu, A. (1984). Robustness Analysis of Process Control Systems: A Case Study of Decoupling Control in Distillation. *Industrial and Engineering Chemistry Process Design and Development* 23(1), 93–101.

Biswas, P.P., Ray, S. & Samanta, A.N. (2009). Nonlinear Control of High Purity Distillation Column under Input Saturation and Parametric Uncertainty. *Journal of Process Control* 19(1), 75–84.

Böling, J.M. & Häggblom, K.E. (1996). Control-Relevant Identification of an Ill-Conditioned Distillation Column. *Proceedings of IEEE Conference on Control Applications*, Dearborn, MI, 570–575.

Bristol, E.H. (1966). On A New Measure of Interactions for Multivariable Process Control. *IEEE Transactions on Automatic Control* 11(1), 133–134.

Chen, J. (1995). Sensitivity Integral Relations and Design Trade-Offs in Linear Multivariable Feedback Systems. *IEEE Transactions on Automatic Control* 40(10), 1700–1716.

Christen, U., Musch, H.E. & Steiner, M. (1997). Robust Control of Distillation Columns: μ- vs. H_∞-Synthesis. *Journal of Process Control* 7(1), 19–30.

Doyle, J.C. (1982). Analysis of Feedback Systems with Structured Uncertainties. *IEE Proceedings* 129(6), 242–250.

Doyle, J.C. & Stein, G. (1981). Multivariable Feedback Design: Concepts for A Classical/Modern Synthesis. *IEEE Transactions on Automatic Control* 26(1), 4–16.

Finco, M.V., Luyben, W.L. & Polleck, R.E. (1989). Control of Distillation Columns with Low Relative Volatilities. *Industrial and Engineering Chemistry Research* 28(1), 75–83.

Georgiou, A., Georgakis, C. & Luyben, W.L. (1988). Nonlinear Dynamic Matrix Control for High-Purity Distillation Columns. *AIChE Journal* 34(8), 1287–1298.

Glover, K. (1984). All Optimal Hankel-Norm Approximations of Linear Multivariable Systems and Their L^∞-Error Bounds. *International Journal of Control* 39(6), 1115–1193.

Grosdidier, P., Morari, M. & Holt, B.R. (1985). Closed-Loop Properties from Steady-State Gain Information. *Industrial and Engineering Chemistry Fundamentals* 24(2), 221–235.

Heath, J.A., Kookos, I.K. & Perkins, J.D. (2000). Process Control Structure Selection Based on Economics. *AIChE Journal* 46(10), 1998–2016.

Hori, E.S. & Skogestad, S. (2007). Selection of Control Structure and Temperature Location for Two-Product Distillation Columns. *Chemical Engineering Research and Design* 85(3), 293–306.

Hurowitz, S., Anderson, J., Duvall, M. & Riggs, J.B. (2003). Distillation Control Configuration Selection. *Journal of Process Control* 13(4), 357–362.

Jacobsen, E.W. & Skogestad, S. (1994). Inconsistencies in Dynamic Models for Ill-Conditioned Plants: Application to Low-Order Models of Distillation Columns. *Industrial and Engineering Chemistry Research* 33(3), 631–640.

Kariwala, V., Skogestad, S. & Forbes, J.F. (2006). Relative Gain Array for Norm-Bounded Uncertain Systems. *Industrial and Engineering Chemistry Research* 45(5), 1751–1757.

Luyben, W.L. (1987). Sensitivity of Distillation Relative Gain Arrays to Steady-State Gains. *Industrial and Engineering Chemistry Research* 26(10), 2076–2078.

Luyben, W.L. (2005). Effect of Feed Composition on the Selection of Control Structures for High-Purity Binary Distillation. *Industrial and Engineering Chemistry Research* 44(20), 7800–7813.

McDonald, K.A., Palazoglu, A. & Bequette, B.W. (1988). Impact of Model Uncertainty Descriptions for High-Purity Distillation Control. *AIChE Journal* 34(12), 1996–2004.

Packard, A. & Doyle, J.C. (1993). The Complex Structured Singular Value. *Automatica* 29(1), 71–109.

Razzaghi, K. & Shahraki, F. (2005). Robust Multivariable PID-Controller Design for a High-Purity Distillation Column Using μ-Synthesis. *Proceedings of the 55th Canadian Chemical Engineering Conference*, Toronto, Ontario, Canada, October 16–19.

Razzaghi, K. & Shahraki, F. (2007). Robust Control of an Ill-Conditioned Plant Using μ-Synthesis: A Case Study for High-Purity Distillation. *Chemical Engineering Science* 62(5), 1543–1547.

Razzaghi, K. & Shahraki, F. (2009). A Survey for the Selection of Control Structure for Distillation Columns Based on Steady State Controllability Indexes. *Iranian Journal of Chemical Engineering* 6(2), 29–36.

Rivera, D.E., Morari, M. & Skogestad, S. (1986). Internal Model Control: 4. PID Controller Design. *Industrial and Engineering Process Design and Development* 25(1), 252–265.

Shin, J., Seo, H., Han, M. & Park, S. (2000). A Nonlinear Profile Observer Using Tray Temperatures for High-Purity Binary Distillation Column Control. *Chemical Engineering Science* 55(4), 807–816.

Shinskey, F.G. (1984). *Distillation Control*. 2nd ed., McGraw-Hill, New York.

Skogestad, S. & Lundström, P. (1990). Mu-Optimal LV-Control of Distillation Columns. *Computers and Chemical Engineering* 14(4–5), 401–413.

Skogestad, S., Lundström, P. & Jacobsen, E.W. (1990). Selecting the Best Distillation Control Configuration. *AIChE Journal* 36(5), 753–764.

Skogestad, S. & Morari, M. (1987a). Control Configuration Selection for Distillation Columns. *AIChE Journal* 33(10), 1620–1635.

Skogestad, S. & Morari, M. (1987b). Implications of Large RGA Elements on Control Performance. *Industrial and Engineering Chemistry Research* 26(11), 2323–2330.

Skogestad, S. & Morari, M. (1988). LV-Control of a High-Purity Distillation Column. *Chemical Engineering Science* 43(1), 33–48.

Skogestad, S., Morari, M. & Doyle, J.C. (1988). Robust Control of Ill-Conditioned Plants: High-Purity Distillation. *IEEE Transactions on Automatic Control* 33(12), 1092–1105.

Sriniwas, G.R., Arkun, Y., Chien, I.-L. & Ogunnaike, B.A. (1995). Nonlinear Identification and Control of a High-Purity Distillation Column: A Case Study. *Journal of Process Control* 5(3), 149–162.

Stichlmair, J. (1995). Conceptual Design of the Control Configurations of Distillation Columns. *Chemical Engineering and Processing* 34(2), 61–69.

Trentacapilli, M., Semino, D. & Brambilla, A. (1997). High-Purity Distillation Control: Some Issues Regarding the Application of Multivariable Control. *Distillation and Absorption '97*, Vol. 1, IChemE, 313–322.

Waller, J.B., Sågfors, M. & Waller, K.E. (1994). Ill-Conditionedness and Process Directionality—The Use of Condition Numbers in Process Control. *Proceedings of IFAC Symposium*, Kyoto, Japan, 465–470.

Zhou, K., Doyle, J.C. & Glover, K. (1996). *Robust and Optimal Control*. Prentice-Hall, Inc., Upper Saddle River, New Jersey.

13

Loop Transfer Recovery for the Grape Juice Concentration Process

Nelson Aros Oñate[1] and Graciela Suarez Segali[2]
[1]*Departamento de Ingeniería Eléctrica, Facultad de Ingeniería, Universidad de La Frontera, Temuco,*
[2]*Departamento de Ingeniería Química, Facultad de Ingeniería, Universidad Nacional de San Juan, San Juan,*
[1]*Chile*
[2]*Argentina*

1. Introduction

It is necessary to ensure the quality of concentrated, because it is highly used in the food industry in juices, drinks, sweets, etc. Its application is in full development because it can compete with any other constituent, it is a natural product, and considering that is a very important regional industry, marketing greatly affects the regional economy. Because of this, that it is extremely important to ensure quality and quantity concentrate.
Argentina is one of the principal producers and exporters of concentrated clear grape juices in the world. They are produced mainly in the provinces of San Juan and Mendoza (Argentine Republic) from virgin grape juice and in the major part from sulfited grape juices. The province of San Juan's legislation establishes that a portion of the grapes must be used for making concentrated clear grape juices. This product has reached a high level of penetration in the export market and constitutes an important and growing productive alternative.
An adequate manufacturing process, a correct design of the concentrate plants and an appropriate evaluation of their performance will facilitate optimization of the concentrated juices quality parameters. The plant efficiency is obtained from knowledge of the physics properties of the raw material and products. These properties are fundamental parameters that are used in the designing and calculations on all the equipment used and also in the control process.
The multi-step evaporation (M-SE) is the most important unit operation used in the food industry to concentrate juices of grapes and apples. Even when the main objective of this process is to produce a concentrated product, it should also possess certain organoleptic properties that are critical with respect to its quality and acceptance grade by the customers.
Product requirements and the complex characteristics of the process such as non-linear behavior, input and output constraints, time delays and loop interactions justify the use of an advanced control system.
The rheological behavior influences directly the heat transfer coefficient (Pilati, 1998; Rubio, 1998) and therefore its knowledge is essential together with the influence of temperature on

its value. The juices (concentrate and intermediate products) physical properties, such as density, viscosity, boiling point elevation, specific heat and coefficient of thermal expansion, are affected by their solid content and their temperature. For this reason, it is necessary to know the physical properties values, as a function of the temperature and the solids content, during the manufacture process.

The principal solids constituents of clear grape juices are sugars and its concentration affects directly the density, viscosity and refraction index. Tables were developed to relate reducing sugar contents, refractometric values and density of pure solutions, at 20°C, for concentrate ranges from 0% to 85% w/w and sucrose solutions for different range concentrations 0% to 70% and a temperature range from 0 to 100°C (AOAC, 1995).

Barbieri (1980) worked with white concentrated clear grape juice in a falling film multiple effect evaporators. They obtained 18.2, 27.3, 38.6, 48.6 and 64.6°Brix samples. They measured density, viscosity and boiling point elevation as a function of soluble solids concentration and temperature. They presented the results in plots with predictive equations for the properties which were studied.

Di Leo (1988) published density, refraction index and viscosity data for a rectified concentrated grape juice and an aqueous solution of a 1:1 glucose/levulose mixture, for a soluble solids concentrate range from 60 to 71% (in increments of 0.1%) and 20°C.

Pandolfi, (1991) studied physical and chemical characteristics of grape juices produced in Mendoza and San Juan provinces, Argentina. They determined density at 20°C in sulfited grape juices of 20–22°Bx and concentrated grape juices of 68–72 °Bx. They obtained no information on intermediate concentrations or other temperatures.

In general, the clarified juice concentrates have a Newtonian behavior (Ibarz 1993; Rao 1984; Saenz, 1986; Saravacos, 1970), although some authors have found a small pseudoplasticity in the flow of grape concentrates, from the variety Concord (Vitis labrusca) for concentrations above 55°Bx. It has been attributed to the presence of some soluble solids, mostly pectins and tartrates (Moressi, 1984; Saravacos, 1970). Other authors consider the juice concentrates as Newtonian, even at high soluble solids concentrations of 60–70°Bx (Barbieri, 1980; Di Leo, 1988; Rao, 1984; Schwartz, 1986).

If we analyze the temperature influence on this product's viscosity, it seems which is directly related with soluble solids concentration; the higher the concentration, the higher is the variation of the viscosity with temperature (Rao, 1984; Saravacos, 1970; Bayindirli, 1992; Crapiste, 1988; Constela, 1989).

Schwartz (1986) determined clear grape juice viscosity at 20, 30, 40 and 50°C, for 30, 40, 50, 60 and 66% soluble solids concentration, but did not publish the experimental data. These authors presented the correlation constants values of the Arrhenius equation for temperature, a potential and an exponential model between viscosity and solids concentration for each temperature studied.

The physical property that represents density change in a material, due to an increase in its temperature at constant pressure, is called the coefficient of thermal expansion. The importance of this parameter can be seen in the effect that density change in the product can have over heat transfer during the process. There is not publish data on the coefficient of the thermal expansion for grape juices and their concentrates. The existing information did not cover all the temperature and concentration ranges that are used in the evaporation process, or else cover to pure sugar solutions, or grape juices of other varieties and/ or originating in other geographical zones.

On the other hand, sensitivity theory, originally developed by Bode (Bode, 1945), has regained considerable importance, due to the recent work developed by many researchers. This research effort has made evident the fundamental role played by sensitivity theory to highlight design tradeoffs and to analyze, qualitatively, control system performance. One of the fields of active research is the analysis of different design strategies from to point of view of sensibility properties. From this perspective one of the richest strategies is the optimal linear quadratic regulator (LQR). It is well known that the sensitivity of a LQR Loop is always less than one (Anderson, 1971). However, it is also known that when the state is not directly fed back but reconstructed through an observer this property is normally lost. In fact, the situation more general, since the recovery problem appears every time a control design based on state feedback design is implemented through observers.

It has been shown that when the plant is minimum-phase, a properly design Kalman filter provides complete recovery of the input sensitivity achieved by LQR will full state feedback (Doyle, 1979). Either full or partial-order filter may be used. On the other hand it is also known that it is generally impossible to obtain LTR if we use observers for a plant with unstable zeros. An exception to this rule arises in MIMO Systems when input directions are orthogonal to non-minimum phase zero directions (Zhang, 1990).

On the other hand, it is known that the only way to obtain full recovery for a general non-minimum-phase plant is to increase the number of independent measurements. This idea has been suggested in conjunction with the use of reduced-order Kalman filters (Friedland, 1989).

The additional independent measurements are used to modify the structure of the open loop transfer function. The standard LTR procedure is applied and it is the implemented combining the resulting full-order Kalman filter with the additional measurements optimally weighted. The idea is obviously to feed back only a subset of the state, for that reason we speak of 'partial' state feedback. The basic approach assumes that all states are available for measurement. However in this paper, it is also shown how to do $\boldsymbol{L}_2$ -optimization on the amount of recovery of the input sensitivity when a given set of measurements is available. This situation is important since in many additional situations there are limitations regarding which variables can be measured and how many additional sensors can be used. This connects the recovery theory with the issue of additional measurements raised in the context of practical ideas for control design, as illustrated in the control of the inverted pendulum; see (Middleton, 1990). The theory supporting the proposal is built on some import technical results which allow for computing the amount of recovery, as a function of frequency (Zhang, 1990).

2. Process description

Figure 1 show the input and output streams in a vertical generic effect evaporator with long tubes. The solution to be concentrated circulates inside the tubes, while the steam, used to heat the solution, circulates inside the shell around the tubes. The evaporator operates in parallel mode. The solution to be concentrated and the steam are fed to the first effect by the bottom and by the upper section of the shell, respectively. Later on, the concentrated solution from the first effect is pumped to the bottom of the second effect, and so on until the fourth effect. On the other hand, the vaporized solvent from each effect serve as heater in the next one. Each effect has a baffle in the upper section that serves as a drops splitter for the solution dragged by the vapor. The vapor from the fourth effect is sent to a condenser and leaves the process as a liquid. Finally, the solution leaving the fourth effect attains the desired concentration and the solution is sent to a storage tank.

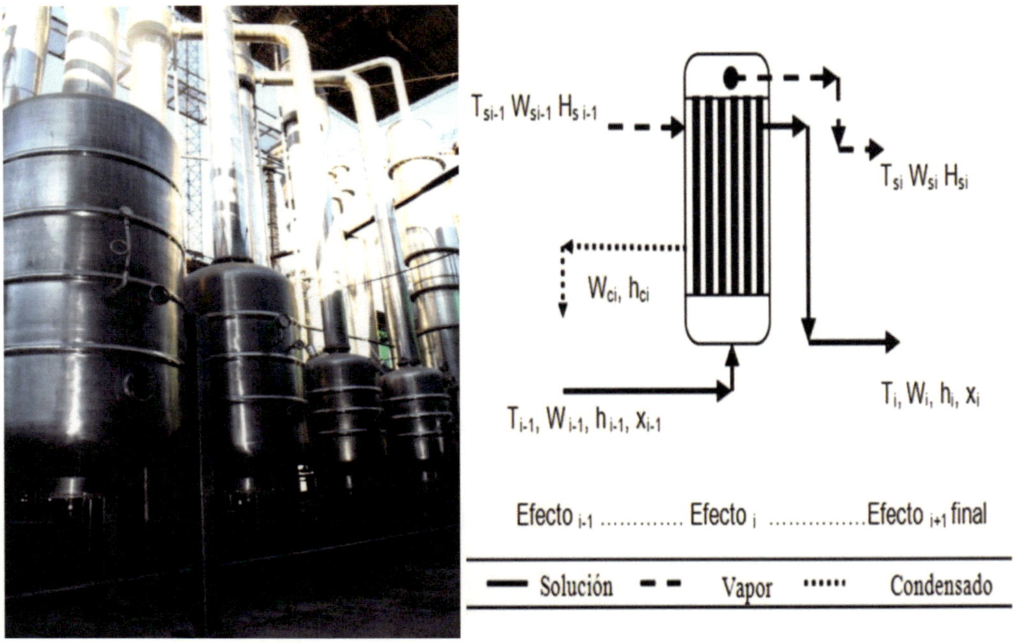

Fig. 1. Photo of evaporator and scheme of effect i in the four-stage evaporator flow sheet. $i = 1, \cdots, 4$.

3. Phenomenological model

Stefanov (2005) has developed a rigorous model with distributed parameters based on partial differential equations for a falling-film evaporator, in which the open-loop stability of the model to disturbances is verified. On the other hand, various methods have been proposed in order to obtain reduced-order models to solve such problems (Armaou, 2002; Camacho, 1999; El-Farra, 2003; Zheng, 2002). However, there is not a general framework yet, which assure an effective implementation of a control strategy in a multiple effect evaporator.

In practice, due to a lack of measurements to characterize the distributed nature of the process and actuators to implement such a solution, the control of systems represented by PDE in the grape juice evaporator, is carried out neglecting the spatial variation of parameters and applying lumped systems methods. However, a distributed parameters model must be developed in order to be used as a real plant to test advance control strategies by simulation.

The mathematical model of the evaporator is obtained by application of the mass and energy balances to each effect:

a. Global mass balances in each effect:

$$\frac{dM_i}{dt} = W_{i-1} - W_{si} - W_i \tag{1}$$

in this equations W_i, $i = 1, \cdots, 4$, are the solution mass flow rates leaving the effects 1 to 4, respectively. W_0 is the input mass flow rate that is fed to the equipment. W_{si}, $i = 1, \cdots, 4$, are the vapor mass flow rates coming from effects 1 to 4, respectively. dM_i/dt, $i = 1, \cdots, 4$, represent the solution mass variation with the time for each effect.

b. Solute mass balances for each effect:

$$\frac{d(M_i X_i)}{dt} = W_{i-1}X_{i-1} - W_i X_i \tag{2}$$

where, X_i, $i = 1, \cdots, 4$, are the concentrations of the solutions that leave the effects 1 to 4, respectively. X_0 is the concentration of the fed solution.

c. Energy balances:

$$\frac{d(M_i h_i)}{dt} = W_{i-1}h_{i-1} - W_i h_i - W_{si}H_{si} + A_i U_i (T_{si-1} - T_i) \tag{3}$$

where, h_i, $i = 1, \cdots, 4$, are the liquid stream enthalpies that leave the corresponding effects, h_0 is the feed solution enthalpy, and H_{si}, $i = 1, \cdots, 4$, are the vapor stream enthalpies that leave the corresponding effects and, A_i represents the heat transfer area in each effect. The model also includes algebraic equations. The vapor flow rates for each effect are calculated neglecting the following terms: energy accumulation and the heat conduction across the tubes. Therefore:

$$W_{si} = \frac{U_i A_i (T_{si-1} - T_i)}{H_{si-1} - h_{ci}} \tag{4}$$

For each effect, the enthalpy can be estimated as a function of temperatures and concentrations (Perry, 1997), where:

$$H_{si} = 2509.2888 + 1.6747 \cdot T_{si} \tag{5}$$

$$h_{ci} = 4.1868 \cdot T_{si} \tag{6}$$

$$C_{\rho i} = 0.80839 - 4.3416 \cdot 10^{-3} \cdot X_i + 5.6063 \cdot 10^{-4} \cdot T_i \tag{7}$$

$$h_i = 0.80839 \cdot T_i - 4.3416 \cdot 10^{-3} \cdot X_i T_i + 2.80315 \cdot 10^{-4} \cdot T_i^2 \tag{8}$$

T_i, $i = 1, \cdots, 4$, are the solution temperatures in each effect, and T_{s0} is the vapor temperature that enters to the first effect. T_{si}, $i = 1, \cdots, 4$, are the vapor temperatures that leave each effect. The heat transfer coefficients are:

$$U_i = \frac{490 \cdot \left(D^{0.57} \cdot W_{si}^{3.6/L}\right)}{\mu_i^{0.25} \cdot \Delta T_i^{0.1}} \tag{9}$$

where, the Arrhenius type equation for the viscosity is:

$$\mu_i = \mu_0 \cdot e^{\frac{A \cdot X_i}{100 - B \cdot X_i}} \tag{10}$$

$$A = C_1 + \frac{C_2}{T_i} \tag{11}$$

$$B = C_3 + C_4 \cdot T_i \tag{12}$$

The global heat-transfer coefficients are directly influenced by the viscosity and indirectly by the temperature and concentration in each effect. The constants C_1, C_2, C_3, C_4 depend on the type of product to be concentrated (Kaya, 2002).
Although the model could be improved, the accuracy achieved is enough to incorporate a control structure.

4. Standard LTR procedure

4.1 The basic approach

Consider a linear time-invariant system with state characterization is given by:

$$\dot{x} = Ax(t) + Bu(t) + v(t) \tag{13}$$

$$y(t) = Cx(t) \tag{14}$$

where $x(t) \in R^n$, $u(t) \in R^m$, $v(t) \in R^m$, $y(t) \in R^l$, A, B and C have consistent dimensions. We further assume that is a wide sense stationary process with covariance matrix Q.
We then have that the system transfer matrix function is given by:

$$G(s) = C(sI - A)^{-1}B \tag{15}$$

If the state feedback law:

$$u(t) = -Fx(t) \tag{16}$$

is applied, we obtain an input sensitivity function is given by:

$$S_0(s) = |I + H(s)|^{-1} \tag{17}$$

where:

$$H(s) = F(sI - A)^{-1}B \tag{18}$$

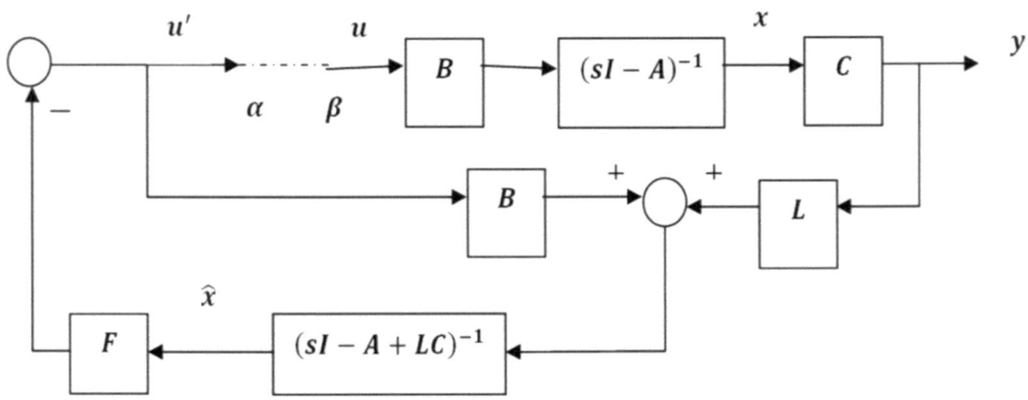

Fig. 2. Optimal LQG Scheme

The complete LQG control system appears in figure 2. The question of loop transfer recovery deals with the problem of keeping the sensitivity given in equation (17) when the control law is implemented with an observer of the state (Stein, 1987), i.e. the control $\boldsymbol{u(t)}$ is generated trough:

$$\boldsymbol{u(t)} = -\boldsymbol{F\hat{x}(t)} \tag{19}$$

where $\hat{x}(\boldsymbol{t})$ is the output of a state observer is given by:

$$\dot{\hat{\boldsymbol{x}}}(\boldsymbol{t}) = \boldsymbol{A\hat{x}(t)} + \boldsymbol{Bu(t)} + \boldsymbol{L}\{\boldsymbol{y(t)} - \boldsymbol{C\hat{x}(t)}\} \tag{20}$$

If the observer is designed using standard Kalman filter theory, then the filter gain $\boldsymbol{L}$ satisfies:

$$\boldsymbol{L} = \sum \boldsymbol{C}^T \tag{21}$$

where $\sum$ is the symmetric nonnegative definite solution of the algebraic Riccati equation:

$$\boldsymbol{A}\sum + \sum \boldsymbol{A}^T - \sum \boldsymbol{C}^T\boldsymbol{C}\sum + \boldsymbol{Q} = \boldsymbol{0} \tag{22}$$

In this case the input sensitivity is given by:

$$\boldsymbol{S_{obs}(s)} = \{\boldsymbol{I} + [\boldsymbol{I} + \boldsymbol{F}(\boldsymbol{sI} - \boldsymbol{A} + \boldsymbol{LC})^{-1}\boldsymbol{B}]^{-1}\boldsymbol{F}(\boldsymbol{sI} - \boldsymbol{A} + \boldsymbol{LC})^{-1}\boldsymbol{LC}(\boldsymbol{sI} - \boldsymbol{A})^{-1}\boldsymbol{B}\}^{-1} \tag{23}$$

After some elementary matrix manipulation, we obtain:

$$\boldsymbol{S_{obs}(s)} = \boldsymbol{S_o(s)}\{\boldsymbol{I} + \boldsymbol{F}(\boldsymbol{sI} - \boldsymbol{A} + \boldsymbol{LC})^{-1}\boldsymbol{B}\} \tag{24}$$

Or

$$\boldsymbol{S_{obs}(s)} = \boldsymbol{S_o(s)}\{\boldsymbol{I} + \boldsymbol{F\phi}[\boldsymbol{B} - \boldsymbol{L}(\boldsymbol{sI} - \boldsymbol{A} + \boldsymbol{LC})^{-1}\boldsymbol{C\phi B}]\} \tag{25}$$

where:

$$\boldsymbol{\phi} = (\boldsymbol{sI} - \boldsymbol{A})^{-1} \tag{26}$$

It becomes then sensible to measure the amount of recovery by the relative sensitivity error (Turan, 1990) given by:

$$\boldsymbol{E(s)} = \boldsymbol{S_o^{-1}}\{\boldsymbol{S_{obs}} - \boldsymbol{S_o}\} \tag{27}$$

Using the equation (25) we obtain:

$$\boldsymbol{E(s)} = \boldsymbol{F}(\boldsymbol{sI} - \boldsymbol{A} + \boldsymbol{LC})^{-1}\boldsymbol{B} \tag{28}$$

It has been shown (Doyle, 1979) that if $\boldsymbol{G(s)}$ is a minimum-phase transfer matrix then complete recovery, i.e. $\boldsymbol{E(s)} = \boldsymbol{0}$, can be achieved provided that:

- We first augment equation (13) to read:

$$\dot{\boldsymbol{x}} = \boldsymbol{Ax(t)} + \boldsymbol{B}\{\boldsymbol{u(t)} + \boldsymbol{Uw(t)}\} + \boldsymbol{v(t)} \tag{29}$$

  where $\boldsymbol{w(t)} \in \boldsymbol{R^m}$ is a wide sense stationary process with covariance matrix $\boldsymbol{q^2I}$ and $\boldsymbol{U}$ is a unitary matrix. We assume that $\boldsymbol{v(t)}$ and $\boldsymbol{w(t)}$ are uncorrelated.
- We then solve the Riccati equation (10) substituting Q by $Q + qBB^T$
- We finally let $\boldsymbol{q} \longrightarrow \infty$. In this case $\boldsymbol{L(q)} \longrightarrow \boldsymbol{qBU}$.

The above procedure yields:

$$\boldsymbol{S_{obs}(s)} \longrightarrow \boldsymbol{S_o(s)} \tag{30}$$

4.2 A factorized form of non-minimum phase plants

It is known that a transfer matrix function $\boldsymbol{G(s)}$ with zeros in $\boldsymbol{C^+}$ (right-half complex plane) can be described as:

$$\boldsymbol{G(s) = G_m(s)B_z(s)} \tag{31}$$

where $\boldsymbol{B_z(s)}$ is a stable all-pass factor with zeros located at the non-minimum-phase zeros of $\boldsymbol{G(s)}$ and satisfies $\boldsymbol{B_z(s)B_z^T(s) = I}$.

One possible way to build the factorization of equation (31) has been proposed by Enns (Enns, 1984; Zhang, 1990). The main result can be stated as follows.

Lemma 2.1 Given a transfer matrix function $\boldsymbol{G(s) = C(sI - A)^{-1}B}$ with $\boldsymbol{l} \in \boldsymbol{C^+}$ zeros (including multiplicity), $z_1, \cdots, z_l$, there exists a matrix $\boldsymbol{B_m}$ such that:

$$\boldsymbol{G(s) = C(sI - A)^{-1}B_m B_z(s)} \tag{32}$$

where $\boldsymbol{G(s) = C(sI - A)^{-1}B_m}$ is minimum-phase and $\boldsymbol{B_z(s)}$ is an all-pass stable matrix factor. We then have that:

$$\boldsymbol{B_z(s) = B_{z_1}(s)B_{z_2}(s) \cdots B_{z_l}(s)} \tag{33}$$

and

$$\boldsymbol{B_m = B_m^l} \tag{34}$$

where, for $i = 1, \cdots, l$:

$$\boldsymbol{B_{z_i}(s) = I - \frac{2\Re e\{z_i\}}{s + z_i^*}\eta_i \eta_i^H} \tag{35}$$

and

$$\boldsymbol{B_m^i = B_m^{i-1} - 2\Re e\{z_i\}\xi_i \eta_i^H} \tag{36}$$

with $\boldsymbol{B_m^0 = B}$, $\boldsymbol{G_m^i(s) := C(sI - A)^{-1}B_m^i}$.

The symbol $\boldsymbol{\Re e\{.\}}$ denotes the real part operator. The vectors $\boldsymbol{\eta_i}$ and $\boldsymbol{\xi_i}$ are solutions of:

$$\begin{bmatrix} z_i\boldsymbol{I - A} & \boldsymbol{-B_m^{i-1}} \\ \boldsymbol{-C} & \boldsymbol{0} \end{bmatrix} \begin{bmatrix} \boldsymbol{\xi_i} \\ \boldsymbol{\eta_i} \end{bmatrix} = \boldsymbol{0} \tag{37}$$

□ □ □

We also have the following useful results:

Lemma 2.2 For SISO systems, the sequence $\{B_m^k\}$ can be alternatively computer as:

$$\boldsymbol{B_m^k = \underbrace{\textstyle\prod_{i=1}^{k}\{I - 2\Re e(z_i)(z_i I - A)^{-1}\}}_{\equiv M_k} B} \tag{38}$$

Proof:

A1.- $\boldsymbol{B_m^0 = B}$

A2.- From (36) $\boldsymbol{B_m^1 = B_m^0 - 2\Re e\{z_1\}\xi_1 \eta_1^H}$.

But for SISO systems $\boldsymbol{\eta_1 = \eta_2 = \cdots = 1}$

and from (37) $\boldsymbol{\xi_1 = (z_1 I - A)^{-1}B_m^0}$

therefore $\boldsymbol{B_m^1 = \underbrace{\{I - 2\Re e(z_1)(z_1 I - A)^{-1}\}}_{\equiv M_1} B}$

A3.- From (36) $\boldsymbol{B_m^k} = \boldsymbol{B_m^{k-1}} - \boldsymbol{2\Re e\{z_k\}}\xi_k$. But for SISO systems $\boldsymbol{\eta_k = 1}$, and from (37) we have that $\boldsymbol{B_m^k} = \{\boldsymbol{I} - \boldsymbol{2\Re e(z_k)(z_k I - A)^{-1}}\}\boldsymbol{B_m^{k-1}}$ then if (37) is satisfied for $k-1$, the result follows.

□ □ □

Corollary:

$$\xi_k = (\boldsymbol{z_k I} - \boldsymbol{A})^{-1}\boldsymbol{M_{k-1}B} \tag{39}$$

$$\boldsymbol{M_0 = I} \tag{40}$$

$$\boldsymbol{M_1 = I - 2\Re e(z_1)(z_1 I - A)^{-1}} \tag{41}$$

□ □ □

Theorem 2.1 Consider a non-minimum-phase system $(\boldsymbol{A}, \boldsymbol{B}, \boldsymbol{C})$, and its minimum-phase counterpart $(\boldsymbol{A}, \boldsymbol{B_m}, \boldsymbol{C})$, with $\boldsymbol{B_m}$ computed according to **lemma 2.1**. Let $\boldsymbol{L}$ and $\boldsymbol{L_m}$ be the optimal observer gains for these two systems, then $\boldsymbol{L_m = L}$.
Proof: See Zhang & Freudenberg (Zhang, 1990).

4.3 Loop transfer recovery and non-minimum phase plants

Assume now that $\boldsymbol{G(s)}$ is a non-minimum phase plant and that it is factorized as in equation (33). If the standard LTR procedure is applied to recover the input sensitivity, then when $\boldsymbol{q} \longrightarrow \infty$ the sensitivity function satisfies:

$$\boldsymbol{S_{obs}(s) = S_o(s)\{I + E(s)\}} \tag{42}$$

where

$$\boldsymbol{E(s) := F(sI - A)^{-1}\{B - qB_m W(I + qC\phi B_m W)^{-1} C\phi B_m B_z(s)\}} \tag{43}$$

then

$$\boldsymbol{E(s) = F(sI - A)^{-1}\{B - B_m B_z(s)\}} \tag{44}$$

It has been also shown that:

$$\boldsymbol{E(s) = E^i(s)} = \sum_{k=1}^{l} \frac{\boldsymbol{2\Re e\{z_k\}}}{\boldsymbol{s+z_k^*}} \boldsymbol{F\xi_k \eta_k^H B_z^{k-1}(s)} \tag{45}$$

From equation (44) it is evident that for this type of plants the amount of recovery at a frequency w depends on the value of $\|\boldsymbol{E(jw)}\|$, where $\|.\|$ is a suitable norm. As in equation (41), $\boldsymbol{E(s)}$ corresponds to the error of the sensitivity in loop with the LTR observer.
The results of the previous two sections can be appreciated if we consider a SISO system with one zero in $\boldsymbol{s = z} \in \boldsymbol{R^+}$. If the standard LTR procedure is applied we have that:

$$\boldsymbol{\lim_{q \to ss} S_{obs}(s) = S_o(s)\left\{1 + \frac{2z}{s+z} H(z)\right\}} \tag{46}$$

where

$$\boldsymbol{H(z) = F(zI - A)^{-1}B} \tag{47}$$

One can then notice that if $|\boldsymbol{H(z)}|$ is small, i.e. when the LQR design bandwidth is small in comparison with the magnitude of the $\boldsymbol{C^+}$ zero, then the recovery is almost complete. This case will be also the situation for high frequencies since the factor $\frac{2z}{s+z}$ is low-pass filter. The sensitivity resulting from a LQR/LTR applied to a non-minimum phase plant is very significant at low frequencies and decreases as the frequency increases. The inability of the LQR/LTR scheme to recover sensitivity is consistent with some fundamental design tradeoffs for non-minimum phase systems (Freudenberg, 1988).

4.4 LTR procedure with partial state feedback (LTR/PSF)

If we assume that, apart from the system output, there is one or more independent measurements, we can implement a control system originally designed to work with state with state feedback via a mixture of observed and measured states (Aros, 1991). We propose to use the scheme shown in figure 3, where $\boldsymbol{\Gamma}$ is a diagonal matrix with nonzero diagonal entries for the corresponding measures states. It then becomes clear that $\boldsymbol{\Gamma} = \mathbf{0}$ corresponds to the standard LTR scheme and $\boldsymbol{\Gamma} = \boldsymbol{I}$ corresponds to the optimal regulator with full state feedback.

A key result to describe the degree of recovery in the non-minimum phase case is given in the next lemma.

Lemma 3.1 Assume that the plant $\boldsymbol{G(s)} = \boldsymbol{C(sI - A)^{-1}B}$ in non-minimum phase with $\boldsymbol{l} \in \boldsymbol{C^+}$ zeros and factorized according to lemma 2.1. Then if we use a LTR/PSF scheme and we let $\boldsymbol{q} \longrightarrow \infty$ the sensitivity function is given by:

$$\boldsymbol{S}_{\Gamma}(\boldsymbol{s}) = \boldsymbol{S}_o(\boldsymbol{s})\{\boldsymbol{I} + \textstyle\sum_{k=1}^{l} \boldsymbol{F}(\boldsymbol{I} - \Gamma)\xi_k \eta_k^H \boldsymbol{W}_k(\boldsymbol{s})\} \tag{53}$$

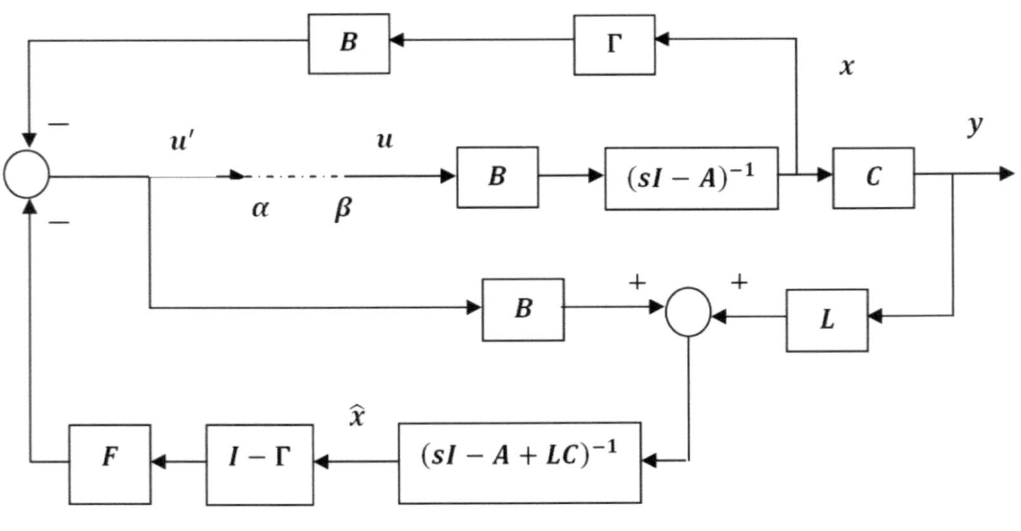

Fig. 3. LTR/PSF scheme.

where:

$$\boldsymbol{W}_k(\boldsymbol{s}) = \frac{2\Re e(z_k)}{s+z_k^*}\boldsymbol{B}_z^{k-1}(\boldsymbol{s}) \tag{54}$$

$$\boldsymbol{B}_z^{k-1}(\boldsymbol{s}) = \textstyle\prod_{i=1}^{k} \boldsymbol{B}_{z_i}(\boldsymbol{s}) \tag{55}$$

$$B_{z_i}(s) = \frac{-s+z_i}{s+z_i^*} \tag{56}$$

Proof: Straightforward on using lemma 2.1 and equation (45).

□ □ □

Remark 3.1 For SISO systems $\eta_k = 1\ \forall k$ and consequently (53) simplifies to:

$$S_\Gamma(s) = S_o(s)\{I + \textstyle\sum_{k=1}^{l} F(I-\Gamma)\xi_k W_k(s)\} \tag{57}$$

Sufficient and necessary conditions for full recovery in SISO systems are given in the next theorem:

Theorem 3.1 Assume that a plant with transfer function $G(s)$ has $l \in C^+$ zeros denoted by $z_1, \cdots, z_l$. We apply the LTR/PSF procedure we obtain $E_\Gamma(s) = 0$, if $F(I-\Gamma)$ is orthogonal to ξ_k for $k = 1, 2, \cdots, l$ when $q \longrightarrow \infty$.

Proof:

i. Sufficiency: straightforward on inspection of equation (53).
ii. Necessity: consider equation (57). We first notice that given the fact that functions $W_k(s)$ form a set of linearly independent functions, the only way to nullify the sum $\forall s$ is that the scalar $F(I-\Gamma)\xi_k$ be made equal to zero $\forall k$.

□ □ □

The user must then choose (if possible) the matrix Γ to satisfy the orthogonality condition in theorem 3.1. Equivalently, Γ must satisfy:

$$FT\zeta_k = F\xi_k \quad k = 1, 2, \cdots, l \tag{58}$$

The computation of Γ is given the next lemma.

Lemma 3.2 Consider a SISO plant as in theorem 3.1. If the LTR/PSF scheme is applied measuring, apart from the output, states $x_1, x_2 \cdots, x_\lambda$ then we obtain full recovery of the sensitivity if there exist $\gamma_1, \gamma_2 \cdots, \gamma_\lambda$, with $\Gamma = diag\{\gamma_1, \gamma_2, \cdots, \gamma_\lambda, 0, \cdots, 0\}$, satisfying:

$$\begin{bmatrix} \alpha_{1_1} & \cdots & \alpha_{\lambda_1} \\ \vdots & \ddots & \vdots \\ \alpha_{1_l} & \cdots & \alpha_{\lambda_l} \end{bmatrix} \begin{bmatrix} \gamma_1 \\ \vdots \\ \gamma_\lambda \end{bmatrix} = \begin{bmatrix} \sum_{i=1}^{n} \alpha_{i_1} \\ \vdots \\ \sum_{i=1}^{n} \alpha_{i_l} \end{bmatrix} \tag{59}$$

where

$$\alpha_{i_k} = f_i \xi_{k_i}; \qquad i = 1,2,\cdots,n \tag{60}$$

$$F = [f_1 \quad f_2 \quad \cdots \quad f_n] \tag{61}$$

$$\xi_K = [\xi_{k_1} \quad \xi_{k_2} \quad \cdots \quad \xi_{k_n}]^T \tag{62}$$

Proof: By straight substitution.

□ □ □

Remark 3.2 From equation (53) it appears that full recovery is obtained $\forall \Gamma$ if η_k and ξ_k are orthogonal. This it shows that the LTR/PSF scheme maintains the standard LTR property claimed in Wall (Wall, 1980) and proved in Zhang (Zhang, 1990).

Remark 3.3 A complete analysis of the conditions the existence of none, one or an infinite number of solutions is out of the scope of this work, but some insight can be gained on analyzing the one RHP zero case and L_2 optimization. Both topics will be addressed below.

Theorem 3.2 Consider a scalar plant with transfer function $G(s)$ and one RHP zero located at $s = z$. If the LTR/PSF is applied measuring the input and one additional state variable (not proportional to the output) then $S_\Gamma(s) \longrightarrow S_o(s)$ when $q \longrightarrow \infty$.
Proof:
From (50) and (46), we can write:

$$\lim_{q\to ss} S_\Gamma(s) = S_o(s)\left\{1 + \frac{2z}{s+z^*}H(z) - \frac{2z}{s+z^*}H'(z)\right\} \tag{63}$$

where $H(s)$ is given in equation (47) and

$$H'(z) = F\Gamma(zI - A)^{-1}B \tag{64}$$

Without loss of generality we can assume that the state variable being fed back is $x_1(t)$. We can thus express Γ as:

$$\Gamma = \gamma\Gamma' \tag{65}$$

where

$$\Gamma' = diag\{1, 0, \cdots, 0\} \tag{66}$$

then full recovery is obtained if γ is chosen to satisfy:

$$\gamma == \frac{F(zI-A)^{-1}B}{F\Gamma'(zI-A)^{-1}B} \tag{67}$$

Remark 3.4 On examining equation (67) we note:
1. If the output is proportional to x_1 there is not solution for γ since then $F\Gamma'(zI - A)^{-1}B = 0$. It certainly agrees with intuition, since nothing can be gained by measuring twice the same variable.
2. There is not solution either when $F\Gamma' = 0$. This case also is intuitive since this situation corresponds to a control law where $x_1(t)$ was not required to be fed back.

These results also apply, mutatis mutandis, to MIMO systems, with the additional complexity which comes from the directionality properties of multivariable systems.

L_2 Optimization

When the designer don't have freedom to choose which state variable can be measured, due either to technical or economical reasons, then the feedback gains for the additional available measurements can be computed by solving an optimization problem. The simplest optimization problem can be posed in L_2. We examine that case for SISO systems.
Assume first that we measure and feed back the state variables $x_1(t), x_2(t), \cdots, x_\lambda(t)$ with gains $[\gamma_1, \gamma_2 \cdots, \gamma_\lambda]^T$. Then the L_2 optimization problem consists in finding a vector γ^o satisfying:

$$\gamma^o = \arg\min_{\gamma\epsilon\Re e^\lambda} J(\gamma) \tag{68}$$

where

$$J = \int_0^\infty |E_\Gamma(jw)|^2\, dw \tag{69}$$

$$J = \int_0^\infty \left\|G_1(jw) + \gamma^T G_2(jw)\right\|^2 dw \tag{70}$$

with

$$G_1(jw) = F(jwI - A + LC)^{-1}B \tag{71}$$

$$\gamma^T G_2(jw) = F\Gamma(jwI - A + LC)^{-1}B \tag{72}$$

This problem has a unique solution if $G(jw)$ is not identical to zero. This unique solution is given by:

$$\gamma^o = -\left[\int_0^\infty G_2(jw)[G_2^*(jw)]^T dw\right]^{-1} \Re e\left\{\int_0^\infty G_1(jw)G_2^*(jw)dw\right\} \tag{73}$$

for $\lambda = 1$ equation (73) becomes:

$$\gamma^o = -\frac{\Re e\left\{\int_0^\infty G_1(jw)G_2^*(jw)dw\right\}}{\int_0^\infty \|G_2(jw)\|^2 dw} \tag{74}$$

5. Simulatinos results

5.1 Open loop

In figure 4, it shows the response of open loop system, when making a disturbance in one of the manipulated variables such as power flow; it represents the temperature of the first effect and concentration effect of the fourth output.

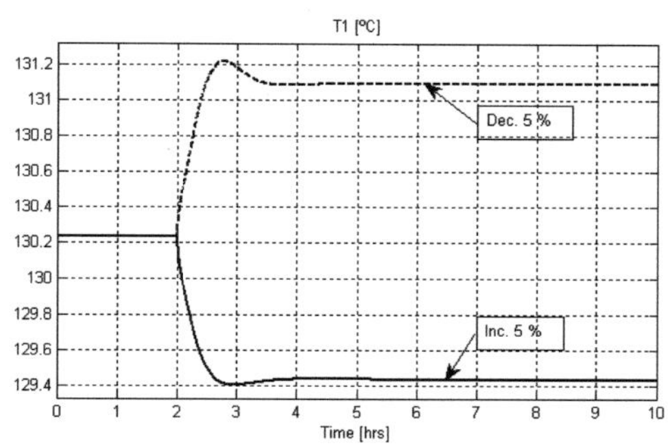

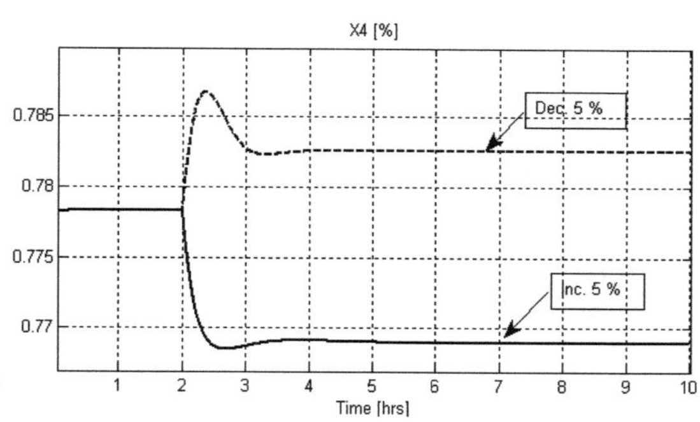

(a) (b)

Fig. 4. Behavior of the concentration in the evaporator to a change of a step in the flow of food (up 5% - decrease of 5%)

In figure 5, it shows the response of open loop system, when making a disturbance in the steam temperature is the other manipulated variable; it represents the temperature of the first effect and concentration effect of the fourth output.

In figure 6, it shows the response of open loop system, when performing a step in one of the shocks as the concentration of power; it represents the temperature of the first effect and concentration effect of the fourth output.

In Figure 7, it shows the response of open loop system, when performing a step in the temperature of the food which is the other perturbations of the system; it represents the temperature of the first effect and concentration effect of the fourth output.

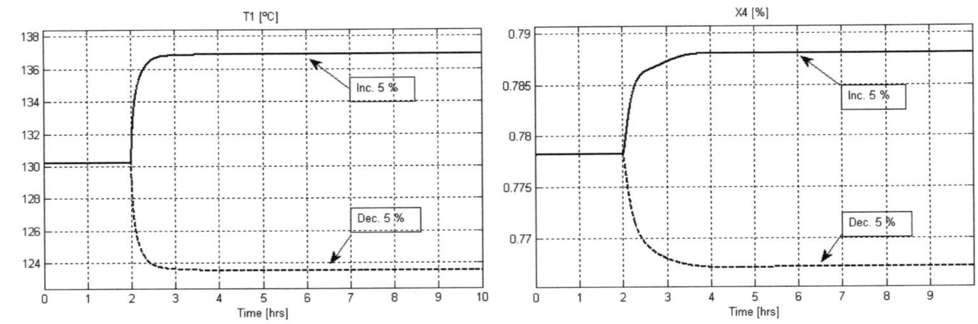

(a) (b)

Fig. 5. Behavior of the concentration in the evaporator to a change of a step in the temperature of steam power (up 5% - decrease of 5%)

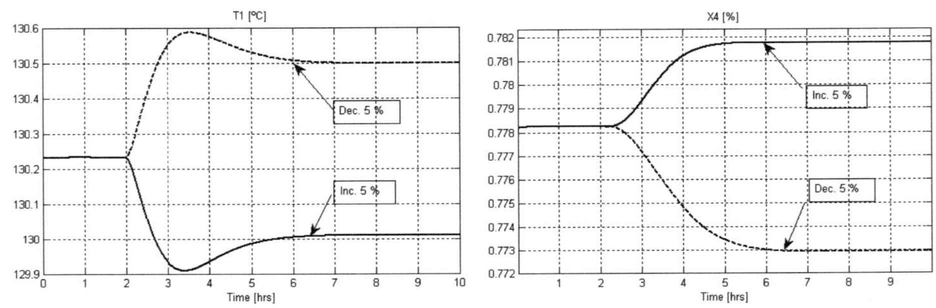

(a) (b)

Fig. 6. Behavior of the concentration in the evaporator to a change of a step in the concentration of power (increase of5% - decrease of 5%)

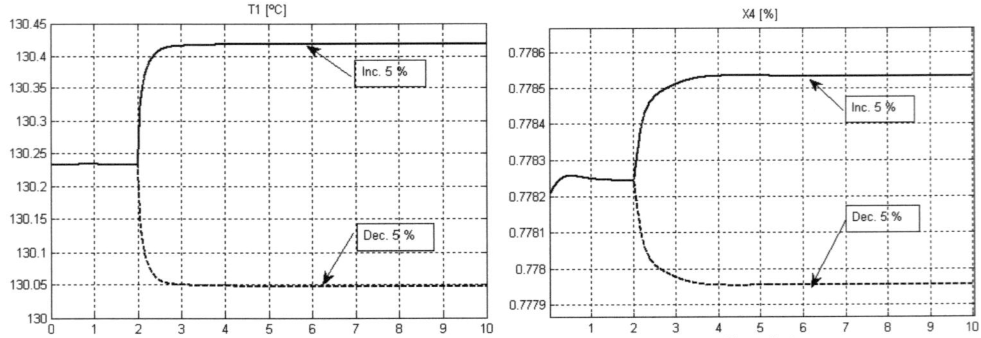

(a) (b)

Fig. 7. Behavior of the concentration in the evaporator to a change of a step in the temperature of the input solution (5% increase - decrease of 5%)

5.2 Close loop

Controlled system response for optimal regulator, whereas white noise disturbances, as well as step-like variation to the inlet concentration to 50, and then a step 100 is added to the feed temperature at the entrance.

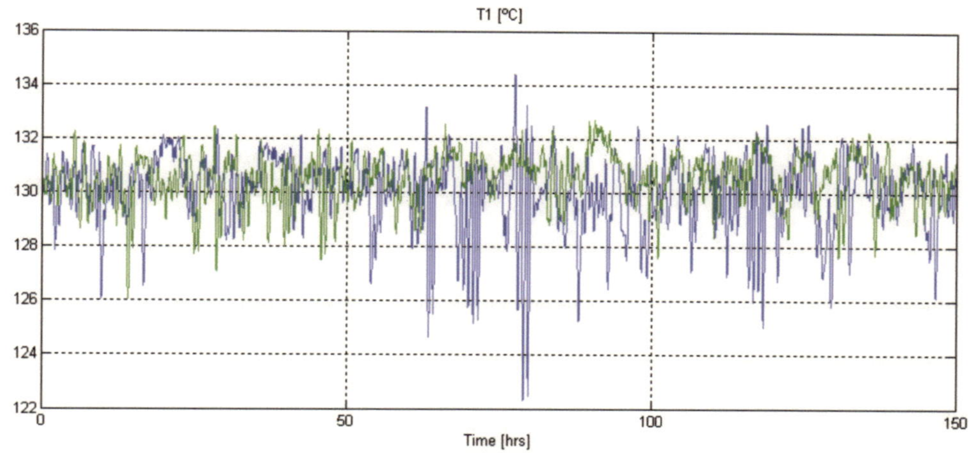

(a)

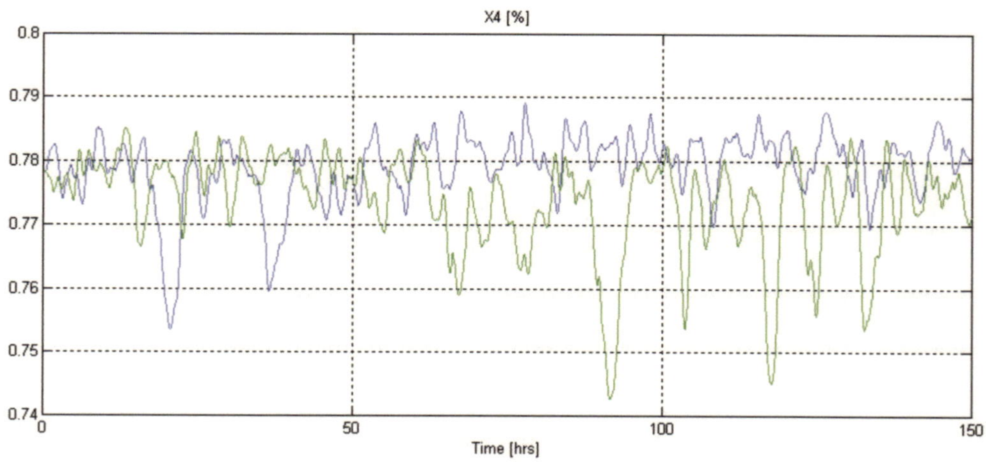

(b)

Fig. 8. Controlled system response to changes in the shocks in type of step and white noise (blue for changes of +5% - green changes -5%)

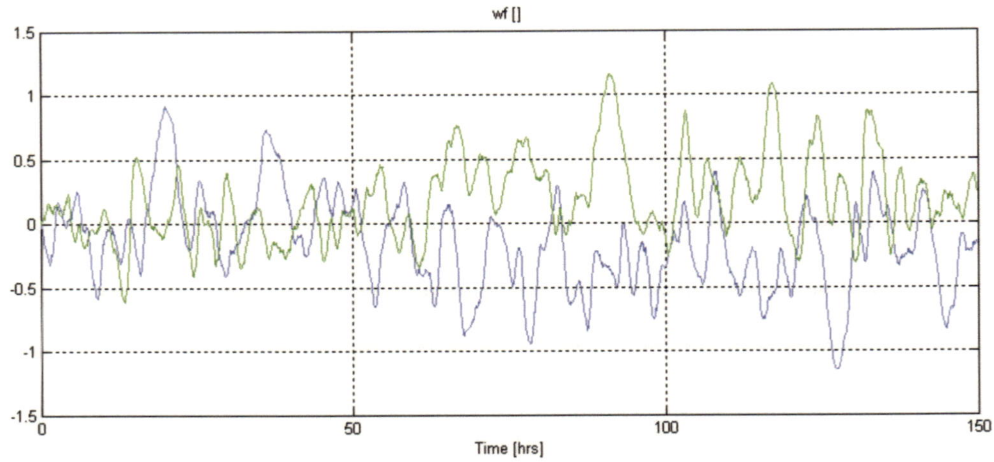

(a)

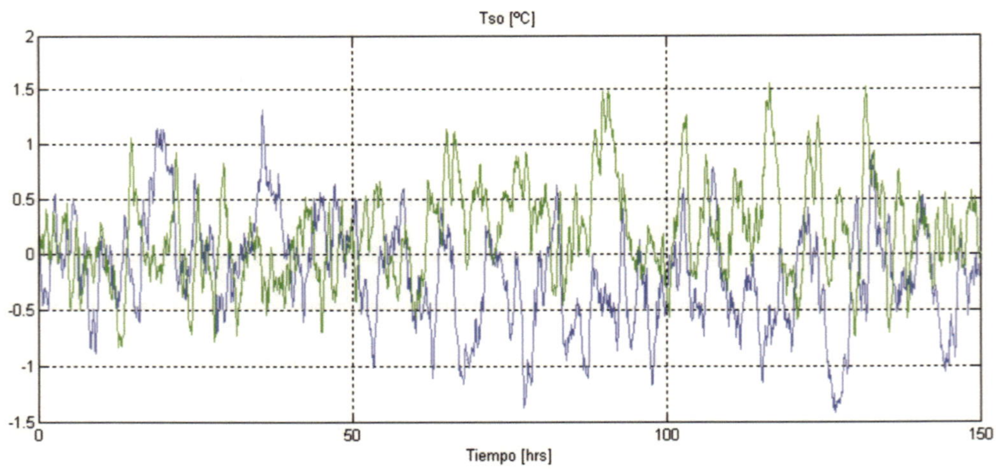

(b)

Fig. 9. Efforts to control the controlled system to changes in the shocks in type of step and white noise (blue for changes of +5% - green changes -5%)

5.3 LQG- design

Controlled system response for optimal regulator, whereas white noise disturbances, as well as like step variation of 5% for the inlet concentration to 50 and then to 100 adds a step is 5% of the feed temperature at the entrance.

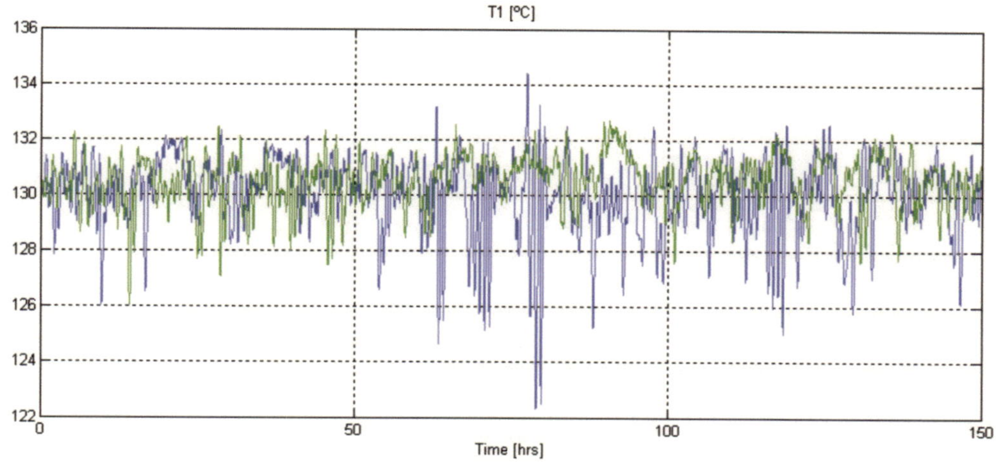

(a)

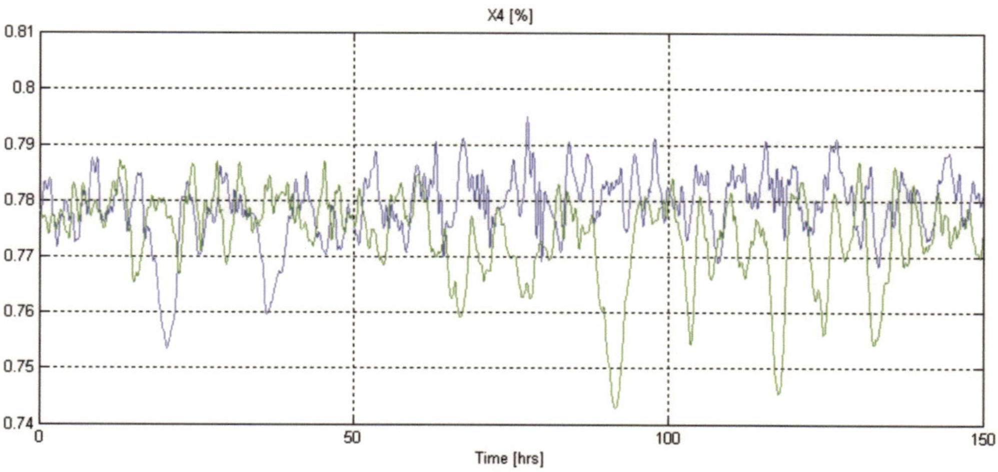

(b)

Fig. 10. LQG controlled system response to changes in the type shocks of step and white noise (blue for changes of +5% - green changes -5%)

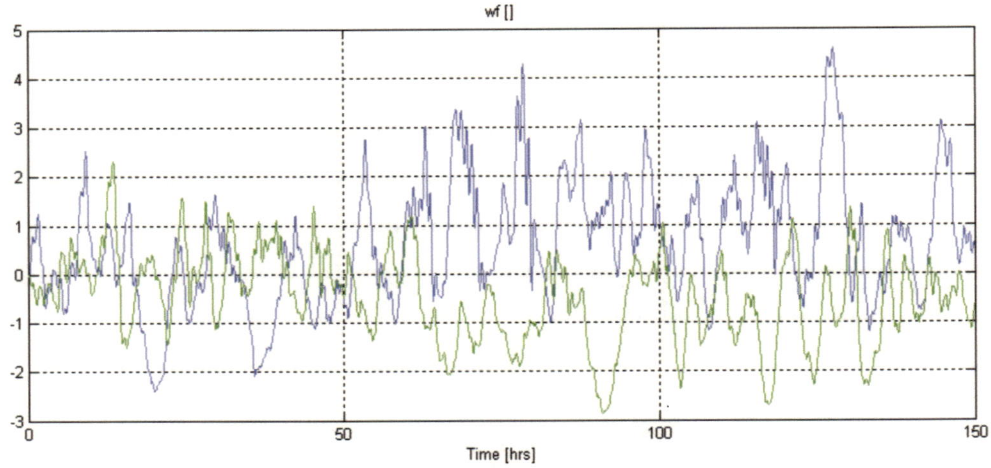

(a)

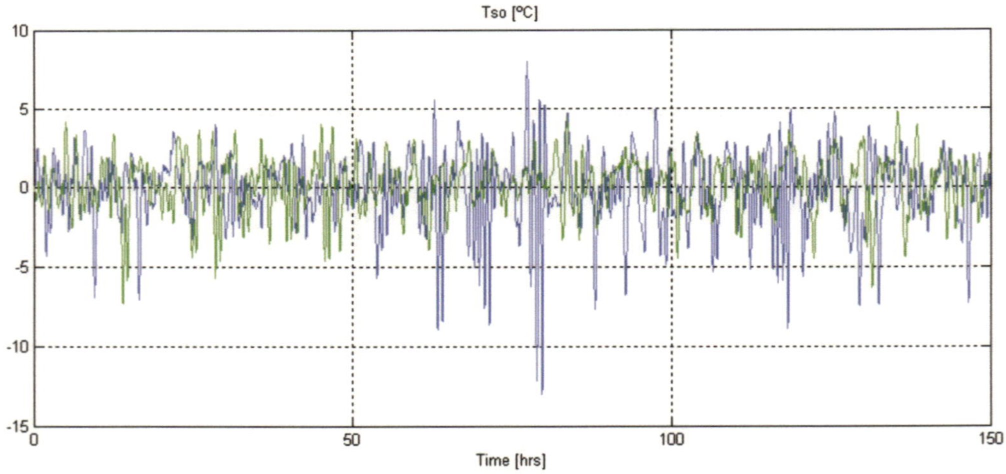

(b)

Fig. 11. Control efforts for the LQG-controlled system to changes in the type shocks of step and white noise (blue for changes of +5% - green changes -5%)

5.4 LQG/LTR

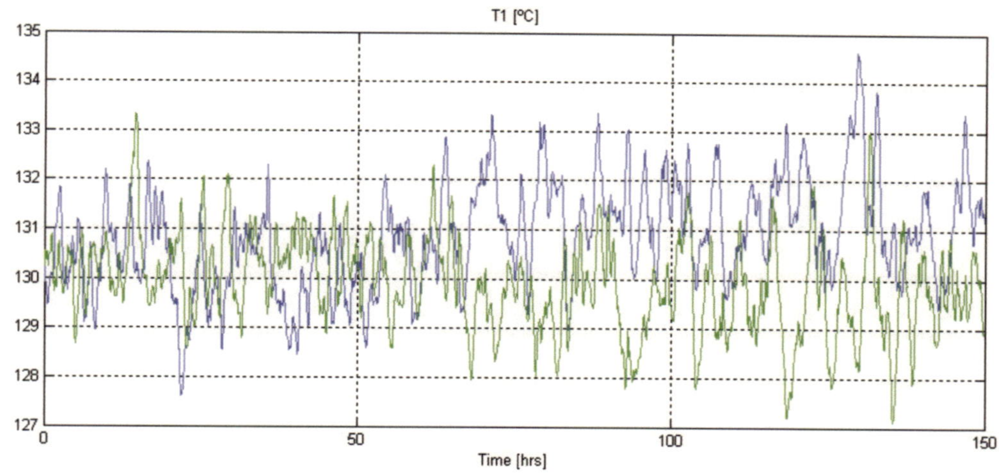

(a)

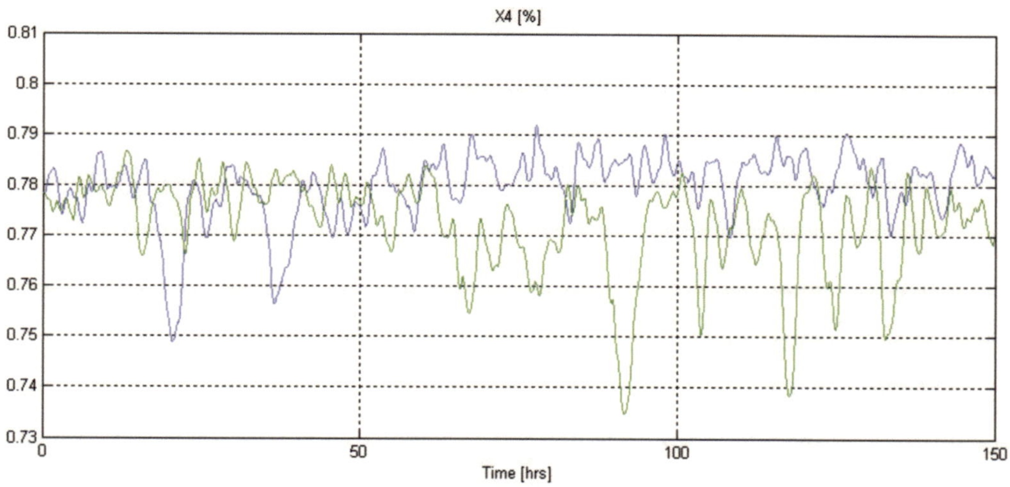

(b)

Fig. 12. Controlled system response LQG / LTR to changes in the type shocks of step and white noise (blue for changes of +5% - green changes -5%)

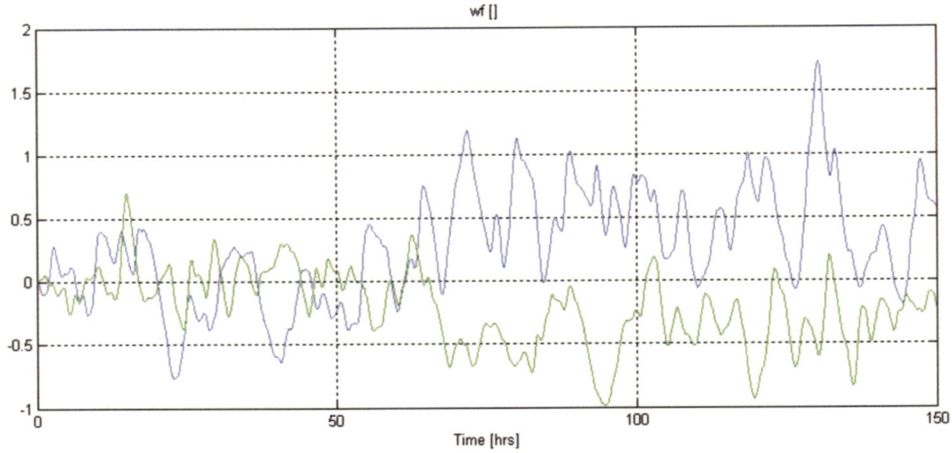

(a)

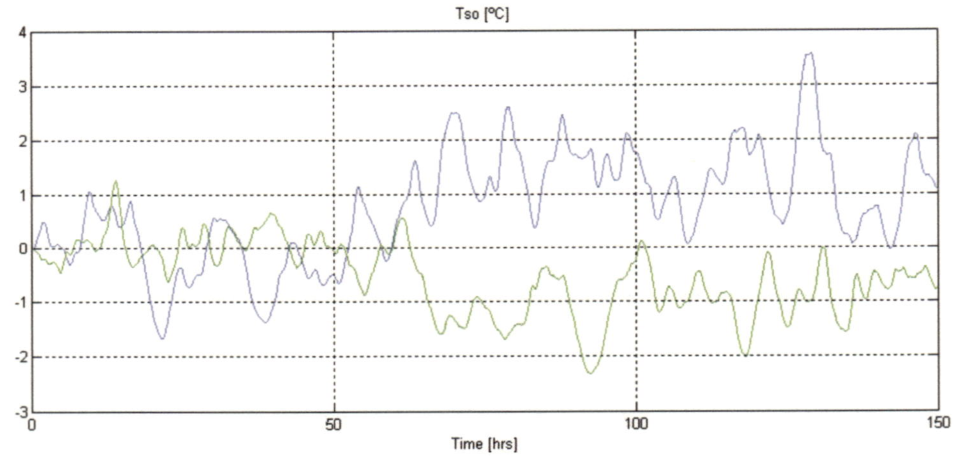

(b)

Fig. 13. Efforts to control the controlled system LQG / LTR to changes in the type shocks of step and white noise (blue for changes of +5% - green changes -5%)

6. Conclusions

Looking at the results presented in figures 4 to 7, show that it is appropriate to consider as manipulated variables steam temperature and feed rate of the solution to concentrate, and as measurable disturbance and characteristic of the system to the concentration the solution concentrated and the inlet temperature of food. You can check the analysis of these figures that the evaporation process presents a complex dynamic, high delay, coupling between the variables, high nonlinearities.

From the results shown in figures 8 to 13, on the behavior of the controlled system verifies that the design LQG/LTR has a better performance especially when control efforts are softer. Partly, it validates the robustness of the proposed control system, despite having analyzed only the rejection of disturbances, since these regulatory systems at the show a good response.

7. Acknowledgment

The authors gratefully acknowledge the financial support of the "Universidad de La Frontera"- Chile DIUFRO DI07-0102, "Universidad Nacional de San Juan"- Argentina, Project FI-I900. They are also grateful for the cooperation of "Mostera Rio San Juan".

8. References

(Anderson, 1971) Anderson B.D.O., Moore J.B., *Lineal Optimal Control Control*. Prentice Hall. Engleewood Cliffs, N.J.

(AOAC, 1995) *Official Methods of Analysis. Reference Tables Appendix C*

(Armaou, 2002) Armaou A., Christofides P.D., "Dynamic Optimization of Dissipative PDE Systems Using Nonlinear Order Reduction". Chemical Engineering Science 57 - 24, pp. 5083-5114.

(Aros, 1991) Aros N.H., *Recuperación de la Transferencia del Lazo por Realimentación Parcial del Estado*. M. Sc. Thesis. Universidad Técnica Federico Santa María, Chile.

(Barbieri, 1980) Barbieri, R., & Rossi, N. "Propietà Fisiche dei Mosti d'uva Concentrati". *Rivista de Viticol. e di Enologia. Conegliano*, No 1, 10–18.

(Bayindirli, 1992) Bayindirli, L. "Mathematical Analysis Of Variation Of Density And Viscosity Of Apple Juice With Temperature And Concentration". *Journal. Of Food Processing and Preservation* (16), 23-28.

(Bode, 1945) Bode H.W., *Network Analysis and Feedback Amplifier Design*. Van Nostrand, New York.

(Camacho, 1999) Camacho E., Bordons F.C., *Model Predictive Control*. Springer-Verlag.

(Crapiste, 1988) Crapiste, G.H. and Lozano, J.E. "Effect Of Concentration And Pressure On The Boiling Point Rise Of Apple Juice And Related Sugar Solutions". *Journal Food Science*. (53), 865-869

(Constela, 1989). Constenla D.T., Crapiste G.H. and Lozano J.E. "Thermophysical Properties Of Clarified Apple Juice As A Function Of Concentration And Temperature". *Journal. Food Science*.(54), 663-669.

(Di Leo, 1988) Di Leo, F. "Caratteristiche Fisico-chimiche dei Mosti Concentrati Rettificati. *Valutazione gleucometrica". Vignevini*, 15(1/2), 43–45.

(Doyle, 1979) Doyle J.C., Stein G., "Robustness with observers". *IEEE Trans. on Auto. Control, Vol. AC-24, April.*

(Enns, 1984) Enns D., *Model reduction for control systems design*. Ph. D. Thesis. Stanford University.

(El-Farra, 2003) El-Farra N.H., Armaou A., Christofides P.D., "Analysis and Control of Parabolic PDE Systems with Input Constraints". *Automatica 39 – 4, pp. 715-725.*

(Freudenberg, 1988) Freudenberg J., Looze D., *Frequency Domain Properties of Scalar and Multivariable Feedback Systems*. Springer Verlag, Berlín.

(Friedland, 1989) Friedland B., "On the properties of reduced-orden Kalman filters". *IEEE Trans. on Auto. Control, Vol. AC-34, March.*

(Ibarz, 1993) Ibarz, A., & Ortiz, J. "Reología de Zumos de Melocotón". Alimentación, Equipos y Tecnología. Octubre, 81–86, Instituto Nacional de Vitivinicultura. *Síntesis básica de estadística vitivinícola argentina, Mendoza*. Varios números.

(Kam, 1999) Kam K.M., Tade M.O., "Case studies on the modelling and control of evaporation systems". *XIX Interamerican Congress of Chemical Engineering COBEQ.*

(Kam, 2000) Kam K.M., Tade M.O., "Simulated Nonlinear Control Studies of Five Effect Evaporator Models". *Computers and Chemical Engineering, Vol. 23, pp. 1795 - 1810.*

(Kaya, 2002) Kaya A., Belibagh K.B., "Rheology of solid Gaziantep Pekmez". *Journal of Food Engineering, Vol. 54, pp. 221-226.*

(Middleton, 1990) Middleton R.H., Goodwin G.C., *Digital Control and Estimation. A Unified Approach.* Prentice Hall, Englewood Cliffs, N.J.

(Moressi, 1984). Moressi, M., & Spinosi, M. "Engineering factors in the production of concentrated fruit juices, II, fluid physical properties of grapes". *Journal of Food Technology*, 5(19), 519–533.

(Niemann, 1995) Niemann H.H., Stoustrup J., Shafai B., Beale S., "LTR design of proportional-integral observers". *Int. Journal Robust Nonlinear Control, pp. 671-693.*

(Pandolfi, 1991) Pandolfi, C., Romano, E. & Cerdán, A. Composición de los Mostos Concentrados Producidos en Mendoza y San Juan, Argentina. Ed. Agro Latino. *Viticultura/Enología profesional* 13, 65–74.

(Perry, 1997) Perry R., *Perry's Chemical Engineers Handbook.* 7TH Edition McGraw Hill.

(Pilati, 1998) Pilati, M. A., Rubio, L. A., Muñoz, E., Carullo, C. A., Chernikoff, R.E. & Longhi, M. F. "Evaporadores Tubulares de Circulación Forzada: Consumo de Potencia en Distintas Configuraciones. III Jornadas de Investigación. FCAI - UNCuyo. Libro de Resúmenes, 40.

(Rao, 1984) Rao, M. A., Cooley, H. J., & Vitali, A. A. "Flow Properties of Concentrated Juices at Low Temperatures. *Food Technology*, 3(38), 113–119.

(Rubio, 1998) Rubio, L. A., Muñoz, E., Carullo, C. A., Chernikoff, R. E., Pilati, M. A. & Longhi, M. F. "Evaporadores Tubulares de Circulación Forzada: Capacidad de Calor Intercambiada en Distintas Configuraciones". *III Jornadas de Investigación. FCAI - UNCuyo. Libro de Resúmenes,* 40.

(Sáenz 1986) Sáenz, C., & Costell, E. "Comportamiento Reológico de Productos de Limón, Influencia de la Temperatura y de la Concentración". *Revista de Agroquímica y Tecnología de Alimentos,* 4(26), 581–588.

(Saravacos, 1970) Saravacos, G. D. "Effect of Temperature on Viscosity of Fruit Juices and Purees". *Journal of Food Science*, (35), 122–125.

(Schwartz, 1986) Schwartz, M., & Costell, E. "Influencia de la Temperatura en el Comportamiento Reológico del Azúcar de Uva (cv, Thompson Seedless)". *Revista de Agroquímica y Tecnología de Alimentos,* 3(26), 365–372.

(Stefanov, 2003) Stefanov Z.I., Hoo K.A., "A Distributed-Parameter Model of Black Liquor Falling Film Evaporators". *Part I. Modeling of Single Plate. Industrial Engineering Chemical Research 42, 1925-1937.*

(Stein, 1987) Stein G., Athans, M., "The LQG/LTR procedure for multivariable feedback control design". IEEE Trans. on Auto. Control, Vol. AC-32, February.

(Suarez, 2010) Suarez G.I., Ortiz O.A., Aballay P.M., Aros N.H., "Adaptive neural model predictive control for the grape juice concentration process". *International Conference on Industrial Technology, IEEE-ICIT 2010, Chile.*

(Turan, 1990) Turan L., Mingori D.L., Goodwin G.C., "Loop transfer recovery design biased and unbiased controllers". Technical Report EE9021, February.

(Wall, 1980) Wall J.E., Doyle J.C, Harvey C.A., "Tradeoffs in the design of multivariable feedback systems". Proc.18th Allerton Conf., October.

(Zang, 1990) Zang Z., Freudenberg J.S., "Loop transfer recovery for nonminimum phase plants". *IEEE Trans. Automatic Control, Vol. 35, pp. 547-553.*

(Zheng, 2002) Zheng D., Hoo K. A., "Low-Order Model Identification for Implementable Control Solutions of Distributed Parameter Systems". *Computers and Chemical Engineering 26 7-8, pp. 1049-1076.*

(Zuritz, 2005) Zuritz C.A., Muñoz E., Mathey H.H., Pérez E.H., Gascón A., Rubio L.A., Carullo C.A., Chemikoff R.E., Cabeza M.S., "Density, viscosity and soluble solid concentration and temperatures". *Journal of Food Engineering, Vol. 71, pp. 143 - 149.*

Part 4

Power Plant and Power System Control

14

A Robust and Flexible Control System to Reduce Environmental Effects of Thermal Power Plants

Toru Eguchi, Takaaki Sekiai, Naohiro Kusumi, Akihiro Yamada, Satoru Shimizu and Masayuki Fukai
Hitachi Ltd.
Japan

1. Introduction

Regulations on environmental effects due to such issues as nitrogen oxide (NOx) and carbon monoxide (CO) emissions from thermal power plants have become stricter[1]; hence the need for compliance with these regulations has been increasing. To meet this need, several technologies with respect to fuel combustion, exhaust gas treatment and operational control have been developed[2-4]. The technologies for the fuel combustion and the exhaust gas treatment include a low NOx burner and an air quality control system, and they are capable of reducing impact on the environment as physical and chemical implementation methods. The operational control technology for the thermal power plants is constantly required to receive changes in operational conditions. It is difficult to realize operational control which responds to combustion properties.

To overcome this issue, the operational control must be able to reduce NOx and CO emissions flexibly in accordance with such changes. Robustness is also required in such control because the measured NOx and CO data often include noise. Therefore, a robust and flexible plant control system is strongly desired to reduce environmental effects from thermal power plants efficiently.

Several studies have proposed plant control technologies to reduce the environmental effects[4-10]. These technologies are classified into two types of methods: model based and non-model based methods. The former methods include an optimization algorithm and a numerical model to estimate plant properties using neural networks (NNs)[11,12] and multivariable model predictive control[13]. The optimization algorithm searches for optimal control signals to reduce NOx and CO emissions using the numerical model. The latter methods have no models and they generates the optimal control signals by fuzzy logic[14]. A fuzzy logic controller outputs the optimal control signals for multivariable inputs using fuzzy rule bases. The fuzzy rule bases are based on *a priori* knowledge of plant control, and they can be tuned by parameters.

These technologies require the measured plant data for initial tuning of the model properties and the parameters of rules when the technologies are installed in plants. It usually takes some time to collect enough plant data. In addition, the search for control

signals can only be made in the past operating range, thus it is difficult to find the optimal control signals of they are located outside the range.
The authors have proposed a new plant control system for reducing environmental effects utilizing numerical calculation technology[15] to shorten the time for initial tuning and search the global optima. The system has one or more calculation databases (DBs) with respect to NOx and CO properties obtained by numerical calculation. Since the model can be tuned using the calculation DBs in advance, it is not necessary to take times for initial tuning when the control system is installed. Moreover, the proposed system obtains better control signals than the conventional technologies because it can model the NOx and CO properties including both inside and outside the operating range by the numerical calculation, which facilitates to search the optimal control signals.
After installation, the proposed control system is capable of tuning its model using the data measured in real time to reduce the model errors. In plant control, the shortest interval for changing operations is every 20 minutes because it often takes about 20 minutes to become static after an operation. The proposed system must be able to calculate the control signals during this interval, hence model tuning and searching for control signals should terminate within 20 minutes.
The proposed system employs radial basis function (RBF) network[16,17] and reinforcement learning (RL)[18]. The RBF network represents the NOx and CO properties to estimate their concentrations according to the control signals. The RL leads to the optimal control signals to achieve the control goals which is to reduce the estimated NOx and CO concentrations. The RBF network is one of the NNs having Gaussian basis functions. The RBF network usually learns the NOx and CO properties faster than ordinary NNs because the learning algorithm of the RBF network can be converted into matrix calculations without iterations. The RL is one of the machine learning methods[19] optimizing action rules of an agent by trial and error. It is preferable to apply the RL to the control system which requires real-time computing because the RL is a single point searching method and its computational cost is relatively small. It is also preferable to use the RL because the control history which can be utilized to improve the control logic can be traced in the RL control system. The proposed control system with the above features is expected to realize robustness, flexibility in control and real-time computing.
However, there are two practical problems to enhance these advantages more efficiently in the proposed control system. The first one is ensuring that the model can achieve enough estimation accuracy within practical computational times. Conventional methods to improve estimation accuracy of the model[11] are to adjust radii parameters of the Gaussian basis functions in RBF networks by calculating the estimation error for regression. However, with the conventional radius adjustment methods it might be difficult to adjust radii parameters within the time restriction because the adjustment of radii by regression requires many iterations. On the other hand, a radius adjustment method without calculation of estimation error has also been proposed[20]. This method determines the radii parameters using an equation considering learning data properties such as size and dimension. Its computational time is fast, but the estimation accuracy is worse than the method with regression. Therefore, it is desired to propose a new radius adjustment method for the plant control to achieve both higher estimation accuracy and faster computation.
The authors propose a novel radius adjustment method to overcome this first problem[21]. The proposed method focuses on the importance of covering input space properly where the model simulates the NOx and CO properties by the Gaussian basis functions to improve

estimation accuracy. This method adjusts radii parameters considering distances among the learning data. Consequently, the Gaussian basis functions can cover the input space properly and both high estimation accuracy and practical computational speed are achieved.

The second problem is to improve flexibility of the learning algorithm. Performance of the RL depends on the definition of a reward function equivalent to an evaluation function. The reward function has to be defined so that the RL algorithm can obtain the desired goal for the problem. As for application of the RL to thermal power plant control, the properties of the model changes in accordance with operational changes, thus the reward function has to be changed flexibly for the operational changes. However, it is quite difficult to prepare the reward functions for all patterns of operational conditions in advance.

To overcome this second problem, the authors introduce a reward function which has variable parameters and they proposed an automatic reward adjustment method[22]. The proposed method adjusts the variable parameters of the reward function automatically based on the NOx and CO emissions obtained in the learning process. As a result, the proposed method can obtain proper reward functions for all kinds of operational conditions.

The following sections outline the proposed control system and its newly proposed methods. Simulations clarify the advantages of the proposed system with respect to the following points: estimation accuracy and computational time of the RBF network, flexibility of the control logic and robustness in control for the noise of data.

2. Proposed plant control system for reducing environmental effects

2.1 Basic structure

Figure 1 shows the basic structure of the proposed control system. This system consists of a plant property estimation part and an operation optimizing part. The plant property estimation part includes a statistical model and measurement and numerical calculation DBs. The statistical model estimates the NOx and CO emission properties in thermal power plants. It is difficult to express these properties as mathematical equations because they have strong nonlinearities. The proposed system employs the RBF network as the statistical model which can estimate NOx and CO emissions for control variables using data stored in the DBs. The measurement DBs store the measured NOx and CO data for some control variables, and the numerical calculation DB stores data consisting of NOx and CO values for control variables calculated by the combustion analysis[15]. The control variables correspond to input of the statistical model, and the estimated NOx and CO emissions correspond to output of it. The statistical model can be modified by measured data obtained during the plant operations.

Conventional studies have been made about the model based control technology to reduce environmental effects from thermal power plants[4,6-8], but none of them considered employing not only the measured DB, but also the numerical calculation DBs. As the model can be tuned using the calculation DBs in advance, it is not necessary to take times for initial tuning at the time of installation. In addition, it is possible to tune the model after the installation by the data in the measured DB.

The operation optimizing part includes a RL agent, a reward calculation module, a reward adjustment module and a learning result DB. The learning procedure is as follows. First, the statistical model calculates and outputs the model outputs for the model inputs changed by the RL agent. Secondly, the reward calculation module calculates a reward using the model inputs and gives it to the RL agent. Thirdly, the RL agent learns its control logic. Learning results are stored in the learning result DB, and they are converted into modification signals. The control signals of the plant are generated by adding the modification signals to original control signals of the basic controller. The reward adjustment module adjusts reward parameters using the model outputs and the calculated reward. Normalized Gaussian network (NGnet)[23] has been employed as the structure of the RL agent. The learning algorithm of the NGnet is an actor-critic learning method[18], and it is appropriate for learning in a continuous environment.

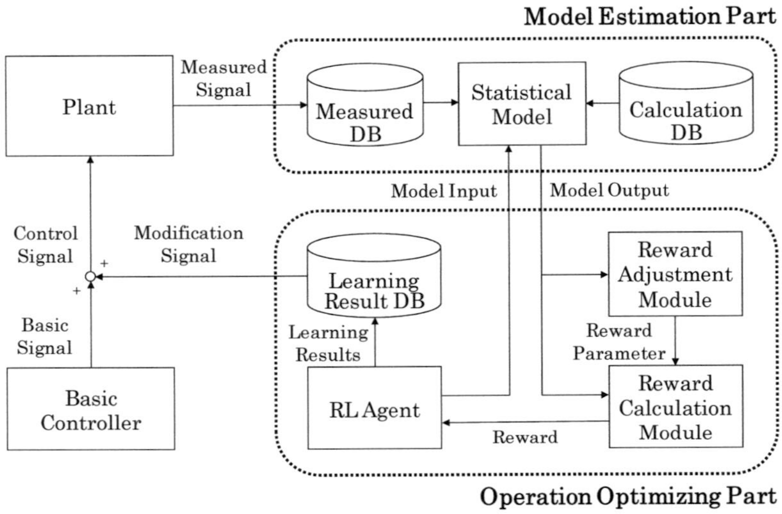

Fig. 1. Basic Structure of the Proposed Plant Control System

2.2 RBF network

The basic structure of the RBF network is shown in Fig. 2. The RBF network has three layers: an input layer, a hidden layer with Gaussian function, and an output layer. First, the J-dimensional vector is input in the input layer. Secondly, Gaussian function values are calculated using the input in the hidden layer. Finally, the P-dimensional vector is calculated by the Gaussian function values and weight parameters in the output layer. The RBF network is preferred for constructing a response surface due to the following properties.

- The RBF network avoids overfitting by the parameter of weight decay[16] to reduce the influences of noise included in the learning data.
- The RBF network does not need iterative calculations for learning of weight parameters like back propagation does[12].

Here, the input and output of the RBF network are denoted as $\mathbf{x}^T = \{x_1, \dots x_j, \dots x_J\}$ $(j \in J)$, $\mathbf{y}^T = \{y_1, \dots y_p, \dots y_P\}$ $(p \in P)$, then the p-th output y_p is calculated by Eqs. (1) and (2).

$$h_l(\mathbf{x}) = \exp\left(-\frac{(\mathbf{x}-\mathbf{c}_l)^T(\mathbf{x}-\mathbf{c}_l)}{r_l^2} \right) \tag{1}$$

$$y_p(\mathbf{x}) = \sum_{l=1}^{N_M} u_{lp} h_l(\mathbf{x}) \tag{2}$$

Here, $h_l(\mathbf{x})$ is the Gaussian function value of the l-th basis function, N_M is the number of basis functions, u_{lp} is the weight parameter between the hidden layer and output layer and $\mathbf{c}_l$, r_l are center coordinates and radius of the l-th basis function, respectively. The parameters $\mathbf{c}_l$ and r_l should be determined appropriately because they have much influence on estimation accuracy. In this chapter, the center coordinates are set to the learning data, and the radii are adjusted by the proposed radius adjusting method described later.

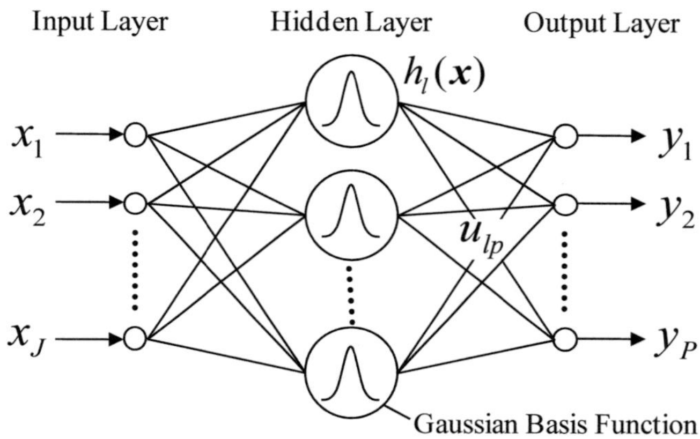

Fig. 2. Basic Structure of RBF Network

Learning of the RBF network corresponds to the determination of the weight parameter u_{lp} to minimize the energy function E_p given by Eq. (3) when the teaching data paired with learning data $\mathbf{x}_q$ are denoted as y_{pq}.

$$E_p = \sum_{q=1}^{N_D} (y_{pq} - y_p(\mathbf{x}))^2 + \lambda \sum_{l=1}^{N_M} u_{lp}^2 \tag{3}$$

Here, N_D is the number of learning data and λ is a weight decay reducing influences of noise included in learning data. The proposed control system can realize a robust control by tuning this parameter in accordance with the learning data. Then, the following matrices are defined.

$$\mathbf{U} = \begin{bmatrix} u_{11} & u_{12} & \cdots & u_{1P} \\ u_{21} & u_{22} & \cdots & u_{2P} \\ \vdots & \vdots & \ddots & \vdots \\ u_{N_M 1} & u_{N_M 1} & \cdots & u_{N_M P} \end{bmatrix} \tag{4}$$

$$\mathbf{H} = \begin{bmatrix} h_1(\mathbf{x}_1) & h_2(\mathbf{x}_1) & \cdots & h_{N_M}(\mathbf{x}_1) \\ h_1(\mathbf{x}_2) & h_2(\mathbf{x}_2) & \cdots & h_{N_M}(\mathbf{x}_2) \\ \vdots & \vdots & \ddots & \vdots \\ h_1(\mathbf{x}_{N_D}) & h_2(\mathbf{x}_{N_D}) & \cdots & h_{N_M}(\mathbf{x}_{N_D}) \end{bmatrix} \quad (5)$$

$$\mathbf{Y} = \begin{bmatrix} y_{11} & y_{12} & \cdots & y_{1P} \\ y_{21} & y_{22} & \cdots & y_{2P} \\ \vdots & \vdots & \ddots & \vdots \\ y_{N_M 1} & y_{N_M 1} & \cdots & y_{N_M P} \end{bmatrix} \quad (6)$$

$$\mathbf{\Lambda} = \lambda \mathbf{I} \quad (7)$$

In Eq. (3), both sides are partially differentiated by u_{lp} and Eqs. (4)-(7) are substituted, then Eq. (8) is obtained[16]. The learning of the RBF network can be described as the calculation of the weight matrix $\mathbf{U}$ given by Eq. (8).

$$\mathbf{U} = \left(\mathbf{H}^T\mathbf{H} + \mathbf{\Lambda}\right)^{-1}\mathbf{H}^T\mathbf{Y} \quad (8)$$

2.3 Reinforcement learning

2.3.1 Basic algorithm

The NGnet for learning of the RL agent learns its action, *i.e.*, control logic, and state value by putting Gaussian basis functions on its state space. Here, the state space is a mapping space to identify its status in the learning environment. The state value is a degree to evaluate how desirable the agent is in its current state. NGnet is known to be able to learn faster than other RL algorithms such as tile coding[18] because of the following features.

- NGnet can learn locally by the Gaussian basis functions.
- NGnet can reduce necessary basis function size by normalization.
- NGnet can add/delete the basis functions and parameter tuning.

Figure 3 shows the basic structure of NGnet. First, NGnet calculates activations of its Gaussian basis functions a_i and normalized activations b_i for the input $\mathbf{x}$ by Eqs. (9)-(11). Next, outputs of actor $\mathbf{m}(\mathbf{x}) = \{m_1, \ldots m_k, \ldots m_K\}$ $(k \in K)$ *i.e.*, action and critic $V(\mathbf{x})$ *i.e.*, state value are calculated by Eqs. (12)-(14).

$$a_i = \exp\left(-\frac{1}{2}(\mathbf{x} - \boldsymbol{\mu}_i)^T \sum\nolimits_i^{-1} (\mathbf{x} - \boldsymbol{\mu}_i)\right) \quad (9)$$

$$\sum\nolimits_i = diag(\boldsymbol{\sigma}_i^2) \quad (10)$$

$$b_i = \frac{a_i}{\sum_{t=1}^{N_L} a_t} \quad (11)$$

$$m_k(\mathbf{x}) = m_k^{\max} f_{sig}\left(\sum_{i=1}^{N_L} w_{ki} b_i + \beta n_k\right) \quad (12)$$

$$f_{sig}(z) = \frac{2}{1-\exp(-z)} - 1 \tag{13}$$

$$V(\mathbf{x}) = \sum_{i=1}^{N_L} v_i b_i \tag{14}$$

Here, i, j, k denote the subscripts of the basis functions of the agent, inputs and actor outputs, respectively. J, K also denote the dimensions of the inputs and actor outputs. In this chapter, the input of the statistical model is defined as becoming equal to that of the RL agent. In other words, the RL agent outputs the control bias to the input condition $\mathbf{x}$. The reward is calculated based on the results of control, *i.e.*, the outputs of the statistical model obtained after the control.

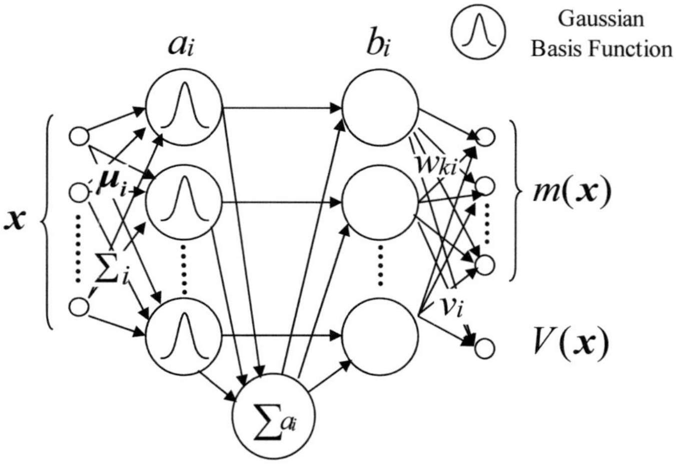

Fig. 3. Basic Structure of NGnet

Here, $\sum_i$ is the covariance matrix of the Gaussian basis function. $\boldsymbol{\mu}_i = \{\mu_{i1}, \ldots \mu_{ij}, \ldots \mu_{iJ}\}$, $\boldsymbol{\sigma}_i^2 = \{\sigma_{i1}^2, \ldots \sigma_{ij}^2, \ldots \sigma_{iJ}^2\}$ are the center and radii vectors, respectively. N_L is the basis function size. w_{ki}, v_i are the weight parameters of actor and critic, respectively.

The procedures to calculate the actor outputs m_k are as follows. First, the sum of the normalized activations b_i is added to a noise component to search for optimal actions. Next, they are converted to the region of [–1.1] by a sigmoid function. Finally, the actor outputs m_k are calculated by multiplying the maximum values of the actor outputs $m_k^{\max}$ and the converted value. Here, n_k is normalized noise whose average is 0 and variance is 1. β is a noise ratio.

2.3.2 Learning algorithm

Learning of NGnet is executed by the following procedures: updating the weight parameters w_{ki}, v_i, adding/deleting of the Gaussian basis functions, and tuning $\boldsymbol{\mu}_i$, $\boldsymbol{\sigma}_i^2$. TD learning[17] is employed to update w_{ki}, v_i. The agent updates its input ($\mathbf{x} \rightarrow \mathbf{x}'$) by its actor outputs m_k, then the model outputs are calculated by the actor outputs. Eq. (15) calculates TD error δ by *reward* calculated by the model outputs and the state value $V(\mathbf{x}')$ calculated by the input $\mathbf{x}'$.

$$\delta = reward + \gamma V(\mathbf{x}') - V(\mathbf{x}) \tag{15}$$

Here, γ is a discount ratio for the future reward. The actor of NGnet learns its actions to improve $V(\mathbf{x})$, and the critic of NGnet also learns to estimate $V(\mathbf{x})$ appropriately. w_{ki}, v_i are updated by Eqs. (16) and (17) using δ.

$$w_{ki} = w_{ki} + \alpha_A b_i \delta n_k \tag{16}$$

$$v_i = v_i + \alpha_C b_i \delta \tag{17}$$

Here, α_A and α_C denote the learning rates of w_{ki} and v_i, respectively.
The other learning procedures execute adding/deleting the Gaussian basis functions and tuning of $\boldsymbol{\mu}_i, \boldsymbol{\sigma}_i^2$ so that the NGnet can obtain enough resolutions to learn its state space. The proposed control system employs the following algorithm: the sizes of basis functions of the NGnet are initialized to 0, and new basis functions are added adaptively in its learning.

Basis Addition Algorithm

Step 1. If the current basis function size N_L satisfies $N_L < N_L^{\max}$, then the algorithm goes to **Step 2**. Otherwise, it terminates.

Step 2. The activations of the agent's current basis functions a_i are calculated for the input $\mathbf{x}$ during its learning.

Step 3. If there is no basis function i which meets $a_i \geq a_{\min}$, then the algorithm goes to **Step 4.** Otherwise, it terminates.

Step 4. If $\delta > \delta_{\min}$ is satisfied, the algorithm goes to **Step 5**. Otherwise, it terminates.

Step 5. A basis function whose center and radius is set to $\mathbf{x}$ and $\boldsymbol{\sigma}_i$ is added to NGnet, then the algorithm terminates.

Here, $N_L^{\max}, a_{\min}$ and $\delta_{\min}$ denote maximum basis function size, threshold value of activation and threshold value of TD error, respectively. This algorithm adds new basis functions in the regions of the state space which are not sufficiently covered with learned basis functions. In addition, the maximum basis function size $N_L^{\max}$ is set because it might be possible to add unnecessary basis functions by increasing variation of the TD error due to the proposed automatic reward adjustment method described later. Therefore, the agent can put only the necessary basis functions in its state space.

2.4 Learning flow of the proposed control system

The learning algorithm flow of the proposed control system consists of the following steps.

Learning Algorithm of the Proposed Control System

Step 1. Initialize learning parameters of the RBF network and RL.

Step 2. Adjust radii of the RBF network.
Step 3. Calculate weight parameters of the RBF network.
Step 4. Determine initial control variables.
Step 5. Change control variables by the RL agent.
Step 6. Calculate model outputs by the RBF network.
Step 7. Calculate reward.
Step 8. Calculate TD error.
Step 9. Update weight parameters of the RL agent.
Step 10. Add new basis functions of the RL agent.
Step 11. If the terminal condition of the episode is reached, go to **Step 12**. Otherwise, return to **Step 5**.
Step 12. Adjust the reward parameters.
Step 13. If the terminal condition of learning is reached, terminate the algorithm. Otherwise, return to **Step 4**.

In the above algorithm, an episode terminates after executing the processes between **Step 5** and **Step 10** for S times, and a trial of learning terminates after executing the processes between **Step 4** and **Step 12** for T times.

3. Adaptive radius adjustment method

3.1 Basic concepts

In the proposed control system, the outputs of the RBF network are calculated by the Gaussian basis functions according to the input space. To obtain high estimation accuracy, the radii should be adjusted so that the basis functions can cover the space sufficiently.
The proposed method focuses on the covering rate of the basis functions on the input space. It adjusts the radii based on the distances between a randomly generated input and the center of the basis functions selected to surround the input, where the learning data are located. As a result, the radii of basis functions whose distances to other data are short become small, and *vice versa*.

3.2 Algorithm of the proposed method

The algorithm of the proposed method consists of the following steps.

Algorithm of the Radius Adjustment Method

Step 1. Initialize the radii and adjusting parameters.
Step 2. Generate an input randomly.
Step 3. Select pairs of learning data by the *k*-SN (*k*-surrounded neighbor) method[24].
Step 4. Exclude the selected data from the data candidates for selection.
Step 5. If there are no data candidates, go to **Step 4.** Otherwise, return to **Step 3.**
Step 6. If there are no selected data, go to **Step 8**. Otherwise, go to **Step 7.**
Step 7. Update radii of the selected data
Step 8. If n reaches N, terminate the algorithm. Otherwise, increment n and return to **Step 2.**

In **Step 1**, the radii are initialized as a small value. In **Step 2**, an input condition $\mathbf{x}_n$ (n : suffix showing the number of iterations) is generated randomly. **In Step 3**, the pairs of learning data ($\mathbf{x}_m^1$, $\mathbf{x}_m^2$) (m : suffix showing the number of pairs) for which the radii are to be adjusted are selected for the generated $\mathbf{x}_n$ using the *k*-SN method. The *k*-SN method is a data extraction method to satisfy the condition of interpolation. It selects the pair of data ($\mathbf{x}_m^1$, $\mathbf{x}_m^2$) so that $\mathbf{x}_n$ is surrounded by them.

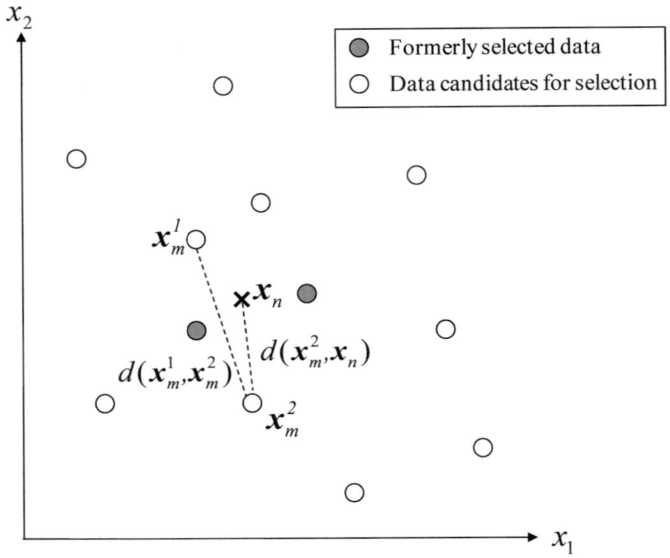

Fig. 4. Mechanism of k-SN method

Figure 4 shows the mechanism of the *k*-SN method in a 2-dimensional input space. The nearest datum $\mathbf{x}_m^1$ to $\mathbf{x}_n$ is selected from the data candidates available for selection, *i.e.*, learning data excluding the formerly selected data. Then the datum $\mathbf{x}_m^2$ paired with $\mathbf{x}_m^1$ is selected according to Eq. (18).

$$\mathbf{x}_m^2 = \arg\min_{z \in Z} d(\mathbf{x}_z, \mathbf{x}_n)$$

$$\textit{subject to } d(\mathbf{x}_z, \mathbf{x}_n) < d(\mathbf{x}_m^1, \mathbf{x}_z) \tag{18}$$

Here, z denotes the suffix of the data candidates available for selection and $d(\mathbf{x}_z, \mathbf{x}_n)$ denotes the distance between $\mathbf{x}_z$ and $\mathbf{x}_n$. In **Step 4**, the selected data ($\mathbf{x}_m^1$, $\mathbf{x}_m^2$) are excluded from the data candidates. If there is no $\mathbf{x}_z$ satisfying Eq. (18), only $\mathbf{x}_m^1$ is excluded. In this way, the radii of basis functions in an interpolative relation with inputs are adjusted, then the basis functions can cover the input space sufficiently. This selection continues until all the data candidates have been selected.

In **Step 7**, the radii (r_m^1, r_m^2) set at the selected data are adjusted by Eqs. (19) and (20).

$$r_m^1 = r_m^1 + \alpha_{rad}\tau^n \left(d(\mathbf{x}_m^1, \mathbf{x}_n) - r_m^1 \right) \tag{19}$$

$$r_m^2 = r_m^2 + \alpha_{rad}\tau^n \left(d(\mathbf{x}_m^2, \mathbf{x}_n) - r_m^2 \right) \tag{20}$$

Here, α_{rad} is an initial step size parameter of radius, and τ is a decay rate of the step size parameter ($0 < \tau < 1$). The second term in the right sides of both Eqs. (19) and (20) decays as iteration n increases, then the radii finally converge to certain values. These steps are iterated until n reaches N, then the radii are adjusted to certain values according to the distribution of learning data.

3.3 Simulations

In this section, some simulations are executed in order to evaluate the performances of the proposed radius adjustment method. The proposed method is compared with two conventional radius adjustment methods with respect to estimation accuracy and computational time using the test function data.

3.3.1 Simulation conditions

Simulations are executed in the following steps: a) determination of radii, b) calculation of weight parameters, and c) evaluation of estimation error. In step a), the proposed method, the Cross Validation (CV) method[11] and the radius equation method[20] are used to determine radii. The CV method adjusts radii with regression, and the radius equation method adjusts radii without regression. (See appendix). In step b), the weight parameters of the RBF network are calculated by Eq. (8). In step c), the estimation errors between the outputs of the RBF network and the test data are evaluated.

In the case of plant control, the shape of the response surface changes according to the plant properties, input dimensions and numbers of learning data. In order to simulate various response surfaces, the learning data are created for different test functions, input dimensions and numbers of data. The test functions $F_1(\mathbf{x})$ and $F_2(\mathbf{x})$ ($\mathbf{x} \in [-5,5]$) described as Eqs. (21) and (22) are used in the simulations. These functions are often used as benchmark problems of RBF networks[20].

$$F_1(\mathbf{x}) = \sum_{j=1}^{J} \left(x_j^4 - 16x_j^2 + 5x_j + 100 \right) \tag{21}$$

$$F_2(\mathbf{x}) = \sum_{j=1}^{J} \left(\sum_{k=1}^{j} x_k \right)^2 \tag{22}$$

Table 1 shows settings of learning data and test data of the RBF network in the simulations. Here, the numbers of learning data and test data are denoted as N_D and N_{Test}, respectively. In simulations, the output dimension P is fixed to 1, while the input dimension J is varied from 2 to 10. The parameters of α_{rad}, τ and N are set to 0.01, 0.999 and 3000, respectively. They are set appropriately based on prior experimental results. The parameters of $r^{\min}$, $r^{\max}$ and Δr used in the CV method are shown in Table 2. The common parameter, λ is set to 0.01. Each simulation is executed for 25 random sequences using a Linux machine (CPU clock: 2.8[GHz]).

Case	Function	Input Dimension J	Data Size N_D	Data Size N_{Test}
1	$F_1(\boldsymbol{x})$	2	25	25
2			50	
3			100	
4		5	100	50
5			300	
6			500	
7		10	100	100
8			500	
9			1000	
10	$F_2(\boldsymbol{x})$	2	25	25
11			50	
12			100	
13		5	100	50
14			300	
15			500	
16		10	100	100
17			500	
18			1000	

Table 1. Specifications of learning data and test data in the simulation cases

Case	r^{min}	r^{max}	Δr
1,2,3,10,11,12	0.1	10	0.1
4,13	5	15	0.1
5,14	5	15	0.5
6,15	5	15	1
7,16	5	20	0.1
8,9,17,18	5	20	1

Table 2. Parameter conditions in CV method

3.3.2 Results and discussions

In order to evaluate estimation accuracy of the proposed method, root mean square error $RMSE_{cn}$ calculated by Eq. (23) is used.

$$RMSE_{cn} = \sqrt{\sum_{t=1}^{N_{Test}} \frac{\left(y_{cn}(\mathbf{x}_t) - F_{cn}(\mathbf{x}_t)\right)^2}{N_{Test}}} \tag{23}$$

Here, $y_{cn}(\mathbf{x}_t)$ and $F_{cn}(\mathbf{x}_t)$ are an output value of the RBF network and the test data for the input $\mathbf{x}_t$ (t : suffix of the test data) in case cn ($cn = \{1,2,\ldots 18\}$), respectively.

First, convergence performance of the proposed method is studied using the RMSE and adjusted radii parameters. Figs. 5 and 6 show the RMSE and several radii parameters for iteration N in case 5 of Table 1. Case 5 is the most suitable condition for real plants with respect to input dimensions and numbers of learning data. The other cases also show the results similar to those of case 5. From Fig. 5, it is confirmed that the RMSE decreases and converges into a certain value with iteration.

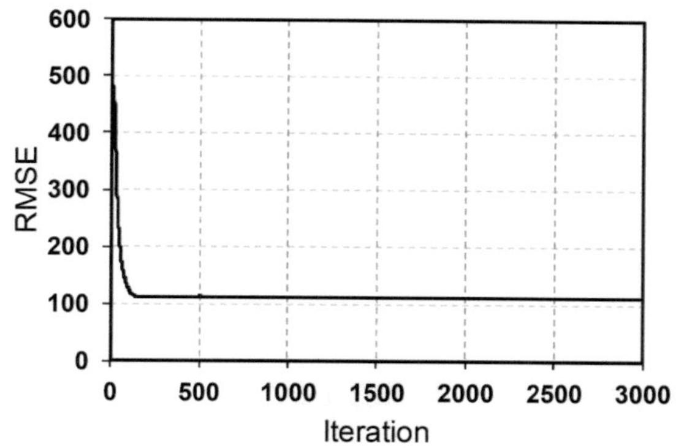

Fig. 5. RMSE curve obtained by the proposed method

Figure 6 shows the adjustment history of 10 typical radii selected from those of 300 Gaussian basis functions corresponding to the numbers of learning data in case 5. In this figure, the radii soon increase with iteration but converge into different values. The reason why the adjusted radii converge into different values is that the proposed method adjusts the radii based on the distribution of learning data. For the data whose distances to other data are short, the distances between the learning data and $\mathbf{x}_n$ become short. Consequently, the radii of the data in the region become shorter than those in the region whose distances are long. It is also confirmed by comparing Figs. 5 and 6 that the convergence of radii due to the decay of τ^n contributes to the convergence of RMSE.

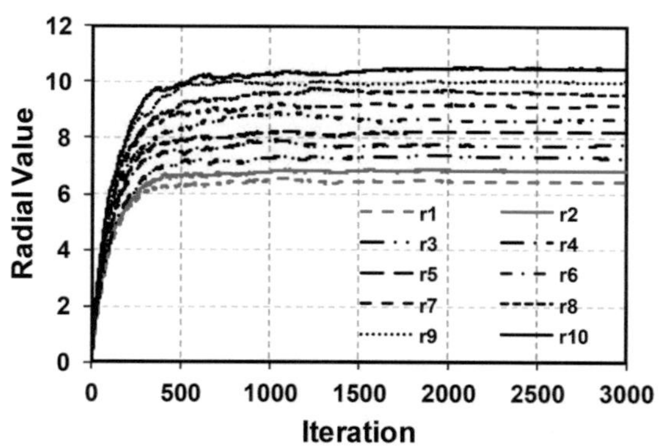

Fig. 6. Adjusting history of typical radial values by the proposed method

Next, Fig. 7 shows the radial values plotted for the crowding index ci of their basis functions calculated by Eqs. (24) and (25). The crowded index ci represents how the center coordinate $\mathbf{c}_i$ of the basis function i is covered with all the basis functions having uniform radii, thus this index of the data whose distances to other data are short usually becomes large.

$$h_t(\mathbf{c}_i) = \exp\left(-\frac{(\mathbf{c}_i - \mathbf{c}_t)^T (\mathbf{c}_i - \mathbf{c}_t)}{r_{ci}^2} \right) \tag{24}$$

$$ci(\mathbf{c}_i) = \frac{1}{N_D} \sum_{t=1}^{N_D} h_t(\mathbf{c}_i) \tag{25}$$

Here, $h_t(\mathbf{c}_i)$ is the Gaussian function value of the basis function whose center is $\mathbf{c}_t$ and r_{ci} is the radius to which a certain constant value is set. In this simulation, r_{ci} is set to 3.89 considering the range of input values.

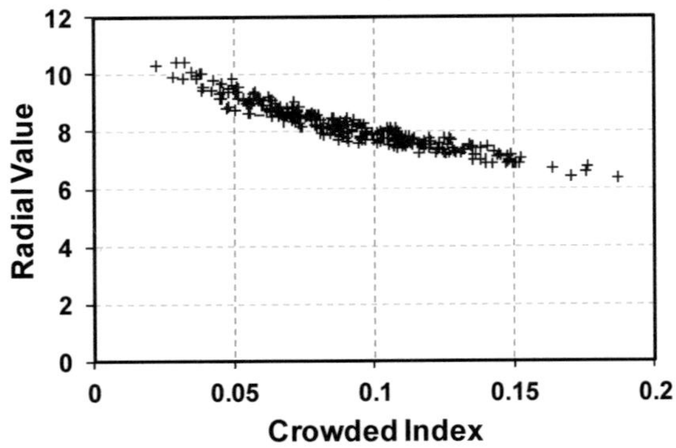

Fig. 7. Relation between the crowded index and radial values

In Fig. 7, the radial values with low crowded index are larger than those with high crowded index because the basis functions in the region where distances to data are long need to cover a wider input space. This result indicates that the proposed radius adjustment algorithm works properly.

Case	Proposed Method	CV Method	Radius Equation
1	76.0	83.5	96.5
2	71.4	70.9	99.7
3	29.5	36.1	63.3
4	138.3	130.3	186.5
5	116.6	115.4	170.1
6	107.2	113.3	153.1
7	201.0	234.1	272.7
8	174.4	166.8	230.7
9	164.0	158.4	239.3
10	7.9	6.1	14.2
11	2.7	2.7	10.5
12	1.5	2.3	9.3
13	39.8	35.3	79.1
14	16.4	12.9	58.5
15	8.9	10.9	48.4
16	268.1	292.2	294.7
17	82.0	58.8	173.5
18	63.9	38.2	175.6
Ave.	87.3	87.1	132.0

Table 3. Comparisons of the RMSEs obtained by the proposed and conventional methods

Table 3 compares the RMSEs of the proposed method and conventional methods. The case values in the table are the averages of 25 simulation results. The RMSEs of the proposed method are smaller than those for the radius equation in each case. The radius equation is usually applied to learning data having a uniform crowded index[20]. Therefore, it is difficult to apply it to plant control where the learning data usually have deviations of crowded index like Fig. 7. The proposed method can adjust the radii considering the distribution of the learning data, thus the RMSEs are an average of 33.9[%] better compared to those from the radius equation. The proposed method also has the same performances as the CV method.

Table 4 compares computational times of the proposed and conventional methods. These case results are also the averages of 25 simulation results. The computational times of the radius equation are enormously short because it spends time only in the calculation of Eq. (34) to adjust the radii. Regarding the CV method, the computational times increase exponentially with the number of data because error evaluations are needed for all learning data. There are some cases where the computational times are well beyond the limitation of practical use (20 minutes). Therefore, it is difficult to apply the CV method to plant control. On the other hand, the computational times of the proposed method in every case are within 20 minutes. These computational times are practical for plant control and it is confirmed that the proposed method is the most suitable for plant control.

These simulation results show that the proposed plant control system can construct a flexible statistical model having high estimation accuracy for various operational conditions of thermal power plants within a practical computational time. It is expected to improve effectiveness in reducing NOx and CO by learning with such a statistical model.

Case	Proposed Method	CV Method	Radius Equation
1	2.8E-02	6.5E-01	7.6E-06
2	9.9E-02	9.2E+00	2.8E-05
3	3.7E-01	1.5E+02	1.1E-04
4	4.6E-01	1.4E+02	1.4E-04
5	3.9E+00	2.6E+03	1.3E-03
6	1.1E+01	1.7E+04	3.6E-03
7	6.6E-01	2.2E+02	2.8E-04
8	1.6E+01	2.3E+04	6.9E-03
9	6.4E+02	6.5E+05	3.1E-02
10	2.7E-02	6.5E-01	7.6E-06
11	9.8E-02	9.2E+00	2.7E-05
12	3.7E-01	1.5E+02	1.1E-04
13	4.6E-01	1.4E+02	1.4E-04
14	3.9E+00	2.6E+03	1.3E-03
15	1.1E+01	1.6E+04	3.6E-03
16	6.6E-01	2.2E+02	2.8E-04
17	1.6E+01	2.3E+04	6.9E-03
18	6.4E+02	6.5E+05	3.1E-02

Table 4. Comparisons of the computational times [s] for the proposed and conventional methods

4. Automatic reward adjustment method

4.1 Basic concepts

When the RL is applied to the thermal power plant control, it is necessary to design the reward so that it can be given to the agent instantly in order to adapt to the plant properties which change from hour to hour. So far, studies with respect to designing reward of the RL have reported[25,26] that high flexibility could be realized by switching or adjusting the reward in accordance with change of the agent's objectives and situations. However, it would be difficult to apply this to thermal power plant control which needs instant reward designing for changes of plant properties because the reward design and its switching or adjusting depend on *a priori* knowledge.

The proposed control system defines a reward function which does not depend on the learning object and proposes an automatic reward adjustment method which adjusts the parameters of the reward function adaptively based on the plant property information obtained in the learning. It is possible to use the same reward function for different operating conditions and control objectives in this method, and the reward function is adjusted in accordance with learning progress. Therefore, it is expected possible to construct a flexible plant control system without manual reward design.

4.2 Definition of reward

The statistical model in the proposed control system has a unique characteristic due to specifications of applied plants, kinds of environmental effects and operating conditions. In case such a model is used for learning, the reward function should be generalized because it is difficult to design unique reward functions for various plant properties in real time. Thus the authors have defined the reward function as Eq. (26).

$$reward = \begin{cases} reward_{\max} \exp\left(\dfrac{\rho - f}{\phi}\right) & (f \geq \rho) \\ reward_{\max} & (f < \rho) \end{cases} \tag{26}$$

Here, $reward_{\max}$ and f are maximum reward value and sum of weighted model outputs calculated by Eq. (27), respectively. ϕ and ρ are the parameters to determine shapes of the reward function.

$$f = \sum_{p=1}^{P} C_p y_p \tag{27}$$

Here, C_p are the weight of the model output y_p, and p is a suffix for model output. In Eq. (26), the conditions $\phi > 0$, $\rho \geq 0$ are satisfied. If ϕ and ρ become larger, a larger reward is gotten for f. In addition, it is possible for f to weight y_p by C_p in accordance with control goals. Fig. 8 shows the shape of the reward function where $reward_{\max} = 1$, $\phi = 10$, $\rho = 20$ are set in Eq. (26).

The reward function defined as Eq. (26) can be applied for various kinds of statistical models where the operating conditions and the control goals are different because it is possible to define the reward only by ϕ, ρ and C_p. C_p is set in accordance with the control goals, and ϕ, ρ are adjusted automatically by the proposed automatic reward adjustment method.

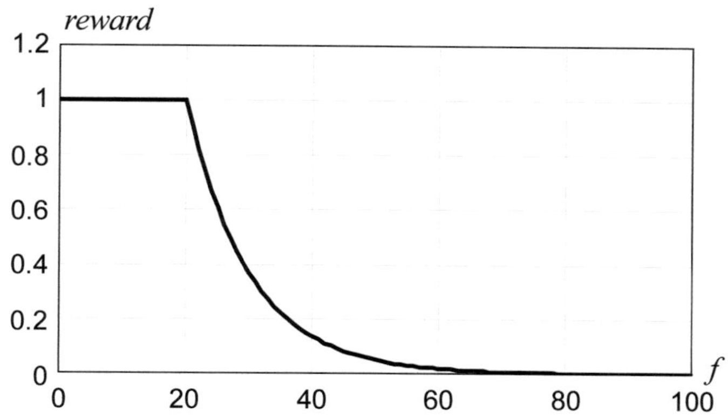

Fig. 8. Schematic of reward function

4.3 Algorithm of the proposed reward adjustment method

The proposed reward adjustment method adjusts the reward parameters ϕ, ρ using the model outputs which are obtained during the learning so that the agent can get the proper reward for (1) characteristics of the learning object and (2) progress of learning. Here, (1) means that this method can adjust the reward properly for the statistical models whose optimal control conditions and NOx/CO properties are different by adjusting ϕ, ρ. (2) means that this method makes it easier for the agent to get the reward and accelerate learning at the early stage, while also making the conditions to get the reward stricter and improving the agent's learning accuracy.

The reward parameters are updated based on the sum of weighted model outputs f obtained in each episode and the best f value obtained during the past episodes. Hereafter, the sum of weighted model outputs and the reward parameters at episode t are denoted as f_t, ϕ_t and ρ_t, respectively.

The algorithm of the proposed method is as follows. First, f_t is calculated by Eq. (28), then its moving average $\overline{f_t}$ is calculated.

$$\overline{f_t} = \varepsilon f_t + (1-\varepsilon)\overline{f_{t-1}} \tag{28}$$

Here, ε is a smoothing parameter of the moving average. The parameter ϕ_t is updated by Eqs. (29) and (30) where $\overline{f_t} > \rho_t$ is satisfied.

$$\phi_{t+1} = \phi_t + \alpha_\phi(\phi'_t - \phi_t) \tag{29}$$

$$\phi'_t = \frac{\rho_t - \overline{f_t}}{\ln(\theta_t / reward_{\max})} \tag{30}$$

Here, ϕ'_t is an updating index of ϕ_t, θ_t is a threshold parameter to determine the updating direction (positive/negative), and α_φ is a step size parameter of ϕ_t. As shown in Fig. 9, ϕ'_t corresponds to the ϕ when the reward value for $\overline{f_t}$ becomes θ_t. The updating direction of ϕ_t becomes positive where θ'_t calculated by Eq. (31) is smaller than θ_t, and *vice versa*.

$$\theta'_t = reward_{\max} \exp\left(\frac{\rho_t - \overline{f}_t}{\phi_t}\right) \tag{31}$$

θ_t is updated by Eq. (32) so that it becomes closer to θ'_t.

$$\theta_{t+1} = \theta_t + \alpha_\theta(\theta'_t - \theta_t) \tag{32}$$

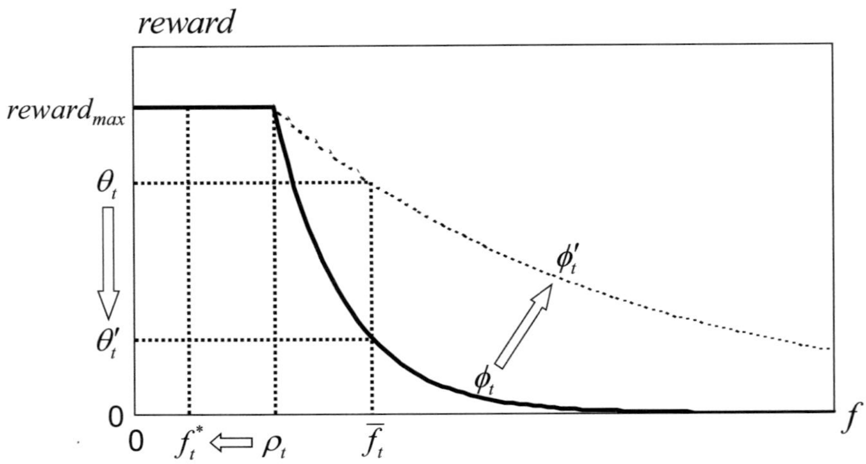

Fig. 9. Mechanism of the proposed method

Here, α_θ is a step size parameter of θ_t. θ_t is initialized to small value. As a result of updating θ_t by Eq. (32), finally ϕ'_t becomes equal to ϕ_t. This means that the reward is given to the agent appropriately for current $\overline{f}_t$. The value of θ_t depends on the learning object and progress, hence it is preferable to acquire empirically in the learning process. That is because θ'_t, the reward value for $\overline{f}_t$ is defined according to the updating index of θ_t.

The parameter ρ_t is updated to approach the f_t^* by Eq. (33) which is the best value of f during past learning.

$$\rho_{t+1} = \rho_t + \alpha_\rho(f_t^* - \rho_t) \tag{33}$$

Here, α_ρ is a step size parameter of ρ_t.

The above algorithm is summarized as the following steps.

Reward Automatic Adjustment Algorithm

Step 1. Calculate $\overline{f}_t$ by Eq. (28).

Step 2. If $\overline{f}_t > \rho_t$ is satisfied, go to **Step 3**. Otherwise, go to **Step 5**.

Step 3. Update ϕ_t by Eqs. (29) and (30).

Step 4. Update θ_t by Eqs. (31) and (32).

Step 5. Update ρ_t by Eq. (33) and terminate the algorithm.

4.4 Simulations

In this section, simulations are described to evaluate the performances of the proposed control system with the automatic reward adjustment method when it is applied to virtual plant models configured on the basis of experimental data. The simulations incorporate changes of the plant operations several times and the data for the RBF network. The evaluations focus on the flexibility in control of the proposed reward adjustment method for the change of the operational conditions. In addition, the robustness in control for the statistical model including noise by tuning the weight decay parameter of RBF network is also studied.

4.4.1 Simulation conditions

Figure 10 shows the basic structure of the simulation. The objective of the simulation is to reduce NOx and CO emissions from a virtual coal-fired boiler model (statistical model) constructed with three numerical calculation DBs. The RL agent learns how to control three operational parameters with respect to air mass flow supplied to the boiler. Therefore, input and output dimensions (J,P) of the control system are 3 and 2, respectively. The input values are normalized into the range of $[0,1]$. The three numerical calculation DBs have different operational conditions, and each DB has 63 data whose input-output conditions are different. These data include some noise similar to the actual plant data.

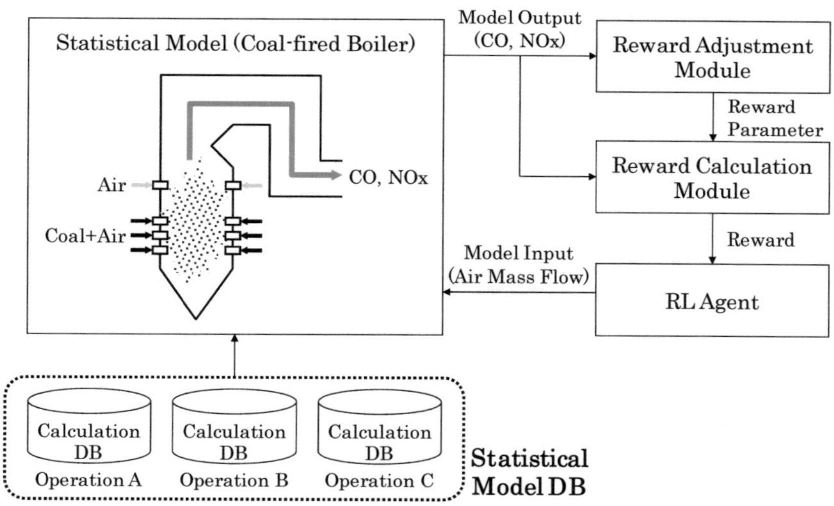

Fig. 10. Basic structure of thermal power plant control simulation

In this simulation, the robustness and flexibility of the proposed control system are verified by implementing the RL agent so that it learns and controls the statistical model which changes in time series. Two kinds of boiler operational simulations are executed according to Table 5. Each simulation case is done for six hours (0:00-6:00) of operation, and it is considered that the statistical model is changed at 0:00, 2:00 and 4:00. One of the simulations considers three kinds of operational conditions (A,B,C) where coal types and power outputs are different, and the other considers three kinds of control goals defined as Eq. (27), where the weight coefficients C_1,C_2 of CO and NOx, respectively in that equation are different.

The simulations are executed by two reward settings: the variable reward for the proposed reward adjustment method (proposed method) and the fixed reward (conventional method). Both reward settings are done under two conditions where the weight decay λ for the RBF network is set to 0, 0.01 to evaluate the robustness of control by λ settings. The RL agent learns at the times when operational conditions or control goals (0:00, 2:00 and 4:00) are changed, and the control interval is 10 minutes. Hence it is possible to control the boiler 11 times in each period.

Parameter conditions of learning are shown in Table 6. These conditions are set using prior experimental results. The parameter conditions of reward are shown in Table 7. The parameters ($\varepsilon, \alpha_\phi, \alpha_\theta, \alpha_\rho$) of the proposed method are also set properly using prior experiments. In the conventional method, the values of ϕ, ρ are fixed to their initial values which are optimal for the first operational condition in Table 5 because their step size parameters (α_ϕ, α_ρ) are set to 0.

Objective	Change of Operational Conditions			Change of Goals		
Time	Ope. Cond.	C_1	C_2	Ope. Cond.	C_1	C_2
0:00 - 2:00	A	0.1	0.9	A	0.1	0.9
2:00 - 4:00	B	0.1	0.9	A	0.9	0.1
4:00 - 6:00	C	0.1	0.9	A	0.001	0.999

Table 5. Time table of plant operation simulation

Parameter		Condition
Radius of Gaussian basis	σ	0.2
Max. output of NGnet	m_k^{max}	0.2
Noise ratio	β	0.2
Discount rate	γ	0.9
Learning rate for actor	α_A	0.1
Learning rate for critic	α_C	0.02
Max. basis num of agent	N_L^{max}	100
Min. a_i for basis addition	a_{min}	0.368
Min. δ for basis addition	δ_{min}	0.01
Max. iteration in 1 episode	S	30
Max. episode	T	10000

Table 6. Parameter conditions of learning

Parameter		Prop. Method	Conv. Method
Max. reward	$reward_{max}$	1	1
Smoothing parameter	ε	0.1	0.1
Step size parameter of ϕ	α_ϕ	0.05	0
Step size parameter of ρ	α_ρ	0.05	0
Step size parameter of θ	α_θ	0.05	0
Initial value of ϕ		0.001	3
Initial value of ρ		0.001	0
Initial value of θ		0	186

Table 7. Reward conditions of each method

4.4.2 Results and discussion

Figure 11 shows the time series of normalized f as a result of controls by the two methods, where the initial value at 0:00 is determined as the base. There are four graphs in Fig. 11 with combinations of the two objectives of simulations and λ settings. The optimal f value in each period is shown as well. The computational time of learning in each case was 23[s].

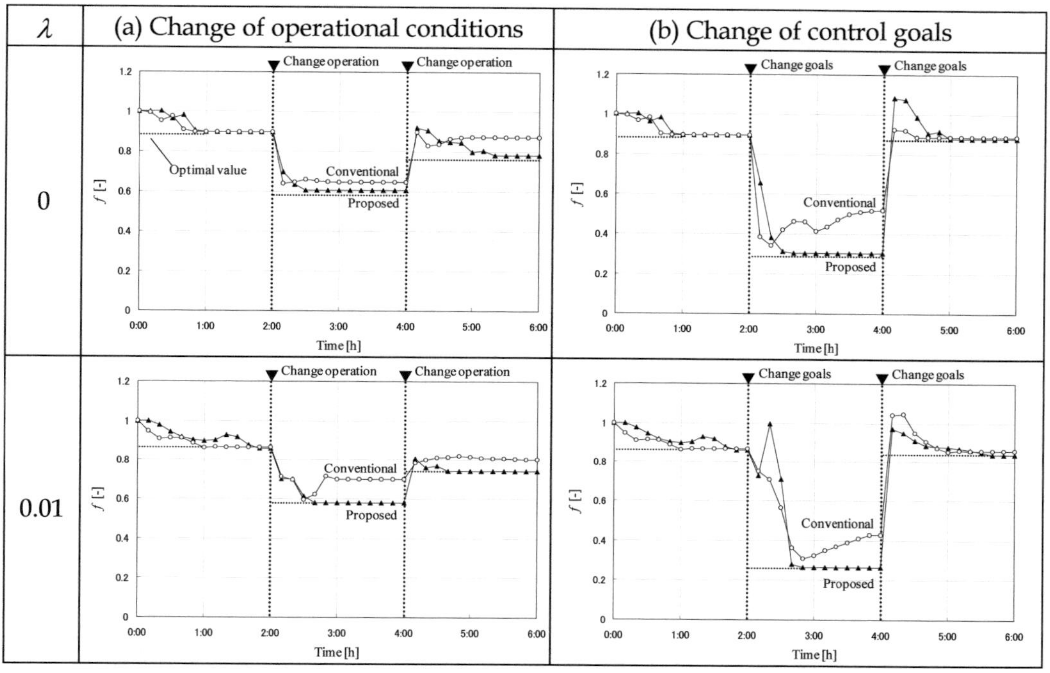

Fig. 11. Time series of normalized f in the boiler operation simulations

To begin with, time series of the normalized f values by the proposed method and conventional method in the case of λ=0.01 are discussed. The initial f values at 0:00 of these methods have offsets with the optimal values, but they are decreased for control and finally converged near the optimal values. This is because the reward functions used in each method are appropriate to learn the optimal control logic. The RL agent relearns its control logic when the statistical model and its optimal f values are changed at 2:00 by the change of operational conditions or control goals. However, the f values of the conventional method after 11 control times still have offsets from the optimal values, while the proposed method can obtain the optimal values after 11 times. The initial reward setting of the conventional method would be inappropriate for the next operational condition. Similar results of control are obtained for the same reason after changing the statistical model at 4:00. As discussed above, the plant control system by the conventional method has a possibility to deteriorate the control performances in thermal power plants for which operational conditions and control goals are changed frequently. Therefore, the proposed reward adjustment method is effective for the plant control, which can adjust the reward function flexibly for such changes.

Next, the robustness of the proposed control system by weight decay (λ) tuning is discussed. In Fig. 11, every f value of the proposed method can reach nearly the optimal value when λ is 0.01, whereas f converges into the values larger than the optimal values when λ is 0 for 2:00-6:00 in (a) and 2:00~4:00 in (b). The RBF network cannot learn with considered the influences of noise included in the learning data when λ is 0[16]. The response surface is created to fit the noised data closely and many local minimum values are generated in it compared with the response surface of $\lambda = 0.01$. This is because the learned control logic is converged each local minimum. The above results show that the RBF network can avoid overfitting by tuning λ properly and the proposed control system can control thermal power plants robustly.

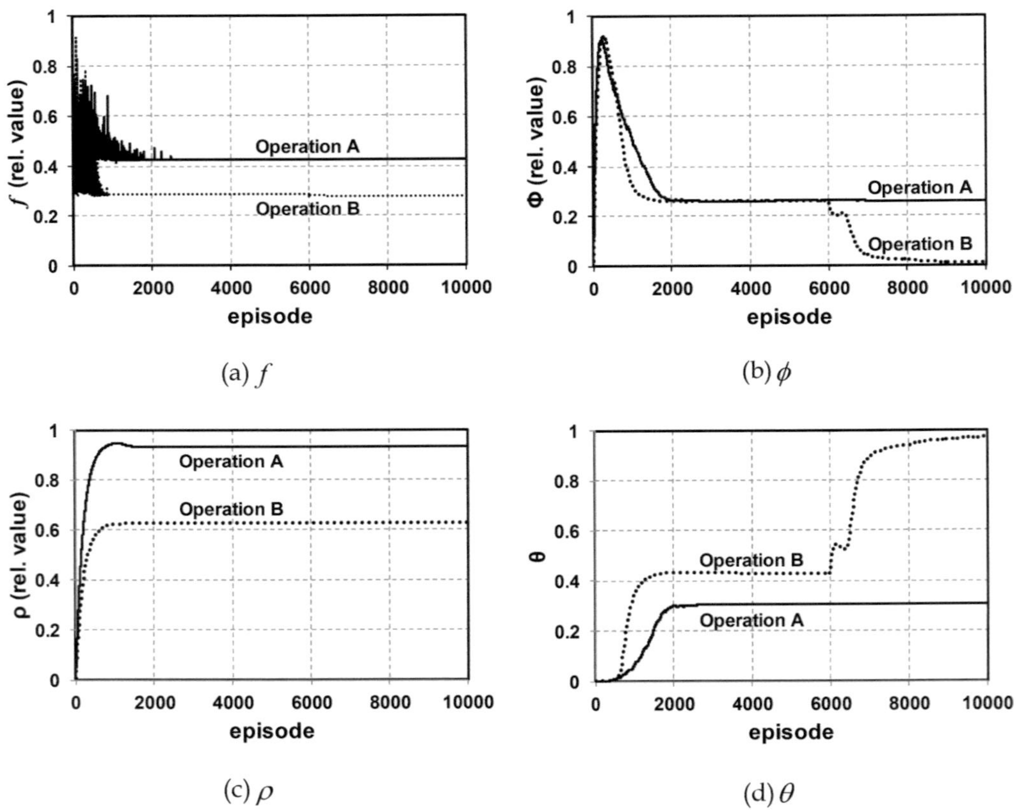

(a) f (b) ϕ (c) ρ (d) θ

Fig. 12. Learning processes of f and reward parameters (ϕ, ρ, θ) of the proposed method

Finally, the learning processes of f and reward parameters of the proposed method are studied. Fig. 12 shows the f, ϕ, ρ, θ values for episodes in learning at the operational changes at 0:00 and 2:00 when λ is 0.01. In the early stage of learning (episodes 1-500), the ϕ parameter in each case increases nearby 0.9 because the f value does not decrease due to insufficient learning of the RL agent. In the next 1000 episodes, ϕ increases and θ decreases simultaneously as the learning progresses. This behavior can be explained by the Eqs. (29)-(32) which are the updating algorithms of ϕ, θ. On the other hand, ρ value in each case converges to certain values by the 2000th episode. This indicates that the optimal f values are found in the learning process. Then the parameters of each case remain stable during the middle stage of learning (episode 2000-6000), but ϕ, θ change suddenly at the 6000th episode only in the case of operation B. This is because the RL agent can learn the control logic to get a better f value, then ϕ, θ are adjusted flexibly in accordance with the change of f used in Eqs. (29) and (30). As a result, these parameters converge into different values.

These adjustment results of reward parameters for different statistical models can be discussed as follows. By analysis of the characteristics of these statistical models, it seems that the gradient of f in operation A is larger than that of operation B because operation A has a larger difference between the maximum and minimum value of f than operation B. When the gradient of f is larger, f will vary significantly for each control thus it is necessary to set ϕ larger so that the agent can get the reward easily. On the other hand, it is useless to set ϕ larger in the statistical model in operation B for which the gradient of f is small. As for the results of adjustment of ϕ, ρ, θ in Fig. 12, the reward function of operation A certainly becomes easier to give the reward due to the larger ϕ than for operation B. Therefore, the above results show that the proposed method can obtain the appropriate reward function flexibly in accordance with the properties of the statistical models.

5. Conclusions

This chapter presented a plant control system to reduce NOx and CO emissions exhausted by thermal power plants. The proposed control system generates optimal control signals by that the RL agent which learns optimal control logic using the statistical model to estimate the NOx and CO properties. The proposed control system requires flexibility for the change of plant operation conditions and robustness for noise of the measured data. In addition, the statistical model should be able to be tuned by the measured data within a practical computational time. To overcome these problems the authors proposed two novel methods, the adaptive radius adjustment method of the RBF network and the automatic reward adjustment method.

The simulations clarified the proposed methods provided high estimation accuracy of the statistical model within practical computational time, flexible control by RL for various changes of plant properties and robustness for the plant data with noise. These advantages led to the conclusion that the proposed plant control system would be effective for reducing environmental effects.

6. Appendix A. Conventional radius adjustment method

A.1 Cross Validation (CV) method

The cross validation (CV) method is one of the conventional radius adjustment methods for the RBF network with regression and it adjusts radii by error evaluations. In this method, a datum is excluded from the learning data and the estimation error at the excluded datum is evaluated. Iterations are repeated until all data are selected as excluded data to calculate RMSE. After the calculations of RMSE for several radius conditions, the best condition is determined as the radius to use. The algorithm is shown as follows.

Algorithm of Cross Validation Method

Step 1. Initialize the radius is initialized to $r^{\min}$.
Step 2. Select an excluded datum.
Step 3. Learn weight parameters of RBF network using all data except the excluded datum.
Step 4. Calculate the output of the RBF network at the point of the excluded datum.
Step 5. Calculate the error between the output and the excluded datum.
Step 6. Go to **Step 7** if all data have been selected. Otherwise, return to **Step 2**.
Step 7. Calculate RMSE by the estimation errors.
Step 8. Increment the radius by Δr.
Step 9. Select the radius with the best RMSE if the radius is over $r^{\max}$ and terminate the
Step 10. algorithm. Otherwise, return to **Step 2**.

A.2 Radius equation

This method is one of the non-regression methods and it adjusts the radius r by Eq. (34).

$$r = \frac{d_{\max}}{\sqrt{J}\left(\sqrt[J]{N_D} - 1\right)} \tag{34}$$

Here, $d_{\max}$ is the maximum distance among the learning data.

7. References

[1] U.S. Environmental Protection Agency, Available from http://www.epa.gov/air/ oaq_caa.html/

[2] Ochi, K., Kiyama, K., Yoshizako, H., Okazaki, H. & Taniguchi, M. (2009), Latest Low-NOx Combustion Technology for Pulverized-coal-fired Boilers, *Hitachi Review*, Vol. 58, No. 5, pp. 187-193.

[3] Jorgensen, K. L., Dudek, S. A. & Hopkins, M. W. (2008), Use of Combustion Modeling in the Design and Development of Coal-Fired Furnaces and Boilers, *Proceedings of ASME International Mechanical Engineering Congress and Exposition*, Boston.

[4] EPRI (2005), *Power Plant Optimization Industry Experience*, 2005 Update. EPRI, Palo Alto.

[5] Rangaswamy, T. R.; Shanmugam J. & Mohammed K. P. (2005), Adaptive Fuzzy Tuned PID Controller for Combustion of Utility Boiler, *Control and Intelligent Systems*, Vol. 33, No. 1, pp. 63-71.

[6] Booth, R. C. & Roland W. B. (1998), Neural Network-Based Combustion Optimization Reduces NOx Emissions While Improving Performance, *Proceedings of Dynamic Modeling Control Applications for Industry Workshop*, pp.1-6.

[7] Radl B. J. (1999), Neural networks improve performance of coal-fired boilers, CADDET Energy Efficiency Newsletter, No.1, pp.4-6.

[8] Winn H. R. & Bolos H. R. (2008), Optimizing the Boiler Combustion Process in Tampa Electric Coal Fired Power Plants Utilizing Fuzzy Neural Model Technology, *Proceedings of Power-Gen International 2008*, Orlando, FL.

[9] Vesel R. (2008), The Million Dollar Annual Payback: Realtime Combustion Optimization with Advanced Multi-Variable Control at PPL Colstrip, *Proceedings of Power-Gen International 2008*, Orlando, FL.

[10] Airikka P. & Nieminen V. (2010), Optimized Combustion through Collaboration of Boiler and Automation Suppliers, *Proceedings of Power-Gen International 2008*, Amsterdam.

[11] Wasserman P. D. (1993), *Advanced Methods in Neural Computing*, Van Nostrand Reinhold.

[12] Rumelhart D. E.; Hinton G. E. & Williams R. J. (1986), Learning Representations of Back-propagation Errors, *Nature*, vol. 323, pp. 533-536.

[13] Camacho, E. F. & Bordons, C. (1999), *Model Predictive Control*. Springer.

[14] Jamshidi, M., Titli, A., Zadeh, L. & Boverie, S. (1997), *Applications of Fuzzy Logic*. Prentice Hall.

[15] Yamamoto K.; Fukuchi T.; Chaki M.; Shimogori Y. & Matsuda J. (2000), Development of Computer Program for Combustion Analysis in Pulverized Coal-fired Boilers, *Hitachi Review*, Vol. 49, No. 2, pp. 76-80.

[16] Orr M. J. L.. Introduction to Radial Basis Function Networks. Available from http://anc.ed.ac.uk/mjo/rbf.html

[17] Maruyama M. (1992), Learning Networks Using Radial Basis Function - New Approach for the Neural Computing. *Trans. of ISCIE*, Vol. 36, No. 5, pp. 322–329. (in Japanese)

[18] Sutton R. S. & Barto A. G. (1998), *Reinforcement Learning-An Introduction*, MIT Press.

[19] Bishop, C., M. (2006). *Pattern Recognition And Machine Learning*. Springer-Verlag.

[20] Kitayama S.; Yasuda K. & Yamazaki K. (2008), The Integrative Optimization by RBF Network and Particle Swarm Optimization, *IEEJ Trans. on EIS*, Vol. 128, No. 4, pp. 636-645. (in Japanese)

[21] Eguchi, T.; Sekiai T.; Yamada, A.; Shimizu S. & Fukai M. (2009), A Plant Control Technology Using Reinforcement Learning Method with Automatic Reward Adjustment, *IEEJ Trans. on EIS*, Vol. 129, No. 7, pp. 1253-1263. (in Japanese)

[22] Eguchi, T.; Sekiai T.; Yamada, A.; Shimizu S. & Fukai M. (2009), An Adaptive Radius Adjusting Method for RBF Networks Considering Data Densities and Its Application to Plant Control Technology, *Proceedings of ICCAS-SICE2009*, pp.4188-4194, Fukuoka, Japan, August 18-21.

[23] Moody J. & Darken C. J. (1989), Fast learning in networks of locally-tuned processing units, *Neural Computation*, Vol.1 , pp. 281-294.

[24] Zhang J.; Yim Y. & Yang J. (1997), Intelligent Selection of Instances for Prediction Function in Lazy Learning Algorithms, *Artificial Intelligence Review*, Vol. 11, pp. 175-191.

[25] Ng. A.; Harada D. & Russell S. (1999), Policy invariance under reward transformations: Theory and application to reward shaping, *Proceedings of 16th International Conference on Machine Learning*, pp.278-287.

[26] Li J. & Chan L. (2006), Reward Adjustment Reinforcement Learning for Risk-averse Asset Allocation, *Proceedings of International Joint Conference on Neural Networks 2006 (IJCNN06)*, pp.534-541.

15

Wide-Area Robust H_2/H_∞ Control with Pole Placement for Damping Inter-Area Oscillation of Power System

Chen He[1] and Bai Hong[2]
[1]*State Power Economic Research Institute, State Grid Corporation of China*
[2]*China Electric Power Research Institute*
China

1. Introduction

The damping of inter-area oscillations is an important problem in electric power systems (Klein et al., 1991; Kundur, 1994; Rogers, 2000). Especially in China, the practices of nationwide interconnection and ultra high voltage (UHV) transmission are carrying on and under broad researches (Zhou et al., 2010), bulk power will be transferred through very long distance in near future from the viewpoints of economical transmission and requirement of allocation of insufficient resources. The potential threat of inter-area oscillations will increase with these developments. If inter-area oscillations happened, restrictions would have to be placed on the transferred power. So procedures and equipments of providing adequate damping to inter-area oscillations become mandatory.

Conventional method coping with oscillations is by using power system stabilizer (PSS) that provides supplementary control through the excitation system (Kundur, 1994; Rogers, 2000; Larsen et al., 1981), or utilizing supplementary control of flexible AC transmission systems (FACTS) devices (Farsangi et al., 2003; Pal et al., 2001; Chaudhuri et al., 2003, 2004). Decentralized construction is often adopted by these controllers. But for inter-area oscillations, conventional decentralized control may not work so well since they have not observability of system level. Maximum observability for particular modes can be obtained from the remote signals or from the combination of remote and local signals (Chaudhuri et al., 2004; Snyder, et al., 1998; Kamwa et al., 2001). Phasor measurement units (PMUs)-based wide-area measurement system (WAMS) (Phadke, 1993) can provide system level observability and controllability and make so-called wide-area damping control practical.

On the other hand, power system exists in a dynamic balance, its operating condition always changes with the variations of generations or load patterns, as well as changes of system topology, etc. From control theory point of view, these changes can be called uncertainty. Conventional control methods can not systemically consider these uncertainties, and often need tuning or coordination. Therefore, so-called robust models are derived to take these uncertainties into account at the controller design stage (Doyle et al., 1989; Zhou et al., 1998). Then the robust control is applied on these models to realize both disturbance attenuation and stability enhancement.

In robust control theory, H_2 performance and H_∞ performance are two important specifications. H_∞ performance is convenient to enforce robustness to model uncertainty, H_2 performance is useful to handle stochastic aspects such as measurement noise and capture the control cost. In time-domain aspects, satisfactory time response and closed-loop damping can often be achieved by enforcing the closed-loop poles into a pre-determined subregion of the left-half plane (Chilali et al., 1996). Combining there requirements to form so-called mixed H_2/H_∞ design with pole placement constrains allows for more flexible and accurate specification of closed-loop behavior. In recent years, linear matrix inequalities (LMIs) technique is often considered for this kind of multi-objective synthesis (Chilali et al., 1996; Boyd et al., 1994; Scherer et al., 1997, 2005). LMIs reflect constraints rather than optimality, compared with Riccati equations-based method (Doyle et al., 1989 ; Zhou et al., 1998), LMIs provide more flexibility for combining various design objectives in a numerically tractable manner, and can even cope with those problems to which analytical solution is out of question. Besides, LMIs can be solved by sophisticated interior-point algorithms (Nesterov et al., 1994).

In this chapter, the wide-area measurement technique and robust control theory are combined together to design a wide-area robust damping controller (WRC for short) to cope with inter-area oscillation of power system. Both local and PMU-provided remote signals, which are selected by analysis results based on participation phasor and residue, are utilized as feedback inputs of the controller. Mixed H_2/H_∞ output-feedback control design with pole placement is carried out. The feedback gain matrix is obtained through solving a family of LMIs. The design objective is to improve system damping of inter-area oscillations despite of the model changes which are caused mainly by load changes. Computer simulations on a 4-generator benchmark system model are carried out to illustrate the effectiveness and robustness of the designed controller, and the results are compared with the conventional PSS.

The rest of this chapter is organized as follows: In Section 2 a mixed H_2/H_∞ output-feedback control with pole placement design based on the mixed-sensitivity formulation is presented. The transformation into numerically tractable LMIs is provided in Section 3. Section 4 gives the benchmark power system model and carries out modal analyses. The synthesis procedures of wide-area robust damping controller as well as the computer simulations are presented in Section 5. The concluding remarks are provided in Section 6.

2. H_2/H_∞ Control with pole placement constrain

2.1 H_∞ mixed-sensitivity control

Oscillations in power systems are caused by variation of loads, action of voltage regulator due to fault, etc. For a damping controller these changes can be considered as disturbances on output **y** (Chaudhuri et al., 2003, 2004), the primary function of the controller is to minimize the impact of these disturbances on power system. The output disturbance rejection problem can be depicted in the standard mixed-sensitivity (**S**/**KS**) framework, as shown in Fig. 1, where sensitivity function $\mathbf{S}(s)=(\mathbf{I}-\mathbf{G}(s)\mathbf{K}(s))^{-1}$.

An implied transformation existing in this framework is from the perturbation of model uncertainties (e.g. system load changes) to the exogenous disturbance. Consider additive model uncertainty as shown in Fig. 2, The transfer function from perturbation **d** to controller output **u**, $\mathbf{T}_{ud}$, equals $\mathbf{K}(s)\mathbf{S}(s)$. By virtue of small gain theory, $\|\mathbf{T}_{ud}\Delta(s)\|_\infty<1$ if and only if $\|\mathbf{W}_2(s)\mathbf{T}_{ud}\|_\infty<1$ with a frequency-depended weighting function $|\mathbf{W}_2(s)|>|\Delta(s)|$. So a system with additive model uncertain perturbation (Fig. 2) can be transformed into a disturbance rejection problem (Fig. 1) if the weighted H_∞ norm of transfer function form **d** to **u** is small than 1, and the weighting function $\mathbf{W}_2(s)$ is the profile of model uncertainty.

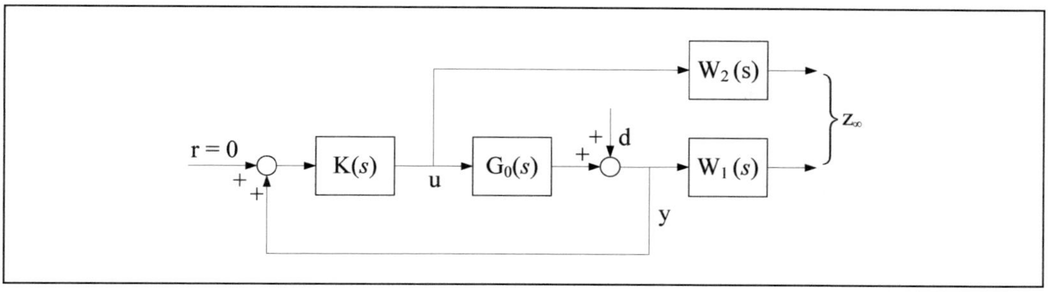

Fig. 1. Mixed sensitivity output disturbance rejection

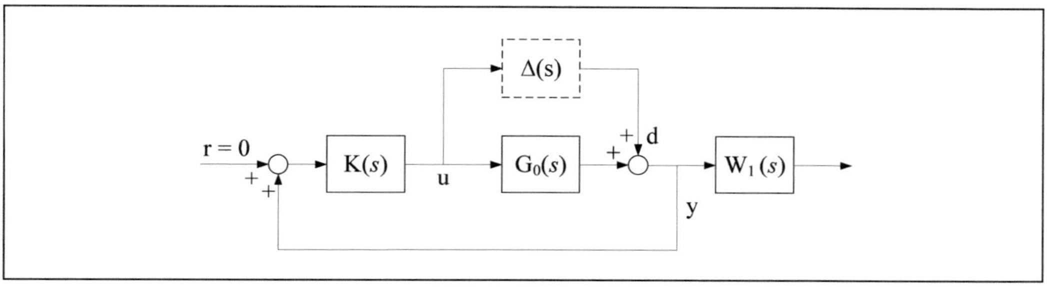

Fig. 2. System with additive model uncertainty

The design objective of standard mixed-sensitivity design problem, shown in Fig. 1, is to find a controller **K**(s) from the set of internally stabilizing controller $\mathbb{S}$ such that

$$\min_{K\in\mathbb{S}} \left\| \begin{bmatrix} \mathbf{W}_1(s)\mathbf{S}(s) \\ \mathbf{W}_2(s)\mathbf{K}(s)\mathbf{S}(s) \end{bmatrix} \right\|_\infty < 1 \tag{1}$$

In (1), the upper inequality is the constraint on nominal performance, ensuring disturbance rejection, the lower inequality is to handle the robustness issues as well as limit the control effort. Knowing that the transfer function from **d** to **y**, $\mathbf{T}_{yd}$, equals **S**(s). So condition (1) is equivalent to

$$\min_{K\in\mathbb{S}} \left\| \begin{bmatrix} \mathbf{W}_1(s)\mathbf{T}_{yd} \\ \mathbf{W}_2(s)\mathbf{T}_{ud} \end{bmatrix} \right\|_\infty < 1 \tag{2}$$

or

$$\min_{K\in\mathbb{S}} \left\| \mathbf{T}_{z_\infty d} \right\|_\infty < 1 \tag{3}$$

The system performance and robustness of controlled system is determined by the proper selection of weighting function $\mathbf{W}_1(s)$ and $\mathbf{W}_2(s)$ in (1) or (2). In the standard H_∞ control design, the weighting function $\mathbf{W}_1(s)$ should be a low-pass filter for output disturbance rejection and $\mathbf{W}_2(s)$ should be a high-pass filter in order to reduce the control effort and to ensure robustness against model uncertainties. But in some cases, there would be a low-pass requirement on $\mathbf{W}_2(s)$ when the open-loop gain is very high by applying standard lower-pass design, which will result in a conflict in the nature of $\mathbf{W}_2(s)$ to ensure robustness and minimize control effort (Pal et al., 2001). So the determination of $\mathbf{W}_2(s)$ should be careful.

2.2 H_2 performance for control cost requirement

It is known that the control cost can be more realistically captured through H_2 norm, see (Pal et al., 2001) and its reference, this enlightens directly adding H_2 performance on controller output $\mathbf{u}$ at the design stage, i.e. consider constraint

$$\left\| \mathbf{W}_3(s)\mathbf{T}_{ud} \right\|_2 < \gamma_2 \tag{4}$$

to constrain the control effort and mitigate the burden of selection of $\mathbf{W}_2(s)$. The weighting function $\mathbf{W}_3(s)$ is used to compromise between the control effort and the disturbance rejection performance, as shown in Fig. 3.

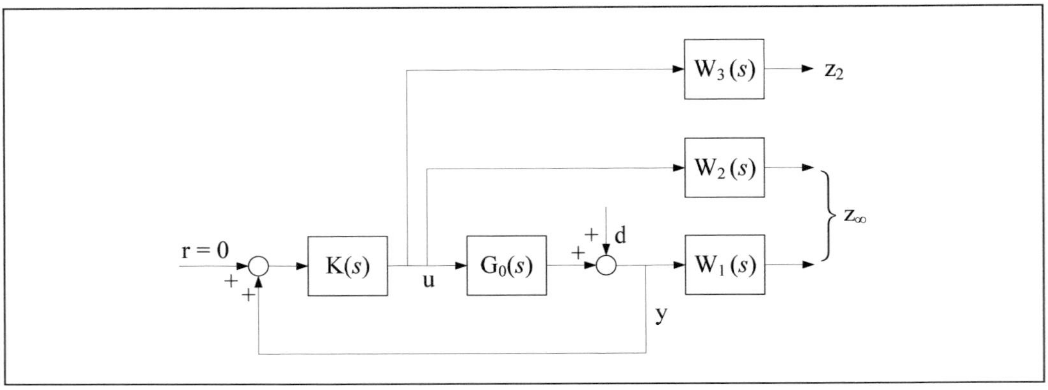

Fig. 3. Mixed sensitivity output disturbance rejection with other constraint

2.3 Pole placement constraint

H_2/H_∞ design deals mostly with frequency-domain aspects and provides little control over the transient behavior and closed loop pole location. Satisfactory time response and closed-loop damping can often be achieved by forcing the closed-loop poles into a suitable subregion of the left-half plane, and fast controller dynamics can also be prevented by prohibiting large closed-loop poles. Therefore, besides H_∞ and H_2 norm constraint, pole placement constraint that confine the poles to a LMI region is also considered.

A LMI region $S(a, r, \theta)$ is a set of complex number $x+jy$ such that $x<-a<0$, $|x+jy|< r$, and $\tan(\theta)x<-|y|$, as shown in Fig. 4. Confining the closed-loop poles to this region can ensure a minimum decay rate a, and minimum damping ratio $\zeta=\cos(\theta)$, and a maximum undamped natural frequency $\omega_d = r\sin(\theta)$. The standard mathematical description of LMI region can be found in (Chilali et al., 1996).

The multiple-objective design including H_∞/H_2 norm and pole placement constrains can be formulated in the LMIs framework and the controller is obtained by solving a family of LMIs.

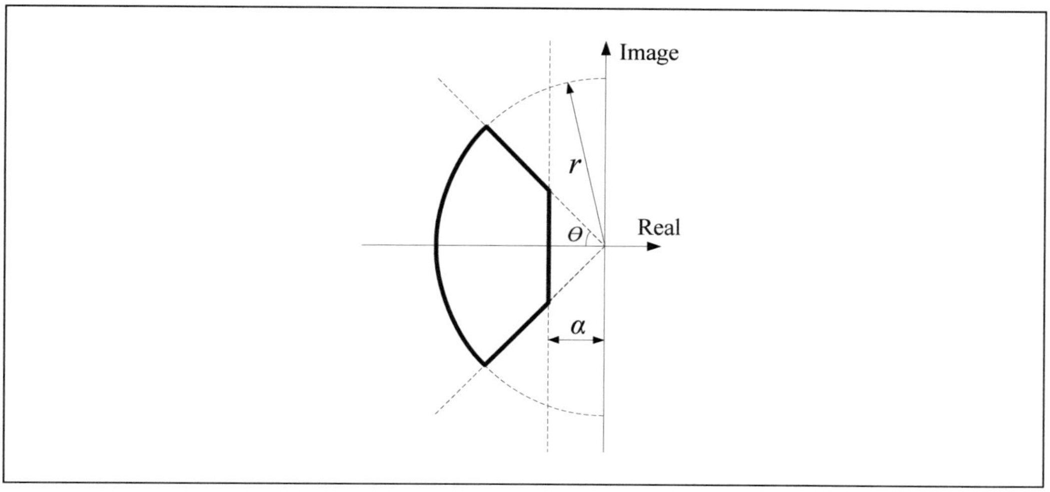

Fig. 4. LMI region $S(a, r, \theta)$

3. Multiple-objective synthesis using LMI method

General mixed H_2/H_∞ control with pole placement scheme has multi-channel form as shown in Fig. 5. **G**(s) is a linear time invariant generalized plant, **d** is vector representing the disturbances or other exogenous input signals, $\mathbf{z}_\infty$ is the controlled output associated with H_∞ performance and $\mathbf{z}_2$ is the controlled output associated with H_2 performance, **u** is the control input while **y** is the measured output.

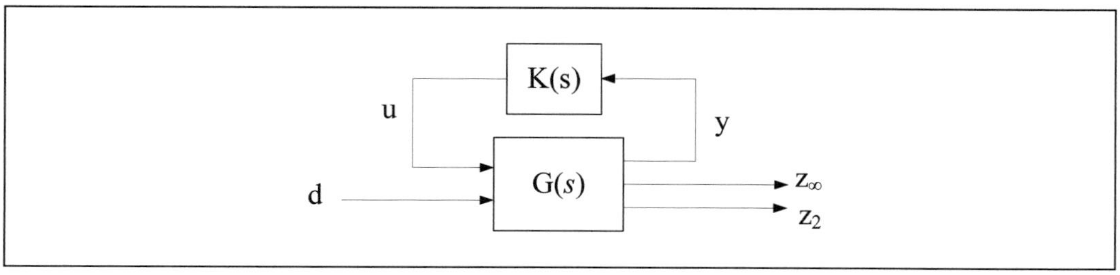

Fig. 5. Multiple-objective synthesis

The state-space description of above system can be written as

$$\left.\begin{aligned} \dot{\mathbf{x}} &= \mathbf{A}\mathbf{x} + \mathbf{B}_w\mathbf{d} + \mathbf{B}_u\mathbf{u} \\ \mathbf{z}_\infty &= \mathbf{C}_\infty\mathbf{x} + \mathbf{D}_{\infty 1}\mathbf{d} + \mathbf{D}_{\infty 2}\mathbf{u} \\ \mathbf{z}_2 &= \mathbf{C}_2\mathbf{x} + \mathbf{D}_{21}\mathbf{d} + \mathbf{D}_{22}\mathbf{u} \\ \mathbf{y} &= \mathbf{C}_y\mathbf{x} + \mathbf{D}_y\mathbf{d} \end{aligned}\right\} \tag{5}$$

The goal is to compute a output-feedback controller $\mathbf{K}(s)$ in the form of

$$\left.\begin{aligned} \dot{\zeta} &= \mathbf{A}_K\zeta + \mathbf{B}_K\mathbf{y} \\ \mathbf{u} &= \mathbf{C}_K\zeta + \mathbf{D}_K\mathbf{y} \end{aligned}\right\} \tag{6}$$

such that the closed-loop system meets mixed H_2/H_∞ specifications and pole placement constraint. The closed-loop system can be written as

$$\left.\begin{aligned} \dot{\mathbf{x}}_c &= \mathbf{A}_c\mathbf{x}_c + \mathbf{B}_c\mathbf{d} \\ \mathbf{z}_\infty &= \mathbf{C}_{c1}\mathbf{x}_c + \mathbf{D}_{c1}\mathbf{d} \\ \mathbf{z}_2 &= \mathbf{C}_{c2}\mathbf{x}_c + \mathbf{D}_{c2}\mathbf{d} \end{aligned}\right\} \tag{7}$$

By virtue of bounded real lemma (Boyd et al., 1994) and Schur's formula for the determinant of a partitioned matrix, matrix inequality condition (3) is equivalent to the existence of a symmetric matrix $\mathbf{X}_\infty>\mathbf{0}$ such that

$$\begin{pmatrix} \mathbf{A}_c\mathbf{X}_\infty + \mathbf{X}_\infty\mathbf{A}_c^{\mathrm{T}} & \mathbf{B}_c & \mathbf{X}_\infty\mathbf{C}_{c1}^{\mathrm{T}} \\ \mathbf{B}_c^{\mathrm{T}} & -\mathbf{I} & \mathbf{D}_{c1}^{\mathrm{T}} \\ \mathbf{C}_{c1}\mathbf{X}_\infty & \mathbf{D}_{c1} & -\mathbf{I} \end{pmatrix} < \mathbf{0} \tag{8}$$

The closed-loop poles lie in the LMI region (see Fig. 4) $S(0, 0, \theta)$ if and only if there exists a symmetric matrix $\mathbf{X}_D$ such that (Chilali et al., 1996):

$$\begin{pmatrix} \sin(\theta)(\mathbf{A}\mathbf{X}_D + \mathbf{X}_D\mathbf{A}^{\mathrm{T}}) & \cos(\theta)(\mathbf{A}\mathbf{X}_D - \mathbf{X}_D\mathbf{A}^{\mathrm{T}}) \\ \cos(\theta)(\mathbf{X}_D\mathbf{A}^{\mathrm{T}} - \mathbf{A}\mathbf{X}_D) & \sin(\theta)(\mathbf{A}\mathbf{X}_D + \mathbf{X}_D\mathbf{A}^{\mathrm{T}}) \end{pmatrix} < \mathbf{0} \tag{9}$$

For H_2 performance, $\|\mathbf{W}_3(s)\mathbf{T}_{ud}(s)\|_2$ does not exceed γ_2 if and only if $\mathbf{D}_{c2}=\mathbf{0}$ and there exist two symmetric matrices $\mathbf{X}_2>\mathbf{0}$ and $\mathbf{Q}>\mathbf{0}$ such that

$$\left.\begin{aligned} &\begin{pmatrix} \mathbf{A}_c\mathbf{X}_2 + \mathbf{X}_2\mathbf{A}_c^{\mathrm{T}} & \mathbf{B}_c \\ \mathbf{B}_c^{T} & -\mathbf{I} \end{pmatrix} < \mathbf{0} \\ &\begin{pmatrix} \mathbf{Q} & \mathbf{C}_{c2}\mathbf{X}_2 \\ \mathbf{X}_2\mathbf{C}_{c2}^{\mathrm{T}} & \mathbf{X}_2 \end{pmatrix} > \mathbf{0}, \mathrm{Trace}(\mathbf{Q}) < \gamma_2^{\,2} \end{aligned}\right\} \tag{10}$$

This condition can be deduced from the definition of H_2 norm (Chilali et al., 1996 ; Scherer et al., 1997). The multiple-objective synthesis of controller is through solving matrix inequality (8) to (10). But this problem is not jointly convex in the variable and nonlinear, for example nonlinear entry $\mathbf{A}_c\mathbf{X}_\infty$ in (8), so they are not numerically tractable. Choosing a single Lyapunov matrix $\mathbf{X}=\mathbf{X}_\infty=\mathbf{X}_2=\mathbf{X}_D$ and linearizing change of variables can cope with this problem. Choosing a single Lyapunov matrix makes the resulting controller not globally optimal, but is not overly conservative from the practical point of view. The linearizing change of variables is important for multiple-objective output feedback robust synthesis based LMIs. The details can be found in (Chilali et al., 1996 ; Scherer et al., 1997) and the references in them. Finally the result can be obtained as

$$\begin{cases} \min \ \mathbf{c}^T\mathbf{x} \\ \text{s.t. linearized LMIs constraints from (8) to (10)} \end{cases} \tag{11}$$

This standard LMI problem (Boyd et al., 1994) is readily solved with LMI optimization software. An efficient algorithm for this problem is available in *hinfmix*() function of the LMI control toolbox for Matlab (Gahinet et al., 1995).

4. A Benchmark system with undamped inter-area oscillation

4.1 Low frequency oscillation in power system

One of the major problems in power system operation is low frequency (between 0.1 and 2 Hz) oscillatory instability. Normally no apparent warning can be identified for the occurrence of such kinds of growing oscillations caused by the changes in the system's operating condition or by improper-tuned sustained excitation.

The change in electrical torque of a synchronous machine following a perturbation can be resolved as $\Delta T_e = T_S\Delta\delta + T_D\Delta\omega$, where $T_S\Delta\delta$ is the component of torque change in phase with the rotor angle perturbation $\Delta\delta$ and is referred as the synchronizing torque component, T_S is the synchronizing torque coefficient. Lack of sufficient synchronizing torque will result in aperiodic drift in rotor angle. $T_D\Delta\omega$ is the component of torque in phase with the speed deviation $\Delta\omega$ and is referred to as the damping torque component, T_D is the damping torque coefficient. Lack of sufficient damping torque will result in oscillatory instability.

In next section, an example will be used to illustrate the low frequency oscillation of a weak-tied system and the design of a wide-robust damping controller (WRC) to effectively increase the damping ratio of inter-area mode.

4.2 System model and modal analysis

A 4-generator benchmark system shown in Fig. 6 is considered. The system parameters is from (Klein et al., 1991) or (Kundur, 1994). However some modifications have been made to facilitate the simulations. The generator G2 is chosen as angular reference to eliminate the undesired zero eigenvalues. Saturation and speed governor are not modeled. Excitation system is chosen by thyristor exciter with a high transient gain. All loads are represented by constant impedance model and complete system parameters are listed in Appendix.

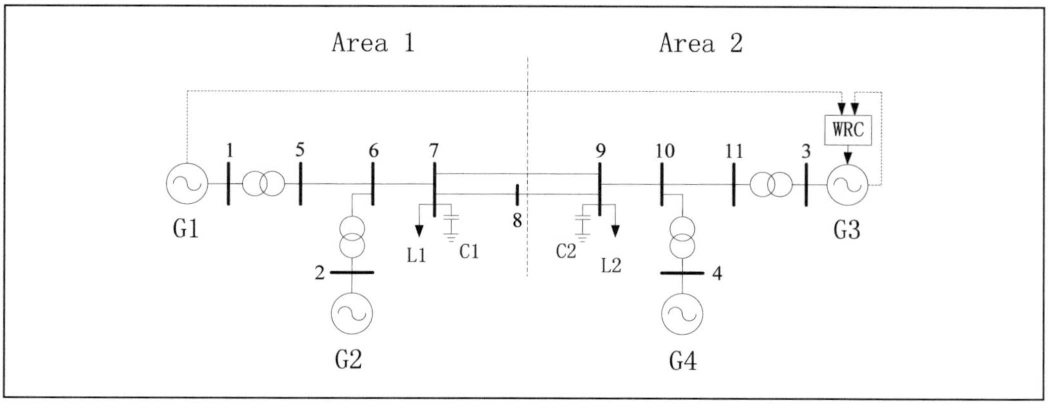

Fig. 6. 4-generator benchmark system model

After linearization around given operating condition and elimination of algebraic variables, the following state-space representation is obtained.

$$\left.\begin{aligned}\dot{\mathbf{x}}&=\mathbf{A}\mathbf{x}+\mathbf{B}_u\mathbf{u}\\ \mathbf{y}&=\mathbf{C}_y\mathbf{x}\end{aligned}\right\} \tag{12}$$

where $\mathbf{x}$ is state vector; $\mathbf{u}$ is input vector, $\mathbf{y}$ is output vector; $\mathbf{A}$ is the state matrix depending on the system operating conditions, $\mathbf{B}_u$ and $\mathbf{C}_y$ are input and output matrices, respectively. The number of the original state variables is 28, since generator 2 has been chosen as angular reference, 2 sates are eliminated, so the number of state variables is 26.

Following the small-signal theory (Kundur, 1994), the eigenvalues of the test system and corresponding frequencies, damping ratios and electromechanical correlation ratios are calculated. The results are classified in Table 1. It can be found that mode 3 is undamped, which means that the disturbed system can not hold transient stability.

The electromechanical correlation ratio in Table 1 is determined by a ratio between summations of eigenvectors relating to rotor angle and rotor speed and summations of other eigenvectors. If the absolute value of one entry (correlation ratio) is much higher than 1, the corresponding mode is considered as electromechanical oscillation.

No.	Mode	Frequency (Hz)	Damping Ratio (%)	Electromechanical Correlation Ratio
1	−0.7412±6.7481	1.0740	0.1092	5.7087
2	−0.7154±6.9988	1.1139	0.1017	5.6918
3	**0.0196±3.9141**	**0.6229**	**−0.0050**	**13.2007**

Table 1. Results of Modal Analysis

A conception named participation phasor is used to facilitate the positioning of controller and the selection of remote feedback signal. Participation phasor is defined in this easy way: its amplitude is participation factor (Klein et al., 1991; Kundur, 1994) and its phase angle is angle of eigenvector. The analysis results are shown in Fig. 7, in which all vectors are originated from origin (0, 0) and vector arrows are omitted for simplicity.

It can be seen that

- Mode 1 is a local mode between G1 and G2. The Participation phasor of G3 and G4 are too small to be identified;
- Mode 2 is a local mode between G3 and G4. The Participation phasor of G1 and G2 are too small to be identified;
- Mode 3 is an inter-area mode between G1, G2 and G3, G4.

Wide-area controller is located in G3, which has highest participation factor than others. Even if using local signal only, the controller locating in G3 will have more effects than locating in other generators.

Often the residue indicates the sensitivity of eigenvalues to feedback transfer function (Rogers, 2000), that is to say if residue is 0 then feedback control have no effects on controlled system, so residue is used to select suitable remote feedback signal provided by PMU. The residue corresponding to the transfer function between rotor speed output of G1 and excitation system input of G3 is 1.58 (normalized value), while the residue corresponding to the transfer function between rotor speed output of G2 and excitation system input of G3 is 1 (normalized value). So the remote signal is chosen from G1.

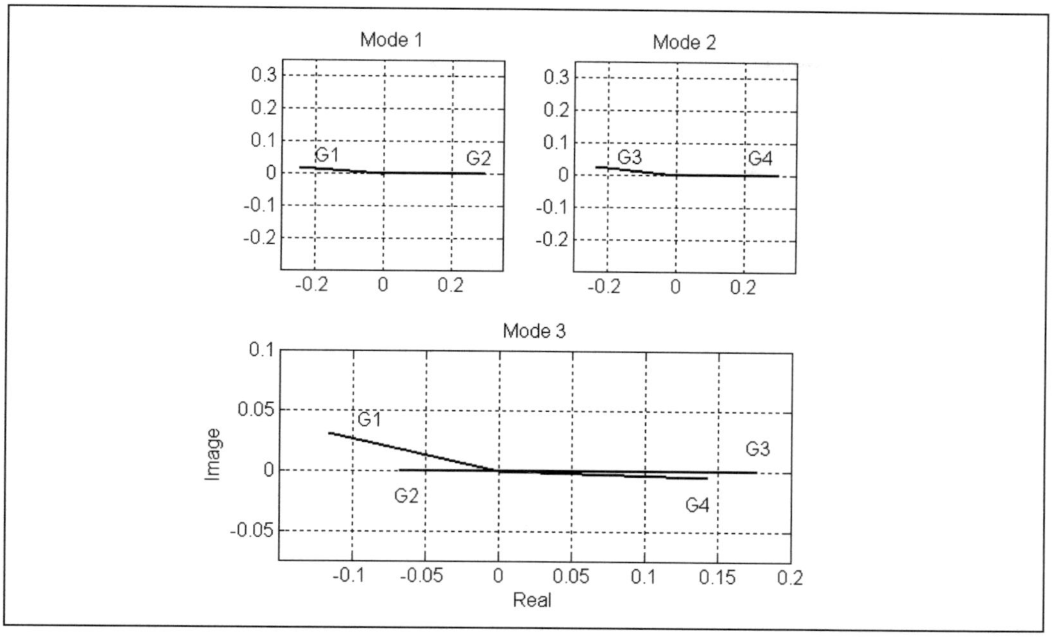

Fig. 7. Participation phasors of considered power system

The positioning of controller and the selection of signals are shown in Fig. 6. Both local and remote feedback signals are rotor speed deviation $\Delta\omega$, in this way the component of torque (see in section 4.1) can be increased directly, and controller output $\boldsymbol{u}$ of WRC is an input to the automatic voltage regulators (AVRs) of G3. The configuration of WRC, excitation system and voltage transducer is shown in Fig. 8.

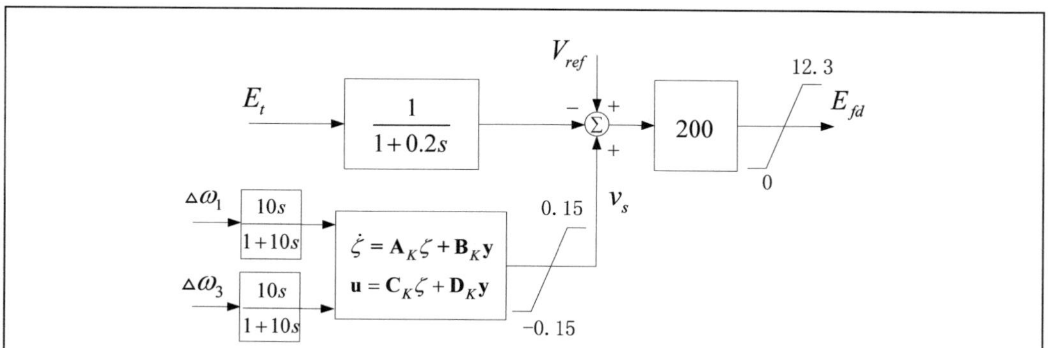

Fig. 8. The configuration of WRC, excitation system and voltage transducer

5. Wide-area robust damping controller design

5.1 Designprocedure

The basic steps of controller design are summarized as below.

(1) Reduce the original system model through Schur balanced truncation technique (Zhou et al., 1998), a reduced 9-order system model can be obtained. The frequency responses of

original and reduced model are compared in Fig. 9, it shows that reduced system has proper approximation to original system within considered frequency range.

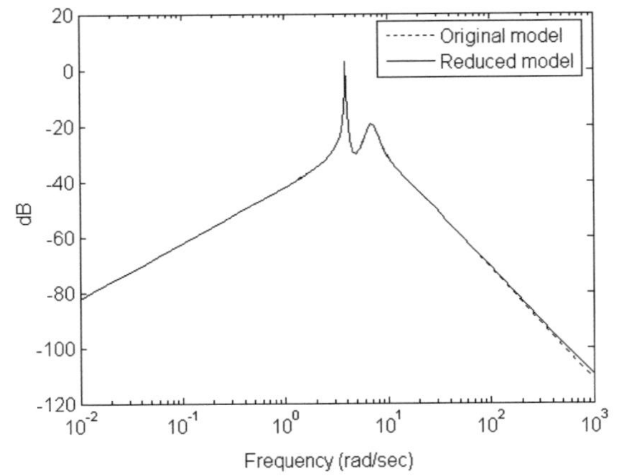

Fig. 9. Frequency response of original system model and reduced system model

(2) Formulate the generalized plant in Fig. 5 using the reduced model and the weighting function. The weighting functions are chosen as follows:

$$\mathbf{W}_1(s) = \frac{80}{s+41}, \quad \mathbf{W}_2(s) = \frac{8.6s+4}{s+4}, \quad \mathbf{W}_3(s) = 1 \tag{13}$$

The weighting functions are in accordance with the basic requirements of mixed-sensitivity design. $\mathbf{W}_1(s)$ is a low-pass filter for output disturbance rejection, $\mathbf{W}_2(s)$ is a high-pass filter for covering the additive model uncertainty, and $\mathbf{W}_3(s)$ is a weight on H_2 performance.
(3) Controller design by using the Robust Control Toolbox in Matlab. The solution is numerically sought using suitably defined objectives in the arguments of the *hinfmix*() function of the Robust Control Toolbox. The LMI region is chosen as a conic sector with inner angle equals 2*acos(0.17) (corresponding damping ratio 17%) and apex at the origin.
(4) Controller reduction through Schur balanced truncation technique. A 4-order 2-input 1-output controller is obtained. The state-space representation of the designed controller is

$$\mathbf{A}_K = \begin{pmatrix} -5.4 & 11.6 & -4.0 & -0.1 \\ -1.9 & -14.2 & 16.1 & -0.5 \\ -9.3 & 25.2 & -14.5 & 0.0 \\ -3.3 & 125.6 & -10.4 & -2.7 \end{pmatrix}, \quad \mathbf{B}_K = \begin{pmatrix} -0.36 & 6.66 \\ -0.36 & -8.82 \\ 0.72 & 14.94 \\ 8.1 & 25.52 \end{pmatrix}$$

$$\mathbf{C}_K = \begin{pmatrix} -19.5 & -25.7 & 60 & 2.1 \end{pmatrix}, \quad \mathbf{D}_K = \begin{pmatrix} 0 \\ 0 \end{pmatrix}.$$

A washout filter 10s/(10s+1) is added in each feedback channel as shown in Fig. 8. This is a standard practice to prevent the damping controllers from responding to very slow variations in the system conditions (Kundur, 1994). A limit of [—0.15, 0.15] (pu) is imposed on the output of the designed controller.

5.2 Computer simulations and robustness validation

Computer simulations are carried out to test the effectiveness and performance of the designed controller and validate the robustness in different operating conditions. The simulation is carried out by Matlab-Simulink.

A 5%-magnitude pulse, applied for 12 cycles at the voltage reference of G1, is used to simulate the modes of oscillation. For comparison, one conventional PSS is also considered. The PSS has one gain, one washout and two phase compensations, the block diagram representation of the conventional PSS is shown in Fig. 10. The parameters are adopted directly from (Kundur, 1994).

$$\Delta\omega \rightarrow \boxed{20} \rightarrow \boxed{\frac{10s}{1+10s}} \rightarrow \boxed{\frac{1+0.05s}{1+0.02s}} \rightarrow \boxed{\frac{1+3.0s}{1+5.4s}} \rightarrow v_s$$

Fig. 10. Block diagram of conventional PSS

Figure 11 shows the tie line (transmission lines between bus 7 and bus 9 in Fig. 6) active power response to the pulse disturbance without any damping controller (with only AVRs in each generator). It shows that the open-loop system oscillates and is unstable.

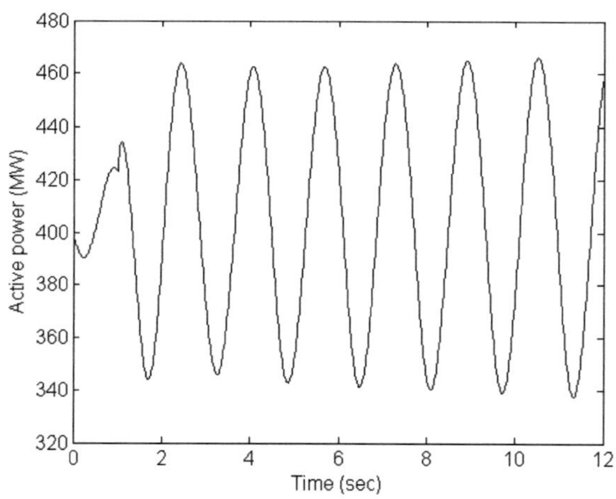

Fig. 11. Tie line active power response with AVRs only

The pulse response with the designed WRC is shown in Fig. 12, which is compared with the response with one conventional PSS located in G3. The state variable is the tie line active power. Both of the damping controllers can ensure the system asymptotic stable but better damping performance is achieved by the WRC.

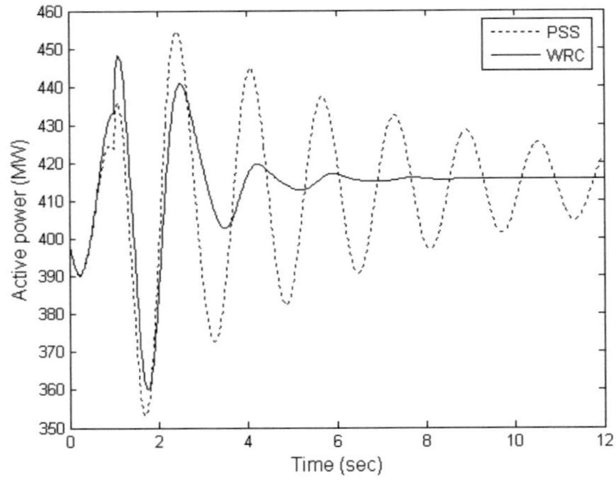

Fig. 12. Tie line active power response with one PSS and the WRC

Figure 13 shows the pulse responses of the system in the cases of open-loop, controlled by one PSS and by the WRC. The state variables in this figure are the rotor speeds of all the generators. The inter-area mode oscillation between G1, G2 and G3, G4 can be clearly identified from the open-loop responses. The rotor speed response of the designed controller shows better damping performance than that of conventional PSS.

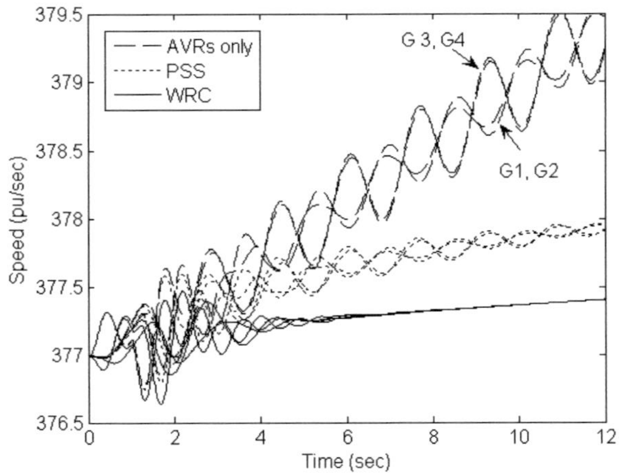

Fig. 13. Rotor speed responses of all the generators with AVRs only, one PSS and the WRC

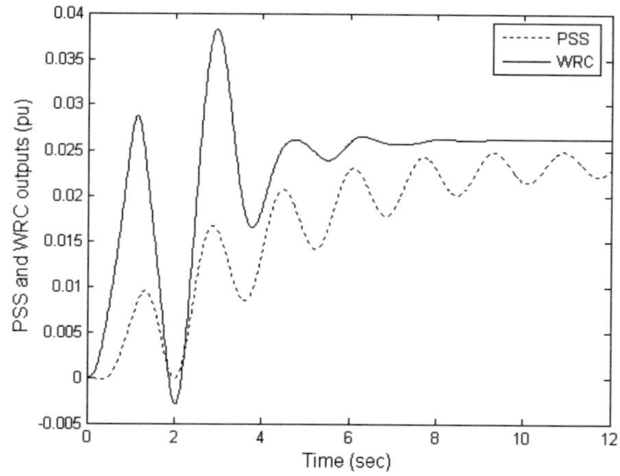

Fig. 14. Outputs of PSS and WRC

Figure 14 shows the outputs of the PSS and the WRC, the WRC show better transient performance and its output is not higher than 0.04 pu.
To test the robustness of the designed controller to changes of operating conditions (or model uncertainties), load changes are considered. Eight different operation conditions are considered, corresponding load L_1 and L_2 in normal conditions and change between ±5%and ±10%, respectively. The load change, making the tie line power change, is the primary factor affecting the eigenvalues of the matrix **A** (also the damping ratios) in system model (12), and also used to select the weighting function $\mathbf{W}_2(s)$. Fig. 15 shows the frequencies and damping ratios corresponding to these changes. The horizontal axis is the load changes

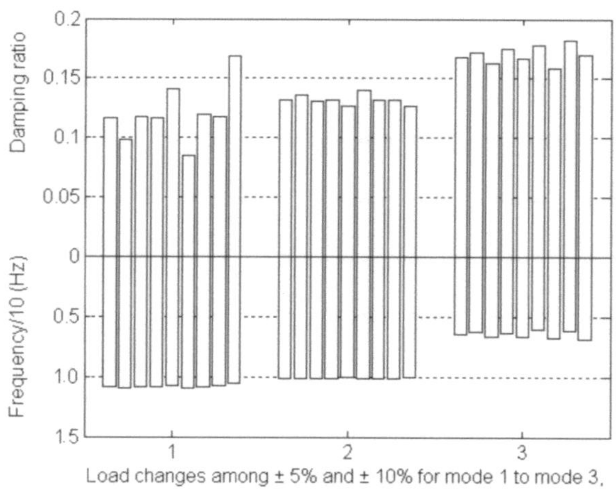

Fig. 15. Damping ratios and frequency corresponding to load change for mode 1 to mode 3

(including the nominal operating condition) for mode 1 to mode 3. The upper vertical axis is the damping ratios corresponding to each load change, the lower vertical axis is the frequencies corresponding to each damping ratio. For inter-area mode, mode 3, the damping ratios are higher than 0.15 in all these load levels. The damping rations of the whole system is higher than 0.08 in all cases. The controlled system has proper damping performance and keeps robustness against the variations of system loads.

6. Conclusion

This chapter applies robust control theory to power system, to design wide-area robust damping controller to cope with inter-area oscillation. Both local signal and suitable chosen PMU-provided remote signal are utilized to construct the feedback loop. A conception named participation vector is used for facilitating the positioning of controller, and the residue is utilized to select suitable remote signal. The controller is designed based on mixed H_2/H_∞ output-feedback control with pole placement, and the controller parameters are obtained through solving a family of linear matrix inequalities. The designed controller is applied on a 4-generator power system model. The computer simulations are performed for pulse disturbance as well as system operating changes. The designed controller shows better damping than conventional PSS and keeps robustness with load variations.

7. Appendix: Benchmark system model parameters

Synchronous machine data (pu)

X_d=1.8, X_d'=0.3, X_d''=0.25, X_q=1.7, X_q'=0.55, X_q''=0.25, X_l=0.2, R_a=0.0025, T_{d0}'=8, T_{d0}''=0.03, T_{q0}'=0.4, T_{q0}''=0.05, H_1=6.5, H_2=6.175.

Transmission system data in per unit

r=0.0001, x_L=0.001, b_C=0.0018, x_T=0.15.

Excitation system data (pu)

K_A=200, T_R=0.01, E_{FMAX}=12.3, E_{FMIN} =0.

Generation (power flow results calculated by Matlab-Simulink) (MW, MVar)

G1: P=700, Q=146.5; G2: P=678.9, Q=137.3;
G3: P=719, Q=138.1; G4: P=700, Q=109.1.

Load model (MW, MVar)

L1: P_L= 967, Q_L=100, Q_C=187;
L2: P_L= 967, Q_L=100, Q_C=187.

Shunt capacitor: (MVar)

C1: Q_C=100; C2: Q_C=250.

8. Acknowledgment

The authors thank Prof. Guangxiong Wang and Prof. Huijun Gao, Harbin Institute of Technology, for their kindly discussions on robust control theory.

9. References

Klein, M.; Rogers, G. J. & Kundur, P. (1991). A Fundamental Study of Inter-Area Oscillations in Power Systems. *IEEE Trans. Power Syst.*, Vol. 6, No. 3, pp. 914-921, ISSN: 0885-8950

Kundur, P. (1994). *Power System Stability and Control.* McGraw-Hill Professional, ISBN 007035958X, New York, USA

Rogers, M. G. (2000). *Power System Oscillations.* Springer, ISBN 978-0-7923-7712-2, Boston, USA

Zhou, X. X.; Yi, J.; Song, R. H.; Yang, X. Y.; Lia, Y. & Tang, H. Y. (2010) et al, An Overview of Power Transmission Systems in China, Energy, Vol. 35, Issue 11, pp. 4302-4312, ISSN: 1540-7977

Larsen, E. V. & Swann, D. A. (1981). Applying Power System Stabilizers, Part I-III. *IEEE Trans. Power Apparat. Syst.*, Vol. PAS-100, Issue.6, (February 1981), pp. 3017 - 3024, ISSN: 0018-9510

Farsangi, M. M.; Song, Y. H. & Tan, M.(2003). Multi-Objective Design of Damping Controllers of Facts Devices via Mixed H_2/H_∞ With Regional Pole Placement. *Electrical Power and Energy Systems*, Vol. 25, pp. 339-346, ISSN: 0142-0615

Pal, B. C.; Coonick, A. H. & Cory, B. J. (2001). Linear Matrix Inequality Versus Root-Locus Approach for Damping Inter-Area Oscillations in Power Systems, *Electrical Power and Energy Systems*, Vol. 23, No. 6, pp. 481-489, ISSN: 0142-0615

Chaudhuri, B.; Pal, B. C.; Zolotas, A. C.; Jaimoukha, I. M. & Green, T. C. (2003). Mixed-Sensitivity Approach to H_∞ Control of Power System Oscillations Employing Multiple Facts Devices, *IEEE Trans. Power Syst.*, Vol. 18, No. 3, pp. 1149-1156, ISSN: 0885-8950

Chaudhuri, B. & Pal, B. C. (2004). Robust Damping of Multiple Swing Modes Employing Global Stabilizing Signals With a TCSC, *IEEE Trans. Power Syst.*, Vol. 19, No. 1, pp. 499-506, ISSN: 0885-8950

Snyder, A. F.; Hadjsaid, N.; Georges, D.; Mili, L.; Phadke, A. G.; Faucon, O. & Vitet, S. (1998). Inter-Area Oscillation Damping With Power System Stabilizers and Synchronized Phasor Measurements, *Proceedings* of *International Conference on Power System Technology*, pp.790-794, ISBN: 0-7803-4754-4, Beijing , China, August 1998.

Kamwa, I.; Grondin, R.; & Hebert, Y. (2001). Wide-Area Measurement Based Stabilizing Control of Large Power Systems —A Decentralized/Hierarchical Approach, *IEEE Trans. Power Syst.*, Vol. 16, pp. 136-153, ISSN: 0885-8950

Phadke, A. G. (1993). Synchronized Phasor Measurement in Power Systems, *IEEE Computer Applications in Power*, Vol. 6, No. 2, pp. 10-15, ISSN: 0895-0156

Doyle, J. C.; Glover, K.; Khargonekar, P. P. & Francis B. A. (1989). State-space solutions to standard H_2 and H_∞ control problems, *IEEE Trans. Automat. Contr.*, Vol. 34, No. 8, pp. 831–847, ISSN: 0018-9286

Zhou, K.; Doyle, J. C. (1998). *Essentials of Robust Control.* Prentice Hall, ISBN 0-13-525833-2, New Jersey, USA

Chilali, M.; Gahinet, P. (1996). H_∞ Design With Pole Placement Constraints: An LMI Approach, *IEEE Trans. Automat. Contr.*, Vol. 41, No. 3, pp. 358-367, ISSN: 0018-9286

Boyd, S.; Ghaoui, L. El; Feron, E. & Balakrishnan, V. (1994). *Linear Matrix Inequalities in System and Control Theory,* Volume 15 of Studies in Applied Mathematics. SIAM, ISBN 0-89871-334-X, Philadelphia, PA, June 1994.

Scherer C. W. & Weiland S (2005). Lecture notes DISC course on linear matrix inequalities in control-2004/2005. Delft University of Technology, Netherlands. Compilation: pp. 70-71. [Online]. Available: http://www.cs.ele.tue.nl/sweiland/lmi.html

Scherer, C. W.; Gahinet, P.; Chilail, M. (1997). Multiobjective Output- Feedback Control via LMI Optimization, *IEEE Trans. Automat. Contr.*, Vol. 42, No. 7, pp. 896-911, ISSN: 0018-9286

Nesterov, Y. & Nemirovski, A. (1994). *Interior Point Polynomial Algorithms in Convex Programming*. SIAM publications, ISBN: 0898715156, Philadelphia, USA

Gahinet, P.; Nemirovskii, A.; Laub, A. J. & Chilali, M. (1995). *LMI Control Toolbox*. The MathWorks Inc, 1995.

Part 5

Selected Issues and New Trends in Robust Control Applications

16

Robust Networked Control

Wojciech Grega
Department of Automatics, AGH University of Science and Technology
Poland

1. Introduction

Most of integrated industrial control systems adopt a multilevel, vertical control hierarchy. Logically, such a system (Fig. 1) is structured in three levels: the direct (device) control level, the supervisory level and the management level (Grega, 2010, Tatjewski, 2007, Grega *at al.*, 2009).

The basic task of the direct (device) control level is to maintain the process states at the prescribed set values. The device controller level provides an interface to the hardware, either separate modules or microprocessors incorporated in the equipment to be controlled. Here, mainly PID digital control algorithms are implemented - in some cases these are more advances control methods such as multivariable control or adaptive functions. A number of embedded control nodes and Programmable Logical Controllers (PLC) are used as the front-ends to take the control tasks. High speed networks and fieldbuses are implemented at the direct control level to exchange in real time the information between front-ends and the device controllers and, vertically, with the supervisory control level. This architecture has the advantage of locating the hard real-time activities as near as possible to the equipment.

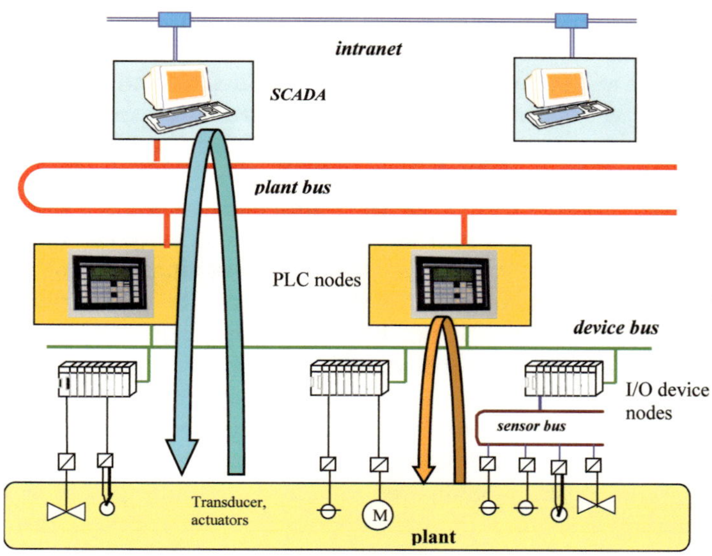

Fig. 1. Multilevel structure of an industrial control system

The supervisory level comprises workstations and industrial PCs providing high-level control support, database support, graphic man-machine interface, network management and general computing resources. Classically, the supervisory level calculates set points for controllers according to the defined criteria. For this purpose more complex mathematical models of the process are employed at this level to find the optimal steady-state, by solving optimisation and identification tasks. Due to the rapid development of control technology, there is growing scope for more advanced close-loop algorithms (predictive control, repetitive control) located at this level. However, increasing computational efficiency of PLCs at the device level supported by high performance networks transferring data and control signals vertically gives more flexibility to the designer. The control loops can be handled by local, device–level controllers, and also by the supervisory controllers (Fig.1). For example, a predictive control algorithm can be handled by a supervisory workstation as well as by a local PLC. It should be noted that upper level loops usually offer shorter computational time due to the higher efficiency of the workstations.

Feedback control systems wherein the control loops are closed through a communication network are referred to as Distributed Control Systems (DSC). They are distributed in the sense that their sensors, actuators and controllers (referred as "nodes") communicate via a shared data transmission network. The behaviour of a networked control system depends on the performance parameters of the underlying network, which include transmission rate and access method to the network transmission medium.

Communication networks were introduced in control in the 1970s. They can be grouped into fieldbuses (e.g. CAN, Profibus, Modbus) and general purpose networks (e.g. IEEE standard LANs), (Zurawski, 2005). Each type of network has its own protocol that is designed for a specific range of applications. Fieldbuses are intended for real-time applications. The most important feature of these industrial networks is that they guarantee bounded transmission delays. More and more popular is application of general-purpose networks, inexpensive and easy to maintain. Ethernet is a solution, which seems to become an industrial standard in the near future (Felsner, 2005).

The advantages of data transmission channels integration into control system are obvious, such as reducing wiring costs and increasing flexibility. Thanks to these important benefits, typical applications of these systems range over various fields, such as automotive, mobile robotics, advanced aircraft, and so on. However, introduction of communication networks in the control loops makes the analysis and synthesis of distributed control systems more complex.

DCS can be considered a special case of digital control systems, as data is sent through the network periodically, in units called packages. Therefore, any signal continuous in time must be sampled to be carried over the network. Real-time assumptions are as important for DCS as for any other computer controlled systems. Hence, there are similarities between DCS and real-time digital control systems due to sampling effects. The most challenging problem with DSC that needs to be properly addressed are time delays. A network induced delays occurs while sending data among nodes connected to the shared data transmission medium of limited throughput. Network-induced delays may vary depending on the network load and Medium Access Protocol (MAC). Lack of access to the communication network is an important constraint compared to lack of computer power or time errors of the real-time operating system. It is well known that time delays can degrade the performance of the control system or even destabilize the system.

Especially, the following effects are observed in DCS:

- variable computation-induced delays,
- variable network induced delays,
- data loss, caused by packet dropouts.

resulting in:

- violation of the assumption that sampling/actuation intervals are evenly spaced,
- violation of the causality principle.

From the point of view of control theory networked control often introduces some additional dynamics and temporal non-determinism. Therefore, novel methodologies should be developed for stability analysis of DCS and optimise the performance. An integrated approach is necessary, that combines data transmission issues (modelling of variable communication delays), sampling theory and control theory.

The notion of robustness of various DCS properties (especially stability) plays an important role in design of control systems, as confirmed by extensive literature discussion (Walsh *et al*, 2002, Gupta and Chow, 2010). Very general formulation of robustness for DCS is illustrated in Fig. 2. As it was mentioned before, DCS can be considered as a special case of digital control systems. Therefore, it is sensitive to the sampling period T_0 variations. For non-networked digital control system the quality of control generally increases while T_0 is getting shorter. This must not be true for DCS. Increasing network traffic results in longer and variable network-induced delays, and leads to the deterioration of control quality. In this case robust design means shifting the DCS quality characteristic as close as possible to the characteristics of digital (non-network) control system.

During last 20 years various methods have been developed to maintain the stability and the performance of DCS with delay problems. In order to enhance robustness of DCS against network induced delays appropriate methods of control theory are supplemented by some methods of network traffic engineering. Therefore, two main research approaches can be distinguished (Gupta and Chow, 2010).

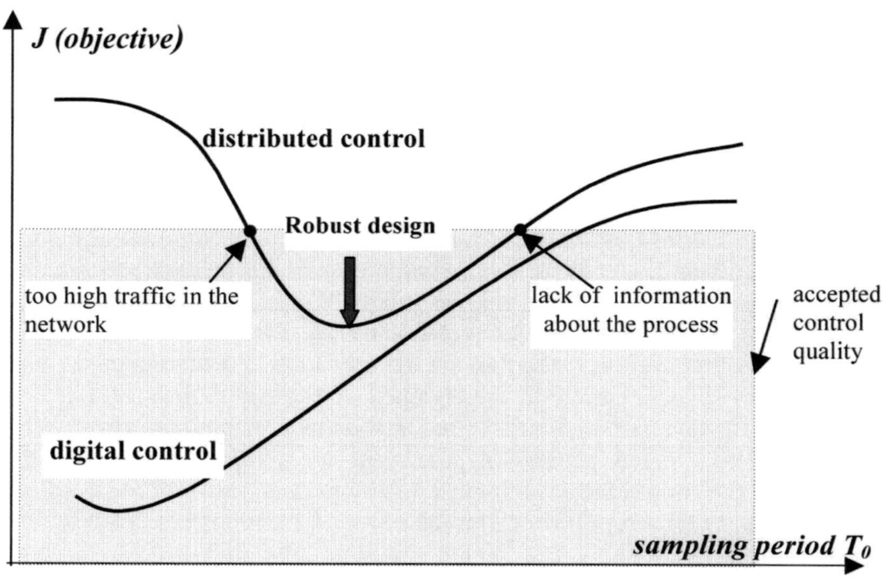

Fig. 2. Control quality versus sampling period

Study and research on communications and networks to make them suitable for real-time DCS, e.g. routing control, real-time protocols, congestion reduction, real-time protocols, codesign of networking and controllers are referred as *Control of network.*

Developing of control strategies and control systems design over the network to minimize the effect of adverse network parameters on DCS performance, such as network delay is referred as *Control over network.* The main advantage of this approach is its simplicity: the designer of DCS can exploit standard control algorithms and make them robust against effects of networking.

Following the *Control of network* approach, effects of the network configuration on the performance of the control system have been studied and different improvements have been proposed. At the physical level the network topology cannot be chosen freely but is subject to many practical constraints such as cost and reliability considerations. For example, the real-time performance of industrial Ethernet network depends strongly on the way the devices are allocated to the individual switches in the network. Therefore, the problem of optimal device allocation in industrial Ethernet networks with real-time constraints remains an important topic (Georges *et al*, 2006).

Another concept was to modify scheduling methods and communication protocols in such a way that data delays are minimized. Several solutions have been proposed. The most interesting of these involve:

- a new scheduling strategies based on a time division (Al-Hammouri *et al*, 2006),
- obtaining a maximum allowable delay bound for DCS scheduling (Walsh *et al*, 2002),
- adjustment of the network parameters (link quality measures) to the control quality,
- measures, by studying impact of frames priorities (Juanole *et al*, 2006).

Desire to incorporate a real-time element into some popular single-network solution has led to the development of different real-time Industrial Ethernet solutions, called *Real-time Ethernet*.

If the second approach is implemented (*Control over network*), the network is considered as a passive component of feedback loop, modeled in a simplified way. In most cases the control theory of delayed systems can be applied to compensate the effects of communication in order to guarantee the Quality of Control (QoC), (Hirai, 1980).

Network delays can be modeled and analyzed in various ways. They can be modeled as a constant delay (timed buffers), independent random delay and delay with known probability distribution, governed by Markov chain model.

One of the first applications taking the randomness of the network into account, either as a constant probability function or as a Markov chain together with time stamping was thesis of Nilson (Nilsson, 1998). Later, the optimal stochastic methods approached the problem as a Linear-Quadratic-Gaussian (LQG) problem where the LQG gain matrix is optimally chosen based on the network delay statistics (Nilsson *et al*, 1998).

One simple idea is that constant delay in the control loop is better than variable delay. Introducing buffers reduces temporal dependency of the individual components of the close-loop model. The data package is delivered as soon as possible, but is hold in the buffer and is implemented to the process in the next sampling intervals. By this way, synchronisation of the control loop is achieved. Constant delay can be compensated using a standard approach, e.g. Smith predictor. It must be noted that constant delay buffer usually creates conservative controller gains. Better solutions give applications of switched or variable delay buffer. The stability analysis of the switched buffer model can be reduced to the problem of stability of the *Asynchronous Dynamical Systems* (ASD) , (Hassibi, 1999).

Smith Predictor-based approach was proposed by several authors (Vatanski *et al.*, 2009) for the control in the case when accurate delay measurements are accessible. In contrast to the robust control-based approach when only the estimate of the upper-bound end-to-end delays are available (Grega, 2002).
Other concept is to increase network utilization by modification of the transmission pattern - by samples grouping. The samples from sensor are transferred through network, however they are grouped together into *M*-element packages before they enter the network. Grouping effects can be compensated by an approximate model of the process ("observer") at the controller side, and by control signal estimator (output to actuators) for some range of the sampling period and modeling errors (Grega and Tutaj, 2007).
Finally, network observers and state observes can be applied. The idea is that the communication delays between the sensor and the controller can be compensated by an approximate (non-exact) model of the process at the controller side, for some range of the sampling period and modelling errors. The performance of the method greatly depends on the model accuracy (Montestruque *et al.*, 2003).
An intelligent control was proposed using fuzzy logic to adaptively compensate network induced time delay in DCS applications (Cao and Zhang, 2005). The advantage of the fuzzy logic compensator is that the existing PI controller needs not to be redesigned, modified, or interrupted for use on a network environment.

2. Control of the network

2.1 Optimizing protocols

The idea is to implement communication protocols and network topology that minimise data delays. Current communication systems for automation implement different protocols. This is a substantial disadvantage, leading to the need to use vendor-specific hardware and software components, which increase installation and maintenance costs. Moreover, presently used fieldbus technologies make vertical communication across all levels of the automation systems difficult. Gateways need to be used to establish connections between different kinds of fieldbus systems used in the lower levels, and Ethernet used in the upper level.
The evolution of industrial communication has moved to Industrial Ethernet networks replacing the proprietary networks (Larson, 2005, ARC Advisory Group, 2007). Ethernet provides unified data formats and reduces the complexity of installation and maintenance, which, together with the substantial increase in transmission rates and communication reliability over the last few years, results in its popularity in the area of industrial communications.
Ethernet, as defined in IEEE 802.3, is non-deterministic and, thus, is unsuitable for hard real-time applications. The media access control protocol, CSMA/CD can not support real-time communication because back-off algorithm for collision resolution is used. With CSMA/CD it can not be determine in advance how long the collision resolution will take. It was explained before, that delays and irregularities in data transmission can very severely affect real-time system operation. Therefore, various techniques and communication protocol modifications are employed in order to eliminate or minimise these unwanted effects and make the data transmission system time invariant.
To employ Ethernet in an industrial environment, its deterministic operation must first be assured. Coexistence of real-time and non-real time traffic on the same network infrastructure remains the main problem. This conflict can be resolved in several ways by:

- embedding a fieldbus or application protocol on TCP(UDP)/IP – the fieldbus protocol is tunneled over Ethernet, and full openness for "office" traffic is maintained,
- using a special Data Link layer for real-time devices – dedicated protocol is used on the second OSI Layer, implemented in every device. The real-time cycle is divided into slots, one of which is opened for regular TCP/IP traffic, but the bandwidth available is limited,
- using application protocol on TCP/IP, direct MAC addressing with prioritization for real-time, and hardware switching for fast real-time.

All these specific techniques allow a considerable improvement in terms of determinism. Different real-time Industrial Ethernet solutions were proposed, called Real-time Ethernet, such as PROFINET, EtherCAT, Ethernet/IP and many more (CoNet, 2011). The conditions for the industrial use of Ethernet are described by international standard IEC 61 784-2 Real Time Ethernet (See Fig. 3). IEC stands for International Electrotechnical Commission.

The following parameters are covered by the network performance metrics:

- latency (delay) – the amount of time required for a frame to travel from source to destination,
- jitter – a measure of the deviation of the latency from its average value,
- loss rate – the probability that an individual packet is lost (dropped) during the transmission,
- throughput – the amount of digital data transferred per time unit.

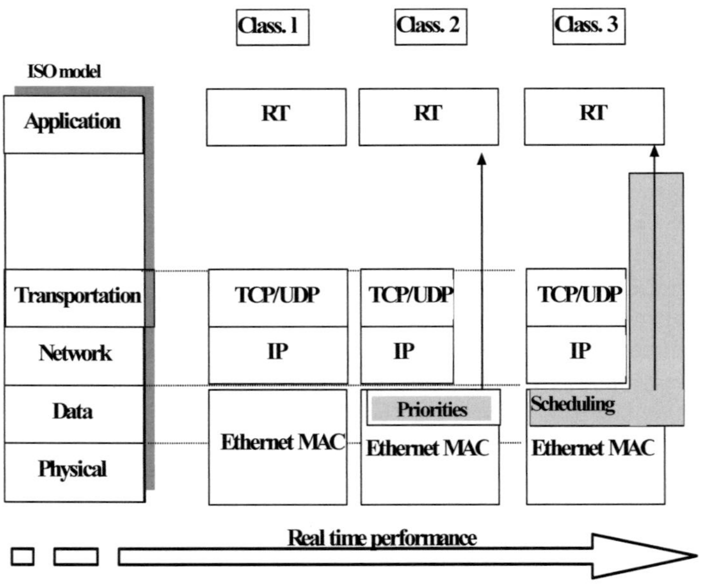

Fig. 3. Classification of industrial Ethernet (IEC 61 784-2)

Class 1 describes the use of standard Ethernet TCP/IP as it is. In this case the different real time protocols and the best-effort protocols, like HTTP, SNMP, FTP etc., uses the services of the TCP/IP protocol suite. This includes examples such as CIP Sync (Ethernet/IP, ModBus/TCP). The class 1 has the largest conformity to the Ethernet TCP/IP standard and can thereby use standard hardware and software components.

Class 2 introduces optimizations, whereby the realtime data bypasses the TCP/IP stack and thus considerably reduces the latency time and increases the achievable packet rate. In Classes 1 and 2, the priority support described by IEEE 802.1Q can also be used depending on the approach. In Class 3 the scheduling on the MAC level is again modified through the introduction of a TDMA method. Class 3 can be used in applications that require maximum latency in the range 1ms and maximum jitter below 1microsec. In this class there are strong restrictions for the use of standard hardware components or the necessity for special components, like dedicated switches. Generally, conformance with the Ethernet standard decreases when ones increase the Class number, while the achievable real-time performance increases.

2.2 Robust codesign

2.2.1 Dynamics of distributed control system

The basic model of the DCS is shown in Fig. 4. The process outputs are measured and control signals are applied through the distance I/O devices. The I/O devices are integrated with A/D and D/A converters.

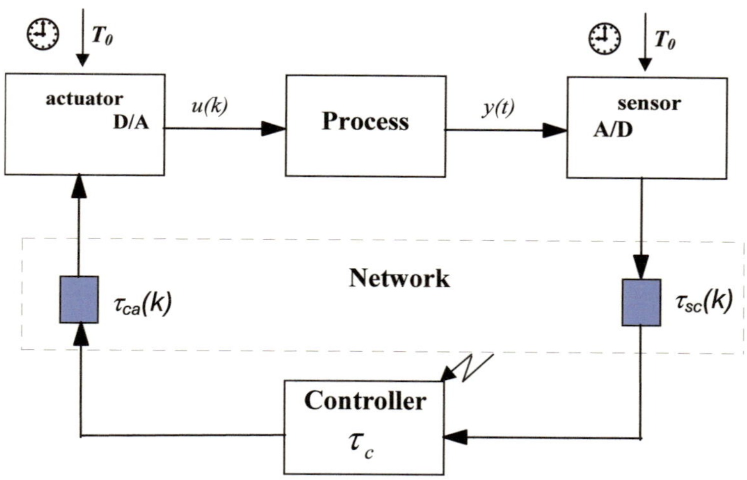

Fig. 4. Basic model of distributed control system

The communication to and from the controller node is supported by a network. From a digital control point of view, it is natural to sample the process with an equal period T_0 and to keep the control delay as short as possible. This suggests that the sensor and actuator (A/D and D/A) converters are time-triggered (sampling period T_0), while the controller is event-triggered, which means that they are triggered by the arrival of the new data. The main complication of this control architecture is the presence of variable time delays. The additional dynamics observed in distributed control system depends on the performance parameters of the underlying network, which include transmission rate and transmission medium access method. Under certain circumstances the network-induced delays can be consider constants, but generally they might be varying from transfer to transfer (Fig.4). Thus, the introduction of a network in the feedback loop violates conventional control theory assumptions such as non-delayed sensing and actuation. This can degrade the performance of the control system or even can destabilise the system.

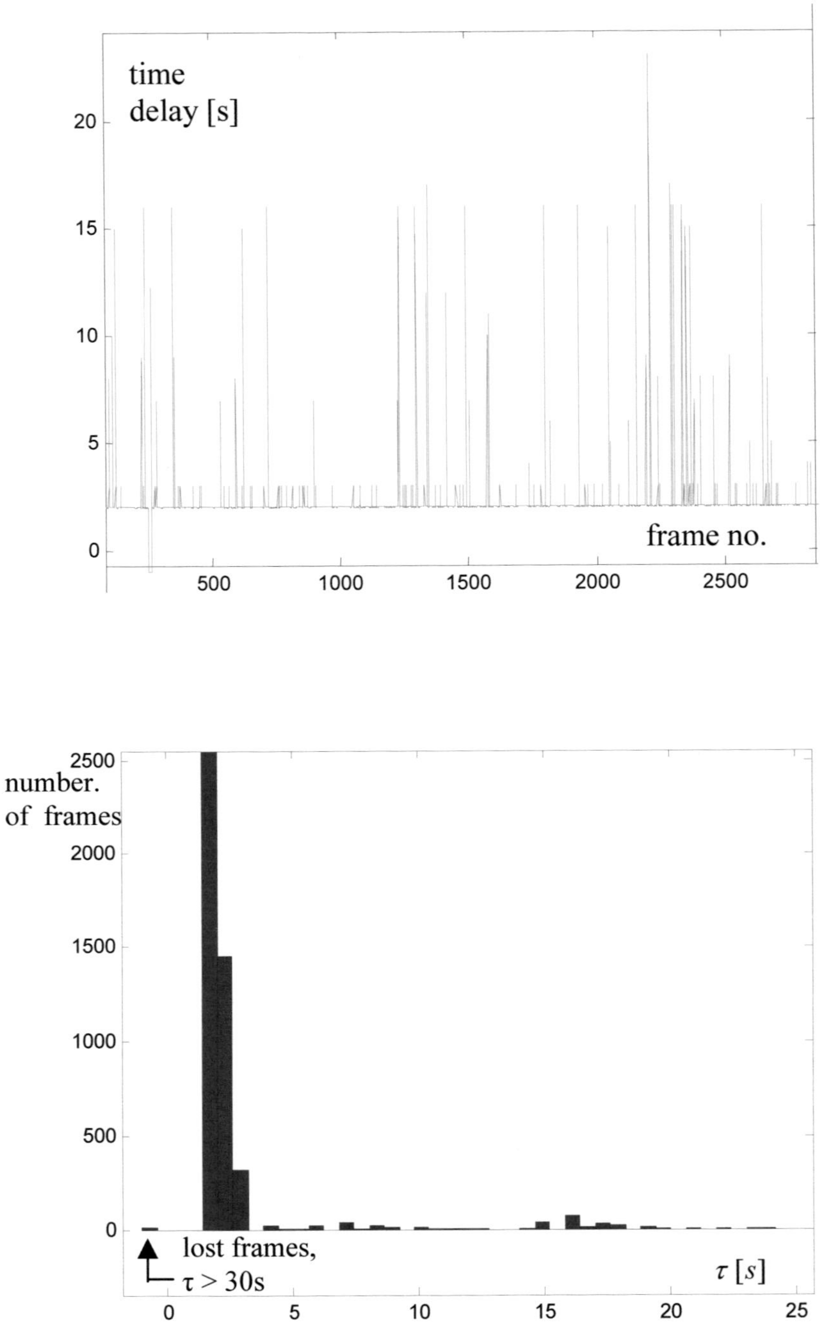

Fig. 5. Example: wireless network data transfer times and histogram of delays ¶

2.2.2 Co-design

Computer implementation of distributed control systems, real-time algorithms, data transmission models and digital control theory methods cannot be developed separately because an unexpected control system performance may occur. Three parameters need particular attention from the distributed control design perspective: sampling and actuation tasks period, controller task period and network parameters (latency and jitter). Due to the close relationships between the network and control parameters the selection of the best sampling period will be a compromise. In this section we will demonstrate the construction of a networked control design chart, which can be used to select proper design parameters.

2.2.3 Sampling and actuation task

We will assume that the control algorithm design is based on correctly identified: model of the process and the model of disturbances (referred to as "nominal models"). We assume that it is possible for the nominal models to estimate a maximal, admissible sampling period, which would guarantee acceptable control performance.

One accepted rule is (Aström and Wittenmark, 1997) that the control task period should be a ($a > 1, a \in N$) times smaller than the period of the cut-off frequency, approximated in some reasonable way for the nominal process model. This upper bound of T_0 is denoted as T_0^u (Fig. 6).

For the design purpose we assume that performance of the closed-loop control system is a strictly monotonic function of T_0: any sampling (actuation) period $T_0 < T_0^u$ improves the control performance. For $T_0 < T_0^l$ improvement is not observed. Finally, the sampling (actuation) task period can be estimated as $T_0 \in [T_0^l, T_0^u]$.

2.2.4 Controller task period

The applied control platforms (processor, peripherals hardware and operating systems) are characterized by a closed - loop execution time, estimated as $\delta_s \in [\delta_s^l, \delta_s^u]$, where δ_s^l - is the lower bound of the execution time for simple control algorithms, δ_s^u - is the execution time of complex control algorithms.

The control algorithm is classified as "simple", if pseudocode of the controller task includes no more than 5-10 operations (loops are excluded). Examples of "simple" algorithms are: incremental PID or state feedback controller. If the pseudocode of the controller includes more than 10 operations or loops are included then the algorithm is classified as "complex".

2.2.5 Network parameters

Presence of networks introduces communication delays and limits the amount of data that can be transferred between nodes. In some cases not all samples from sensor or to actuator (produced with period T_0) can be sent, because the network requires intervals longer than T_0 between the transfers of two consecutive packets. Therefore, constraints on the process data availability, introduced by the communication channel are defined.

The average communication delay between the sensor node and the controller node is denoted as τ^{sc}, τ^{ca} is average communication delay between the controller node and the actuator node, $\Delta(k)$ represents a total jitter in the feedback loop, k – is the number of the control step.

Actually, the communication delays and jitters can be added to the controller execution time creating an estimation of delays and uncertainty in the control loop. The total delay in the control loop is

$$\tau(k) = \tau_{sc} + \tau_{ca} + \delta_s + \Delta(k)$$

It will also be assumed that the jitter is bounded by $0 \le \Delta(k) \le \Delta^u$.

2.2.6 Robust codesign

In the previous section we have introduced a number of parameters that need special attention from the perspective of real-time digital control: T_0 - sampling period defining the temporal granularity related to the process dynamics, δ_s - execution time describing the efficiency of the hardware and software application platform and $\tau_{sc}, \tau_{ca}, \Delta$ - communication delays and jitter. Now, we will demonstrate, how these parameters interacts one to another, how to select the application platforms and how to set closed-loop execution times in such a way, that process dynamics and communication network properties are balanced.

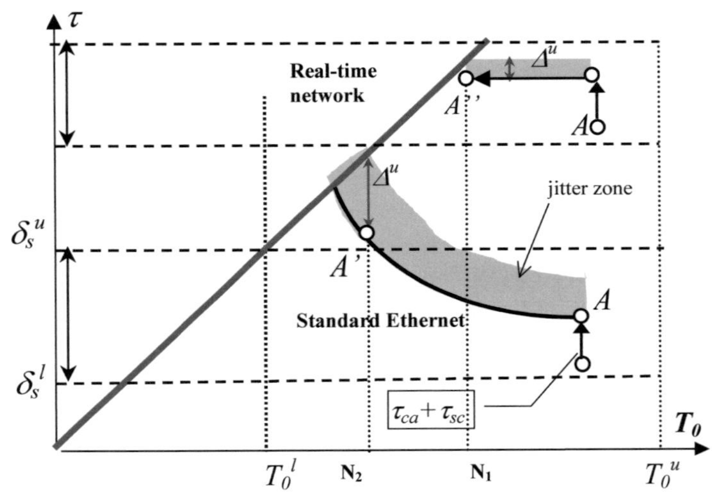

Fig. 6. Distributed control system design chart

The operating point of the distributed control system should be located in the area between T_0^l and T_0^u in Fig. 6. The operating must lie below the line separating "time critical" solution, which simply means that control loop execution time must be less than sampling period. Points A, A' in Fig. 6 also represent a situation where the design is robust against possible variations (jitter) of the task execution and data transfer times (shadowed area in Fig. 6).

Let us assume that Ethernet network is implemented. Computational delay of the controller δ_s is fixed, but for Ethernet network the transmission time delay increases linearly with increasing load - in same case exponentially, when the load on the network exceeds 35 - 40%.

It means, that a faster sampling rate for guaranteeing better control performance will saturate the network traffic load, and eventually increase the data transmission time. For the example given in Fig. 6, the best operating point for Ethernet network is A′ and is constrained by the process data availability introduced by transmission time delays of the communication channel.

If communication can be supported by high-speed real-time - network, e.g. ProfiNet, Class 2 (Amiguet *et al.* 2008) the constraint of this kind is not active. However, another constraint becomes active and critical. Control loop execution time can not be longer than the sampling period (A″ in Fig.6), including the jitter $\Delta(k)$. The reason is that cycles of the control loop do not accept intervals between transfers of the two consecutive packets shorter, than N_1. The time diagram for this situation is given in Fig. 7. For the model from Fig.6 we must assume that

$$\tau_{sc} + \tau_c + \tau_{sc} + \Delta(k) \le T_0 = N_1$$

It means, that the operating point (A″) must be located below the line separating "time-critical" zone, including the jitter zone (Fig.6).

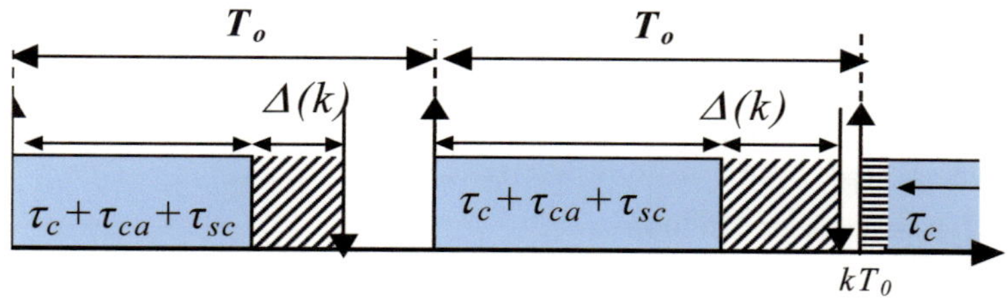

Fig. 7. Timing model that can be used for a regularly sampled process

3. Control over the network: Increasing the robustness

One commonly used approach to increase the robustness of DCS stability with respect to the network effect is extension of the standard control algorithms by new components.

3.1 Buffering

The idea is to reduce temporal dependency of the individual parts of the model from Fig. 4 by introducing buffers at the actuator (Tutaj, 2006). Buffering can be easily implemented using PLCs' or embedded controller at the device level. In digital control this operation can be handled by use of a zero-order holds on the control signal.

First approach presented in this section incorporates one-step buffer introduced at actuator side to compensate variable time delays. Let τ be the overall delay (round trip latency time, $\tau = \tau^{sc} + \tau^{ca} + \tau^{ca}$). The controlled process model is assumed to be linear, in the form

$$\frac{dx}{dt} = Ax(t) + Bu(t-\tau),\ x(t) \in R^n,\quad u(t) \in R^1 \tag{1}$$

For applied delayed linear control law

$$u(t-\tau) = Kx(t-\tau)$$

the closed loop model takes the form

$$\frac{dx}{dt} = Ax(t) + BKx(t-\tau) = Ax(t) + A_1(t-\tau)$$

The maximum tolerable time delay for given K (or bounds for K under some assumptions on τ) can be computed from the solution of LMI optimisation problem. We should notice, that generally the network induced delays are different from the process delays, because they are time varying and unknown. One solution proposed in (Yi and Hang, 2002) determines condition for exponential stability of system (1) for $\tau(t) \in C_b^0$ - nonnegative, continuous and bounded at $[0,+\infty)$

$$\frac{A^T + A}{2} + \sqrt{\lambda_{\max}(A_1 A_1^T)} I < 0$$

where $\lambda_{\max}$ - is the maximum eigenvalue.
Several authors have pointed out (Fujioka, 2009) that the above stability condition is usually conservative.
Assuming that:

- signal transmission is with a single packet (or frame),
- the sensor and actuator are time driven, the controller is event driven. The clocks operate at time period T_0 and are synchronized,
- the process dynamics is controllable,

then discrete time model can be introduced. For brevity in the ensuing text notation $x(k)$ will be used in place of $x(kT_0)$.
If the actuation period is selected as T_0, than $u(t-\tau)$ is piecewise constant over the actuation period and only changes value at $(kT_0 + \tau)$. Integration of (1) over the sampling period gives a discrete-time, finite dimensional approximation of the delayed model (1)

$$x[k+1] = \Phi_0 x[k] + \Gamma_1 u[k-q-1] + \Gamma_0 u[k-q]$$

where

$$\tau = qT_0 + \gamma,\ q \geq 1,\ \Gamma_1 = \int_{T_0-\gamma}^{T_0} e^{As} B ds, \Gamma_0 = \int_0^{T_0-\gamma} e^{As} B ds, \Phi_0 = e^{AT_0}$$

We define new state variables

$$\begin{aligned} z_1[k] &= u[k-q-1] \\ z_2[k] &= u[k-q] \\ &\vdots \\ z_{q+1}[k] &= u[k-1] \end{aligned}$$

For the assumed considered timing method and the condition on total network delay

$$\tau(k) \leq T_0 \tag{2}$$

fulfilled, the model

$$\begin{bmatrix} x(k+1) \\ z(k+1) \end{bmatrix} = \begin{bmatrix} \Phi_0 & \Gamma_1(\tau) \\ 0 & 0 \end{bmatrix} \begin{bmatrix} x(k) \\ z(k) \end{bmatrix} + \begin{bmatrix} \Gamma_0(\tau) \\ 1 \end{bmatrix} u(k) \tag{3}$$

$$y(k) = \begin{bmatrix} 1 & 0 \end{bmatrix} \begin{bmatrix} x(k) \\ z(k) \end{bmatrix}$$

describes behaviour of closed-loop system.
It is known, that for a discrete linear system with time-varying parameters location of the system eigenvalues in a stable region for all admissible values of the parameters does not imply stability of the system. The buffer can be used at actuator side to eliminate the delay variability in the loop, thereby enabling more effective use of delay compensation algorithms (e.g. Smith predictor). Generally, the buffered control loop can take advantage of more deterministic loop delay, and in consequence the controller can be design more "aggressively" - if only a good process model is available.
The augmented state model with one-step, constant length buffer is obtained in the form

$$\begin{bmatrix} x(k+1) \\ z_1(k+1) \end{bmatrix} = \begin{bmatrix} \Phi_0 & \Gamma_1 \\ 0 & 0 \end{bmatrix} \begin{bmatrix} x(k) \\ z_1(k) \end{bmatrix} + \begin{bmatrix} 0 \\ 1 \end{bmatrix} u(k)$$

The data package is delivered as soon as possible to the actuator, but is hold in the buffer and is implemented to the process in the next sampling intervals. As long as (2) is fulfilled, the "buffered" loop delay is constant and is equal to the buffer length ($\tau_B = T_0$).
If the control strategy is assumed as linear feedback

$$u(k) = -\begin{bmatrix} K_x & 0 \end{bmatrix} \begin{bmatrix} x(k) \\ z_1(k) \end{bmatrix}, \quad u(k) \in R^1 \tag{4}$$

the closed-loop system can be written as

$$\begin{bmatrix} x(k+1) \\ z(k+1) \end{bmatrix} = \begin{bmatrix} \Phi_0 - \Gamma_0(\tau)K & \Gamma_1(\tau)) \\ -K & 0 \end{bmatrix} \begin{bmatrix} x(k) \\ z(k) \end{bmatrix} \tag{5}$$

If the condition (2) is not fulfilled for some kT_0, the two-step, constant length buffer can be applied ($\tau \le 2T_0$), Fig.8. For this case the model takes the form ($q = 1,\ \tau_B = 2T_0$).

$$\begin{bmatrix} x(k+1) \\ z_1(k+1) \\ z_2(k+1) \end{bmatrix} = \underbrace{\begin{bmatrix} \Phi_0 & \Gamma_1 & 0 \\ 0 & 0 & 1 \\ -K_x & 0 & 0 \end{bmatrix}}_{\Phi 2} \begin{bmatrix} x(k) \\ z_1(k) \\ z_2(k) \end{bmatrix} \tag{6}$$

If the loop delay $\tau(k)$ is time varying between *[0, 2T₀]*, it is reasonable to switch between T_0 and $2T_0$ buffers. The stability analysis of this model is the problem of stability of the *Asynchronous Dynamical Systems* (ASD) (Hasibi *et al*, 1999).
The model (5) can be rewritten in the equivalent form, as

$$\begin{bmatrix} x(k+1) \\ z_1(k+1) \\ z_2(k+1) \end{bmatrix} = \underbrace{\begin{bmatrix} \Phi_0 & 0 & \Gamma_1 \\ 0 & 0 & 0 \\ -K_x & 0 & 0 \end{bmatrix}}_{\Phi 1} \begin{bmatrix} x(k) \\ z_1(k) \\ z_2(k) \end{bmatrix}$$

The following result applies in this case (Zhang, 2001). If for the linear DCS model

$$w((k+1)) = \Phi_{s(k)} w(k), \; w(k) = \begin{bmatrix} x(k) \\ z_1(k) \\ z_2(k) \end{bmatrix}, \; s(k) = (1,2)$$

for a given rate r of the frames transmission there exists the Lyapunov function such that

$$V(w(k)) = w^T(k) P w(k)$$

and scalars α_1, α_2 such that

$$\alpha_1^r \alpha_2^{1-r} > 1$$

$$\Phi_1^T P \Phi_1 \le \alpha_1^{-2} P, \; \Phi_2^T P \Phi_2 \le \alpha_2^{-2} P \tag{7}$$

than the system is exponentially stable. The rate r represents the fraction of time that each discrete state transition matrix (Φ_1, Φ_2) occurs. Assuming the transmission rate, the problem (7) can be solved as the LMI problem.

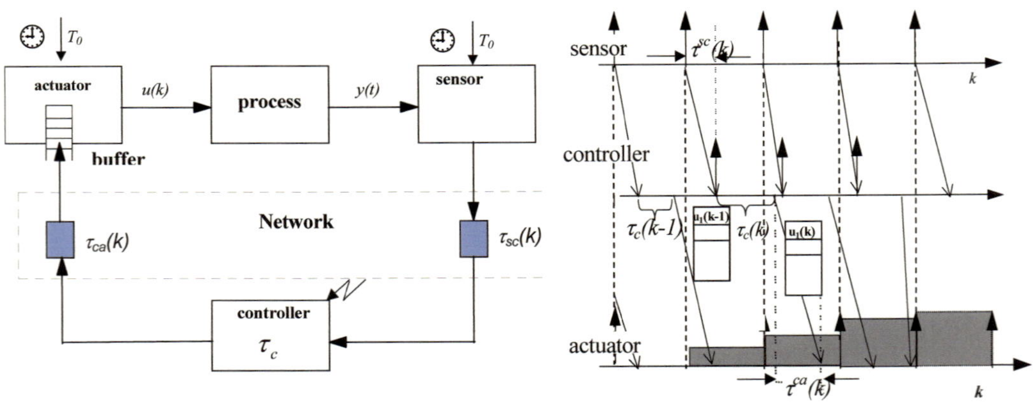

Fig. 8. Time diagram of buffering for $\tau_B = 2T_0$

Clearly, adding any delay to a closed-loop system generally degraders the performance. Therefore, once must investigate:

- proper buffer length for assumed model of delay distribution,
- design of controller that takes advantage of an effectively more deterministic loop delay.

A natural extension of this approach is application of variable length (adaptive) buffers (Tutaj, 2006). It is assumed that frames order can not be changed, frames are not lost or doubled. The initial length of the buffer is T_0. The buffer length is adapted according to the following formula:

$$\tau_B(k) = T_0 + \alpha T_0 (p - \varphi(k))$$

where:

α - adaptation parameter, $\alpha > 0$

p - assumed rate of frames delivered to the buffer in time (during the time interval no longer than $\tau_B(k)$), , $0 < p < 1$

$$\varphi(j) = \begin{cases} 1 & \textit{if the frame was delired in time} \\ 0 & \textit{otherwise} \end{cases}$$

If the frame is not delivered in time (at $t = kT_0$ the buffer is empty) than $\varphi(k)$ is set to 1. First frame delivered to the buffer is released immediately.
After $k+1$ steps the buffer length can be calculated as (Tutaj, 2006)

$$\tau_B(k+1) = T_0 + \alpha T_0 \sum_{j=0}^{k} (p - \varphi(j))$$

Such a model implements a kind of "filtration" of delays effects (Fig. 9).

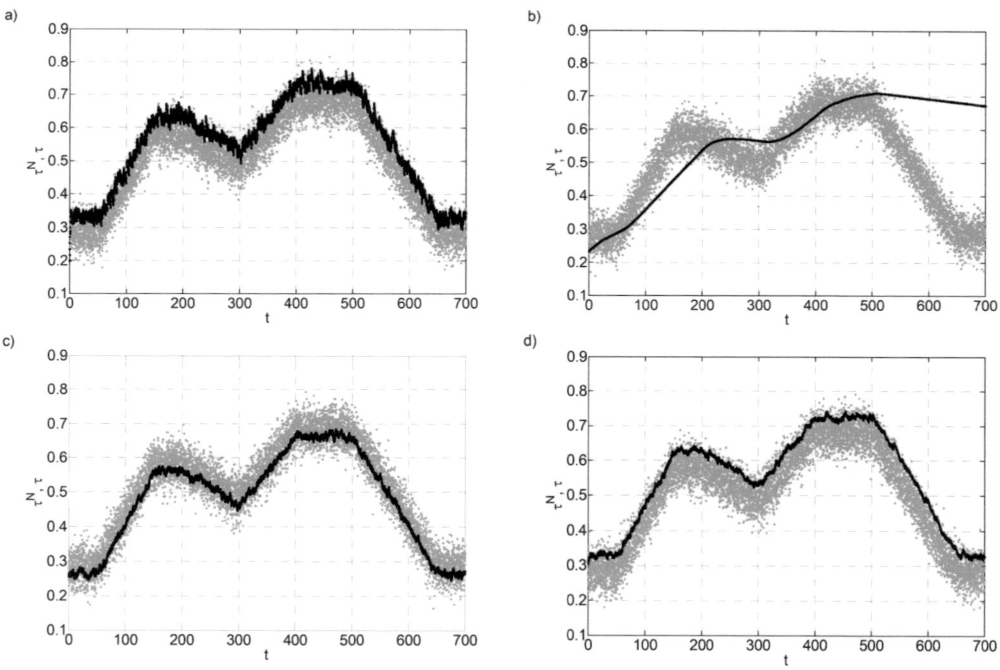

Fig. 9. Example of adaptive filter operation (Tutaj, 2006): a) α = 0,2; p = 0,9, b) α = 0:002; p = 0,9, c) α = 0,05; p = 0,3, d) α = 0,05; p = 0,9 (black – after buffer, grey – before buffer)

3.2 Robust stability of the buffered DCS

Application of variable length buffer simplifies analysis of DCS. It can be assumed that the control delays are constant but not exactly known. In this case the problem of stability analysis of the DCS can be formulated as a parametric robust control problem. This allows using the mapping theorem (Bhattacharyya *et al.* 1995) to develop an effective computational technique to determine robust stability. The advantage of this approach over the stochastic method is that it is not necessary to identify the stochastic model of the delay.

3.2.1 Time-invariant delays in DCS

We assume that the total delay is slowly varying and known only with some precision

$$\tau(k)=\tau,\quad \tau_{\min}\le\tau\le\tau_{\max}$$

In such a case we could design a controller stable for some range of slowly varying delay. The solution of this problem gives answer to the basic question "how much delay can the system tolerate"?

The state matrix of the closed loop system (5) can be next rewritten in the form

$$\begin{aligned} M(\Psi) &= \begin{bmatrix} \Phi_0-\Gamma_0(\tau)K & \Gamma_1(\tau)) \\ -K & 0 \end{bmatrix} = \\ & M_0+M_1(\varphi_1)+M_2(\varphi_2)+..+M_n(\varphi_n)= \\ &= M_0+\varphi_1(\tau)\hat{M}_1+\varphi_2(\tau)\hat{M}_2+...+\varphi_n(\tau)\hat{M}_n \end{aligned} \tag{8}$$

where

$$\Psi^T(\tau)=A^{-1}\Phi_0\Phi(\tau)B=\left[\varphi_1(\tau)\quad \varphi_2(\tau)\mid \varphi_n(\tau)\right]$$

$$\Phi(\tau)=e^{-A\tau},\qquad M_0=\begin{bmatrix}\Phi_0+A^{-1}BK & A^{-1}\Phi_0 B\\ -K & 0\end{bmatrix}$$

$$\begin{aligned}\hat{M}_i &= \begin{bmatrix} 0_1 & 0_2 \\ -K & -1 \\ \hline 0 & 0 \end{bmatrix}, \\ & \dim 0_1=(i-1)\times n,\ \dim 0_2=(i-1)\times 1\end{aligned} \tag{9}$$

The uncertain delay enters affinely into the state matrix of the closed loop system. If $\tau_{\min}\le\tau\le\tau_{\max}$, then we could obtain the boundaries

$$\Psi_{\min}\le\Psi(\tau)\le\Psi_{\max} \tag{10}$$

The following stability problem is important for the model formulated above: determine if matrix (9) remains Schur-stable as φ_i parameters ranges over the bounds given by (10)? The structure of the closed loop state matrix (8) is a special case of the interval matrix family and we are free to use results of the robust theory solutions for checking stability (Bhattacharyy *et al.* 1995). Under the assumption $rank(M_i)=1$ for $i=1..n$ the coefficients of the characteristics polynomial of $M(\Psi)$ are multilinear function of φ. The following theorem applies in this case:

Let the matrix M_0 *be Schur stable. If the* $rank(\hat{M}_i)=1$ *for* $i=1..n$, *than the family of the matrices* $M(\Psi),\ \ \Psi\in\Theta$ *defined by (8)-(9) is robust Schur-stable if the testing function*

$$F(y)>0 \quad \forall y\in Y,$$

where the testing function is defined as

$$F(y)=\pi-\alpha(y) \tag{11}$$

$$\alpha(y)=\max\{|\arg(\tilde{p}_r(f(y))-\arg(\tilde{p}_k(f(y))|\},$$
$$r,k=1,2,...K,\quad r\neq k$$

$$\tilde{p}_k(z)=\frac{p_k(z)}{w_0(z)},\ k=1,2,...,K$$

$$p_k(z)=\det(zI-M(\varphi_k))$$

$$w_0(z)=\det(zI-M_0)\ \ K=2^n.$$

The function $f(y)=\exp(j\pi y),\ \ y\in Y=[0,2]$ is a parametric description of the unit circle, $M(\varphi_k)$ is a vertex matrix calculated for each φ_k - the vertex of the set Θ, K is the number of the vertex matrices. The testing function (11) checks the maximal phase differences of the vertex polynomials over parameter box corresponding to the vertices given by (10).

3.2.2 Example: Distributed control of a tank system

Let us consider a problem of distributed control of a tank system. The process consists of the upper tank having constant cross section and the lower cylindrical tank, so having variable cross section. Liquid is pumped into the top tank by DC motor driven pump. The liquid outflows of the tanks only due to gravity. The orifices C_1 and C_2 determine the outflow of the liquid. The general objective of the control is to reach and stabilise the level in the lower tank by adjustment of the pump operation. The levels in the tanks are measured with pressure transducers (S). The appropriate interfaces (I) enabling distance transmission of the control signals to the pump were installed, creating a distributed control system from Fig.10.

If levels in the tanks are introduced as the states variables, the nonlinear model could be linearized at $H^0\ =[H_1^0,H_1^2]^T$ giving finally (Grega, 2002)

$$\frac{d}{dt}\begin{bmatrix}h_1\\h_2\end{bmatrix}=\begin{bmatrix}a_1 & 0\\a_3 & a_4\end{bmatrix}\begin{bmatrix}h_1\\h_2\end{bmatrix}+\begin{bmatrix}b_1\\0\end{bmatrix}u$$

where: $\quad H_1=H_1^0+h_1,\quad H_2=H_2^0+h_2,\quad q=q^0+u,\quad b_1=\frac{1}{S},\ a_1=\frac{-C_1}{2S\sqrt{H_1^0}},$

$$a_3 = \frac{C_1\sqrt{H_1^0}}{4H_1^0 \cdot w\sqrt{r^2-(r-H_2^0)^2}}, \quad a_4 = \frac{-C_2\sqrt{H_2^0}}{4H_2^0 \cdot w\sqrt{r^2-(r-H_2^0)^2}}$$

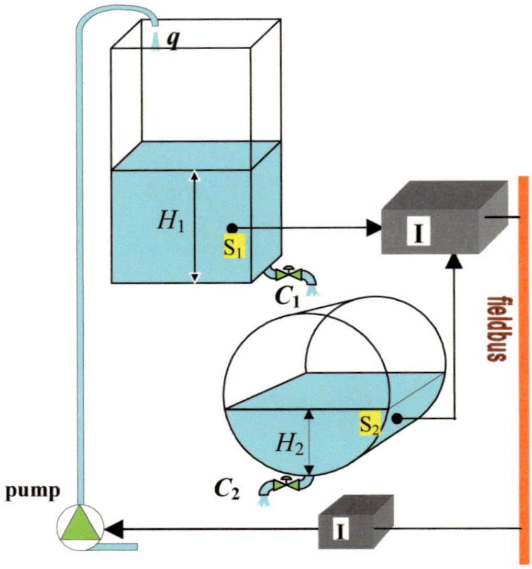

Fig. 10. Distributed control of tank system

For an assumed sampling period T_0 the equivalent discrete model is ($\tau = 0$)

$$\begin{bmatrix} h_1(k+1) \\ h_2(k+1) \end{bmatrix} = \begin{bmatrix} e^{a_1 T_0} & 0 \\ \frac{a_3}{a_1 - a_4}(e^{a_1 T_0} - e^{a_4 T_0}) & e^{a_4 T_0} \end{bmatrix} \begin{bmatrix} h_1(k) \\ h_2(k) \end{bmatrix} + $$

$$+ \begin{bmatrix} \frac{b_1}{a_1}(e^{a_1 T_0} - 1) \\ \frac{b_1 a_3}{a_1 a_4}[\frac{a_1(e^{a_1 T_0} - e^{a_4 T_0})}{a_1 - a_4} - (e^{a_1 T_0} - 1)] \end{bmatrix} u(k)$$

Linear feedback control law is in the form

$$u(k) = -K\, h(k)$$

It was assumed that the controller has been design ignoring the network, hence the state matrix of (3) is stable for $\tau = 0$. The assumed parameters of the tank model were: $C_1 = 10$, $C_2 = 15$, $T_0 = 80s$, giving the LQ controller gains: $K_1 = 0.7167$ $K_2 = 3.0950$. Fig. 11 demonstrates the LQ optimal output of the model (simulation). Figure 13 illustrates observed perturbation of data transmission times, when Ethernet protocol was applied and some additional traffic in the network was generated.

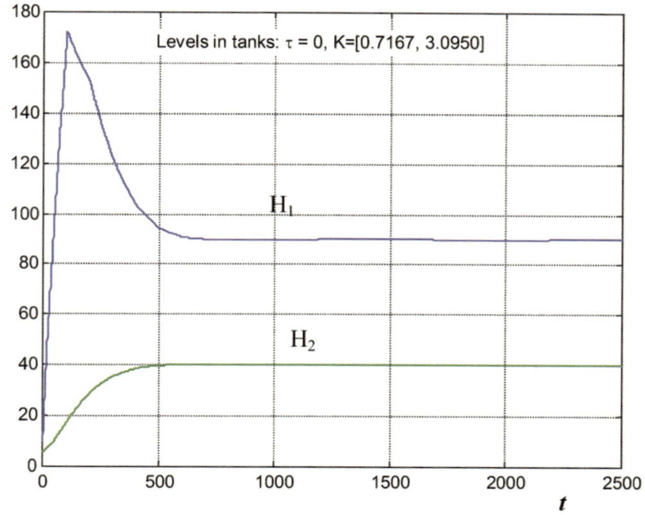

Fig. 11. LQ - optimal control of the tank system

The delay in the control loop reduces the stability margin of the system. Figure 12 shows how the fixed feedback delay ($\tau = 80s$) degrades the performance of the tank system control. Notice that this is equivalent to implementation of the fixed size buffer ($\tau_B = T_0$). So, to increase the stability margin and improve stability it is necessary to tune the feedback gains.

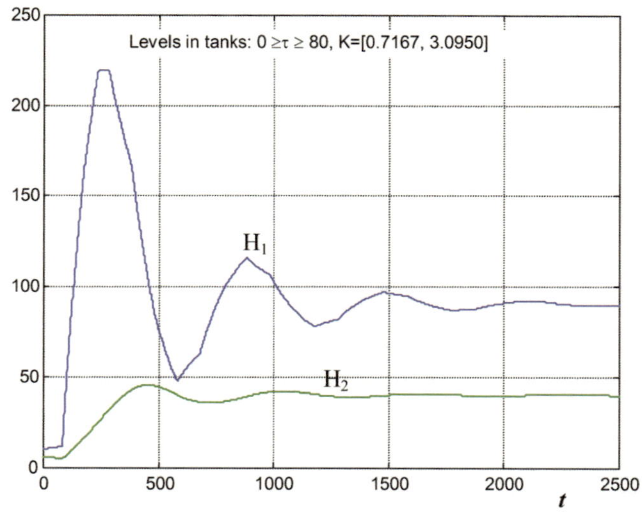

Fig. 12. System performance degradation

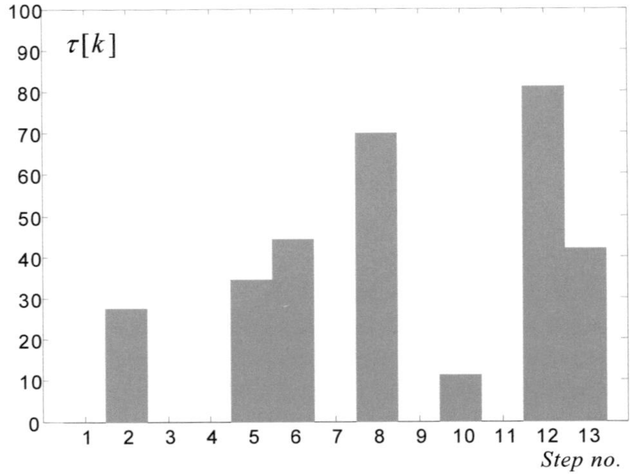

Fig. 13. Network delays - perturbation of data transmission times, $\tau \in [0,100]s$

If the variable-length buffer is introduced, stability of the distributed digital control system for the assumed controller gains and $0 \le \tau \le \tau_{\max}$ can be verified using the methodology described above. It is assumed now, that the control delays are constant but not exactly known. The LQ optimal robust gains (giving the stable matrix M_0) were calculated as: $K_1 = 0.3745 \quad K_2 = 0.3420$.

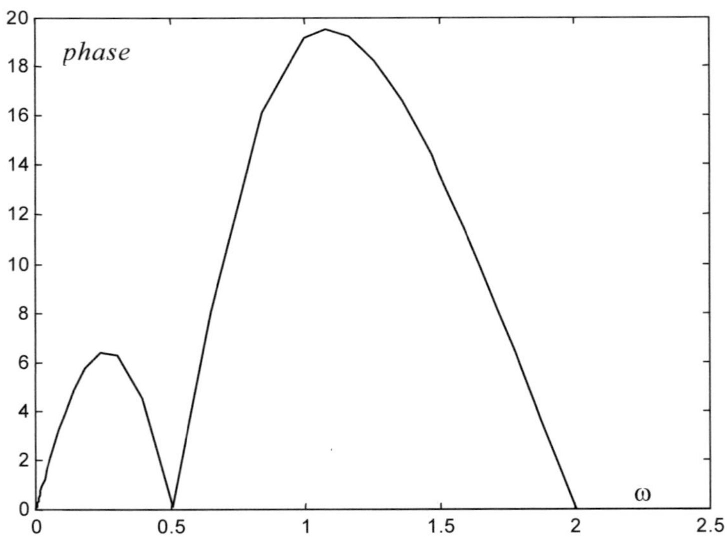

Fig. 14. Maximum phase diffences of vertex matrices

The next step is verification of the testing function *F(y)*. The appropriate testing function is given in Fig. 14. The maximum phase difference over all vertices at each $\omega \in [0, 2\pi)$ is less than 180^0. Figure 15 shows operation of the LQ controller for the above set of controller parameters and network delays , as given in Fig.13.

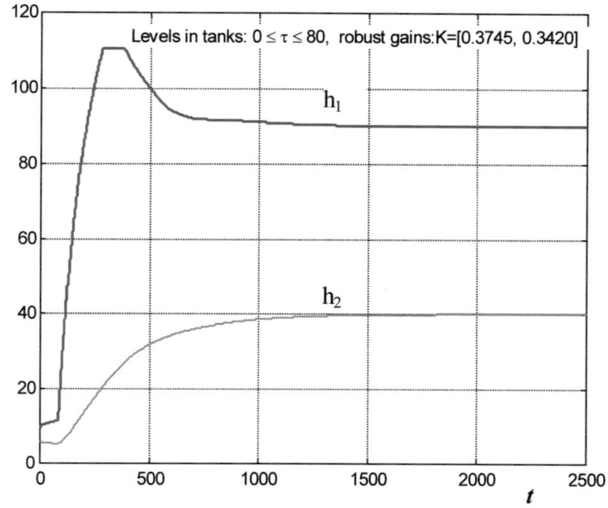

Fig. 15. Robust, LQ - optimal control of the tank system

4. Final remarks

The introduction of networks, limited throughput of data transmission channels, combined with non-optimised hardware and software components introduce non-determinism in the distributed control system. For multilevel industrial systems this problem becomes even more complex. Some control loops can be handle by local, device - level controllers, but also by the supervisory controllers all them implementing data transmission networks. Special care must be taken when the communication channel generates sampling - actuation jitters or other kinds of run time violation of the closed-loop timing assumptions. It means that the introduction of data transmission networks into the feedback loop in many cases violates conventional control theories assumptions such as non-delayed or evenly spaced sampling and actuation. It is now reasonable to redesign controllers improving the temporal robustness of the distributed control system.

Control engineers do not care very much about real-time or distributed control implementations of control algorithms. In many cases they do not understand control timing constraints. The typical solutions proposed are: "buy a faster computer" or "install a more efficient data transmission network". Basic control theory does not advise them on how to redesign controllers to take network limitation into account.

It was demonstrated in this chapter, that robust design it is not only a proper selection and tuning of control algorithms, but also study on communications protocols and networks, to make them suitable for real-time DCS.

We have proposed an integrated design approach combining several components: process dynamics, controller parameters and network constraints, and resulting in better quality of control systems.

Finally, it was shown how the extension of the standard controller with a buffer improves robustness of distributed control system. The model was formulated as variable parameter linear discrete-time model, where variability of parameters was introduced by the time varying delays. The variable length buffer was used at actuator side to eliminate high speed delay variability in the loop, thereby enabling more effective use of delay compensation algorithms. A water tank control example has shown how implementation of variable-length buffer algorithm and application of some results of interval matrices theory increases robustness of the control loop.

5. Acknowledgment

This work was supported by the European Regional Development Fund Grant no. UDA-POIG.01.03.01-12-171/08/00. Special thanks are given to dr Andrzej Tutaj from Department of Automatics, AGH University of Science and Technology for his valuable experiments results included in this work.

6. References

Al-Hammouri, A.T., Branicky, M.S., Liberatore, V. & Phillips, S.M.(2006). Decentralized and dynamic bandwidth allocation in networked control systems, in: *20th International Parallel and Distributed Processing Symposium*, 25-29 April 2006

Amiguet, A., Etienne, L. & Jaggi A. (2008). Performance analysis of PROFINET networks, Available: http://lamspeople.epfl.ch/decotignie/Midterm08_profinet_slides.pdf

ARC Advisory Group (2007). *Industrial Ethernet Infrastructure Worldwide Outlook* (2007), Market Analysis and Forecast through 2011 ed. by Cisco

Aström, K.J. & Wittenmark, B. (1997). *Computer Controlled Systems,* Prentice Hall, London

Bhattacharyya, S.P., Chapellat, H. & Keel, L.H. (1995). *Robust Control: The Parametric Approach*, Prentice Hall, New Jersey, 1995

CoNeT (Cooperative Network Training) project website. (2011). http://www.conet-eu.net/

Felsner, F. (2005). Real-Time Ethernet – Industry Prospective, in: *Proceedings of the IEEE,* Vol. 93, June 2005, pp. 1118- 1129

Fujioka, H. (2009). Stability analysis of systems with aperiodic sample-and-hold devices, *Automatica*, vol. 45, no.3, pp. 771–775

Georges, G.P., Krommenacker, N., Divoux T. & Rondeau, E. (2006). A design process of switched Ethernet architectures according to real-time application constraints, *Engineering Applications of Artificial Intelligence*, vol. 19, no.3, pp. 335-344

Grega, W. (1998). Time-optimal Control of N-tank System, in: *Proceedings of 1998 IEEE International Conference on Control Applications,* Triest, pp. 532-536

Grega, W. (2002). Stability of Distributed Control Systems with Uncertain Delays, in: *8th IEEE International Conference on Methods and Models in Automation and Robotics,* Międzyzdroje 2002, pp. 303 – 307

Grega, W. & Tutaj A. (2007). Network traffic reduction by sample grouping for distributed control systems, in: 3rd *Proceedings International Workshop on: Networked control systems: tolerant to faults*, June 20–21, 2007, Nancy, France, pp. 2-8

Grega, W. (2009). Codesign of distributed and real-time control systems, in: *14th International Conference on Methods and Models in Automation and Robotics,* Międzyzdroje, August 2009, pp. 85-88

Grega, W., Byrski, W. & Duda J. (2009). *InStePro – Integrated Production Control.* Project Description, available from: http://www.InStePro.agh.edu.pl

Grega, W. (2010). Multilevel Control under Communication Constraints, In: *Proceedings of 2010 IEEE International Symposium on Computer-Aided Control System Design,* Yokohama, Japan, September 8-10, 2010, pp. 1176 - 1181, IEEE Xplore Digital Library, Digital Object Identifier: 10.1109/CACSD.2010.5612686

Gupta, R. A. & Chow, M.-Y. (2010). Networked Control System: Overview and Research Trends, *IEEE Transaction on Industrial Electronics*, vol. 57, no.7, pp. 2527-2535

Hassibi A., Boyd, S.P. & How, J.P. (1999). Control of Asynchronous Dynamical Systems with Rate Constraints on Events, in: *Proceedings of 37 IEEE Conference on Decision and Control*, pp. 1345–1351

Hirai, K. & Satoh, Y. (1980). Stability of a System with Variable Time Delay, In: *IEEE Transaction on Automatic Control*, vol. AC-25, pp. 552-554

Juanole, G. & Mouney, M. (2006). Real time distributed systems: QoS and impact on the performance of process control applications. In: *17th International Symposium on Mathematical Theory of Networks and Systems (MTNS'06)*, July 2006, Kyoto, Japan, pp. 1739-1746

Larsson, L. (2005). *Fourteen Industrial Ethernet solutions under the spotlight*, The Industrial Ethernet Book, vol.28, Available from: http://ethernet.industrialnetworking.com /articles/articledisplay.asp?id=854

Montestruque, L.A. & Antsaklis, P.J. (2003). On the model-based control of networked systems", *Automatica*, vol. 39, pp. 1837-1843

Nilson, J. (1998). *Real-time Control systems with Delays*, Ph. D. Dissertation, Lund Institute of Technology, Sweden, 1998

Nilsson, J., Bernhardsson, B. & Wittenmark, B. (1998). Stochastic analysis and control of real-time systems with random time delays, *Automatica*, vol. 34, pp. 57-64

Tatjewski, P. (2007). *Advanced Control of Industrial Processes,* Springer-Verlag, London

Tutaj, A. (2009). Packets buffering in network traffic in distributed control systems, in: *Proceedings of 12th IEEE international conference on Methods and Models in Automation and Robotics*, 28–31 August 2006, Międzyzdroje, pp. 27–28

Vatanski, N., Georges, J-P., Aubrun, Ch., Rondeau, E. & Jämsä-Jounela, S-L. (2009). Networked Control with Delay Measurement and Estimation, *Control Engineering Practice* vol.17, no.2, pp. 231-244

Walsh, G.C., Ye, H., & Bushnell, L.G. (2002) Stability analysis of networked control systems, *IEEE Transactions on Control Systems Technology*, vol. 10, no. 3, pp. 438-446

Yi, Z. & Heng, P. A. (2002). Stability of Fuzzy Control Systems with Bounded Uncertain Delays, IEEE *Transactions on Fuzzy Systems*, vol. 10, no. 1 (Feb. 2002), pp. 92-97

Cao, Y. and & Zhang, W. (2005). Modified Fuzzy PID Control for Networked Control Systems with Random Delays, in: *World Academy of Science, Engineering and Technology*, vol. 12, pp. 520-523

Zhang, W. (2001). *Stability Analysis of Networked Control Systems*, PhD Thesis, Case Western Reserve University

Zurawski, R. (2005). *The Industrial Communication Systems. Handbook*. CRC Press, ISBN: 9780849330773

17

An Application of Robust Control for Force Communication Systems over Inferior Quality Network

Tetsuo Shiotsuki
Tokyo Denki University
Japan

1. Introduction

The developments of computer and network technologies have provided a virtual reality environment and ubiquitous network systems. Especially audio-visual devices play an important role in communication. For example, voice communication by telephone, audio-visual communication, streaming technology, digital television system and so on. However, we know that the human makes communication not only by audio-visual information but also by using all five-senses (touch, taste, hearing, eyesight, and smell). The realization of the five-senses communication system is one of the prospected technologies.
Especially force communication is a hopeful application in the coming e-world. Several kinds of gimmicks can be considered for transmitting or exchanging the sense of touch, haptic, tactile, force and kinesthetic. In the area of the wearable computing technologies some force-like communication system is realized by using pressure, tension, bending, stress sensors and vibration or pressure actuators, which give the illusion of force communication. On the other hand, robotic researchers have discussed on bilateral tele-operation systems, which realizes remote-manipulations with the sense of reaction forces caused by collision or touching of remote objects and environments. An aim of the technology is that the communication channel between two terminals simulates as if a rigid rod or tight rope. In this article, we consider the bilateral tele-operation systems as a force communication device. It is a well-known that the computer network has inevitable time-delay and jitter in the transmission of the data. And in control engineering deterioration of the stability and performance of the closed loop systems is a well-known fact. Control researchers have proposed several kinds of approaches to overcome the problems. The rest of the chapter is composed as follows. In Section 2, a characterization of the computer network from the view point of transmission delay is discussed. In Section 3, control systems of force sensorless bilateral tele-operation system and the problems caused by transmission delay are examined with a brief historical review. Section 4 presents a procedure how to design a robust control system over the uncertain time-delay network. In section 5 a simulation result is introduced, and some discussions are presented. In section 6 experimental results over the real broadband computer network are introduced. And the results of experiments and investigation are explained in detail. Section 6 concludes the article.

2. Communication network and time delay

Fusion of computer and tele-communication technologies has provided the revolution of the computer network such as the Internet. Before the revolution usual tele-communication is established in two steps. Firstly, according to the request from the sender the system searches the receiver and establishes a communication channel by reserving network resources exclusively. Secondly, the session starts on the reserved real communication channel. After the end of the session, the reserved resources are released. In this case, the time-delay over the communication channel is so small as can be ignored.

On the other hand, communication on the computer network between two terminal nodes is realized as a set of the exchange process of datagrams (frame, packet, cell,Ąc). For example, the information is converted into digital data and divided into datagrams. These datagrams are put on the node and travel along the path while looking for appropriate next node until they reach to the destination. In general the data exchange process includes the huge number of data processing such as encoding, storing and (route) switching. And the length of the processing time depends on the size of datagram and transmitting rate the busyness of the equipments. Especially the network routers are shared by multi-users. Since the practically implemented algorithm is almost trying and error type, the data buffer sometimes overflowed and fails data(packet loss). In order to ensure the reproducibility of the data several kinds of data processing algorithms are implemented according to transmission protocols. TCP/IP(Transmission Control Protocol/Internet Protocol) provides confirming of receiving data (acknowledge), control of window size, and data retransmission and so on. Because of the complexity of the mechanism and sharing of the resource of the network the time-delay is greater than the circuit channel type communications. And the jitter, variation of the time delay occurs frequently.

Fig.1 shows an example of time delay during a day between two campuses(Kumamoto and Fukuoka) in 1998. The left graph (a) indicates the time series from midnight to midnight, and the right(b) is the histogram of the number of packets with respect to transmission delay. It is too difficult to construct the prediction model of time-delay because of the randomness and chaos. Here we adopt a statistical model as a rectangular distribution as follows.

$$0 < L \leq L_{max} \tag{1}$$

In practice it is possible to set L_{max} such that 95% of packets are travels in the time interval $[0, L_{max}]$.

3. Historical review of tele-operation systems

3.1 Master slave system

Suppose the situation in which an operator manipulates (push/ pull/ lift/ put on and so on) some object through the communication network. Such a kind of system is called a master slave system or *tele-operation system*. Usually the terminals for the operator and the object are called the *master* and the *slave* equipment each other. Operators motion is converted to the motion data by the master mechanism, transmitted to the slave side and realized as a motion of the slave equipment. If the system can transmit the force information caused in the slave side to the master side it called *bilateral tele-operation systems*. Several kind of mechanisms are proposed for the control of bilateral tele-operation systems.

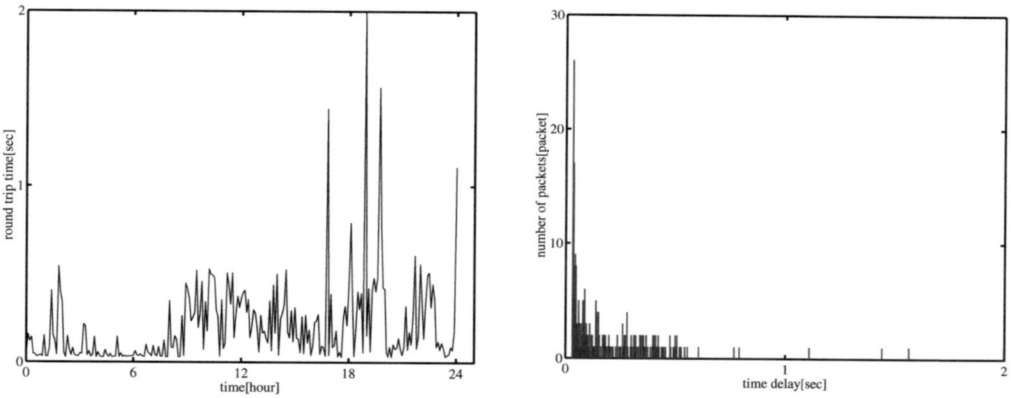

Fig. 1. An example of transmission delay of the Internet: (left)time history(time vs. delay), (right) Histogram (delay vs. frequency)

PE(position error) type is the most simple one. Master and slave exchange the position data each other. Both controllers compensate the deviation of the positions independently. It means that the system is a combination of two position feedback control systems. When the master and the slave equipments have the same characteristics the structure of the system is completely symmetric. In FR(Force reflection) type, master equipment transmits the position data and receives the force data from the slave equipment. On the other hand FRP(Force reflection with passivity) type exchanges the velocity data and the force data respectively.

3.2 Delay and instability

It is well-known that the time delay in the loop deteriorates the stability, performance and robustness of the feedback systems. Fig.2(b) shows a demonstration of the tarde-off of gain K and time-delay L. If (L, K) is chosen in the range of *stable* region the closed loop system depicted in Fig. 2(a) becomes stable, and vice versa. This trade-off curve is identical to the contour of the H_∞ norm of the transfer function

$$\gamma := \left\| \frac{K}{s^2 + K}(1 - e^{-Ls}) \right\|_\infty \tag{2}$$

with $\gamma = 1$. It is easy to calculate that the transfer function matches to one at the cutting point a in Fig.2.

This plot says the following facts.

(1) If $L = 0$ then the gain margin is infinite.

(2) The gain margin decreases rapidly as the time-delay grows, that is, the robustness to time-delay deteriorate as the time-delay grows.

According to the considerations the necessity of careful investigation to time-delay is required.

3.3 Scattering and wave variable method

Anderson and Spong (6) introduced a new communication architecture for tele-operation over the network with time-delay. Their method is based on the passivity and scattering representation of the network. Thus the strictly passivity of master and slave systems

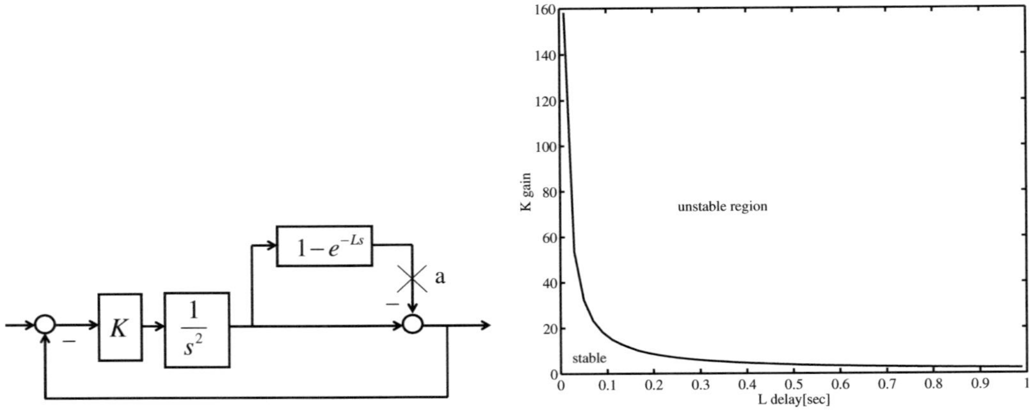

Fig. 2. Trade-off between time-delay and loop gain (delay vs. loop gain)

and stationary time-delay are assumed, which are strong constraints for design. Moreover Niemeyer and Slotine (7) extended their method by using wave variables. Since it is a generalization of Anderson-Spong method, it has the same constraints and difficulties in practice. On the other hand Leung, Francis and Apkarian (8) proposed a controller designed via $\mu-$Synthesis. The proposed method based on robust control theory can deal fluctuation of time-delay and has strong practicability. But all the above methods have the same configuration in which the master and slave system exchanges the velocity and the force variables (v, f) through the network. This means that the position, integral of velocity v, of the master and the slave systems are depend on the initial conditions, and the stability is ensured not in the sense of position but velocity. Moreover the necessity of force sensors makes the systems configuration sophisticated. The more simple architecture is prefer for the practical application.

4. Robust control approach

4.1 Paradigm of robust control

There are several kind of strategies to overcome the problem of time delay. Assuming the rectangular (uniform) distribution of time delay H_∞ control theory can be applied as follows. Fig.4 shows the correspondence between time-delay and multiplicative uncertainty. Now let define a 1-st order high-pass filter $W_D(s; L)$ as

$$W_D(s; L) = \frac{As}{s + \dfrac{1}{L}} \tag{3}$$

where $A = 2.102904074495...$ It is easy to verify that the norm of $(e^{-Ls} - 1)$ holds the following inequality for any frequency(on the imaginary axis) and any time delay L with $0 < L \leq L_{max}$

$$|1 - e^{-j\omega L}| < W_D(j\omega; L) \leq W_D(j\omega; L_{max}), \quad \forall j\omega \in j\mathbf{R}, \tag{4}$$

where L_{max} is the upper bound of the estimated time-delay. This means that the uncertainty caused by the variation of time-delay between $[0, L_{max}]$ can be covered by the weighting function $W_D(s; L_{max})$ as a high-pass filter with cut-off frequency $1/L_{max}$[rad/sec]

$$W_D(s, L_{max}) = \frac{2.1s}{s + \frac{1}{L_{max}}}. \tag{5}$$

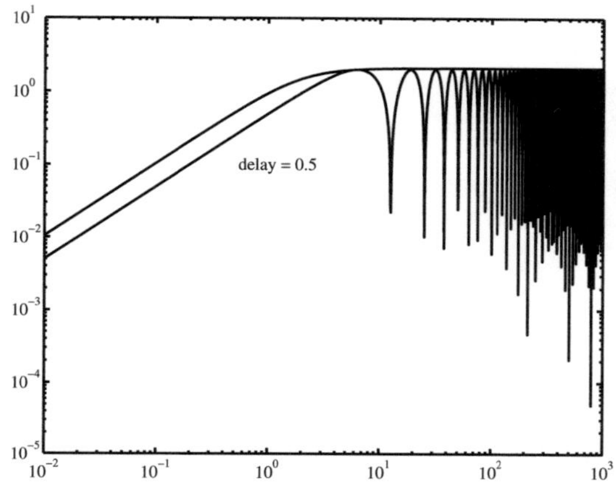

Fig. 3. variation of the gain casued by time delay and inequality in (4)

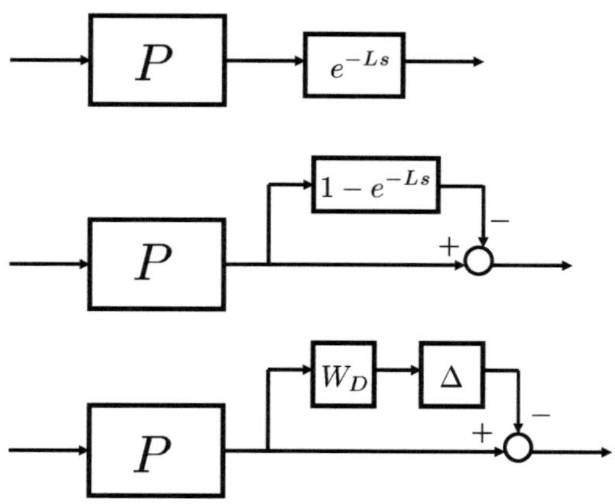

Fig. 4. correspondence between time-delay and multiplicative uncertainty

By using the function $W_D(s; L_{max})$ and uncertain bounded function Δ ($|\Delta| < 1$) the inequality 4 can be replaced as

$$e^{-Ls} = 1 + W_D(s; L_{max})\Delta. \tag{6}$$

4.2 PE type bilateral tele-operation system

Here we introduce a simple PE type bilateral tele-operation system designed with robust control technique. Two joystick mechanisms, corresponds to master and slave, are considered.

Each joystick has a DC-servo motor for torque generation and a position sensor for measurement of the angle of the joy-stick. Force sensors attached to the joysticks are not for use of the control but for the performance evaluation of force communications. They are controlled by computers which are connected to the computer network (see Fig.5). These joysticks are assumed to be modeled as

$$J_m \ddot{x}_m(t) + D_m \dot{x}_m(t) = K_m u_m(t) + f_m(t) \tag{7}$$
$$J_s \ddot{x}_s(t) + D_s \dot{x}_s(t) = K_s u_s(t) - f_s(t) \tag{8}$$

where x , f and u indicate the variables of position of the joystick, external force and input voltage for motor torque generator each other. J and D indicate the physical parameters of inertia and friction each other. The suffixes m, s indicate the master and the slave respectively.

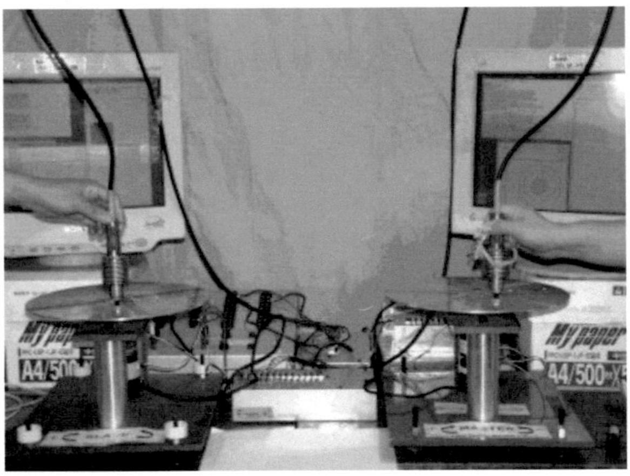

Fig. 5. A view of experimental equipments: Two joysticks controlled by computers connected to the network

J_m	0.0140	[Kgm^2]
J_s	0.0379	[Kgm^2]
D_m	0.0110	[Nms]
D_s	0.0250	[Nms]
K_m	0.2557	[Nm/V]
K_s	0.2557	[Nm/V]

Table 1. Parameters of master and slave joysticks

If there is no time delay between master and slave sides the deviations of the joysticks are evaluated as

$$e_{m0}(t) = x_m(t) - x_s(t) \tag{9}$$
$$e_{s0}(t) = x_s(t) - x_m(t) = -e_{m0}(t). \tag{10}$$

When the master and the slave controllers exchange the information through the network, as stated in the previous section, the time-delay must be considered. Let us assume that the time-delay $L > 0$ exists between master and slave controllers symmetrically. The evaluated deviations (9)(10) at each controller might be computed as follows.

$$e_{mL}(t) = x_m(t) - x_s(t - L) \tag{11}$$
$$e_{sL}(t) = x_s(t) - x_m(t - L) \tag{12}$$

In the rest of the chapter $e_{m0}, e_{s0}(e_{mL}, e_{sL})$ are called as errors in ideal (computed) deviation. The Laplace transform of computed deviation e_{mL} (11) is written as

$$\begin{aligned} E_{mL}(s) &= X_m(s) - e^{-Ls}X_s(s) \\ &= X_m(s) - X_s(s) + X_s(s) - e^{-Ls}X_s(s) \\ &= E_{m0}(s) + (1 - e^{-Ls})X_s(s). \end{aligned} \tag{13}$$

In the same way the Laplace transform of e_{sL} can be written as

$$E_{sL}(s) = E_{s0}(s) + (1 - e^{-Ls})X_m(s). \tag{14}$$

This means that the minimization of computed deviations $(E_{mL}(s), E_{sL}(s))$ is acomplished by the simultaneous minimization of $E_{m0}(s)(= -E_{s0}(s))$, $(1 - e^{-Ls})X_s(s)$ and $(1 - e^{-Ls})X_m(s)$ from the inequality as

$$\begin{aligned} |E_{mL}(j\omega)| &\leq |E_{m0}(j\omega)| + |(1 - e^{-j\omega L})X_s(j\omega)| \\ &\leq |E_{m0}(j\omega)| + |W_D(j\omega; L)X_s(j\omega)|. \\ &\leq |E_{m0}(j\omega)| + |W_D(j\omega; L_{max})X_s(j\omega)|. \end{aligned} \tag{15}$$

$$|E_{sL}(j\omega)| \leq |E_{s0}(j\omega)| + |W_D(j\omega; L_{max})X_m(j\omega)|. \tag{16}$$

As mensioned in previous section the time delay L includes uncertainty. But if the upper bound of L is obtained as L_{max} according to 4and 5 the minimization problem can be acomplished by the minimization of $E_{m0}(s)$,$W_D(s; L_{max})X_s(s)$ and $W_D(s; L_{max})X_m(s)$.
H_∞ theory gives a design method to obtain an appropriate feedback gain to keep stability and robustness against the type of model uncertainty.

4.3 Plant model

Let us consider the two joystick mechanisms as a system with two inputs and two outputs plant

$$\begin{bmatrix} X_m(s) \\ X_s(s) \end{bmatrix} = \begin{bmatrix} P_m(s) & 0 \\ 0 & P_s(s) \end{bmatrix} \begin{bmatrix} U_m(s) \\ U_s(s) \end{bmatrix}. \tag{17}$$

where,

$$P_m(s) = \frac{K_m}{s(J_m s + D_m)}, \quad P_s(s) = \frac{K_s}{s(J_s s + D_s)}. \tag{18}$$

The purpose is the design of a controller

$$\begin{bmatrix} U_m(s) \\ U_s(s) \end{bmatrix} = \begin{bmatrix} C_{mm}(s) & C_{ms}(s) \\ C_{sm}(s) & C_{ss}(s) \end{bmatrix} \begin{bmatrix} X_m(s) \\ X_s(s) \end{bmatrix} \tag{19}$$

which satisfies the requirements specified as follows. The schematic diagram is depicted in Fig.6

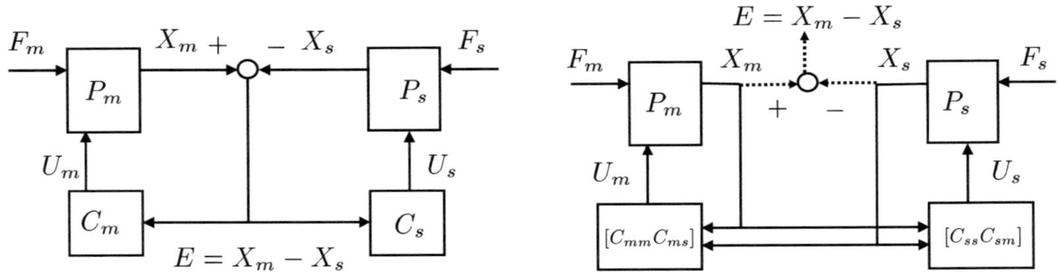

Fig. 6. scheme of PE type master-slave system (a)ideal (non time-delay) scheme,(b) practical(implementable) scheme

tracking performance

The closed loop system is a kind of regulator which makes the deviation $e = X_m - X_s \to 0$ as time goes. For the robust control systems design the deviation e is generalized to a criteria for tracking performance as

$$z_1 = W_{11}e, \quad e = x_m - x_s. \tag{20}$$

In general W_{11} has to be chosen as high gain at low frequency and low gain at high frequency.

stability augmentation

More over in order to obtain a adequate local feedback gain which improves the stability and robustness of closed loop system, the criteria for stability is formulated as

$$z_2 = W_{12}x_m, \quad z_3 = W_{13}x_s. \tag{21}$$

properness of controller

In order to ensure the properness of the controller $C(s)$ in (19) the input variables for the plant are added to the criteria for design as

$$z_4 = W_{21}(w_1 + u_1), \quad z_5 = W_{22}(w_2 + u_2) \tag{22}$$

where w_1, w_2 are exogenous inputs or exerted external forces(torques) as in Fig.7.

robust stability against time-delay

As mentioned in the previous section the robustness corresponds to the minimization of $W_D(s; L_{max})X_s(s)$ and $W_D(s; L_{max})X_m(s)$. Thus we introduce new two output variables

$$z_6 = W_D x_m, \quad z_7 = W_D x_s. \tag{23}$$

with two more variables w_3, w_4 which come from the uncertainty (6) as

$$w_3 = \Delta z_6 \ , \ w_4 = \Delta z_7. \tag{24}$$

4.4 Construction of generalized plant

Let's define the exogenous input w and evaluated output z as

$$w = \begin{bmatrix} w_1 \ w_2 \ w_3 \ w_4 \end{bmatrix}^T \tag{25}$$

$$z = \begin{bmatrix} z_1 \ z_2 \ z_3 \ z_4 \ z_5 \ z_6 \ z_7 \end{bmatrix}^T. \tag{26}$$

Moreover defining the following vectors

$$U = \begin{bmatrix} U_m \ U_s \end{bmatrix}^T \tag{27}$$

$$X = \begin{bmatrix} X_m \ X_s \end{bmatrix}^T \tag{28}$$

the generalized plant is obtained as follows (see Fig.7).

$$\begin{bmatrix} z \\ X \end{bmatrix} = \begin{bmatrix} G_{11} & G_{12} \\ G_{21} & G_{22} \end{bmatrix} \begin{bmatrix} w \\ U \end{bmatrix} \tag{29}$$

$$\begin{bmatrix} G_{11} & G_{12} \\ G_{21} & G_{22} \end{bmatrix} = \left[\begin{array}{cccc|cc} W_{11}P_m & -W_{11}P_s & W_{11} & -W_{11} & W_{11}P_m & -W_{11}P_s \\ W_{12}P_m & 0 & W_{12} & 0 & W_{12}P_m & 0 \\ 0 & W_{13}P_s & 0 & W_{13} & 0 & W_{13}P_s \\ W_{21} & 0 & 0 & 0 & W_{21} & 0 \\ 0 & W_{22} & 0 & 0 & 0 & W_{22} \\ W_{D1}P_m & 0 & 0 & 0 & W_{D1}P_m & 0 \\ 0 & W_{D2}P_s & 0 & 0 & 0 & W_{D2}P_s \\ \hline P_m & 0 & 0 & 0 & P_m & 0 \\ 0 & P_s & 0 & 0 & 0 & P_s \end{array}\right] \tag{30}$$

By applying the controller (19) to the above system a transfer function matrix from w to z

$$G_{zw}(s) = G_{11} + G_{12}(I - CG_{22})^{-1}CG_{21} \tag{31}$$

can be obtained. By using the design procedure based on the H_∞ control theory a controller is obtained such that

$$\|G_{zw}(s)\|_\infty < \gamma \tag{32}$$

where γ is a design parameter chosen as small as possible (5).

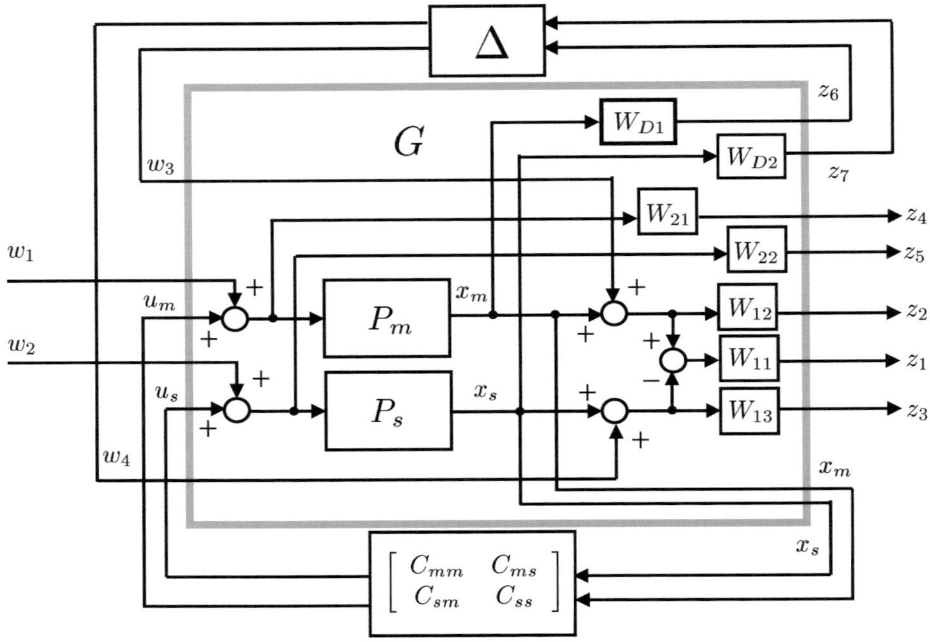

Fig. 7. generalized plant for master-slave systems

5. Simulation and estimates of robustness

In order to demonstrait the robustness w.r.t. time delay computer simulations on the stability and tracking analysis are shown here. The time-delay is assumed as $L_{max} = 0.1$[sec]. According to the specifications discussed above weighting functions $\{W_i\}, i = 1, 2, 3$ are set as follows.

$$W_{D1}(s) = W_{D2}(s) = \frac{2.1s}{s+10} \tag{33}$$

$$W_{11}(s) = \frac{1.0 \times 10^5 s + 2.0 \times 10^5}{1.0 \times 10^4 s + 1.0} \tag{34}$$

$$W_{12}(s) = W_{13}(s) = \frac{0.01s+1}{0.1s+1} \tag{35}$$

$$W_{21}(s) = W_{22}(s) = 10 \tag{36}$$

Fig.8 shows the bode diagrams of the transfer functions.
Table 2 shows simulation results for analysis of stability and tracking performance.
The values in the left column indicate actual time-delay L_{act} in simulation. When the actual time-delay is not greater than the assumed maximum one, that is $L_{act} \leq L_{max}$, the stability and tracking performance are kept well. On the other hand if the time-delay exceeds estimated value , $L_{act} > L_{max}$, the performance of the system becomes worse. It points that the importance of the estimate of maximum time delay L_{max}.

<table>
<tr><th>L[sec]</th><th>stability</th><th>tracking</th></tr>
<tr><td>0.00</td><td>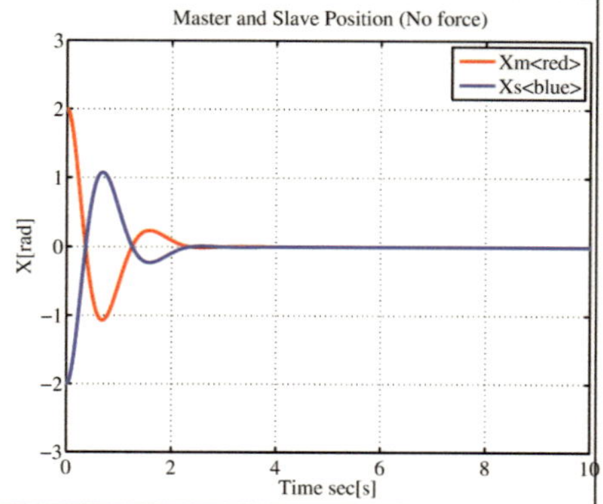
</td><td>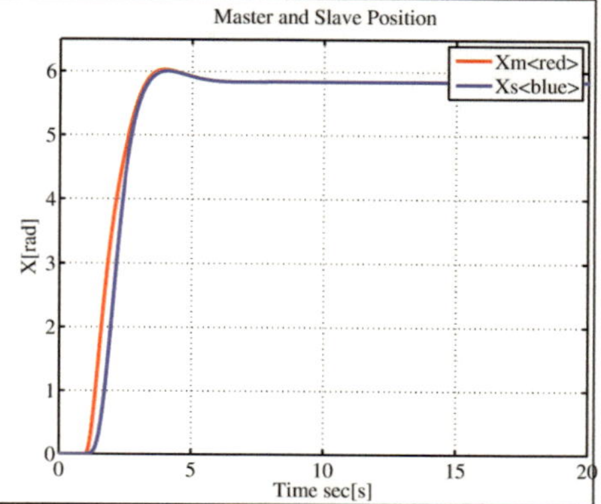
</td></tr>
<tr><td>0.10</td><td>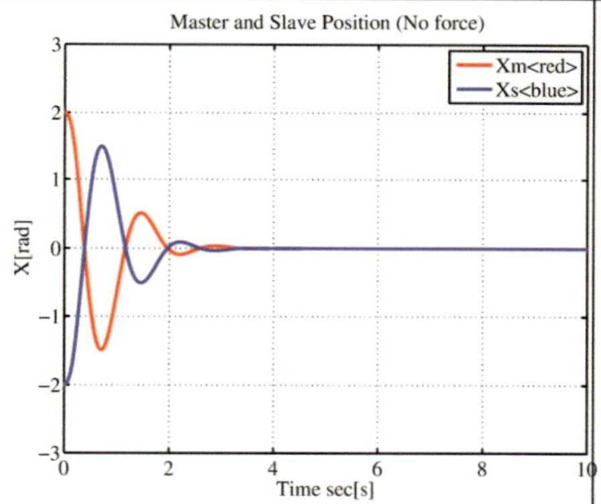
</td><td>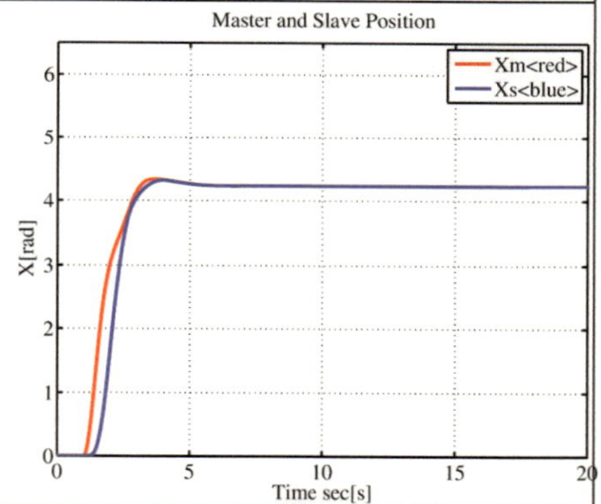
</td></tr>
<tr><td>0.50</td><td>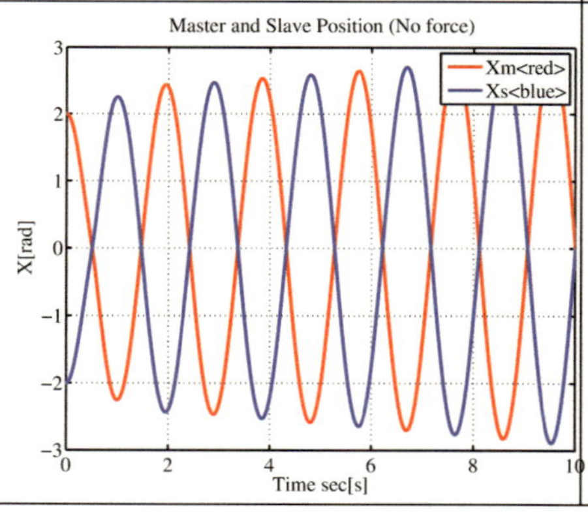
</td><td>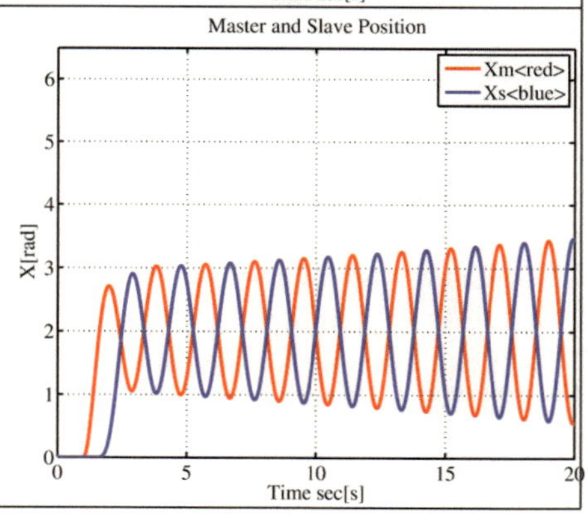
</td></tr>
</table>

Table 2. Analysis of stability and tracking performance w.r.t. time-delay

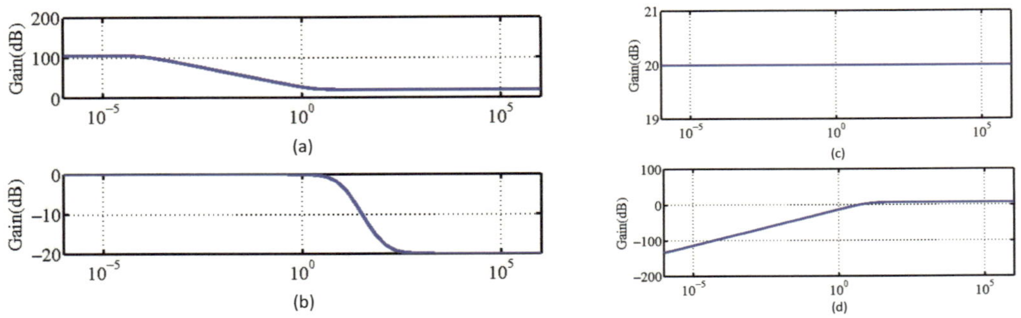

Fig. 8. Bode (gain) diagrams (a) W_{11} ,(b) W_{12}, W_{13}, (c) W_{21}, W_{22}, (d) W_{D1}, W_{D2}

Analysis of designed controller

Fig. 9 shows bode diagrams of controller $C(s)$. It can be observed that roughly seeing of the controller is a kind of integrator, but it works constant gain in the middle range $1 \sim 10^3$[rad/sec].

5.1 Analysis via hybrid matrix

In order to investigate the force communication ability and transperency of master-slave system hybrid matrix is defined as

$$\begin{bmatrix} F_s \\ X_m \end{bmatrix} = \begin{bmatrix} h_{11} & h_{12} \\ h_{21} & h_{22} \end{bmatrix} \begin{bmatrix} X_s \\ F_m \end{bmatrix}. \tag{37}$$

Table 10 shows the bode daigram of hybrid matrix. $h_{12} = F_s/F_m$ and $h_{21} = X_m/X_s$ indicate that the tracking ability of position and force communication are expected to work in the range from DC upto 1 rad/sec.

6. Experiments over the network with time-delay

In order to demonstrate the robustness of the proposed control systems a networked control system is constructed as in Figure 11.

Master and slave mechanisms and their local controllers (C_m, C_s) are located on the same cite (at Kumamoto city), and another computer (C_T) is located beyond the network (at Kitakyushu city, 150km far from Kumamoto city). These three computers are connected to the network JGN , which was Japanese broadband network as an experimental testbed administrated by TAO [1]. The controllers C_m and C_s can communicate each other by way of relay computer C_T, but not admitted to communicate directly. The transmission capacity of the network is about 100Mbps. The control period at C_m and C_s is 5[msec] and that of communication period between C_m and C_s is 10[msec]. The communication protocol UDP/IP is adopted.

[1] Telecommunications Advancement Organization of Japan; reorganized to NICT(National Institute of Information and Communications Technology) in 2004 (http://www.nict.go.jp/)

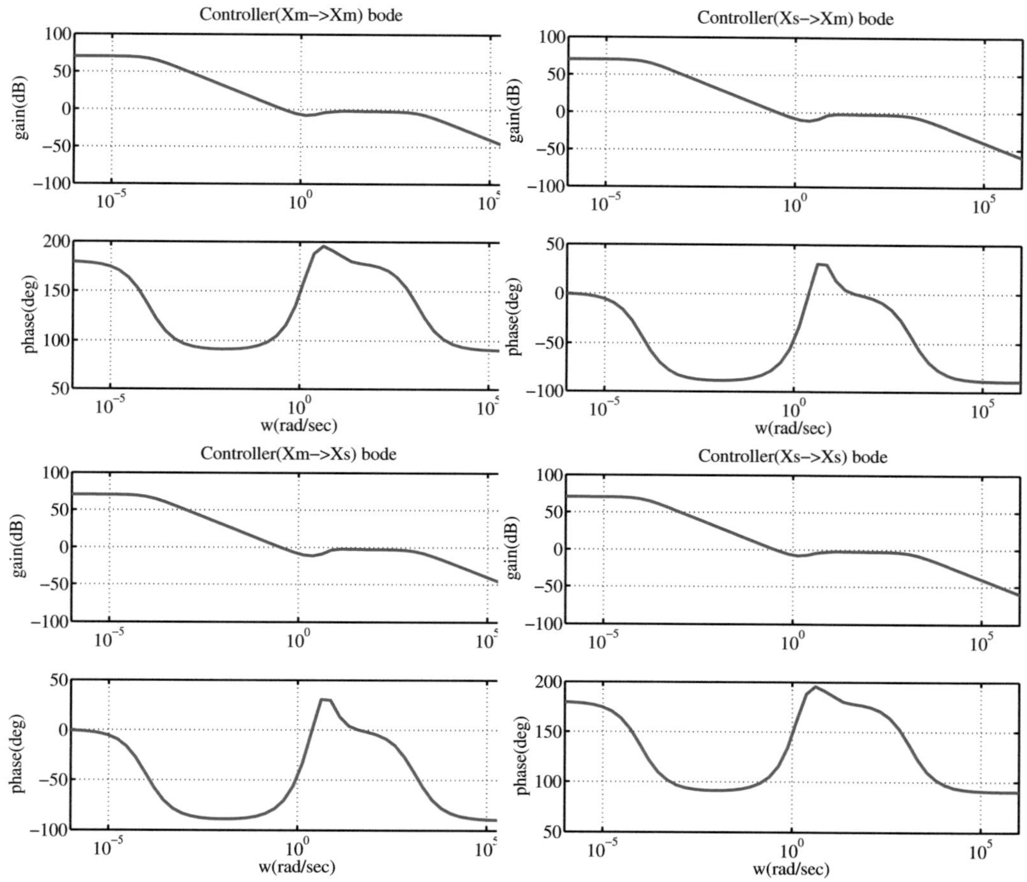

Fig. 9. Bode diagram of the controller $C(s)$

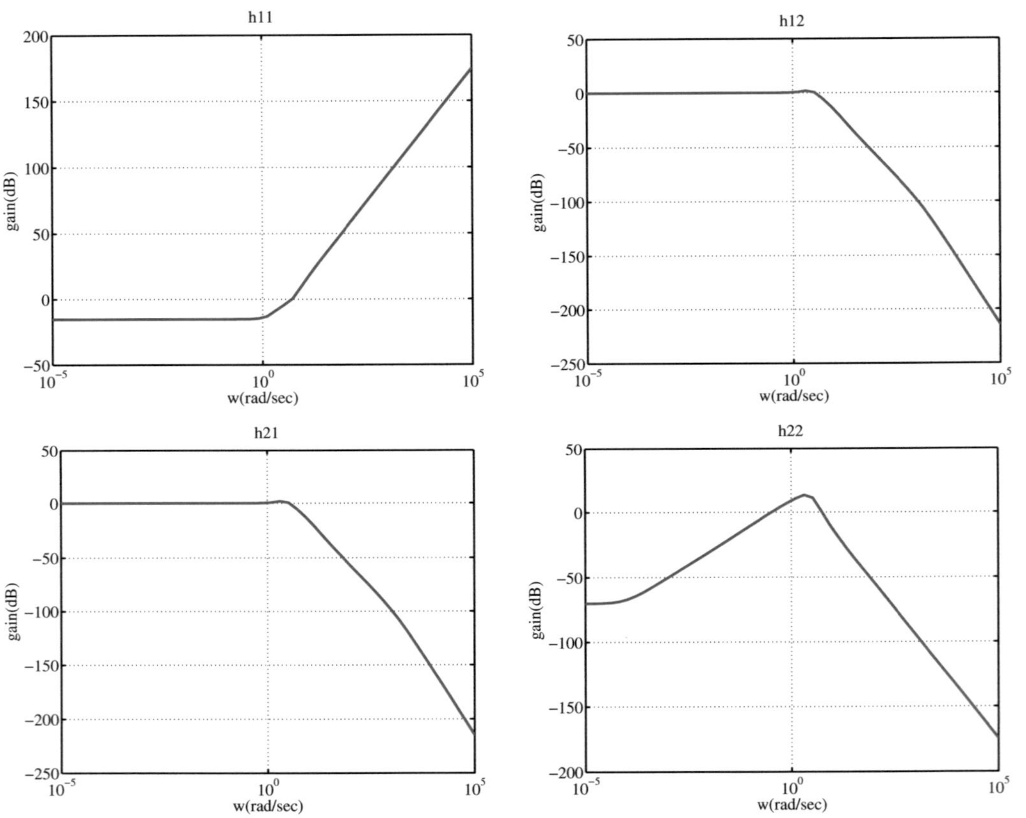

Fig. 10. Bode (gain) diagram of hybrid matrix (37)

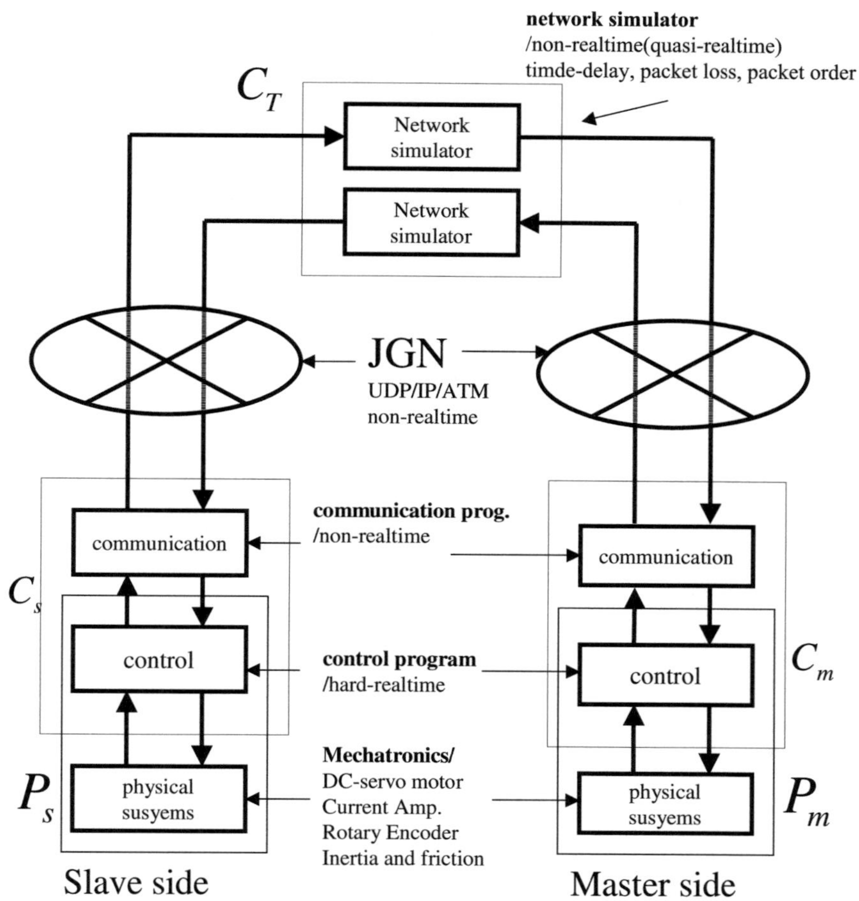

Fig. 11. experimental network system for robust control on JGN

design parameters

Weighting functions in generalized plant (29) are specified as

$$W_{11} = \frac{0.3s + 20 \times 10^7}{3.9 \times 10^6 s + 10^5}, \tag{38}$$

$$W_{12} = W_{13} = \frac{1}{1.5}, \tag{39}$$

$$W_{21} = W_{22} = 0.2, \tag{40}$$

$$W_D = \frac{2.1s + \epsilon}{s + \frac{1}{L_{max}}} \tag{41}$$

By specifying the allowable time-delay L_{max} and upper bound of H_∞ norm γ in (32) the controller $C(s)$ is obtained by using MATLAB [2].

[2] MATLAB is a product of The MathWorks, Inc.

implementation issues

The controllers are implemented in personal computers. The algorithms are coded by C-language with RT-Linux formats and embedded as a kernel modules of Linux system. The control period is set at 5 [msec] and data exchange rate is set at 10 [msec]. Thus the data processing sequences must be synchronized.

Fig.12 shows a sequence diagram. The time goes from left to right and the datagram travels from master side(top) to slave side(bottom) through the network(middle). Because of the control period is a half of the communication period a copied value of the oposit side is used once every two control calculation. The right half part of the diagram assumes the case of long time-delay. In this case copied value is used over and over again until the new datagram reaches again.

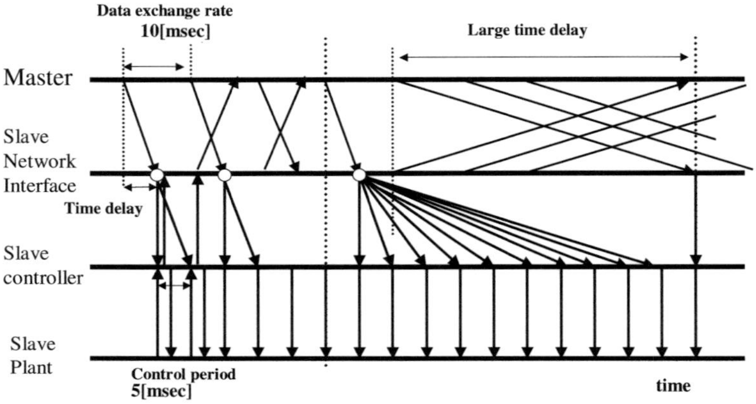

Fig. 12. Sequence diagram of data exchanges between master and slave

MASTER	CPU	AMD Duron 600MHz
	NIC	100/10 Base T
	OS	RT-Linux 3.1 on Linux 2.2.19
SLAVE	CPU	Pentium 75+ 166MHz
	NIC	100/10 Base T
	OS	RT-Linux 2.2 on Linux 2.2.14
RELAY	CPU	Pentium 600MHz
	NIC	100/10 Base T
	OS	Linux 2.2.14

Table 3. Parameters of master and slave controllers

emulation of network with poor quality

C_T is a computer located beyond the network to emulate various kind of qualities. It can emulate various kind of probability distribution of transmission delay, packet loss, packet shuffling and so on. Here we set the maximum time delay 1.0[sec] and the jitter in the Pareto distribution . The design parameter for robustness w.r.t. time-delay is set at $L_{max} = 1.5$[sec] and $\epsilon = 6.6 \times 10^{-3}$ Here the $\epsilon > 0$ is selected to reduce the effects of integrator.

experimental result

Fig. 13 indicates an experimental result for evaluation of tracking performance. In the first half master(solid line) leads to the slave(dashed line). And last half slave leads to the master. This means that the master-slave tele-operation system works symmetricaly well.

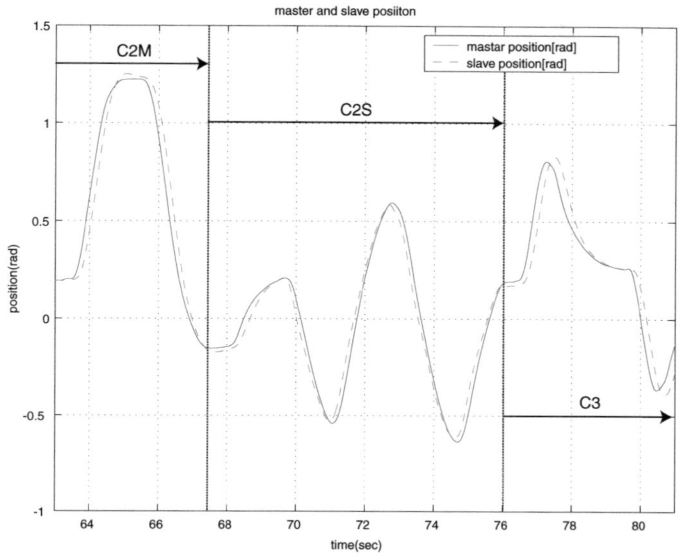

Fig. 13. experimental result of position tracking: master (- slid line), slave(– dashed line)

7. Summary

In this chapter we discussed on an application of robust control for force communication systems over inferior quality network. According to the investigation of the experiments the effectiveness of bilateral tele-operation system for force communication is confirmed. Especially the most important problem of the robustness w.r.t. time-delay is improved by H_∞ control systems theory.

8. References

[1] Shiotsuki,T., Force communication over the computer network, *3rd IFAC Symposium on Mechatronic Systems, Manly Beach, Sydney, Australia*, (2004)353-358

[2] Shiotsuki,T., Nasu,T., A case study of tele-operation system with time-delay, *SICE02* (2002)

[3] Kim,J., Kim,H., Tay,B.K.,Muniyandi, M., Srinivasan,M.A., Jordan, J., Mortensen,J., Oliveira,M. and Slater,M., Transatlantic Touch: A Study of Haptic Collaboration over Long Distance, *PRESENCE*, 13-3, 328-337 (2004)

[4] Yashiro,D., Tian,D., and Ohnishi,K., Centralized Controller based Multilateral Control with Communication Delay, *Proceeding of The IEEE International Conference on Mechatronics*, ICM 2011, Istanbul, Turkey, 13th(2011)

[5] Doyle , Francis and Tannenbaum(1992), *Feedback Control Theory*, Macmillan Publish Company

[6] R. J. Anderson and M. W. Spong, Asymptotic stability for force reflecting teleoperators with time delay *IEEE Trans. Automatic. Contr.* (1988) 1618–1625

[7] Niemeyer,G and Slotine,J.J, Stable adaptive tele-operation, *IEEE Journal of Ocean Eng.* 16-1,(1991)152–162

[8] Leung,G.M.H. Francis,B.A. and Apkarian,J., Bilateral controller for teleoperators with time delay via μ-synthesis, *IEEE Trans.of Robotics and Automation*, RA-11-1,(1995) 105–116

[9] N.L.Johnson A.Kotz and N.Balakrishnan, Continuous uninvariate distributions: Weily series in probability and mathematical statistics, 1, John Weily & Sons, Inc.(1994)

18

Robust Control for Single Unit Resource Allocation Systems

Shengyong Wang, Song Foh Chew and Mark Lawley
University of Akron, Southern Illinois University Edwardsville, and Purdue University
USA

1. Introduction

Supervisory control for deadlock-free resource allocation has been an active area of manufacturing systems research. Most work, however, assumes that allocated resources do not fail. Little research has addressed allocating resources that may fail. Automated manufacturing systems have many types of components that may fail unexpectedly. We develop robust controllers for single unit resource allocation systems with unreliable resources (Chew et al., 2008; Chew et al., 2011; Chew & Lawley, 2006; Lawley, 2002; Lawley & Sulistyono, 2002; Wang et al., 2008; Wang et al., 2009). These controllers guarantee that when unreliable resources fail, parts requiring failed resources do not block the production of parts not requiring failed resources. Further, while resources are down, the system is controlled so that when repair events occur, the system is in a safe and admissible state.

There is little manufacturing research literature on robust supervision. Reveliotis (1999) considers the case where parts requiring a failed resource can be re-routed or removed from the system through human intervention. Park & Lim (1999) address existence questions for robust supervisors. Hsieh (2004) develops methods that determine the feasibility of production given a set of resource failures modelled as the extraction of tokens from a Petri net. In contrast, our work models the failure of the workstation server while assuming that buffer space remains accessible after the failure event. We assume that when the server of a workstation fails, we can continue allocating its buffer space up to capacity, but that none of the waiting parts can be processed and thus cannot proceed along their routes until the server is repaired. We further assume that server failure does not prevent finished parts occupying the workstation's buffer space from being moved away from the workstation and proceeding along their routes. Finally, we assume that server failure does not damage or destroy the part being processed and that failure can only occur when the server is working. The last two assumptions are made for notational efficiency and presentation clarity. They can be easily relaxed by adding appropriate events and state variables to our treatment.

Our objective is to control the system so that failure of an unreliable resource does not prevent processing of parts not requiring the failed resource. When a resource fails, all parts in the system requiring the failed resource for future processing are unable to complete until the failed resource is repaired. Because these parts occupy buffer space, they can block production of parts not requiring the failed resource. Thus, we want to assure that, when unreliable resources fail, the buffer space allocation can evolve under normal operation so that parts not requiring failed resources can continue production. Operation must continue to obey part routings and must assure that when a failed resource is repaired, the system is not in an unsafe state. We refer to supervisors guaranteeing this as *robust*.

The remainder of the chapter comprises the following sections. Most briefly, Section 2 discusses the way we model our systems. An example system is presented in this section to motivate properties that robust controllers must possess. In Section 3, we develop robust controllers for systems with multiple unreliable resources where each part type requires at most one unreliable resource. Specifically, Subsection 3.1 develops two robust controllers using a neighbourhood policy, a modified version of banker's algorithm, and a single step look ahead policy. Subsection 3.2 uses a resource order policy to construct another robust controller; Subsection 3.3 employs a notion of shared buffer capacity to develop a robust controller. Relaxing the restriction, Section 4 builds robust controllers for systems for which part types may require multiple unreliable resources. Finally, Section 5 concludes the chapter and discusses future research directions.

2. Modelling of robust control

There are two subsections in this section. Specifically, we will discuss the way we model our systems in Subsection 2.1. Subsection 2.2 will provide examples to motivate properties that robust controllers must possess.

2.1 The discrete event system

We model our systems using the approach of Ramadge & Wonham (1987). This is necessary to define the properties that we want our supervisors to enforce. The following model is similar to that developed by Lawley & Sulistyono (2002), but differs in that now we have more complex failure scenarios and thus some of the underlying formalism has to be generalized. Figure 1 provides an example for the following development.

The system is defined as a 9–tuple vector $S = \langle R,C,P,\rho,Q,Q_0,\Sigma,\xi,\delta\rangle$. In S, R is the set of system resource types, with $R=R^R\cup R^U$, $R^R\cap R^U=\varnothing$, where R^R is the set of reliable resource types, not subject to failure and R^U is the set of unreliable resource types, subject to failure. Let $C=\langle C_i : i=1\ldots|R|\rangle$ where C_i is the capacity of the buffer space associated with system resource type $r_i\in R$.

The set P of part types is produced by the system with each part type $P_j\in P$ representing an ordered set of processing stages, $P_j=\langle P_{j1}\ldots P_{j|P_j|}\rangle$, where P_{jk} represents the kth processing stage of P_j. Also, let $RP_{jk}=\langle P_{jk}\ldots P_{j|P_j|}\rangle$ be the *residual* part stages. We will use p_{jk} to represent a part instance of P_{jk}. Let $\rho:P_j\to R$ such that $\rho(P_{jk})$ returns the resource type required by P_{jk}. Thus, the route of P_j is $T_j=\langle\rho(P_{j1})\ldots\rho(P_{j|P_j|})\rangle$, and the residual route $RT_{jk}=\langle\rho(P_{jk})\ldots\rho(P_{j|P_j|})\rangle$. Finally, let $\Omega_i=\{P_{jk}:\rho(P_{jk})=r_i\in R\}$, the set of part type stages associated with resource $r_i\in R$.

We will suppose that our resource types are workstations with buffer space for staging and storing parts and a processor or server for operating on parts. We will use the standard assumption from queuing theory that the server is not idle so long as there are unfinished parts in a workstation's buffer space. The resource units that we are concerned with allocating are instances of the workstation's buffer space. The controllers that we design are not intended to allocate the server among parts waiting at the workstation. We assume this to be done by some local queuing discipline.

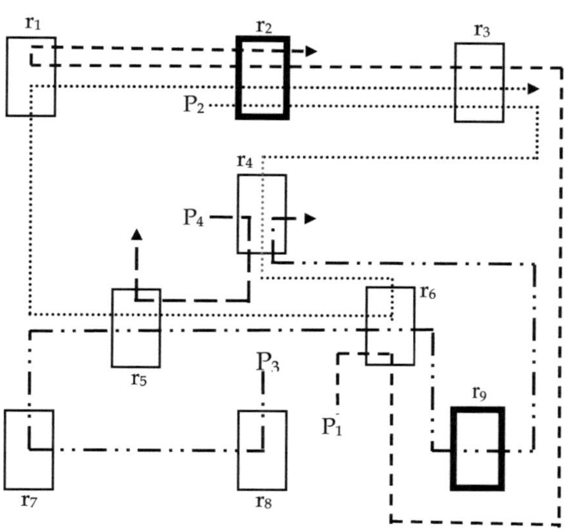

$$
\begin{aligned}
R &= \{r_1, r_2, r_3, r_4, r_5, r_6, r_7, r_8, r_9\} \\
R^R &= \{r_1, r_3, r_4, r_5, r_6, r_7, r_8\} \\
R^U &= \{r_2, r_9\} \\
C &= \langle C_1, C_2, C_3, C_4, C_5, C_6, C_7, C_8, C_9 \rangle \\
&= \langle 1,1,1,2,1,1,1,1,1 \rangle \\
P &= \{P_1, P_2, P_3, P_4\} \\
P_1 &= \{P_{11}, P_{12}, P_{13}, P_{14}, P_{15}\} \\
T_1 &= \langle r_6, r_3, r_2, r_1, r_2 \rangle \\
P_2 &= \{P_{21}, P_{22}, P_{23}, P_{24}, P_{25}, P_{26}, P_{27}, P_{28}\} \\
T_2 &= \langle r_2, r_3, r_4, r_6, r_5, r_1, r_2, r_3 \rangle \\
P_3 &= \{P_{31}, P_{32}, P_{33}, P_{34}, P_{35}, P_{36}\} \\
T_3 &= \langle r_8, r_7, r_5, r_6, r_9, r_4 \rangle \\
P_4 &= \{P_{41}, P_{42}\} \\
T_4 &= \langle r_4, r_5 \rangle \\
\Sigma_c &= \{\alpha_{1i} : i=1...6\} \cup \{\alpha_{2i} : i=1...9\} \cup \\
&\quad \{\alpha_{3i} : i=1...7\} \cup \{\alpha_{4i} : i=1...3\} \\
\Sigma_{u1} &= \{\beta_{1i} : i=1...5\} \cup \{\beta_{2i} : i=1...8\} \cup \\
&\quad \{\beta_{3i} : i=1...6\} \cup \{\beta_{4i} : i=1,2\} \\
\Sigma_{u2} &= \{\kappa_2, \kappa_9, \eta_2, \eta_9\} \\
\psi^{\{r2\}} &= \{\alpha_{1i} : i=1...6\} \cup \{\alpha_{2i} : i=1...9\} \cup \\
&\quad \{\beta_{1i} : i=1...5\} \cup \{\beta_{2i} : i=1...8\} \\
\psi^{\{r9\}} &= \{\alpha_{3i} : i=1...7\} \cup \{\beta_{3i} : i=1...6\} \\
\psi^{\{r2,r9\}} &= \psi^{\{r2\}} \cup \psi^{\{r9\}}
\end{aligned}
$$

Fig. 1. An example system with two unreliable resources

Workstation *failure* will imply the failure of the workstation's server, not any of its buffer space. We will assume that when the server of a workstation fails, we can continue to allocate its buffer space up to capacity, but that none of the waiting parts can be processed and thus cannot proceed along their respective routes until the server is repaired. We further assume that server failure does not prevent finished parts occupying the workstation's buffer space from being moved away from the workstation and proceeding along their respective routes. Finally, we assume that server failure does not damage or destroy the part being processed and that failure can only occur when the server is working.

We are now in a position to define the system states and events. Let Q represent the set of system states, where $Q \ni q = \langle sv_i, y_{jk}, x_{jk} : i=1\ldots|R|, j=1\ldots|P| \text{ and } k=1\ldots|P_j| \rangle$, with sv_i being the status of the server of workstation i (0 if failed, 1 if operational), y_{jk} being the number of unfinished units of P_{jk} (parts waiting or in-process) located in the buffer space of $\rho(P_{jk})$, and x_{jk} being the number of finished units of P_{jk} located in the buffer space of $\rho(P_{jk})$. Q_0 is the set of initial states with $q_0 \in Q_0$ being the state in which no resources are allocated and all servers are operational. The dimension of q is $\sum_{j=1}^{|P|} 2|P_j| + |R|$.

Let $\Sigma = \Sigma_c \cup \Sigma_u$, where $\Sigma_c = \{\alpha_{jk} : j=1\ldots|P| \text{ and } k=1\ldots|P_j|+1\}$ is the set of controllable events with α_{jk} representing the allocation of $\rho(P_{jk})$ to a part instance of P_{jk}; that is, α_{jk} is the event that a part instance of a part type P_j advances into the buffer space of a workstation that will perform its kth operation. Then, $\alpha_{j,|P_j|+1}$ represents a finished part of type P_j leaving the system. We assume that the supervisor controls the occurrences of these events through resource allocation decisions.

We have $\Sigma_u=\Sigma_{u1}\cup\Sigma_{u2}$ being the set of uncontrollable events where $\Sigma_{u1}=\{\beta_{jk} : j=1\ldots|P| \text{ and } k=1\ldots|P_j|\}$ represents the completion of service for P_{jk}. Then, $\Sigma_{u2}=\{\kappa_i,\eta_i : r_i\in R^U\}$ represents the failure (κ_i) and repair (η_i) of the server of unreliable resource $r_i\in R^U$. Service completions, failures and repairs are assumed to be beyond the controller's influence.

Let $\xi:Q\rightarrow 2^{\Sigma}$ be a function that, for a given state, returns the set of enabled events. This function is defined for a state, $q\in Q$, as follows:

1. For $P_{j1}\in\Omega_i$, if $C_i - \sum_{p_{jk}\in\Omega_i}(y_{jk}+x_{jk})>0$, then $\alpha_{j1}\in\xi(q)$.

   Events that release new parts into the system are enabled when space is available on the first required workstation in the route.
2. For $P_{jk}\in\Omega_i$, if $y_{jk}>0$ and $sv_i=1$, then $\beta_{jk}\in\xi(q)$.
   If a part is at service, then the corresponding service completion event is enabled.
3. For $r_i\in R^U$, $P_{jk}\in\Omega_i$ and $\beta_{jk}\in\xi(q)\Rightarrow\kappa_i\in\xi(q)$.
   If the server is busy with a part, then the corresponding failure event is enabled.
4. For $r_i\in R^U$, if $sv_i=0$, then $\eta_i\in\xi(q)$ and $\beta_{jk}\notin\xi(q)\ \forall P_{jk}\in\Omega_i$.
   If the server is failed, the corresponding repair event is enabled and the corresponding service completion events are disabled.
5. For $P_{jk}\in\Omega_i$, $1<k\le|P_j|$, if $x_{j,k-1}>0$ and $C_i - \sum_{p_{jk}\in\Omega_i}(y_{jk}+x_{jk})>0$, then $\alpha_{jk}\in\xi(q)$.

   When a part finishes its current operation and buffer space becomes available at the next required workstation in its route, the event corresponding to advancing the part is enabled.
6. For $P_{j,|Pj|}\in\Omega_i$, if $x_{j,|Pj|}>0$, then $\alpha_{j,|Pj|+1}\in\xi(q)$.
   If a part has finished all of its operations, the event corresponding to unloading it from the system is enabled.

The state transition function is now defined as follows. The transition function, δ, is a partial function from the cross product $Q\times\Sigma$ to the set Q of system states. Specifically, let $\delta:Q\times\Sigma\rightarrow Q$ such that

$\delta(q,\alpha_{jk})=q-e_{x_{j,k-1}}+e_{y_{jk}}$, advancing a part $p_{j,k-1}$;

$\delta(q,\beta_{jk})=q-e_{y_{jk}}+e_{x_{jk}}$, service completion of a part p_{jk};

$\delta(q,\kappa_i)=q-e_{sv_i}$, failure of server i;

$\delta(q,\eta_i)=q+e_{sv_i}$, repair of server i;

where $e_{x_{j,k-1}}$, $e_{y_{jk}}$, $e_{x_{jk}}$, and e_{sv_i} are the standard unit vectors with components corresponding to $x_{j,k-1}$, y_{jk}, x_{jk} and sv_i being 1, respectively. Note that, $e_{y_{j,|Pj|+1}} = e_{x_{j0}} = \mathbf{0}$, the zero vector with the same dimension, and that p_{j0} represents a raw part of P_j waiting to be released into the system.

We assume that $|R^U|\ge 1$. In this case, any subset of the unreliable resources can be simultaneously in a failed state. Thus, if one of the $\binom{|R^U|}{i}$ subsets of size i, $i=1\ldots|R^U|$, is down, we want the remaining resources to continue producing parts not requiring any of the failed resources without human intervention to remove or rearrange the parts requiring failed resources. Further, when one of the failed resources is repaired, we want production of parts requiring that resource to resume. A robust controller must possess certain properties in order to accomplish the above-mentioned characteristics.

2.2 Motivating examples for properties of robust supervisory control

This subsection motivates a set of desired properties for a robust controller based upon an example production system. Figure 1 presents an example manufacturing system with two unreliable resources. The stages, routes, and resource capacities are given, as is the complete discrete event model. This model enumerates the resources, capacities, events, and so forth. For now, we will constrain our discussion to the system states presented in Figures 2-4. We recall that, by definition, a resource allocation state is safe if, starting from that state, there exists a sequence of resource allocations/deallocations that completes all parts and takes the system to the empty and idle state, the state in which no resources are allocated and no servers are busy. Our underlying assumption is that if a resource allocation state is safe, then, under correct supervision and starting from that state, it is possible to produce all part types indefinitely.

We have several control objectives for the system of Figure 1. First, we desire that the controller guarantee deadlock-free operation, i.e., that it keeps the system producing all part types. Second, in the event that r_2 fails, we want to continue producing part types not requiring r_2, $\{P_3,P_4\}$, without having to intervene by clearing the system of parts requiring r_2. Similarly, in the event that r_9 fails, we want to continue producing part types not requiring r_9, $\{P_1,P_2,P_4\}$, again without having to intervene by clearing the system of parts requiring r_9. Further, if both r_2 and r_9 are in the failed state, we want to continue producing part types not requiring r_2 or r_9, $\{P_4\}$, again without explicit intervention.

Consider for example the state given in Figure 2. This state is safe; however, if r_2 fails while processing part p_{27} in this state, the production of both P_3 and P_4 will be blocked by two p_{23}'s at r_4. Note that if we advance a p_{23} from r_4 to r_6, then production of P_4 can proceed. However, production of P_3 will now be blocked. Thus, this state does not satisfy our condition that after the failure of r_2, we should be able to continue producing both P_3 and P_4.

As another example, consider the state of Figure 3. Again, we see that this state is safe. However, if r_9 fails while processing part p_{35} in this state, production of part types P_1 and P_2 will be blocked by p_{34} at r_6, although the production of P_4 is unaffected.

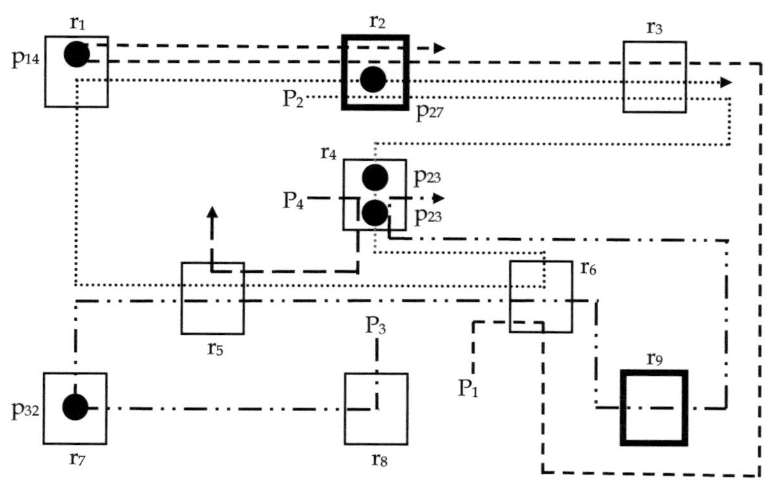

Fig. 2. An undesirable system state since unreliable resource r_2 may fail while processing part p_{27}

Thus, these examples illustrate that parts requiring a failed resource can prevent the system from producing parts not requiring the failed resource through propagation of blocking. Our objective is to develop supervisory controllers that avoid this by guaranteeing that if an unreliable resource fails, it is possible to redistribute the parts requiring that resource so that part types not requiring that resource can continue to produce.

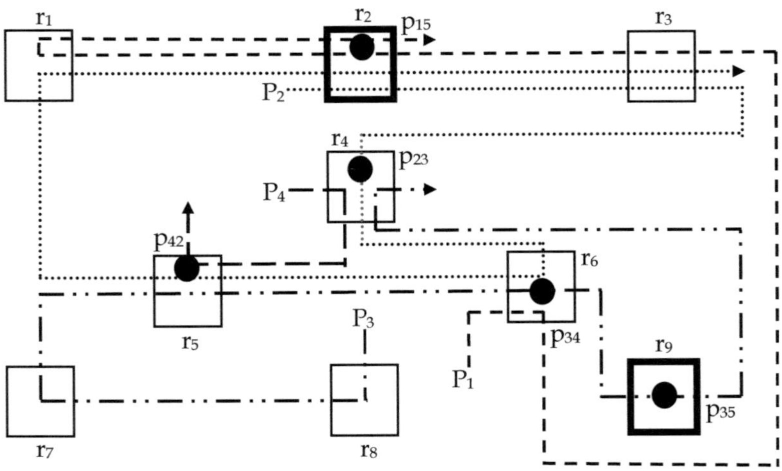

Fig. 3. An undesirable system state since unreliable resource r_9 may fail while processing part p_{35}

For the third objective, consider the state of Figure 4. Again, we see that the state is safe. If r_2 fails, production of P_3 is blocked by p_{11} at r_6. Further, production of P_4 is blocked by p_{33} at r_5. We note that by advancing p_{11} from r_6 to its next required resource, r_3, the blockages of P_3 and P_4 can now be resolved and thus the system can continue producing both P_3 and P_4, as desired. However, when r_2 is repaired, the system is no longer safe since resources r_2 and r_3 are now involved in deadlock. This illustrates that our controller must guarantee that any redistribution of parts requiring the failed resource does not result in system deadlock when the resource is repaired.

The above discussion lays a foundation for a robust supervisory controller. In summary, a supervisory controller is said to be robust to resource failures of R^U if the supervisory controller satisfies Property 2.2.

Property 2.2:

2.2.1: The supervisory controller ensures continuing production of part types not requiring failed resources, given that additional failures/repairs do not occur.

2.2.2: The supervisory controller allows only those states that serve as feasible initial states if an additional resource failure occurs.

2.2.3: The supervisory controller allows only those states that serve as feasible initial states if a failed resource is repaired and becomes operational.

We say that a state is a *feasible initial state* if, starting from that state, it is possible to produce all part types not requiring failed resources. The formal development and definition of this property using language theory is presented in Chew and Lawley (2006).

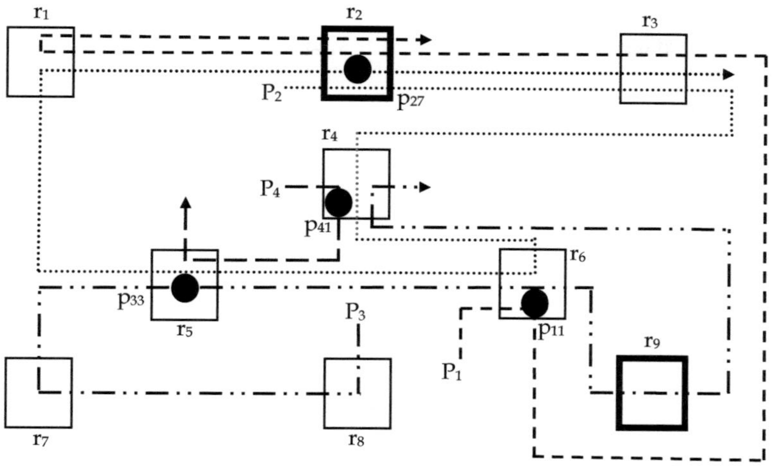

Fig. 4. An undesirable system state since r_2 may fail while processing part p_{27}

3. Robust control for systems with multiple unreliable resources

This section endeavours to delve into robust control for single unit resource allocation systems with unreliable resources.

3.1 Robust control using a neighbourhood policy

This subsection develops controllers that satisfy Property 2.2 above, while maintaining polynomial complexity. Each controller is a conjunction of a modified deadlock avoidance policy and a set of neighbourhood constraints. The deadlock avoidance policy guarantees deadlock-free operation, while the neighbourhood constraints control the distribution of parts that require unreliable resources. Subsection 3.1.1 develops the neighbourhood constraints, NHC. Subsection 3.1.2 constructs a supervisor based on a modified Banker's Algorithm and NHC, while subsection 3.1.3 develops a supervisor based on single-step look-head (SSL) and NHC. The complete proofs can be found in Chew and Lawley (2006).

3.1.1 A neighbourhood policy

In this subsection, we discuss neighbourhood constraints based on the notion of *failure dependency*. Informally, a resource is failure-dependent if every part that enters its buffer space requires some future processing on an unreliable workstation. Thus, all unreliable resources are failure-dependent. Some reliable resources may also be failure-dependent if they only process parts that require future processing on an unreliable resource. This is defined more precisely later. For each failure-dependent resource, we generate a neighbourhood. The neighbourhood of a failure-dependent resource is a virtual space of finite capacity that is used to control the distribution of parts requiring that failure-dependent resource. Again, this is formalized in the following, where we extend the neighbourhood concepts presented by Lawley & Sulistyono (2002) for systems with multiple unreliable resources. We first discuss and illustrate neighbourhood concepts, and then illustrate how neighbourhood constraints are constructed for failure-dependent resources.

Recall that R^U is the set of unreliable resources in the system S, and that $\Omega_i=\{P_{jk}:\rho(P_{jk})=r_i\in R\}$ is the set of part type stages supported by resource r_i. If $r_i\in R^U$, then $r_v\in R$ is said to be *failure-dependent* on r_i if $\forall P_{jk}\in\Omega_v$, $\exists P_{j,k+c}\in\Omega_i$ with $c\geq 0$. In other words, r_v is failure-dependent on r_i if every part that enters the buffer of r_v requires future processing on unreliable resource r_i (note that r_i is failure-dependent on itself). For $r_i\in R^U$, let $R_i^{FD} = \{r_v : r_v\in R \text{ and } \forall P_{jk}\in\Omega_v, \exists P_{j,k+c}\in\Omega_i \text{ with } c\geq 0\}$ be the set of failure-dependent resources on r_i, and let $R^{FD} = \bigcup_{r_i\in R^u} R_i^{FD}$ and $R^{NFD} = R \setminus R^{FD}$.

For each failure-dependent resource of R_i^{FD}, we construct a neighbourhood. The neighbourhood of $r_v \in R_i^{FD}$, NH_v^i, is defined as the set of part type stages that require r_v now or later in their processing and have no intervening failure-dependent resources of R_i^{FD}. Formally, $NH_v^i = \Omega_v \cup \{P_{jk}: \exists c>0 \text{ with } P_{j,k+c}\in\Omega_v \text{ and } \forall d\in[0,c), \rho(P_{j,k+d}) \notin R_i^{FD}\}$. Thus, if $\rho(P_{j,k+c}) = r_v \in R_i^{FD}$, $\rho(P_{j,k-1}) = r_w \in R_i^{FD}$, with $r_v\neq r_w$, and $\{\rho(P_{jk}), \rho(P_{j,k+1})\ldots\rho(P_{j,k+c-1})\} \cap R_i^{FD} = \varnothing$, then $\{P_{jk},P_{j,k+1} \ldots P_{j,k+c-1},P_{j,k+c}\} \subseteq NH_v^i$, and $P_{j,k-1} \notin NH_v^i$. Let $NH^i = \{NH_v^i : r_v\in R_i^{FD}\}$ be the neighbourhood set for $r_i\in R^U$, and let $NH=\{NH^i : r_i\in R^U\}$.

For example, the system of Figure 1 has two unreliable resources, $R^U=\{r_2,r_9\}$. Note that anytime r_1 appears in a route, r_2 appears later in the route, and thus, $R_2^{FD} = \{r_1,r_2\}$. Also, anytime r_7 or r_8 appear in a route, r_9 appears later in the route, so, $R_9^{FD} = \{r_7,r_8,r_9\}$. Thus, $NH^2 = \{NH_1^2, NH_2^2\}$ and $NH^9 =\{NH_7^9, NH_8^9, NH_9^9\}$, where the neighbourhoods are as follows: $NH_1^2 =\{P_{14},P_{22},P_{23},P_{24},P_{25},P_{26}\}$, $NH_2^2 = \{P_{11},P_{12},P_{13},P_{15},P_{21},P_{27}\}$, $NH_7^9 = \{P_{32}\}$, $NH_8^9 = \{P_{31}\}$, $NH_9^9 = \{P_{33},P_{34},P_{35}\}$.

To understand this construction, consider NH_1^2 and NH_2^2. Note that $\Omega_1= \{P_{14},P_{26}\}$ and $\Omega_2=\{P_{13},P_{15},P_{21},P_{27}\}$. Since $\Omega_v \subseteq NH_v^i$, $\{P_{14},P_{26}\} \subseteq NH_1^2$, and $\{P_{13},P_{15},P_{21},P_{27}\} \subseteq NH_2^2$. Now consider $T_1 = \{\rho(P_{11}),\rho(P_{12}),\rho(P_{13}),\rho(P_{14}),\rho(P_{15})\} = \{r_6,r_3,r_2,r_1,r_2\}$. Since $\{r_6,r_3\} \cap R_2^{FD} = \varnothing$, $\{P_{11},P_{12}\} \subseteq NH_2^2$. Similarly, $T_2 = \{\rho(P_{21}),\rho(P_{22}),\rho(P_{23}),\rho(P_{24}),\rho(P_{25}),\rho(P_{26}),\rho(P_{27}),\rho(P_{28})\} = \{r_2,r_3,r_4,r_6,r_5,r_1,r_2,r_3\}$. Since $\{r_3,r_4,r_6,r_5\} \cap R_2^{FD} = \varnothing$, $\{P_{22},P_{23},P_{24},P_{25}\} \subseteq NH_1^2$. Thus, we get $NH_1^2 = \{P_{14},P_{22},P_{23},P_{24},P_{25},P_{26}\}$ and $NH_2^2 = \{P_{11},P_{12},P_{13},P_{15},P_{21}, P_{27}\}$.

Although all parts supported by r_6 later need an unreliable resource, r_6 is shared by r_2 and r_9, and thus it is not failure-dependent on either. This implies that failure-dependent sets are disjoint, i.e., $R_2^{FD} \cap R_9^{FD} = \varnothing$. Furthermore, we observe that no part stage is in more than one neighbourhood, i.e., $NH_1^2 \cap NH_2^2 \cap NH_7^9 \cap NH_8^9 \cap NH_9^9 = \varnothing$. These and other important neighbourhood properties are established in Chew and Lawley (2006).

We restrict the number of parts allowed in a neighbourhood. Our intention is to guarantee that every part in the neighbourhood of a failure-dependent resource has capacity reserved at that resource. That is, we want to be able to advance every part requiring an unreliable resource into its associated failure-dependent resource in the event of a resource failure so that it will not block production of parts not requiring the failed resource. In the example, for a permissible state, we want, for example, every part in $NH_9^9 = \{P_{33},P_{34},P_{35}\}$ to have a reserved unit of buffer at r_9. As a consequence, we will reject a state if this constraint is violated. For instance, a state is not admissible if, at this state, the sum of parts in $NH_9^9 > 1$; recall that r_9 has a single unit of capacity. To see this, at this inadmissible state, if r_9 fails, at least one part of NH_9^9 must reside at r_5 or r_6. Although P_1, P_2, and P_4 do not require failed r_9 in their processing, this distribution of parts may in turn block production of some of these part types. Our objective is to develop supervisory controllers capable of rejecting these undesirable states.

We now construct neighbourhood constraints to enforce the above intention. The constraint for a neighbourhood, say NH_v^i, is an inequality of the form $Z_v^i \le C_v$ where $Z_v^i = \sum_{P_{jk} \in NH_v^i} (x_{jk} + y_{jk})$. Recall that x_{jk} is the number of finished instances, and y_{jk} is the number of unfinished instances, of P_{jk} located in the buffer of $\rho(P_{jk})$; and that the right hand side C_v is the capacity of r_v. NH_v^i is said to be *capacitated* if $Z_v^i = C_v$ and *over-capacitated* if $Z_v^i > C_v$. Define the set of all possible neighbourhood constraints with respect to $r_i \in R^U$ as:

$$NHC_1^i = \{ Z_v^i \le C_v : NH_v^i \in NH^i \}.$$

In the example, we have

$$NHC_1^2 = \{ Z_1^2 = \sum_{P_{jk} \in NH_1^2} (x_{jk} + y_{jk}) \le C_1,\ Z_2^2 = \sum_{P_{jk} \in NH_2^2} (x_{jk} + y_{jk}) \le C_2 \};$$

$$NHC_1^9 = \{ Z_7^9 = \sum_{P_{jk} \in NH_7^9} (x_{jk} + y_{jk}) \le C_7,\ Z_8^9 = \sum_{P_{jk} \in NH_8^9} (x_{jk} + y_{jk}) \le C_8,$$

$$Z_9^9 = \sum_{P_{jk} \in NH_9^9} (x_{jk} + y_{jk}) \le C_9 \}.$$

Constraints of NHC_1^i assure that no neighbourhood of NH^i becomes over-capacitated.
As Lawley & Sulistyono (2002) discuss, NHC_1^i may induce deadlock among failure-dependent resources of R_i^{FD}, since if all neighbourhoods are capacitated, parts cannot move from one neighbourhood to another without over-capacitating a neighbourhood. In the example, a state may satisfy both $1 = Z_1^2 \le C_1 = 1$ and $1 = Z_2^2 \le C_2 = 1$. But, a part moves from one of these associated neighbourhoods to another must over-capacitate the other neighbourhood. To resolve this dilemma, we develop an additional set of constraints, NHC_2^i.
It is first necessary to compute the set of strongly connected neighbourhoods for NHC_2^i. To do this, for each $r_i \in R^U$, we construct a directed graph (NH^i, A^i) where $A^i = \{(NH_g^i, NH_h^i) : \exists P_{jk} \in NH_g^i \text{ with } P_{j,k+1} \in NH_h^i\}$. Thus, in operation, there will be part flow from NH_g^i to NH_h^i. We then compute the set of strong components of (NH^i, A^i) using standard polynomial graph algorithms (Cormen et at., 2002). For example, we see that NH_1^2 and NH_2^2 are strongly connected, since $\{P_{14}, P_{26}\} \subseteq NH_1^2$ and $\{P_{13}, P_{27}\} \subseteq NH_2^2$. Therefore, in operation, there is flow from NH_1^2 to NH_2^2 and from NH_2^2 to NH_1^2. Let SC^i be the set of strongly connected components of (NH^i, A^i). Then, $SC^2 = \{SC_1^2 = \{NH_1^2, NH_2^2\}\}$, and $SC^9 = \{SC_1^9 = \{NH_7^9\}, SC_2^9 = \{NH_8^9\}, SC_3^9 = \{NH_9^9\}\}$. Then, NHC_2^i is stated as follows:

$$NHC_2^i = \{ Z_g^i + Z_h^i < C_g + C_h : \{NH_g^i, NH_h^i\} \subseteq SC_m^i \in SC^i,\ m = 1 \ldots |SC^i| \}.$$

Hence, for every strongly connected component of (NH^i, A^i), NHC_2^i guarantees that at most one neighbourhood can be capacitated at a time. In the example, we have the following:

$$NHC_1^2 = \{Z_1^2 + Z_2^2 < C_1+C_2\},\ NHC_2^2 = \varnothing.$$

NHC_1^2 guarantees that NH_1^2 and NH_2^2 are not simultaneously capacitated.
To summarize, $NHC^i = NHC_1^i \cup NHC_2^i$ guarantees that no neighbourhood is over capacitated, and that neighbourhoods with mutual flow dependencies are not simultaneously capacitated. The complete set of neighbourhood constraints is defined as:

$$NHC = \{NHC^i : r_i \in R^U\}.$$

Note that in the worst case, we generate one constraint for each pair of resources and thus the size of NHC is of $O(|R|^2)$.
Chew and Lawley (2006) establish several important properties of NHC. These properties are required to establish robustness of the two supervisors that we develop later. We next modify two deadlock avoidance policies that we use in conjunction with NHC to develop robust supervisors.

3.1.2 Banker's algorithm

In this subsection, we configure Banker's Algorithm (BA) (Habermann, 1969) to work with NHC. BA is perhaps the most widely known deadlock avoidance policy (DAP), and its underlying concepts have influenced the thinking of numerous researchers. BA is a suboptimal DAP in the sense that it achieves computational tractability by sacrificing some safe states. BA avoids deadlock by allowing an allocation only if the requesting processes can be ordered so that the terminal resource needs for the ith process, P_i, in the ordering can be met by pooling available resources and those released by completed processes P_1, P_2 … P_{i-1}. The ordering is essentially a sequence in which all processes in the system can complete successfully. BA is of $O(mn\log n)$ where m is the number of resource types and n is the number of requests. Other manufacturing related work also uses BA (Ezpeleta et al., 2002; Lawley et al., 1998; Reveliotis, 2000).
Our modifications are straightforward and are a generalization of those undertaken by Lawley & Sulistyono (2002). Our objective is to search for an ordering of parts that advances failure-dependent parts (those requiring unreliable resources) into the resource of their current neighbourhood, and non-failure-dependent parts (those not requiring unreliable resources) out of the system. Again, the ordering is such that the resources required by the first part are all available, those required by the second part are all available after the first part has finished and released the resources held by the part, and so forth. If the system can be cleared in this way (all failure-dependent parts are advanced into failure-dependent resources and all non-failure-dependent parts are advanced out of the system), then we can guarantee that if any unreliable resource fails, the system can continue producing parts that do not require this failed resource.
In the following, let $\Lambda=\Lambda^{NFD}\cup\Lambda^{FD}$ be the set of part type stages instantiated in q *whose parts hold non-failure-dependent resources*, where Λ^{NFD} is the set of non-failure-dependent part type stages (those that do not require failure-dependent resources in the residual route) and Λ^{FD} is the set of failure-dependent part type stages (those that do require failure-dependent resources in the residual route). We now present our modified version of BA as Algorithm A1 as follows.

Algorithm A1:

Query: Is state q admissible?
Input: state q;
Output: ACCEPT / REJECT
Step 1: Initialization
For P(u)
  For r_v R
    ALLOCATION[u][v]=0
    NEED[u][v]=0
    AVAILABLE[v]=C_V
  End For
End For
For P(u)=P_{jk} and r_v = (P_{jk})
  ALLOCATION[u][v]=$x_{jk}+y_{jk}$
End For
For P(u)=P_{jk} NFD
  For r_v RT_{jk}\{ (P_{jk})}
    NEED[u][v]=1
  End For
End For
For P(u)=P_{jk} FD
  For r_v RT_{jk}
    NEED[u][v]=1
  End For
  For r_v RT_{jk} R^{FD}
    NEED[u][v]=0
  End For
  For r_s= (P_{jc}) and r_t= ($P_{j,c+1}$), c=k….|P_j| 1

  NEED[u][t]=NEED[u][s]*NEED[u][t]
  End For
  For r_v RT_{jk}
    If r_v= (P_{jk})
      NEED[u][v]=0
    End If
  End For
End For
For r_v R^{NFD}
  AVAILABLE[v]=C_V $_u$ALLOCATION [u][v]
End For
 =

Step 2: Test and Evaluation
While
  Find P(u) such that NEED[u][v] AVAILABLE[v] for all r_v R^{NFD}
  If no such P(u) exists
    Return REJECT
  Else
     = \{P(u)}

    AVAILABLE[v]=AVAILABLE[v]+ALLOCATION[u][v] for r_v= (P(u))
  End If
End While
Return ACCEPT

Step 1 of the algorithm configures the data structures required. For every part type stage represented in the system, these capture the current resource holding and the future processing need. These structures also capture the resource availability of resources in the state being tested. Three additional comments regarding the algorithm are in order. First, the need for every failure-dependent resource is explicitly set to zero, so this version looks only at the availability of non-failure-dependent resources. Second, for non-failure-dependent part type stages (those not requiring unreliable resources), the need for every resource in the residual route (except the one held) is set to one. Finally, for failure-dependent part type stages (those requiring unreliable resources), the need for every resource in the residual route (except the one held) up to the one immediately preceding the first encountered failure-dependent resource is set to one, all others are set to zero. Note that these are the resources such a part will need to advance into the failure-dependent resource of its current neighbourhood. Step 2 then executes the usual Banker's logic.

Algorithm A1 is not correct by itself, since it does not handle allocation of failure-dependent resources (for detailed examples the reader is referred to the work by Lawley & Sulistyono

(2002)). However, A1 and NHC together form a robust controller, that is, if we allow the system to visit only those states acceptable to both A1 and NHC, then the system operation will satisfy the requirements of Property 2.2. The detailed proofs for this are given in Chew and Lawley (2006). The supervisor is defined as follows:

Definition 3.1.1: Supervisor Δ_1 = A1 $\wedge$ NHC.

Supervisor Δ_1 permits a system state that satisfies both A1 and NHC in runtime. Consider Figure 5, which illustrates a state, say q, in which r_4 holds p_{23} and p_{14}; and r_5 holds p_{33}. It is clear, at q, that there exists an admissible sequence by A1; that is, p_{33} can advance into failure-dependent resource r_9; p_{23} can advance into failure-dependent resource r_1; and finally, p_{41} can be advanced out of the system. In addition, q satisfies NHC since

$$NHC_1^2 = \{ Z_1^2 = 1 \leq 1, \text{(there is one } P_{23}) \quad Z_2^2 = 0 \leq 1\};$$

$$NHC_1^9 = \{ Z_7^9 = 0 \leq 1, \quad Z_8^9 = 0 \leq 1, \quad Z_9^9 = 1 \leq 1 \text{ (there is one } P_{33})\};$$

$$NHC_2^2 = \{1 + 0 < 1 + 1 = 2\}; \; NHC_2^9 = \varnothing.$$

Therefore, q is an admissible state by Δ_1. Supervisor Δ_1 will prohibit, at q, advancing p_{23} into r_6 (where it becomes p_{24}) because p_{24} and p_{33} will block, causing the resulting state to violate A1 although not NHC. Loading a p_{11} into r_6 at q is also precluded by Δ_1 since the resulting state violates NHC_2^2 , although not A1. However, advancing p_{33} one step into r_6 or loading a p_{31} into r_8 will result in an admissible state.

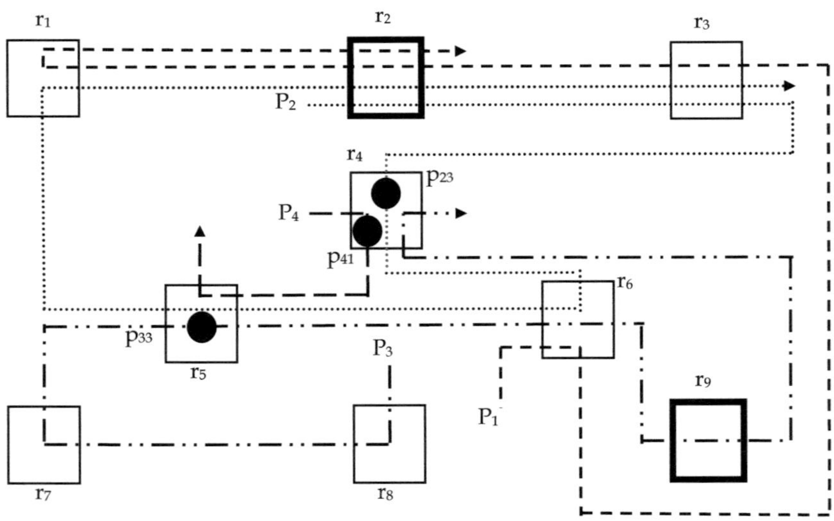

Fig. 5. An admissible system state by supervisor Δ_1

Supervisor Δ_1 is of polynomial complexity since both A1 and NHC require polynomial time for runtime implementation. Chew and Lawley (2006) formally establish that Δ_1 yields a robust supervisor for systems where every part type requires in its route at most one unreliable resource.

3.1.3 A Single step look ahead policy

It is well known that certain system structures, such as a central buffer, input/output bins, and non-unit buffer capacities, eliminate the possibility of deadlock-free unsafe states (Lawley & Reveliotis, 2001). In these systems, every state is either deadlock or safe, and therefore, a single-step look-ahead policy (SSL) is a correct and optimal deadlock avoidance policy. Further, it is of polynomial complexity, and thus ideal for runtime applications in real systems. In the following, we will modify the SSL presented by Lawley (1999) so that it works with systems with multiple unreliable resources.

A resource allocation graph (RAG) is a digraph that encodes the resource requests and allocations of parts (Lawley, 1999). For our purposes, let RAG=($R\backslash R^{FD}$,E) where $R\backslash R^{FD}$ is the set of system non-failure-dependent resource types and E={(r_u,r_v): r_u,$r_v \in R\backslash R^{FD}$ and r_u is holding a part p_{jk} with $\rho(p_{j,k+1})=r_v$}. A subdigraph of RAG, say (R,E), is *induced* when R $\subseteq R\backslash R^{FD}$ and E ={(r_u,r_v):(r_u,r_v)$\in$E and r_u,$r_v \in$ R}. A subdigraph, (R,E), forms a *knot* in RAG if $\forall r_u \in R$, $\Gamma(r_u)$= R, where $\Gamma(r_u)$ is the set of all nodes reachable from r_u in RAG. In other words, a set of nodes, R, forms a knot in RAG when, for every node in R, the set of nodes reachable along arcs in RAG is exactly R. Further, we define a *capacitated knot* to be a knot in which every resource in the knot is filled to capacity with parts requesting other resources in the knot. It is commonly known that a capacitated knot in RAG is a necessary and sufficient condition for deadlock in these types of sequential resource allocation systems. We now provide an algorithm, Algorithm A2, below to detect a capacitated knot in RAG = ($R\backslash R^{FD}$,E). This algorithm has the same polynomial complexity as that given by Lawley (1999).

Algorithm A2:

Input: RAG=($R\backslash R^{FD}$,E)

Output: DEADLOCK, NO DEADLOCK

Step 1: Compute the set of strongly connected components of RAG: C={$C_1 \ldots C_q$}

Step 2: Construct digraph (C,E_c) such that C={$C_1 \ldots C_q$} and E_c={(C_i,C_j):(r_u,r_v)$\in$E with $r_u \in C_i$ and $r_v \in C_j$ for $i \neq j$}

Step 3: For every strongly connected component $C_i \in C$ such that $(C_i,C_j) \notin E_c$ $\forall j=1 \ldots q$

If C_i is a capacitated knot

Return DEADLOCK

End If

End For

Step 4: Return NO DEADLOCK

We note that, for our present work, this version of deadlock detection algorithm operates only on non-failure-dependent resources and parts held by these resources. In A2, Step 1 computes the set of strongly connected components in RAG. As mentioned earlier, this is a standard digraph operation. Step 2 constructs a digraph that defines the reachability relationship between these components. Step 3 looks for a component with no outgoing arc. If such a component is filled to capacity with parts requesting other resources in the component, then it is a capacitated knot, and deadlock exists. If no such capacitated knot exists then the RAG is deadlock-free.

Note that A2 is not correct by itself since it considers only the non-failure-dependent resources. Failure-dependent resources can easily deadlock themselves. However, when A2 is taken in conjunction with NHC, it guarantees Property 2.2 and thus assures that the system will continue to operate even when multiple unreliable resources are down.

Definition 3.1.2: Supervisor $\Delta_2 = A2 \wedge NHC$.
Supervisor Δ_2 accepts a system state that contains no deadlock and satisfies NHC. For example, in Figure 1, suppose that every non-failure-dependent resource has non-unit capacity; that is, $C_i>1$, $\forall r_i \in R \backslash R^{FD} = \{r_3, r_4, r_5, r_6\}$. Then, A2 permits any state in which no subset of parts residing on $\{r_3, r_4, r_5, r_6\}$ is deadlocked on $\{r_3, r_4, r_5, r_6\}$. If the state also satisfies NHC, then Property 2.2 is guaranteed.
Note that $\Delta_2 = A2 \wedge NHC$ is suited for real-time implementation since both A2 and NHC are of polynomial complexity. Chew and Lawley (2006) formally establishes that Δ_2 yields a robust supervisor for systems where every part type requires in its route at most one unreliable resource.

3.2 Robust control using a resource order policy

This subsection configures a deadlock avoidance policy, resource order policy (RO). We will employ this configured resource order policy in conjunction with the neighbourhood constraints of Subsection 3.1 to develop a robust controller. Consider, for configuration purposes, Figure 1. Define $RCO = R \backslash R_i^{FD}$ as the set of non-failure-dependent resources. Since $R_2^{FD} = \{r_1, r_2\}$ and $R_9^{FD} = \{r_7, r_8, r_9\}$, thus $RCO = \{r_3, r_4, r_5, r_6\}$. Let $\omega : RCO \rightarrow \aleph$ (the set of natural numbers) be a one to one mapping of non-failure-dependent resources (ω orders the non-failure-dependent resources so that RO can be applied); $P^{FD} = \{P_j : \rho(P_{jk}) \in R^U \text{ for some } k\}$ (P^{FD} is the set of part types requiring unreliable resources; thus, in Figure 1, $P^{FD} = \{P_1, P_2, P_3\}$); and $P^{NFD} = P \backslash P^{FD}$ (P^{NFD} is the set of part types not requiring any unreliable resources; hence, in Figure 1, $P^{NFD} = \{P_4\}$). For each $P_j \in P^{FD}$, determine all maximal subsequences in the route of P_j that do not contain failure-dependent resources. For instance, in Figure 1, $P_3 \in P^{FD}$ where $P_3 = \langle P_{31}, P_{32}, P_{33}, P_{34}, P_{35}, P_{36} \rangle$ with route $\langle r_8, r_7, r_5, r_6, r_9, r_4 \rangle$, the maximal subsequences in $\langle r_8, r_7, r_5, r_6, r_9, r_4 \rangle$ that do not contain failure-dependent resources are $\langle r_5, r_6 \rangle$ and $\langle r_4 \rangle$.
To express this formally, for each $P_j \in P^{FD}$, break the route of P_j into subroutes as follows: for $P_j = \langle P_{j1} \ldots P_{j,k_1-1}, P_{jk_1}, P_{j,k_1+1} \ldots P_{j,k_2-1}, P_{jk_2}, P_{j,k_2+1} \ldots P_{j,k_{h_j}-1}, P_{jk_{h_j}}, P_{j,k_{h_j}+1} \ldots \rangle$, $\{P_{jk_1}, P_{jk_2} \ldots P_{jk_{h_j}}\}$ being precisely the set of part type stages of P_j that is processed on failure-dependent resources (that is, $\{\rho(P_{jk}) : k=k_1, k_2 \ldots k_{hj}\} \subseteq R^{FD}$ and $\{\rho(P_{jk}) : k \neq k_1, k_2 \ldots k_{hj}\} \cap R^{FD} = \varnothing$), let $P_j^1 = \langle P_{j1} \ldots P_{j,k_1-1} \rangle$, $P_j^2 = \langle P_{j,k_1+1} \ldots P_{j,k_2-1} \rangle$, $P_j^3 = \langle P_{j,k_2+1} \ldots P_{j,k_3-1} \rangle, \ldots,$ $P_j^{h^j} = \langle P_{j,k_{(h_j-1)}+1} \ldots P_{j,(k_{h_j}-1)} \rangle$ and $P_j^{h^j+1} = \langle P_{j,(k_{h_j}+1)} \ldots P_{j|P_j|} \rangle$. For each $P_j \in P^{NFD}$, rename P_j P_j^0. Finally, let $P' = \{P_j^0 : P_j \in P^{NFD}\} \cup \{P_j^k : k = 1 \ldots h_j \text{ and } P_j \in P^{FD}\}$. Note that in P', a part type $P_j \in P^{FD}$ is replaced by a set of part types $\{P_j^1, P_j^2 \ldots P_j^{h^j}\}$ each having a route that is a maximal segment of the route of P_j not containing a failure-dependent resource.
In Figure 1, for example, P_3 is replaced by $P_3^1 = \langle P_{33}, P_{34} \rangle$ with route $\langle r_5, r_6 \rangle$ and $P_3^2 = \langle P_{36} \rangle$ with route $\langle r_4 \rangle$, and P_4 is renamed P_4^0. Thus, the revised set of part types is $P' = \{P_4^0\} \cup \{P_1^1, P_2^1, P_2^2, P_3^1, P_3^2\}$. Note that none of the routes of part types in P' contains any failure-dependent resources.
We now use P' and RCO to construct a set of RO constraints as follows. For each $P_j^i = \langle P_{j,k_{(i-1)}+1} \ldots P_{j,(k_i-1)} \rangle \in P'$ and for each $P_{jk} \in P_j^i$, consider the inclusive remaining route, $\langle \rho(P_{jk}) \ldots \rho(P_{j,(k_i-1)}) \rangle$, and its mapping, $\langle \omega(\rho(P_{jk})) \ldots \omega(\rho(P_{j,(k_i-1)})) \rangle$. (Recall that to implement RO, the resources must be ordered. ω represents the ordering function.) If the mapping of the inclusive remaining route is strictly increasing (decreasing), then P_{jk} is classified as 'right' ('left'); if the mapping of the inclusive remaining route switches direction at some point,

then P_{jk} is classified as 'undirected.' If P_{jk} is terminal, it is ignored. For $r_m \in RCO$, let Π_m^{RU} represent the set of right and undirected part type stages associated with r_m; and Π_m^{LU}, the set of left and undirected part type stages associated with r_m. In the example, consider that $\omega(r_1)=1$, $\omega(r_2)=2$, $\omega(r_3)=3$ $\omega(r_4)=4$, $\omega(r_5)=5$, $\omega(r_6)=6$, $\omega(r_7)=7$, $\omega(r_8)=8$ and $\omega(r_9)=9$. We now have that the inclusive remaining route of P_{33}, $\langle r_5,r_6\rangle$ supporting $\langle P_{33},P_{34}\rangle \subseteq P_3^1$, is strictly increasing for ω, thus P_{33} is classified as 'right' and hence $\Pi_5^{RU} = \{P_{33}\}$. In the meantime, since $P_{25} \in P_2^1$ is the terminal part type stage for P_2^1, P_{25} is ignored. Clearly, $\Pi_5^{RU} = \varnothing$. On the other hand, the inclusive remaining route of P_{11}, $\langle r_6,r_3\rangle$ supporting $\langle P_{11},P_{12}\rangle \subseteq P_1^1$, is strictly decreasing for ω, hence P_{11} is classified as 'left.' The inclusive remaining route of P_{24} is $\langle r_6,r_5\rangle$ supporting $\langle P_{24},P_{25}\rangle \subseteq P_2^1$, which is strictly decreasing for ω, hence P_{24} is classified as 'left.' Therefore, $\Pi_6^{LU} = \{P_{11},P_{24}\}$. Meanwhile, since $P_{34} \in P_3^1$ is the terminal part type stage for P_3^1, P_{34} is ignored. It is obvious that $\Pi_6^{LU} = \varnothing$. After all the part type stages are classified in this way, a constraint is generated for each pair of non-failure-dependent resources, yielding RO constraints. We now define RO constraints formally as follows.

Definition 3.2.1: RO^{RCO} is the set of constraints:

$$\forall r_m, r_n \in RCO \text{ such that } \omega(r_m) < \omega(r_n), \quad \sum_{P_{jk}\in\Pi_m^{RU}} (x_{jk} + y_{jk}) + \sum_{P_{jk}\in\Pi^{LU}} (x_{jk} + y_{jk}) < C_m + C_n$$

where C_m and C_n are the respective buffer capacities of r_m and r_n.

In the example, for r_5, $r_6 \in RCO$, we have $(x_{33}+y_{33}) + (x_{11}+y_{11}+x_{24}+y_{24}) < 2$, recalling that $C_5=C_6=1$. This constraint assures that for every resource allocation state that the system is allowed to visit, the number of 'right' and 'undirected' parts occupying buffer space at r_5 plus the number of 'left' and 'undirected' parts occupying buffer space at r_6 will be less than the combined capacity of the two resources. Similar constraints are generated for the resource pairs $\{r_3,r_4\}$, $\{r_3,r_5\}$, $\{r_3,r_6\}$, $\{r_4,r_5\}$ and $\{r_4,r_6\}$.

We are now in the position to establish that the conjunction of RO^{RCO} and NHC, call it supervisor Δ_3, satisfies Property 2.2. Supervisor Δ_3 is a control policy such that it disables $\alpha_{jk}\in\xi(q)$ if $\delta(q,\alpha_{jk})$ violates either RO^{RCO} or NHC. Formally, it is stated as follows.

Definition 3.2.2: Supervisor $\Delta_3 = RO^{RCO} \wedge NHC$.

Chew et al. (2011) establish that Δ_3 is a robust controller for systems where every part type requires at most one unreliable resource.

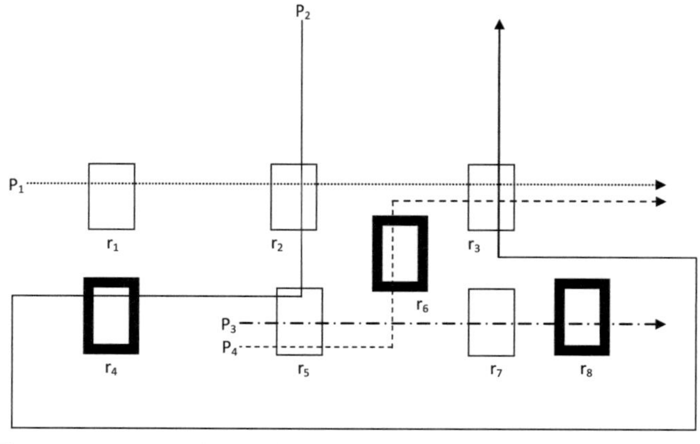

$R=\{r_1,r_2,r_3,r_4,r_5,r_6,r_7,r_8\}$
$P=\{P_1,P_2,P_3,P_4\}$
$R^U=\{r_4,r_6,r_8\}$
$P_1=\{P_{11},P_{12},P_{13}\}$
$T_1=\langle r_1,r_2,r_3\rangle$
$P_2=\{P_{21},P_{22}, P_{23},P_{24}\}$
$T_2=\langle r_2,r_5,r_4,r_3\rangle$
$P_3=\{P_{31},P_{32},P_{33}\}$
$T_3=\langle r_5,r_7,r_8\rangle$
$P_4=\{P_{41},P_{42},P_{43}\}$
$T_4=\{r_5,r_6,r_3\}$
$C=\langle C_1,C_2,\ldots,C_8\rangle=\langle 2,2,2,2,2,2,2,2\rangle$

Fig. 6. An example production system with three unreliable resources

3.3 Robust control using shared resource capacity

The robust supervisory control policies presented in sections 3.1-3.2 assume that that parts requiring failed resources can be advanced into FD buffer. We refer this type of control policies as "absorbing" policies. This subsection relaxes this assumption because, in some systems, providing FD buffer space might be too expensive or it might be desirable to load the system more heavily with FD parts. A "distributing" type of control policy is developed and presented in this subsection. This policy distributes parts requiring failed resources throughout the buffer space of shared resources so that these distributed parts do not block the production of part types that are not requiring failed resources.

Now, the development of the "distributing" control policy, namely, RO^4 policy is discussed in details. First, based on the definitions of resource sets in the previous sections, we further define three resource regions: (1) the region of continuous operation, $RCO=R^{PFD}\cup R^{NFD}$, (2) the region of failure dependency, $RFD=R^{FD}$, and (3) the region of distribution, $ROD=RFD\backslash R^U = R^{FD}\backslash R^U = R^R\backslash R^{NFD}$. In the example system in Figure 6, we have $RCO=\{r_1,r_2,r_3\}$; $RFD=\{r_2,r_4,r_5,r_6,r_7,r_8\}$; $ROD=\{r_2,r_5,r_7\}$. RO^4 policy is the conjunction of four modified RO policies applied to different resource regions. We now define the RO constraints as follows.

Definition 3.3.1: RO^{RCO} is the set of constraints:

$$\sum_{P_{jk}\in\Omega_g} z_{jk} + \sum_{P_{uv}\in\Omega_h} z_{uv} < C_g + C_h$$
$$where \quad z_{st} = x_{st} + y_{st}, \quad r_g, r_h \in RCO \; and \; g \neq h.$$

RO^{RCO} admits states that exhibit at most one capacitated resource in RCO.

Definition 3.3.2: RO^{RFD} is the set of constraints

$$\sum_{P_{jk}\in\Omega_g\cap P_i^{FD}} z_{jk} + \sum_{P_{uv}\in\Omega_h\cap P_i^{FD}} z_{uv} < C_g + C_h \quad for \quad r_i \in R^U$$
$$where \quad z_{st} = x_{st} + y_{st}, \quad r_g, r_h \in RFD, \quad and \quad g \neq h.$$

RO^{RFD} admits states for which at most one resource of RFD is capacitated with P_i^{FD} parts for each $r_i \in R^U$. Note that it does not place any constraint on the total number of RFD resources capacitated.

Definition 3.3.3: RO^{RFD^2} is the set of constraints

$$\sum_{P_{jk}\in\Omega_g\cap P^{FD}} z_{jk} + \sum_{P_{mn}\in\Omega_h\cap P^{FD}} z_{mn} + \sum_{P_{uv}\in\Omega_j\cap P^{FD}} z_{uv} < C_g + C_h + C_j$$
$$where \quad z_{st} = x_{st} + y_{st}, \quad r_g, r_h, r_j \in RFD \; and \; g \neq h \neq j.$$

RO^{RFD^2} admits states for which at most two resources of RFD are capacitated with FD parts, but does not place any constraint on the total number of RFD resources capacitated.

Definition 3.3.4: RO^{ROD} is the set of constraints

$$\sum_{P_{jk}\in\Omega_g\cap P^{FD}} z_{jk} + \sum_{P_{uv}\in\Omega_h\cap P^{FD}} z_{uv} < C_g + C_h$$
$$where \quad z_{st} = x_{st} + y_{st}, \quad r_g, r_h \in ROD \; and \; g \neq h.$$

RO^{ROD} admits states for which at most one resource of ROD=RFD\ R^U is capacitated with FD parts, although it places no constraint on the number of unreliable resources that are capacitated.

As in the example system in Figure 6, the set of constrains are as follows.

RO^{RCO}	r_1r_2:	$z_{11}+z_{12}+z_{21}<4$	r_2r_3:	$z_{12}+z_{21}+z_{13}+z_{24}+z_{43}<4$
	r_1r_3:	$z_{11}+z_{13}+z_{24}+z_{43}<4$		
RO^{RFD}	r_2r_4:	$z_{21}+z_{23}<4$	r_5r_7:	$z_{31}+z_{32}<4$
	r_2r_5:	$z_{21}+z_{22}<4$	r_5r_8:	$z_{31}+z_{33}<4$
	r_4r_5:	$z_{23}+z_{22}<4$	r_7r_8:	$z_{32}+z_{33}<4$
	r_5r_6:	$z_{41}+z_{42}<4$		
RO^{RFD^2}	$r_2r_4r_5$:	$z_{21}+z_{23}+z_{22}+z_{31}+z_{41}<6$	$r_4r_5r_6$:	$z_{23}+z_{22}+z_{31}+z_{41}+z_{42}<6$
	$r_2r_4r_6$:	$z_{21}+z_{23}+z_{42}<6$	$r_4r_5r_7$:	$z_{23}+z_{22}+z_{31}+z_{41}+z_{32}<6$
	$r_2r_4r_7$:	$z_{21}+z_{23}+z_{32}<6$	$r_4r_5r_8$:	$z_{23}+z_{22}+z_{31}+z_{41}+z_{33}<6$
	$r_2r_4r_8$:	$z_{21}+z_{23}+z_{33}<6$	$r_4r_6r_7$:	$z_{23}+z_{42}+z_{32}<6$
	$r_2r_5r_6$:	$z_{21}+z_{22}+z_{31}+z_{41}+z_{42}<6$	$r_4r_6r_8$:	$z_{23}+z_{42}+z_{33}<6$
	$r_2r_5r_7$:	$z_{21}+z_{22}+z_{31}+z_{41}+z_{32}<6$	$r_4r_7r_8$:	$z_{23}+z_{32}+z_{33}<6$
	$r_2r_5r_8$:	$z_{21}+z_{22}+z_{31}+z_{41}+z_{33}<6$	$r_5r_6r_7$:	$z_{22}+z_{31}+z_{41}+z_{42}+z_{32}<6$
	$r_2r_6r_7$:	$z_{21}+z_{42}+z_{32}<6$	$r_5r_6r_8$:	$z_{22}+z_{31}+z_{41}+z_{42}+z_{33}<6$
	$r_2r_6r_8$:	$z_{21}+z_{42}+z_{33}<6$	$r_5r_7r_8$:	$z_{22}+z_{31}+z_{41}+z_{32}+z_{33}<6$
	$r_2r_7r_8$:	$z_{21}+z_{32}+z_{33}<6$	$r_6r_7r_8$:	$z_{42}+z_{32}+z_{33}<6$
RO^{ROD}	r_2r_5:	$z_{21}+z_{22}+z_{31}+z_{41}<4$	r_5r_7:	$z_{22}+z_{31}+z_{41}+z_{32}<4$
	r_2r_7:	$z_{21}+z_{32}<4$		

We are now in the position to establish that RO^4 policy (the conjunction of RO^{RCO}, RO^{RFD}, RO^{RFD^2}, and RO^{ROD}), call it supervisor Δ_4, satisfies Property 2.2. Supervisor Δ_4 is a control policy such that it admits the enabled controllable event α if and only if $\delta(q,a)$ satisfies $RO^{RCO} \wedge RO^{RFD} \wedge RO^{RFD^2} \wedge RO^{ROD}$. Formally, it is stated as follows.

Definition 3.3.5: Supervisor $\Delta_4 = RO^{RCO} \wedge RO^{RFD} \wedge RO^{RFD^2} \wedge RO^{ROD}$.

The intuition behind this control policy is that it ensures that if a shared resource (i.e., a PFD resource) is filled with FD parts, at least one can be advanced out of the shared resources and, thus, out of RCO, which can then operate under RORCO. Furthermore, clearing RCO of this part will not create problems in the FD resources. To summarize, RO^{RFD} allows states with at most one FD resource filled with parts that are FD on the same unreliable resource. RO^{RFD2} allows states for which at most two FD resources are capacitated with FD parts. RO^{ROD} admits states for which at most one resource of ROD is capacitated with FD parts. Wang et al. (2008) establish that Δ_4 is a robust controller for systems where every part type requires at most one unreliable resource.

4. Robust control for product routings with multiple unreliable resources

In Section 3, we develop robust controllers for the single unit resource allocation systems with multiple unreliable resources. These guarantee that if any subset of resources fails, parts in the system requiring failed resources do not block production of parts not requiring failed resources. To establish supervisor correctness, we assume that each part type requires at most one unreliable resource in its route. We now relax this assumption using a central buffer and present robust controllers that guarantee robust operation without assumptions on route structure. To this end, we will construct new robust controllers in conjunction with the robust controllers, Δ_1 and Δ_2, developed in Subsection 3.1. The following three subsections will demonstrate the way we use a central buffer to extend Δ_1 and Δ_2 for systems where parts may require multiple unreliable resources.

4.1 Route partitioning algorithm

We now show how to use a central buffer to extend Δ_1 and Δ_2 for systems where parts may require multiple unreliable resources. We partition routes with multiple unreliable resources into subroutes, each of which contains one unreliable resource. A part in the last stage of a subroute can move to the first resource of the succeeding subroute or into the central buffer. With this partition, the system resembles one with at most one unreliable resource per route, allowing us to apply Δ_1 and Δ_2.

The route partitioning algorithm (RPA) performs this operation. It starts with the last stage and builds the subroute backwards. A subroute is extended until two unique unreliable resources are detected. Then, a new subroute is begun. We demonstrate below on P_1 of Figure 7.

Route Partitioning Algorithm (RPA)

Algorithm Notation: j, q, u are indices and counters; ε is the empty list; Ψ is a temporary set.

for j=1…$|P|$

  let $u=|P_j|$, q=1, $SP_{j1}=\varepsilon$, $\Psi=\varnothing$

  while $u\neq 0$

    (a) if $\rho(P_{ju})\in R^U\backslash\Psi$, $\Psi=\Psi\cup\{\rho(P_{ju})\}$

      (b) if $|\Psi|<2$, $SP_{jq}=push(P_{ju},SP_{jq})$, u=u−1

      (Note: The function 'push' takes two parameters, an object and an ordered list of objects, and inserts the object into the head of the list.)

    (c) else $\Psi=\varnothing$, q=q+1, $SP_{jq}=\varepsilon$

  end while

  $NS_j = q$ *(Number of Segments for P_j)*

For j=1, $u=|P_1|=8$, q=1, $SP_{11}=\varepsilon$, $\Psi=\varnothing$. Then, $\rho(P_{18})=r_1\notin R^U\backslash\Psi=\{r_2,r_4,r_5,r_7\}$, execute (b): $SP_{11}=\langle P_{18}\rangle$, u = 7.

Next, $\rho(P_{17})=r_7\in R^U\backslash\Psi=\{r_2,r_4,r_5,r_7\}$, execute first if: $\Psi=\Psi\cup\{r_7\}=\{r_7\}$. Since $|\Psi|<2$, execute (b) $SP_{11}=\langle P_{17},P_{18}\rangle$, u=6.

Next, $\rho(P_{16})=r_6\notin R^U\backslash\Psi=\{r_2,r_4,r_5\}$, execute (b): $SP_{11}=\langle P_{16},P_{17},P_{18}\rangle$ and u=5.

Next, $\rho(P_{15})=r_5\in R^U\backslash\Psi=\{r_2,r_4,r_5\}$, execute (a): $\Psi=\Psi\cup\{r_5\}=\{r_5,r_7\}$. Since $|\Psi|=2$, execute (c): $\Psi=\varnothing$, q=2, $SP_{12}=\varepsilon$. This completes the first subroute $SP_{11}=\langle P_{16},P_{17},P_{18}\rangle$.

Next, u=5, $\rho(P_{15})=r_5\in R^U\backslash\Psi=\{r_2,r_4,r_5,r_7\}$, execute (a): $\Psi=\Psi\cup\{r_5\}=\{r_5\}$. Since $|\Psi|<2$, execute (b): $SP_{12}=\langle P_{15}\rangle$, u = 4.

Next, $\rho(P_{14})=r_4\in R^U\backslash\Psi=\{r_2,r_4,r_7\}$, execute (a): $\Psi=\Psi\cup\{r_4\}=\{r_4,r_5\}$. Since $|\Psi|=2$, execute (c): $\Psi=\varnothing$, q=3, $SP_{13}=\varepsilon$. This completes the second subroute $SP_{12}=\langle P_{15}\rangle$.

Continuing as shown, RPA partitions P_1 into four subpart types (the remaining two are $SP_{13}=\langle P_{13},P_{14}\rangle$ and $SP_{14}=\langle P_{11}, P_{12}\rangle$) with subroutes $TS_{11}=\langle r_6,r_7,r_8\rangle$, $TS_{12}=\langle r_5\rangle$, $TS_{13}=\langle r_3,r_4\rangle$, and $TS_{14}=\langle r_1,r_2\rangle$. Note that each subroute requires at most one unreliable resource, although the frequency of that resource is not limited. RPA does not affect part types whose routes require at most one unreliable resource.

The maximum number of iterations of the RPA while loop is bounded by the number of part type stages, and thus RPA is no worse than $O(CRL=\sum_{P_j \in P} |P_j|)$, which is polynomial in cumulative route length (CRL).

4.2 Central buffer constraints

The central buffer (CB) will be used to clear workstation buffer space of failure-dependent parts that have finished a subroute. If such parts have completely finished their original routes, they exit the system. Otherwise, they must have available space in the CB. This will ensure that they do not block the production of other part types.

For example, suppose the system of Figure 7 is in a state as follows: r_7 is failed with p_{17} waiting for processing; r_5 is holding a completed p_{15}; and r_4 is holding a completed p_{14}. Because of the blocking effect of p_{14} and p_{15}, it is not possible to produce all other part types. However, if we relocate p_{14} and p_{15} to the CB, the system can continue producing P_2, P_3, and P_4. CB constraints are necessary to achieve this. For P_1, we state the linear inequality: $(x_{11}+y_{11})+(x_{12}+x'_{12}+y_{12})+(x_{13}+y_{13})+(x_{14}+x'_{14}+y_{14})+(x_{15}+x'_{15}+y_{15}) \leq B_1$, where x_{jk} and y_{jk} are the number of finished and unfinished p_{jk}'s at $\rho(P_{jk})$, x'_{jk} is the number of finished p_{jk}'s relocated to the CB, and B_j the CB space reserved for P_j.

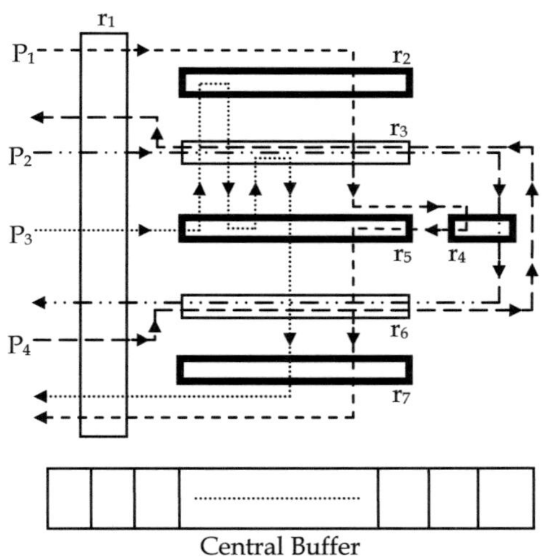

$R=\{r_1,r_2,r_3,r_4,r_5,r_6,r_7\}$ $P=\{P_1,P_2,P_3,P_4\}$

$R^U=\{r_2,r_4,r_5,r_7\}$

$P_1=\{P_{11},P_{12},\ldots,P_{18}\}$ $T_1=\langle r_1,r_2,r_3,r_4,r_5,r_6,r_7,r_1\rangle$

$P_2=\{P_{21},P_{22},\ldots P_{25}\}$ $T_2=\langle r_1,r_3,r_4,r_6,r_1\rangle$

$P_3=\{P_{31},P_{32},\ldots,P_{3,11}\}$ $T_3=\langle r_1,r_5,r_3,r_2,r_3,r_5,r_3,r_5,r_6,r_7,r_1\rangle$

$P_4=\{P_{41},P_{42},P_{43},P_{44}\}$ $T_4=\{r_1,r_6,r_3,r_1\}$

$C=\langle C_1,C_2,\ldots,C_7\rangle=\langle 1,1,1,1,1,1,1\rangle$

Fig. 7. Example with four unreliable resources

With this constraint, finished parts p_{12}, p_{14}, and p_{15}, for subpart types SP_{14}, SP_{13}, and SP_{12}, respectively, can be moved to the CB. Thus, in the example, we can transfer the finished p_{14} and p_{15} to the CB, allowing P_2, P_3, and P_4 to continue production. In the meantime, we decrement x_{14} and x_{15} by 1, and increment x'_{14} and x'_{15} by 1. As an aside, we decrement x'_{14} by 1 and increment y_{15} by 1 when p_{14} advances from the CB into the buffer of r_5.

We now state the CB constraint, CBC. Let $P^*=\{P_j:P_j \in P \wedge |T_j \cap R^U| > 1\}$ be the set of part types that require multiple unreliable resources, and B the total capacity of the CB. For a part type $P_j \in P^*$, let

$$Z_j = \sum_{P_{jk} \in P_j \setminus SP_{j1}} (x_{jk} + y_{jk}) + \sum_{P_{jk} \in LP_j} x'_{jk}$$

where LP_j is the set of "last" part type stages in the subparts of P_j (except SP_{j1}, the final stage of P_j). For example, $LP_1=\{P_{12},P_{14},P_{15}\}$ and $LP_3 = \{P_{32},P_{34},P_{36},P_{38}\}$. In general,

$$LP_j = \{P_{j,|SP_{j,NS_j}|}, P_{j,|SP_{j,NS_j}|+|SP_{j,NS_j-1}|}, \ldots, P_{j,|SP_{j,NS_j}|+\ldots+|SP_{j2}|}\} .$$

Z_j keeps track of the total number of instances of part type stages of $P_j \in P^*$ that are in the system. CBC is defined as:

$$\text{(i)}\ \ Z_j \le B_j,\ \ P_j \in P^* \qquad \text{(ii)}\ \sum_{P_j \in P^*} B_j \le B$$

CBC ensures that every part in the system requiring multiple unreliable resources has capacity reserved on the CB. CBC has no more than CRL*|P| constraints and thus checking CBC computation is no worse than O(CRL*|P|), which is polynomial in stable measures of system size.

The level of B_j for $P_j \in P^*$ can be fixed, in which case B_j does not change; or state-based, where we periodically reallocate CB across all $P_j \in P^*$. Although we cannot preempt CB space from parts that have it reserved, we can reallocate CB space that is not reserved. One simple approach is to let $B_j=Z_j$ as long as (ii) holds. This represents a first-come-first-serve rule. Alternatively, we can solve the following assignment problem:

$$\min \sum_{i=1}^{B} \sum_{j=1}^{|P^*|} C_{ij} X_{ij} \tag{1}$$

$$\text{st.}\quad B_j = \sum_{i=1}^{B} X_{ij},\quad j=1\ldots|P^*| \tag{2}$$

$$Z_j \le \sum_{i=1}^{B} X_{ij},\quad j=1\ldots|P^*| \tag{3}$$

$$\sum_{i=1}^{B} \sum_{j=1}^{|P^*|} X_{ij} \le B \tag{4}$$

$$X_{ij} \in \{0,1\},\ i=1\ldots B,\ j=1\ldots|P^*| \tag{5}$$

Here, X_{ij} is 1 if the ith unit of CB is assigned to $P_j \in P^*$, 0 otherwise. The objective (1) minimizes assignment cost; (2) counts the assignment to each $P_j \in P^*$; (3) assures no preemption from parts in the system; and (4) assures the CB is not over allocated. C_{ij} is the cost of assigning CB space to $P_j \in P^*$. This cost could reflect production priorities or failure probabilities. This problem can be solved in polynomial time using the Hungarian Algorithm (Papadimitriou, 1982). The solution frequency is a topic for future research.

4.3 Robust controllers with CBC

We now define two supervisory controllers. The first is the conjunction of Δ_1 and CBC; and the second is the conjunction of Δ_2 and CBC. Recall that Δ_1 and Δ_2 are the controllers of Subsection 3.1. Formally, the extended supervisors are stated as follows.

Definition 4.3.1: Supervisor $\Delta_5 = \Delta_1 \wedge$ CBC.
Definition 4.3.2: Supervisor $\Delta_6 = \Delta_2 \wedge$ CBC.
The following theorems establish that these supervisors ensure robust operation.
Theorem 4.3.1: Δ_5 is robust to failure of R^U.
Proof: The structure of the proof is as follows. We assume the system to be in an admissible state with parts requiring multiple unreliable resources, with some failed. We show that these parts can advance into the CB or into the buffer space of failure-dependent resources, where they do not block production of parts not requiring failed resources. Let $P_j \in P^*$. The subpart types of P_j constructed by RPA are $\{SP_{j,NS_j}, SP_{j,(NS_j-1)}, \ldots, SP_{j1}\}$. Assume that in the current state, q, unreliable resources in the subroutes of P_j have failed and that q satisfies Δ_5. In the following, we want to show that under Δ_5 parts of type P_j do not block other part types from producing. We ignore parts of type P_j in the final subroute since it is covered by Δ_1. That is, Δ_1 guarantees that parts in the final subroute can be advanced into the buffer space of the last resource and completed and removed from the system if the resource is operational or stored there, out of the way of part types not requiring failed resources, if it is not.
Let $\zeta_{qj}=\{p_{jk} \mid P_{jk} \in SP_{jq}, q = NS_j, (NS_j-1),\ldots,2\}$ be the set of parts of P_j in the state q. Let $\wp_{qj}=\{p_{jk} \mid P_{jk} \in LP_j\}$ be the set of parts of P_j in the final stage of a subroute. By the definition of LP_j, $\wp_{qj} \subseteq \zeta_{qj}$. Now, Δ_1 guarantees that all parts in $\zeta_{qj} \setminus \wp_{qj}$ can be advanced, perhaps through several processing steps, into the buffer spaces of resources required by stages of LP_j. That is, Δ_1 guarantees a sequence of part movements such that the system reaches a new state, say t, where $\zeta_{tj} = \wp_{tj}$. In state t, all instances of P_j are at the end of a subroute.
The left hand side of CBC does not change in moving from state q to state t. To see this, note that CBC is only affected by parts in P*. Since we allow no new parts to be admitted and no part of P* is required to move from one subroute to another (only to the end of the current subroute), the left-hand-side of CBC does not change magnitude. Thus, the part advancement under Δ_1 does not violate CBC.Now, CBC guarantees that every part of ζ_{tj} has capacity reserved on the CB, and any finished part of this set can be moved to the CB. Further, any unfinished part of ζ_{tj} can be finished and moved to the CB if its resource is operational. If the associated resource is not operational, the part can be stored at its failed resource where it will not block the production of part types not requiring failed resources. Thus, all operational resources can be cleared of parts of type P_j. Under Δ_1, the resulting state is a feasible initial state if resource repairs or additional failures occur.
Theorem 4.3.2: Δ_6 is robust to failure of R^U.
Proof: The proof follows the same construction as Theorem 4.3.1. The main difference is in how BA and SSLA operate. Thus, Δ_5 and Δ_6 guarantee robust operation for systems where parts can require multiple unreliable resources. Note that if every resource is unreliable, both theorems continue to hold.

5. Conclusion and future research

Supervisory control for manufacturing systems resource allocation has been an active area of research. Significant amount of theories and algorithms have been developed to allocate resources effectively and efficiently, and to guarantee important system properties, such as system liveness, traceability, deadlock-free operations. However, a major assumption these research works are based on is that resources never fail. While resource failures in automated

manufacturing systems are inevitable, we investigate such system behaviours and control dynamics. First, we developed the notion of robust supervisory control for automated manufacturing systems with unreliable resources. Our objective is to allocate system buffer space so that when an unreliable resource fails the system can continue to produce all part types not requiring the failed resource. We established properties that such a controller must satisfy, namely, that it ensure safety for the system given no resource failure; that it constrain the system to feasible initial states in case of resource failure; that it ensure safety for the system while the unreliable resource is failed; and that during resource repair it constrain the system to states that will be feasible initial states when the repair is completed.
We then developed a variety of control policies that satisfy these robust properties.

Taxonomy for Future Research Directions		
System Structure	S1	at most one unreliable resource for each part type
	S2	random number of unreliable resources for each part type
Central Buffer Capacity	C1	without central buffer
	C2	with central buffer
Flexible Routing	FR1	every part type stage can be performed by exactly one resource
	FR2	every part type stage can be performed by exactly two resources
	…	
	FRj	every part type stage can be performed by exactly j resources
Robustness Level	RB1	no resource failures
	RB2	at most one resource failure at any time
	RB3	at most two resource failures at any time
	…	
	RBi	at most i resource failures at any time
Unreliable Resource Condition	RC1	unreliable resources fail at any time
	RC2	unreliable resource failure characteristics can be estimated
Application Areas	AA1	Manufacturing Systems
	AA2	Business Processes and Workflow Management
	AA3	E-Commerce
	AA4	Supply Chain Management
	AA5	Internet Resource Mangement
	AA6	Transporation Systems
	AA7	Healthcare Systems

Table 1. Taxonomy for future research directions

Specifically, supervisory controllers Δ_1-Δ_4 are for systems with multiple unreliable resources where each part type requires at most one unreliable resource. Supervisory controllers Δ_5-Δ_6 control systems for which part types may require multiple unreliable resources. Another classification of the controllers is based on the underlying control mechanism: controllers Δ_1-Δ_3 'absorb' all parts requiring failed resources into the buffer space of failure-dependent resources, controller Δ_4 distribute' parts requiring failed resources among the buffer space of shared resources, and controllers Δ_5-Δ_6 utilize central buffer to achieve robust operations. These robust controllers assure different levels of robust system operation and impose very different operating dynamics on the system, thus affecting system performance in different ways. An extensive simulation study has been conducted and a set of implementation guidelines for choosing the best robust controller based on manufacturing system characteristics and performance objectives are developed in Wang et al. (2009).

A taxonomy is developed and presented in Table 1 to help guide future research in the area of robust supervisory control. By combining the different system structures, the presence/absence of central buffer, flexible routing capability, system robust level requirements, and unreliable resource failure characteristics, a significant amount of future research and development need to be done to address a variety of system control and performance requirements. And, although automated manufacturing systems are the context in which we develop the robust supervisory control research. We expect to expand our research to other application areas due to the similarity in resource allocation requirement and complexity in workflow management. The robust controllers we developed so far only address a small subset of the research taxonomy. For example, controller Δ_1 falls in the category in the taxonomy of (S1, C1, FR1, RB2, RC1, AA1). Especially, it would be interesting and challenging to develop supervisory control policies for systems with flexible routing and for systems where the failure characteristics of resources are dynamically evolving and can be estimated through sensor monitoring and degradation modelling.

6. References

Chew, S. & Lawley, M. (2006). Robust Supervisory Control for Production Systems with Multiple Resource Failures. *IEEE Transactions on Automation Science and Engineering,* Vol.3, No.3, (July 2006), pp. 309-323, ISSN 1545-5955

Chew, S.; Wang, S. & Lawley, M. (2008). Robust Supervisory Control for Product Routings with Multiple Unreliable Resources. *IEEE Transactions on Automation Science and Engineering,* Vol.6, No.1, (January 2009), pp. 195-200, ISSN 1545-5955

Chew, S.; Wang, S. & Lawley, M. (2011). Resource Failure and Blockage Control for Production Systems. *International Journal of Computer Integrated Manufacturing,* Vol.24, No.3, (March 2011), pp. 229-241, ISSN 0951-192X

Cormen, T.; Leiserson, C. & Rivest, R. (2002). *Introduction to Algorithms* (Second Edition), McGraw-Hill, ISBN 0072970545, New York, USA

Ezpeleta, J.; Tricas, F.; Garcia-Valles, F. & Colom, J. (2002). A Banker's Solution for Deadlock Avoidance in FMS with Flexible Routing and Multiresource States. *IEEE Transactions on Robotics and Automation*, Vol.18, No.4, (August 2002), pp. 621–625, ISSN 1042-296X

Habermann, A. (1969). Prevention of System Deadlocks. *Communications of the ACM*, Vol.12, No.7, (July 1969), pp. 373–377, ISSN 0001-0782

Hsieh, F. (2004). Fault-tolerant Deadlock Avoidance Algorithm for Assembly Processes. *IEEE Transactions on Systems, Man and Cybernetics, Part A,* Vol.34, No.1, (January 2004), pp. 65-79, ISSN 1083-4427

Lawley, M. (1999). Deadlock Avoidance for Production Systems with Flexible Routing. *IEEE Transactions on Robotics and Automation*, Vol.15, No.3, (June 1999), pp. 497-510, ISSN 1042-296X

Lawley, M. (2002). Control of Deadlock and Blocking for Production Systems with Unreliable Resources. *International Journal of Production Research,* Vol.40, No.17, (November 2002), pp. 4563-4582, ISSN 0020-7543

Lawley, M. & Reveliotis, S. (2001). Deadlock Avoidance for Sequential Resource Allocation Systems: Hard and Easy Cases. *International Journal of Flexible Manufacturing Systems,* Vol.13, No.4, (October 2001), pp. 385-404, ISSN 0920-6299

Lawley, M.; Reveliotis, S. & Ferreira, P. (1998). Application and Evaluation of Banker's Algorithm for Deadlock-free Buffer Space Allocation in Flexible Manufacturing Systems. *International Journal of Flexible Manufacturing Systems,* Vol.10, No.1, (February 1998), pp. 73–100, ISSN 0920-6299

Lawley, M. & Sulistyono, W. (2002). Robust Supervisory Control Policies for Manufacturing Systems with Unreliable Resources. *IEEE Transactions on Robotics and Automation,* Vol.18, No.3, (June 2002), pp. 346-359, ISSN 1042-296X

Papadimitriou, C. (1982). *Combinatorial Optimization: Algorithms and Complexity,* Prentice-Hall, ISBN 0486402584, New Jersey, USA

Park, S. & Lim, J. (1999). Fault-tolerant Robust Supervisor for Discrete Event Systems with Model Uncertainty and Its Application to a Workcell. *IEEE Transactions on Robotics and Automation,* Vol.15, No.2, (April 1999), pp. 386–391, ISSN 1042-296X

Ramadge, P. & Wonham, W. (1987). Supervisory Control of a Class of Discrete Event Processes. *SIAM Journal on Control and Optimization,* Vol.25, No.1, (March 1985), pp. 206–230, ISSN 0363-0129

Reveliotis, S. (1999). Accommodating FMS Operational Contingencies through Routing Flexibility. *IEEE Transactions on Robotics and Automation,* Vol.15, No.1, (February 1999), pp. 3–19, ISSN 1042-296X

Reveliotis, S. (2000). Conflict Resolution in AGV Systems. *IIE Transactions,* Vol.32, No.7, (July 2000), pp. 647-659, ISSN 0740-817X

Wang, S.; Chew, S. & Lawley, M. (2008). Using Shared-Resource Capacity for Robust Control of Failure-Prone Manufacturing Systems. *IEEE Transactions on Systems, Man and Cybernetics, Part A,* Vol.38, No.3, (May 2008), pp. 605-627, ISSN 1083-4427

Wang, S.; Chew, S. & Lawley, M. (2009). Guidelines for Implementing Robust Supervisors in Flexible Manufacturing Systems. *International Journal of Production Research,* Vol.47, No.23, (December 2009), pp. 6499-6524, ISSN 0020-7543

19

Design of Robust Policies for Uncertain Natural Resource Systems: Application to the Classic Gordon-Schaefer Fishery Model

Armando A. Rodriguez[1], Jeffrey J. Dickeson[2], John M. Anderies[3] and Oguzhan Cifdaloz[4]
Arizona State University
USA

1. Introduction

Introduction. A critical challenge faced by sustainability science is to develop robust strategies to cope with highly uncertain social and ecological dynamics. The increasing intensity with which human societies utilize (limited) natural resources is fueling the global debate and urging the development of resource management methodologies/policies to effectively deal with very demanding socio-bio-economical issues. Unfortunately, despite concerted efforts by governments, many natural resources continue to be poorly managed. The collapse of many fisheries worldwide is the most notable example (Clark, 2006; Clark et al., 2006; Holland, Gudmundsson; Myers, Worm 2003; Sethi et al., 2005) but other examples include forests (Moran, Ostrom), groundwater basins (Shah, 2000), and soils (ISRIC, 1990). The suggested causes are varied but (Clark, 2006) highlights two: (1) lack of consideration of economic incentives actually faced by economic agents and (2) uncertainty associated with the dynamics of biological populations. In the case of fisheries, Clark notes that "complexity and uncertainty will always limit the extent to which the effects of fishing can be understood or predicted" (Clark, 2006, p. 98). This suggests that we need policies capable of effectively managing natural resource systems despite the fact that we understand them poorly at best.

Real-World Management Issues. Real-world resource management must address three components: goal setting, practical (robust) implementation, and learning. Clark and others (Clark, 2007; 2006; Clark et al., 2006) have recently noted that practical implementation issues are frequently at the root of fishery management failures. For most fisheries, the necessary institutional contexts exist (Wilen, Homans) and we know what to do, yet management efforts fail. This suggests a need to focus on the actual *process* of resource management. For example, how can managers make decisions with incomplete information concerning how the resource and the resource users will respond to management actions?

[1]Electrical Engineering, Ira A. Fulton School of Engineering
[2]Electrical Engineering, Ira A. Fulton School of Engineering
[3]School of Human Evolution and Social Change, School of Sustainability
[4]ASELSAN, Inc. Microelectronics, Guidance and Electro-Optics Division, Turkey

When managers can't learn fast enough, yet still must make decisions, how should they proceed?

Stochastic Optimization. A common approach to such policy[1] problems is stochastic optimization. Examples include studies of the performance of management instruments in the face of a single source of specific uncertainty such as in the size of the resource stock (Clark, Kirkwood; Koenig, 1984), the number of new recruits (Ludwig, Walters; Weitzman, 2002), or price (Andersen, 1982). Unfortunately, because they require assigning probabilities to possible outcomes, the insights from stochastic optimization techniques can be somewhat restricted. As Weitzman puts it, "The most we can hope to accomplish with such an approach is to develop a better intuition about the direction of the pure effect of the single extra feature being added...when the rest of the model is isolated away from all other forms of fisheries uncertainty" (Weitzman, 2002, p. 330). Such models generate interesting insights regarding how uncertain resources *should* be managed, but they contribute little to improving actual resource management practice. In our presentation, we attempt to provide some guidance through the development and application of a set of tools for practical (robust) policy implementation decisions in situations with multiple sources of uncertainty. While our approach is fundamentally deterministic, we show how probabilistic information can be accommodated within our framework.

Literature Survey. Several different threads concerning practical policy implementation challenges have emerged in the literature. Adaptive management (Walters, 1986) and resilience-based management (Holling, Gunderson; 1986; 1973; Ludwig et al., 1997) are examples from ecology. In parallel, robust control ideas from engineering (Zhou, Doyle) have begun to permeate macroeconomics (Hansen, Sargent; Kendrick, 2005) and there is recent work on resource management problems in the engineering literature (Belmiloudi, 2006; 2005; Dercole et al., 2003). A concept of robust optimization has also been developed in the operations research and management science literature (Ben-Tal, Nemirovski; Ben-Tal et al., 2000; Ben-Tal, Nemirovski) with some specific applications of these ideas to environmental problems (Babonneu et al., 2010; Lempert et al., 2006; 2000). The overarching theme of robust optimization is to select the best solution from those "immunized" against data uncertainty, i.e. solutions that remain feasible for all realizations of the data (Ben-Tal, Nemirovski).

Our Approach: Exploiting Concepts from Robust Control. This chapter presents a sensitivity-based robustness-vulnerability framework for the study of policy implementation in highly uncertain natural resource systems in which uncertainty is characterized by parameter bounds (not probability distributions). This approach is motivated by the fact that probability distributions are often difficult to obtain. Despite this, it is shown how one might exploit distributions for uncertain model parameters within the presented framework. The framework is applied to parametric uncertainty in the classic Gordon-Schaefer fishery model to illustrate how performance (income) can be sacrificed (traded-off) for reduced sensitivity, and hence increased robustness, with respect to model parameter uncertainty. Our robustness-vulnerability approach provides tools to systematically compare policy uncertainty-performance properties so that policy options can be systematically discussed.

More specifically, within this chapter, we exploit concepts from robust control in order to analyze the classic Gordon-Schaefer fishery model (Clark, 1990). Classic maximization of net present revenue is shown to result in an optimal control law that exhibits limit

[1] We use the terms "policies" and "control laws" interchangeably in this presentation.

cycle behavior (nonlinear oscillations) when parametric uncertainty is present. As such, it cannot be implemented in practice (because of prohibitively expensive switching costs). This motivates the need for robust policies that (1) do not exhibit limit cycle behavior and (2) offer performance (returns) as close to the optimal perfect information policy as model parameter (and derived fishery biomass target) uncertainty permits. Given the state of most world fisheries, our presentation focuses on a fishery that is nominally (i.e. believed to be) biologically over exploited (BOE); i.e. the optimal equilibrium biomass lies below the maximum sustainable yield biomass (Clark, 2006; Clark et al., 2006; Clark, 1990; Holland, Gudmundsson; Myers, Worm 2003; Sethi et al., 2005). By so doing, we directly address a globally critical renewable resource management problem. As in our prior work (Anderies et al., 2007), (Rodriguez et al., 2010), we do not seek "a best policy." Instead, we seek families of policies that are robust with respect to uncertainties that are likely to occur. Such families can, in principle, be used by a fishery manager to navigate the many tradeoffs (biological, ecological, social, economic, political) that must be confronted. More specifically, our effort to seek robust performance focuses on reducing the worst case downside performance; i.e. maximizing returns when we have the worst case combination of parameters. Such worst case (conservative) planning is critical to avoid/minimize the possibility of major regional/societal economical shortfalls; case in point, the recent "Great Recession." It is important to note that the simplicity of our model (vis-a-vis our performance objective of maximizing the net present value of returns) permits us to readily determine the worst case combination of model parameters (i.e. growth rate, carrying capacity, catchability, discount rate, price, cost of harvesting). Given this, we seek control laws that do not exhibit limit cycle behavior and whose returns are close (modulo limitations imposed by uncertainty) to that of the worst case perfect information optimal control policy - the best we could do in terms of return if we knew the parameters perfectly. Other design strategies are also examined; e.g. designing for the best case set of parameters. "Blended strategies" that attempt to do well for the worst case downside perturbation (i.e. minimize the economic downside) as well as the best case upside perturbation (i.e. maximize the economic upside) are also discussed. Such strategies seek to flatten the return-uncertainty characteristics over a broad range of likely parameters. The above optimal control (derived) policies are used as performance benchmarks/targets for the development of robust control policies. While our focus is on bounded deterministic parametric uncertainty, we also show how probability distributions for uncertain model parameters can be exploited to help in the selection of benchmark (optimal) policies. After targeting a suitable optimal (benchmark) policy, we show how robust policies can be used to approximate the benchmark (as closely as the uncertainty will permit) in order to achieve desired performance-robustness-vulnerability tradeoffs; e.g. have a return that is robust to worst case parameter perturbations.

While the presentation is intended to provide an introduction into how concepts from optimal and robust control may be used to address critical issues associated with renewable resource management, the presentation also attempts to shed light on challenges for the controls community. Although the presentation builds on the prior work presented in (Anderies et al., 2007), (Rodriguez et al., 2010), the focus here is more on defining the problem, describing the many issues, and sufficiently narrowing the scope to permit the presentation of a design methodology (framework) for robust control policies.

Finally, it must be noted that the robust policies that we present are not intended to be viewed as final policies to be implemented. Rather, they should be viewed as policy targets -

providing guidance to resource managers for the development of final implementable policies (based on taxes, quotas, etc. (Clark, 1990, Chapter 8)) that will (in some sense) approximate our robust policies. While our focus has been on parametric uncertainty, it must be noted that robustness to unmodeled dynamics (e.g. lags, time delays) is also important. While some discussion on this is provided, this will be examined in future work.

Contributions of Work. The main contributions of this chapter are as follows:

- *Benefits of Robust Control in Renewable Resource Management.* The chapter shows how robust control laws can be used to eliminate the limit cycle behavior of the optimal control law while increasing robustness to parametric uncertainty and achieving a return that is close (modulo limitations imposed by uncertainty) to the perfect information optimal control law. Special attention is paid to minimizing worst case economic downside. As such, the policies presented shed light on fundamental performance limitations in the presence of (parametric) uncertainty. The policies presented are intended to serve as targets/guidelines that fishery managers may try to approximate using available tools (e.g. taxes, quotas, etc. (Clark, 1990, Chapter 8).
- *Tutorial/Introductory Value.* The chapter serves as an introduction for the controls community to a very important resource management problem in the area of global sustainability. As such, the chapter offers a myriad of challenging problems for the controls community to address in future work.

Organization of Chapter. The remainder of the chapter is organized as follows.

- Section 2 describes the classic Gordon-Schaefer nonlinear fishery model (Clark, 1990) to be used.
- Section 3 describes the optimal control law and its properties. The latter motivates the need for robust control laws for fishery management - laws that try to achieve robust near optimal performance in some sense.
- Section 4 describes a class of robust control laws to be examined.
- Section 5 contains the main results of the work - comparing the properties of the optimal policy to those of the robust policies being considered.
- Finally, Section 6 summarizes the chapter and presents directions for future research.

2. Nonlinear bioeconomic model

In this section, we describe the nonlinear bioeconomic model to be used for control design. The model is then analyzed.

2.1 Description of bioeconomic model

The nonlinear Gordon-Schaefer bioeconomic model (Clark, 1990; Gordon, 1954; Schaefer, 1957) is now described.

Nonlinear Gordon-Schaefer Bioeconomic Model.

The nonlinear model to be used is as follows:

$$\dot{x} = F(x) - qxu_p \qquad x(0) = x_o, \tag{1}$$

where

$$F(x) = rx\left(1 - \frac{x}{k}\right) \tag{2}$$

represents the natural regeneration rate of the resource and x, x_o, and u_p represent resource biomass, initial resource biomass, and harvesting effort, respectively. The parameters r, k, and q, retain their traditional definitions of intrinsic growth rate, carrying capacity, and catchability, respectively. Table 1 in Section 2.5 summarizes model parameter definitions, units, nominal values, and ranges. Model uncertainty will be addressed in Section 2.6.

Saturating Nonlinearity. Typically, effort is bounded above by some maximum and below by zero, i.e. $u_p \in [0, u_{\max}]$. Typically, this physical constraint is implicitly taken into account when the optimal control problem is solved. However, a more general family of controls may generate control signals outside the allowable range, and it is important to be explicit about how these signals are "clipped" by physical constraints. We thus define the saturation function

$$\text{sat}(x; x_{\min}, x_{\max}) \stackrel{\text{def}}{=} \begin{cases} x_{\min} & -\infty < x < x_{\min} \\ x & x_{\min} \leq x \leq x_{\max} \\ x_{max} & x_{\max} < x < \infty. \end{cases} \tag{3}$$

The feasibility condition can then be written in terms of (3), i.e.

$$u_p \in [0, u_{\max}] \Leftrightarrow u_p = \text{sat}(u; 0, u_{max}) \tag{4}$$

where u is the control signal. When there is no risk of confusion, we will write $u_p = \text{sat}(u)$.

Performance Measure. The fishery performance measure to be used, denoted J, is the net present value of future returns:

$$J(u_p) \stackrel{\text{def}}{=} \int_0^T e^{-\delta\tau}(pqx - c)u_p \, d\tau \tag{5}$$

where price p, cost per unit effort c, discount rate δ, and planning horizon T are assumed constant. (We will use $T = \infty$ to develop the optimal control law.)

2.2 Equilibrium analysis of bioeconomic model

One of the desired control objectives will be for the fishery to operate at specific equilibrium (set) points. Given this, the set of equilibria for the nonlinear model are as follows:

$$x_e = o \qquad u_e = 0 \tag{6}$$

$$when \quad u_e \in (0, 1] \qquad x_e = k\left(1 - \frac{q}{r}u_e\right). \tag{7}$$

Observe that as the equilibrium effort increases, the equilibrium biomass decreases (as expected).

2.3 LTI small signal model

To further understand the local characteristics of the above nonlinear model, we can linearize it about equilibria. Doing so yields the following small signal linear time invariant (LTI) model:

$$\delta\dot{x} = a\,\delta x + b\,\delta u \tag{8}$$

$$f(x) = rx\left(1 - \frac{x}{k}\right) - qxu \tag{9}$$

$$a = \left[\frac{\partial f}{\partial x}\right]_{(x_e,u_e)} = r - \frac{2rx_e}{k} - qu_e = -\left(\frac{r}{k}\right)x_e \qquad b = \left[\frac{\partial f}{\partial u}\right]_{(x_e,u_e)} = -qx_e \tag{10}$$

$$\delta x(t) = x(t) - x_e \qquad \delta x(0) = x(0) - x_e = x_o - x_e \qquad \delta u(t) = u(t) - u_e \tag{11}$$

The associated transfer function from δu to δx is given by:

$$P(s) = \frac{b}{s-a} \tag{12}$$

Since $a = -\left(\frac{r}{k}\right)x_e < 0$, it follows that the equilibrium point (x_e, u_e) is asymptotically stable with the rate of convergence (pole) being proportional to the equilibrium biomass x_e, the fishery growth rate r, and inversely proportional to the fishery's carrying capacity k. The dc gain associated with P is $P(0) = -\frac{kq}{r}$; the minus sign implying that fishing reduces the equilibrium biomass.

Utility of LTI Small Signal Model. The above LTI model can be used to approximate the response x of the nonlinear model. If the response of the LTI model is denoted

$$\hat{x} = x_e + \delta x \tag{13}$$

then $\hat{x} \approx x$ when $u \approx u_e$ (i.e. $\delta u(t) \approx 0$) and $x_o \approx x_e$ (i.e. $\delta x(0) = x_o - x_e \approx 0$).

2.4 Control objectives

The control objectives for the fishery may be summarized (roughly) as follows:

1. Maximize the net present value of future returns

$$\text{maximize } J \stackrel{\text{def}}{=} \int_0^\infty e^{-\delta t}(pqx - c)u_p\,dt \tag{14}$$

   Note: We would be willing to give up some return for increased robustness.

2. Closed loop stability
   (a) Limit cycle behavior is not acceptable because it can have an prohibitively expensive implementation cost. While this is not captured in J, it could be addressed by introducing an additional $\dot{u}_p$ term within J.
   (b) Closed loop responses should be "relatively smooth" (continuous) when we have nearly continuous sampling of the biomass x. It is understood that sampling is inevitable in practice; i.e. continuous sampling is prohibitively expensive and hence impossible. As such, closed loop responses should be robust with respect to some discrete sampling.

3. Follow (achievable) step biomass commands issued by the fishery manager in the steady state

4. Reject additive step input and output disturbances in the steady state

5. Attenuate high frequency sensor noise so that it does not significantly impact control action
6. Ensure that the fishery biomass overshoot to step reference biomass commands is suitably bounded
7. Robustness with respect to model parametric uncertainty

2.5 Nominal model parameters

Nominal parameter values to be used are given below in Table 1.

Symbol θ	**Description**	**Unit**	**Nominal** θ_o	**Range**
Biological Parameters				
x_o	Initial resource biomass	*Kilotons, KT*	varies	$[0.5x_o, 1.5x_o]$
u_{min}	Minimum harvesting effort	$fleet \cdot power \cdot year/year$	0	-
u_{max}	Maximum harvesting effort	$fleet \cdot power \cdot year/year$	1	-
r	Intrinsic growth rate	$1/year$	0.3	[0.15,0.45]
q	Catchability	$1/fleet \cdot power \cdot year$	0.3	[0.15,0.45]
k	Carrying capacity	*KT*	100	[50,150]
Economic Parameters				
p	Resource market price	*M$ per kiloton*	10	[5,15]
c	Cost of harvesting per effort	*M$ per year*	13.24	$[6.62, 19.86]$
δ	Annual discount rate	*1/year*	0.1	[0.05,0.15]
T	Planning horizon	*years*	50	N/A

Table 1. Nominal Parameter Values Used

A planning horizon of $T = 50$ years was selected because the nominal discount rate is $\delta = 0.1$ and in roughly $T = \frac{5}{\delta} = 50$ years, the integrand within J is negligible.

Focus of Work: Biologically Exploited (BOE) Fishery. The focus of our presentation will be on a fishery that biologically overly expoilted (BOE) as opposed to biologically under exploited (BUE). This is because most of the world's critical fisheries are overly exploited (Clark, 1990).

- *BOE with the 'low cost'* $c = 13.24$. BOE occurs when the cost is sufficiently small. For the parameters indicated, it can be shown that:

$$x_e^* = 0.75 \cdot x_{\text{MSY}} = 37.5 < x_{\text{MSY}} = \frac{k}{2} = 50$$

  i.e. the optimal equilibrium biomass is below the maximum sustainable yield biomass.

2.6 Model uncertainty and scope of presentation

Within this presentation, we focus on uncertainty associated with the nominal model parameters: r, k, q, p, c, δ. The following uncertainty will not be addressed in this presentation but it is duly noted:

1. The structure of F may be different than considered above. For example, if F has the form $F(x) geq 0$ for $x \in [k_c, k]$ and $F(x) < 0$ for $x \in (0, k_c)$ where $F(0) = 0$ and $F(k) = 0$, then we say that the fishery exhibits *critical depensation* (Clark, 1990, p. 17). In short, this implies that if x ever drops below the critical depensation parameter $k_c > 0$, then x will decrease toward zero regardless of u; i.e. the fishery will be lost.

2. All plant parameters are uncertain. They may even change with time. Moreover, the plant contains additional dynamics; e.g. it takes time for the fishery workers to mobilize. This can contribute additional lags, time delays, and rate limiters within the plant. One can use a decentralized or distributed model in order to capture the decision making made by individual fisher people (Clark, 1990, Ch. 8 & 9).
3. Input and output disturbances are uncertain.
4. Measurement noise is uncertain.
5. The biomass is not known; it must be estimated
6. The output (biomass) is sampled at some rate; if this rate is not sufficiently high, it could cause aliasing (Ogata, 1995); the sampling rate should be (as a rule-of-thumb)greater than ten times the control system bandwidth.

In contrast to many control applications where the "controller" is implemented with great fidelity, fishery controllers are implemented by an organization. As such, there can be considerable implementation issues/uncertainty. This will be discussed further below.

3. Optimal control law and properties

Within this section, we present the optimal control problem, the associated solution (optimal control law), and the properties of the optimal control law.

3.1 Optimal control law

We begin with a brief derivation of the classical optimal control policy stated in a way that will facilitate comparison to the class of LTI policies described later in this section.
The solution of the traditional optimal control problem:

$$\text{maximize } J \stackrel{\text{def}}{=} \int_0^\infty e^{-\delta t}(pqx(t) - c)u_p(t)\,dt \tag{15}$$

$$\text{s.t. } \dot{x}(t) = F(x(t)) - qx(t)u_p(t) \qquad x(0) = x_o \tag{16}$$

$$u_{min} = 0 \leq u_p(t) \leq u_{max} \tag{17}$$

is obtained by forming the Hamiltonian:

$$\mathcal{H}(x,u,\lambda) \stackrel{\text{def}}{=} e^{-\delta t}(pqx - c)u + \lambda\,[F(x) - qxu] = G(x,t)u - \lambda F(x) \tag{18}$$

where $G(x,t) \stackrel{\text{def}}{=} e^{-\delta t}(pqx - c) - \lambda qx$ and λ is the co-state variable. *Pontryagin's Maximum Principle* then implies that an optimal control policy will satisfy:

$$u(t) = \begin{cases} -\infty \text{ when } & G(x,t) < 0 \\ \infty \quad \text{when } & G(x,t) > 0. \end{cases} \tag{19}$$

Because the objective functional is linear, the Maximum Principle says nothing about the case when $G(x,t) = 0$. However, using the co-state variable relationship $\dot{\lambda} = -\frac{\partial \mathcal{H}}{\partial x}$, the well-known implicit formula for the singular control path can be determined (Clark, 1990):

$$F'(x) + \frac{cF(x)}{x(pqx - c)} = \delta \tag{20}$$

Optimal Steady State Equilibrium Biomass. When $F(x) = rx(1 - x/k)$, the above equation can be used to determine the optimal (steady state) equilibrium biomass x_e^*:

$$x_e^* = \frac{\left[\frac{x_\infty}{2} - x_{MSY}\left(\frac{\delta}{r}\right) + x_{MSY}\right] + \sqrt{\left[\frac{x_\infty}{2} - x_{MSY}\left(\frac{\delta}{r}\right) + x_{MSY}\right]^2 + 4x_{MSY}x_\infty\left(\frac{\delta}{r}\right)}}{2}. \quad (21)$$

where

$$x_{MSY} = \frac{k}{2} \quad (22)$$

is the maximum sustainable yield biomass and

$$x_\infty = \frac{c}{pq} \quad (23)$$

is the optimal equilibrium when $\delta = \infty$; i.e. open-access equilibrium (Clark, 1990). The above shows that the optimal biomass x_e^* depends on the three independent parameters x_∞, x_{MSY}, and $\frac{\delta}{r}$. It can be shown that

$$x_\infty \quad \leq \quad x_e^* \quad \leq \quad x_{MSY} \; + \; \frac{x_\infty}{2} \quad (24)$$

for all $\delta \in [0, \infty]$ where the quantity $x_{MSY} + \frac{x_\infty}{2}$ is the optimal x_e^* for $\delta = 0$. The associated optimal (steady state) equilibrium control is given by:

$$u_e^* \stackrel{\text{def}}{=} \frac{r}{q}\left(1 - \frac{x_e^*}{k}\right). \quad (25)$$

Optimal Control Policy. Define the *tracking error* as the difference between the desired (reference) state and the actual state, i.e.

$$e \stackrel{\text{def}}{=} x_{ref} - x. \quad (26)$$

Setting $x_{ref} = x_e^*$ and combining (19) with (25) yields following expression for the control law:

$$u(t) = \begin{cases} -\infty \text{ when } & e > 0 \\ u_e^* \text{ when } & e = 0 \\ \infty \text{ when } & e < 0. \end{cases} \quad (27)$$

The saturation function is then applied to this control signal to capture the physical constraints on the system, i.e. $u_p(t) = \text{sat}(u(t))$. This control law implies the following:

- If $e > 0$, set $u_p(t) = u_{min} = 0$, allow $x(t)$ to increase until $x(t) = x_e^*$, then set $u_p(t) = \text{sat}(u_e^*)$.
- If $e < 0$, set $u_p(t) = u_{max}$ until $x(t)$ decreases to x_e^*, then set $u_p(t) = \text{sat}(u_e^*)$.
- If $e = 0$, set $u_p(t) = \text{sat}(u_e^*)$.

Below, we show that this policy (in general) exhibits limit cycle behavior in the presence of parameter uncertainty (see Figure 6).

3.2 Nominal optimal control policy numerics

The numerics for the nominal optimal perfect information control law are summarized in Table 2.

Case	x_e^*	u_e^*	x_∞	u_∞	Optimal Control Law	Optimal Return, J_o^* small IC	Optimal Return, J_o^* large IC
BOE	37.5	0.625	4.41	0.987	$u(t) = \begin{cases} 0 & x(t) < 37.5 \\ u_e^* \stackrel{\text{def}}{=} \frac{F(x_e^*)}{qx_e^*} = 0.625 & x(t) = 37.5 \\ u_{max} = 1 & x(t) > 37.5 \end{cases}$	451	782

Table 2. Summary of Optimal Control Policy Numerics. BOE corresponds to $c = 13.24$ which yields $x_e^* = 0.75 \cdot x_{\text{MSY}}$. Small IC corresponds to $x_o = 0.5 \cdot x_e^* < x_e^*$. Large IC corresponds to $x_o = 1.5 \cdot x_e^* > x_e^*$. $x_\infty = \frac{c}{pq}$ and $u_\infty = \frac{r}{q}(1 - \frac{x_\infty}{k})$ correspond to infinite discounting (open-access); i.e. $\delta = \infty$.

3.3 Properties of the optimal control law

In this section, we describe the properties of the optimal control law assuming perfect information (i.e. model parameters are known with no error) and imperfect information (i.e. model parameters are not perfectly known). Understanding the properties of the optimal policy is very important for several reasons. We wish to understand (1) the fundamental robustness properties (e.g. economic inefficiency) of an optimal policy (e.g. one based on nominal, worst case, or best case parameters); (2) implementation issues associated with the optimal control policy; (3) how the robustness properties for our robust policies compare to those of a particular optimal control policy; (4) how x_e^* depends on parameter perturbations. The latter is important because we will using x_e^* as the reference command x_{ref} for our robust control law policies. This is an issue because the optimal x_e^* (in general) is uncertain; i.e. x_e^* is only known for specific value selections (e.g. nominal, worst case, best case). As such, we will have to address this uncertainty to clearly understand what our robust control policies (with built-in command following) will be driving the state of the fishery to.

In short, we show below that: (1) Since x_e^* is, in general, uncertain, if x_e^* is the desired (reference) state, then we have a major issue in that we will be driving the fishery to the incorrect state. This can have severe economic as well as biological implications (e.g. driving x below the critical depensation parameter k_c, will destroy the fishery). (2) The optimal policy exhibits limit cycle behavior when x_e^* is uncertain. Moreover, it is very sensitive to any discrete sampling. As such, the (imperfect information) optimal policy is prohibitively expensive to implement (see Figure 6).

Optimal Perfect Information Control Law Sensitivity: Single Parameter Results ($x_o = x_e^*$). The following shows how the the performance of the optimal perfect information control law changes with parameter perturbations. Results for our BOE fishery when $x_o = x_e^*$ are as follows:

1. (J_e^*, x_e^*, u_e^*) increase with increasing k or increasing r.
2. J_e^* increases while (x_e^*, u_e^*) decrease with increasing q.
3. J_e^* decreases while x_e^* increases and u_e^* decreases with increasing δ.
4. J_e^* increases while x_e^* decreases and u_e^* increases with increasing p.

5. (J_e^*, x_e^*) decrease while u_e^* increases with increasing c.

Robustness with Respect to Parametric Uncertainty: Imperfect Versus Perfect Information. Figure 1 shows how the optimal control law performs in the presence of parametric uncertainty. $x_o = x_e^*$ for the perfect information policy. x_o is at the unperturbed/nominal x_e^* for the imperfect information policy. The plots compare the performance of the optimal control law with imperfect parameter knowledge to that with perfect parameter knowledge. The perfect information optimal control law (by definition) results in the maximum achievable return. While it represents a suitable benchmark to compare with, it must be emphasized that x_e^* is always uncertain. This is particularly crucial when x_e^* is being used as the target biomass (reference command) for a robust control law (see Sections 4, 5) because an incorrect reference command x_{ref} will fundamentally limit the achievable performance. Moreover, no (inner loop) robust policy can address this. To properly address this, one needs some combination of parameter estimation, system identification, and learning coupled with some adaptive outer loop policy that adjusts the target based on collected information. While this is challenging and exciting to pursue, it is beyond the scope of our presentation.
Figure 1 specifically shows the maximum theoretical (perfect information) return on the left in blue. The return associated with the imperfect information optimal policy (designed for nominal parameter values) is shown on the left in red. On the right in blue, we see how much the imperfect information optimal control law under performs the perfect information optimal control law. When k is perturbed by -30%, the imperfect law under performs the perfect information law by nearly 10%. Figure 1 shows that for a similar perturbation in r, the imperfect policy under performs by nearly 2%. It can be shown (figures not provided) that for a similar perturbation in δ, the imperfect policy under performs by less than 1%. It can be shown (figures not provided) that for similar perturbations in p, c, or q the imperfect policy under performs by a very small percentage. Why is it that the biological parameters k and r matter more in closing the perfect-imperfect information performance gap than δ, p, c, or q? This is because x_e^* is more sensitive to uncertainty in k and r for the BOE case under consideration. In short, the plots show that we should be concerned primarily with uncertainty in k. More generally, we seek (robust) policies that perform closer to the perfect information optimal policy for likely parametric modeling errors. Imperfect information obviously limits how close we can get. This and associated issues will be addressed below.
Impact of Extremal Parameter Uncertainty on Perfect Information (J_e^*, x_e^*, u_e^*) - At Optimal Equilibrium. In what follows, x_e^* will be used as a reference command x_{ref} to a robust control law with good command following properties. Since x_e^* is uncertain, it is important to understand how commanding an incorrect target will limit achievable performance. Given this, suppose that $x_o = x_e^*$.
We now ask, what is the worst case combination of perturbations for the model parameters (r, k, q, p, c, δ)? While an analytical proof is difficult, it can be shown (numerically) that

J_e^*, in general, decreases when (k, r, p, q) are decreased and/or (c, δ) are increased.

This result is independent of the initial condition x_o for the BOE case under consideration. Given uncertainty bounds for each of the model parameters, this observation permits us to readily determine the worst case set of parameter perturbation - something that, in general, is very difficult to do.

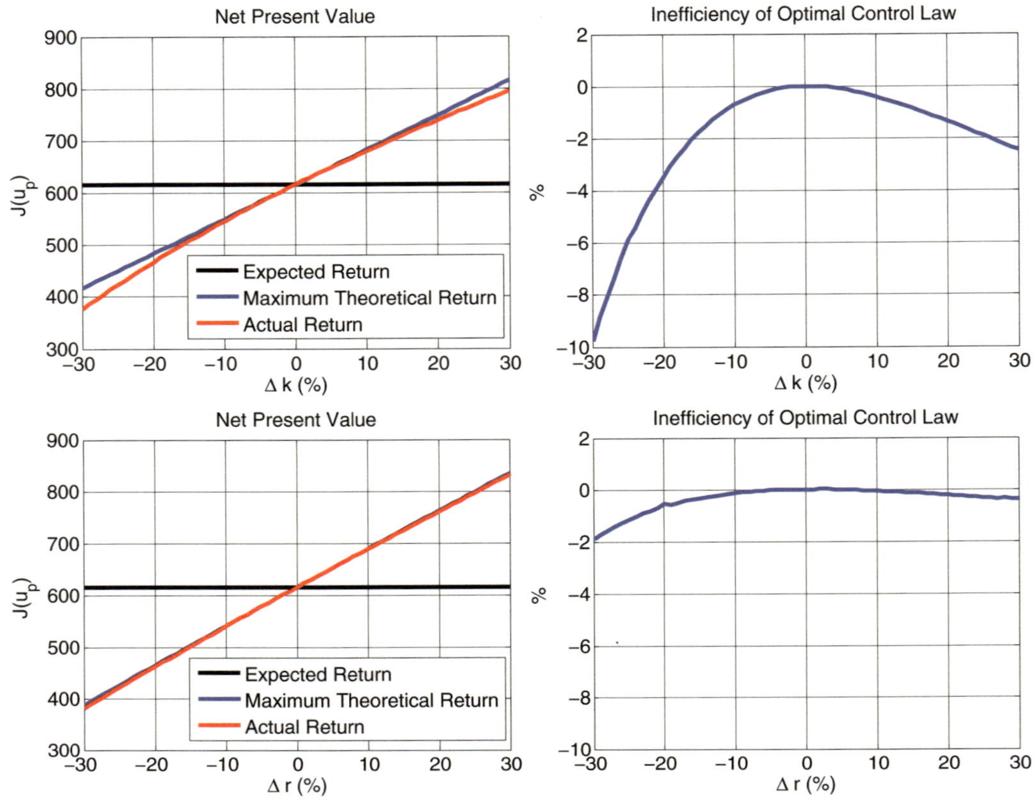

Fig. 1. Economic Inefficiency for Imperfect Information (Nominal) Optimal Control Law: Capacity & Growth Rate Uncertainty

Consider figures 2-3 for (J_e^*, x_e^*, u_e^*), respectively. Within these figures, $x_o = x_e^*$ and perfect information is assumed. The figures show the dependence of the perfect information optimal control law on worst case and best case (extremal) parameter perturbations as defined below.

- *Worst Case Extremal Parameter Perturbations.* Within figures 2-3, negative (worst case extremal) parameter perturbations correspond to

$$\frac{\Delta r}{r_o} = \frac{\Delta k}{k_o} = \frac{\Delta q}{q_o} = \frac{\Delta p}{p_o} < 0 \quad \text{and} \quad \frac{\Delta c}{c_o} = \frac{\Delta \delta}{\delta_o} > 0 \tag{28}$$

  i.e. equal parametric perturbations that result in a smaller return. Here, $\Delta\theta \overset{\text{def}}{=} \theta - \theta_o$ represents a perturbation in the parameter θ with respect to the nominal parameter θ_o.

- *Best Case Extremal Parameter Perturbations.* Within figures 2-3, positive (best case extremal) parameter perturbations correspond to

$$\frac{\Delta r}{r_o} = \frac{\Delta k}{k_o} = \frac{\Delta q}{q_o} = \frac{\Delta p}{p_o} > 0 \quad \text{and} \quad \frac{\Delta c}{c_o} = \frac{\Delta \delta}{\delta_o} < 0 \tag{29}$$

  i.e. equal parametric perturbations that result in a larger return.

The green curves within figures 2-3 represent actual optimal perfect information values. The blue curves give the percent deviation with respect to the nominal value.

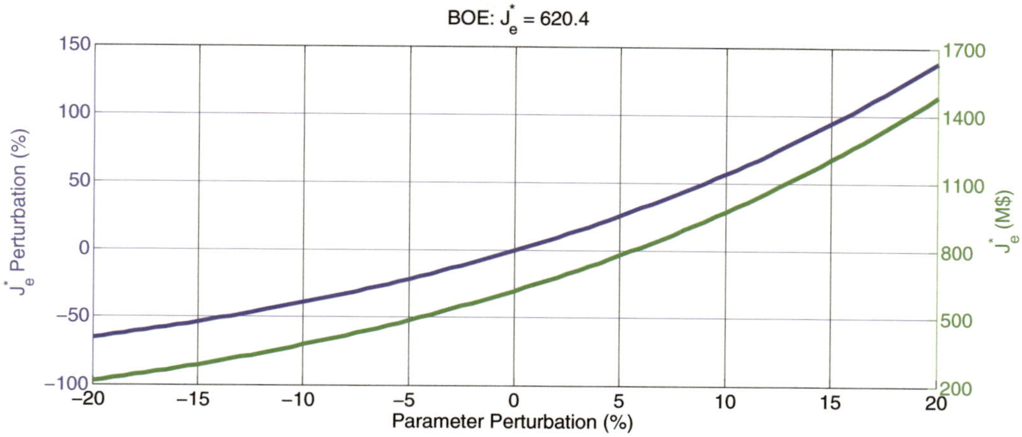

Fig. 2. Perfect Information Optimal Control Law Returns: Extremal Percent Parameter Perturbations, $x_o = x_e^*$

Assuming ±20% uncertainty for each nominal parameter value, figure 2 shows that the worst case perfect information optimal return is \$215.6 M (65.25% below the nominal of $J_e^* = 620.4M$). In contrast, the best case perfect information optimal return is \$1482M (138.95% above the nominal of $J_e^* = 620.4M$) - a 687% improvement with respect to the worst case perfect information optimal return. Also note from figure 3 that the worst case parameter combination results in a 20% reduction in x_e^* with respect to the nominal. From figure 3, we see that u_e^* is increased by less than 1%.

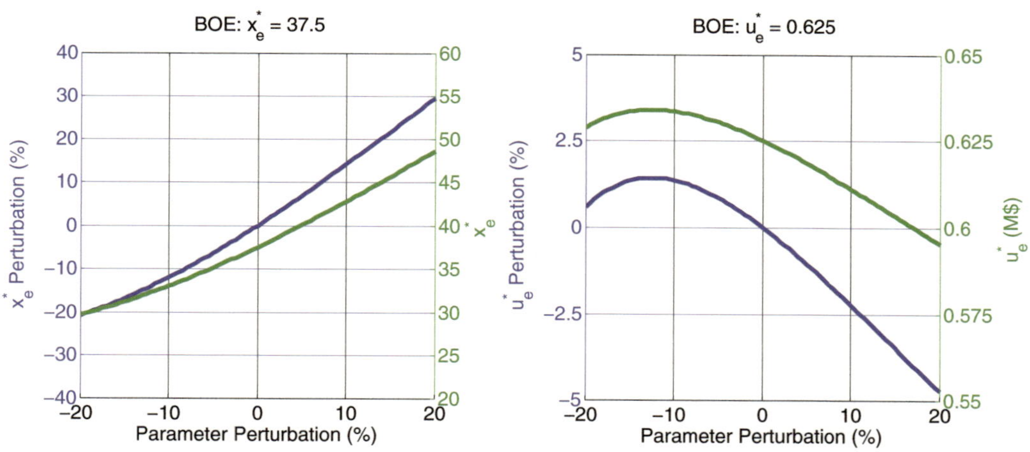

Fig. 3. Perfect Information Optimal Control Law (x_e^*, u_e^*): Extremal Percent Parameter Perturbations, $x_o = x_e^*$

Dealing with Uncertain x_e^* and J_e^*. Let x_{ref} denote the reference biomass at which the fishery manager wishes to operate the fishery [2]. How does a manager choose the target fishery biomass x_{ref}? A biologically conservative manager may wish to keep the fishery at the maximum sustainable yield $x_{ref} = x_{MSY} = \frac{k}{2}$. A financially aggressive manager may chose to operate the fishery at the infinite discount ($\delta = \infty$)) optimal value $x_{ref} = x_{\infty} = \frac{c}{pq}$. More generally, a manager could use the optimal value $x_{ref} = x_e^*$ as the point at which to operate. Given that x_e^* is known to within a percentage $\frac{\Delta x_e^*}{x_e^*}$, it follows that a fishery manager might try to operate at (1) $x_e^* - \Delta x_e^*$ if economic aggression is desired, or at (2) $x_e^* + \Delta x_e^*$ if biological conservatism is desired. The x_e^* concept gives the fishery manager a way to systematically think about fishery biomass targets.

Uncertainty In (x_o, x_e^*, x_{ref}): 6 Cases. In general, x_o and x_e^* are uncertain. How does one choose the target x_{ref}. We'd ideally like $x_{ref} = x_e^*$, but x_e^* is uncertain. What can a manager do? The table below contains the six possible relations that can exist amongst the three scalars (x_o, x_e^*, x_{ref}) - from smallest to biggest. In general, we would (ideally) like the state to move from x_o toward $x_{ref} = x_e^*$. Since x_e^* is uncertain, it follows that x_{ref} (in general) will differ from x_e^*. As such, it follows that we may issue reference commands x_{ref} that move the state x in an incorrect direction. Since the state moves from x_o toward x_{ref}, it follows from the table below that in two cases the state moves in the incorrect direction. In the four other cases, the state moves in the correct direction.

Smallest	$\rightarrow$	Biggest	Direction	
x_{ref}	x_o	x_e^*	Incorrect	Way too much fishing (Way Too Aggressive)
x_{ref}	x_e^*	x_o	Correct	Too much fishing (Very Aggressive)
x_o	x_{ref}	x_e^*	Correct	Too much fishing (Moderately Aggressive)
x_o	x_e^*	x_{ref}	Correct	Too little fishing (Moderately Conservative)
x_e^*	x_{ref}	x_o	Correct	Too little fishing (Very Conservative)
x_e^*	x_o	x_{ref}	Incorrect	Way too little fishing (Way Too Conservative)

Table 3. Six Possible Inequality Relations for (x_o, x_e^*, x_{ref})

To select x_{ref}, we offer the following approaches.

1. **Best-Worst Case Approach.** Assume that we have good bounds on parametric uncertainty (not necessarily tight, but encompassing) for the 6 model parameters under consideration: k, r, q, p, c, δ. Suppose that we design for the best worst case scenario; i.e. try to approach the return of the perfect information optimal policy when the worst case parameter perturbations occur; i.e. $\frac{\Delta k}{k_o} = \frac{\Delta r}{r_o} = \frac{\Delta q}{q_o} = \frac{\Delta p}{p_o} < 0$ and $\frac{\Delta c}{c_o} = \frac{\Delta \delta}{\delta_o} > 0$. We assume worst case maximal parameter perturbations. (For simplicity, we assume that all parameters are perturbed by their maximum worst case percentage and that this percentage is the same for all of the parameters.)

   One could, for example, pick a worst case percentage which bounds all of the parameters. Doing so can be conservative. Parameter estimation can be used to narrow tighten this worst case percentage. If we have fixed percentage bounds for each of the parameters, our approach remains the same. (Recall: Determining the worst case perturbation in our problem is easy. This is not true in most practical scenarios.)

[2] It is understood that x_{ref} can change with time. For now, we assume x_{ref} is fixed.

Now choose x_{ref} equal to the associated worst case x_e^*; i.e. the x_e^* that results from choosing the worst case parameters. By so doing, the actual x_e^* will be greater than x_{ref}. As such, only cases 1-3 can occur; i.e. cases 4-6 cannot occur. The only way, cases 4-6 can occur is if our uncertainty bounds were not truly encompassing.

2. **Best-Best Case Approach.** Assume that we have good bounds on parametric uncertainty (not necessarily tight, but encompassing)for the 6 model parameters under consideration: $k,\ r,\ q,\ p,\ c,\ \delta$. Suppose that we design for the best best case scenario; i.e. try to approach the return of the perfect information optimal policy when the best case parameter perturbations occur; i.e. $\frac{\Delta k}{k_o} = \frac{\Delta r}{r_o} = \frac{\Delta q}{q_o} = \frac{\Delta p}{p_o} > 0$ and $\frac{\Delta c}{c_o} = \frac{\Delta \delta}{\delta_o} < 0$. We assume best case maximal parameter perturbations. (For simplicity, we assume that all parameters are perturbed by their maximum best case percentage and that this percentage is the same for all of the parameters.)

   Now choose x_{ref} equal to the associated best case x_e^*; i.e. the x_e^* that results from choosing the best case parameters. By so doing, the actual x_e^* will be smaller than x_{ref}. As such, only cases 4-6 can occur; i.e. cases 1-3 cannot occur. The only way, cases 1-3 can occur is if our uncertainty bounds were not truly encompassing.

3. **Blended Best-Worst-Best-Best Approach.** One can also try to offer a blended approach that attempts to offer decent returns when either worse case or best case parameter perturbations occur. We shall illustrate this below.

4. **Probabilistic Approach.** If a probability density function for the parameter percentage θ is available, it can be used to determine where to operate. Let f_θ denote a density function for θ. This can be used to derive the density function f_J for J. Given this, the expected value for J is given by $E[J] = \int_J J\ f_J(J)\ dJ \ = \ \int_\theta J(\theta)\ f_J(J(\theta))\ J'(\theta)\ d\theta$. The density function for θ can be used to reflect what parameter perturbations are most likely to occur. The above expectation can then be used to choose x_{ref} to maximize the expectation.

To illustrate the above ideas, consider figures 4-5 for small and large initial conditions, respectively under extremal parameter perturbations. The figures show results for the perfect information designs (black), best worst case design (blue), best best case design (red), and the nominal design (green).
To summarize, the following specific optimal control laws were implemented:

1. Perfect Information Optimal Designs: $x_{ref} = x_e^*,\ u_{ref} = u_e^*$
2. A Best-Worst Case Design: $x_{ref} = 29.7,\ u_{ref} = 0.629$
3. A Nominal Design Based on the Nominal Parameters: $x_{ref} = 37.5,\ u_{ref} = 0.625$
4. A Best-Best Case Design: $x_{ref} = 48.6,\ u_{ref} = 0.595$

The performance of the perfect information designs are always best (by definition). The performance of the best-worst case design (blue) duplicates that of the perfect information design for 20% worst case perturbations since it is based on the worst case parameter model and x_e^*. The performance of the best-best case design (blue) duplicates that of the perfect information design for 20% best case perturbations since it is based on the best case parameter model and x_e^*. The following key observations are in order within figure 4 (small IC case):

1. The best-worst case design does better than the best-best case design when its parameter assumptions are maximally incorrect; falling by less than 20% (with respect to perfect information optimal return) while the best best falls by more than 40% (with respect

to perfect information optimal return) when its parameter assumptions are maximally incorrect.

2. The nominal design can be viewed as a nice compromise or blend between the two prior policies. It is based on the nominal parameter model and x_e^*. Its returns deteriorates by a little more than 10% for worst case parameter uncertainty and by a little more than 5% for best case parameter uncertainty. In short, the returns associated with this nominal (blended) policy offers flatter returns over a wider range of extremal parameter perturbations.

Each of the above three approaches offer a specific design model (to base the control design upon) and a specific x_e^* to use as a target. Control laws are always evaluated with the true (nonlinear) plant. In what follows, we will use the above as benchmarks whose performance we shall target via robust control laws. Similar patterns are observed for the large IC case in figure 5.

Sensitivity Analysis: Extremal Perturbations, Small Initial Condition. The expected value for each of the design cases considered are as follows:

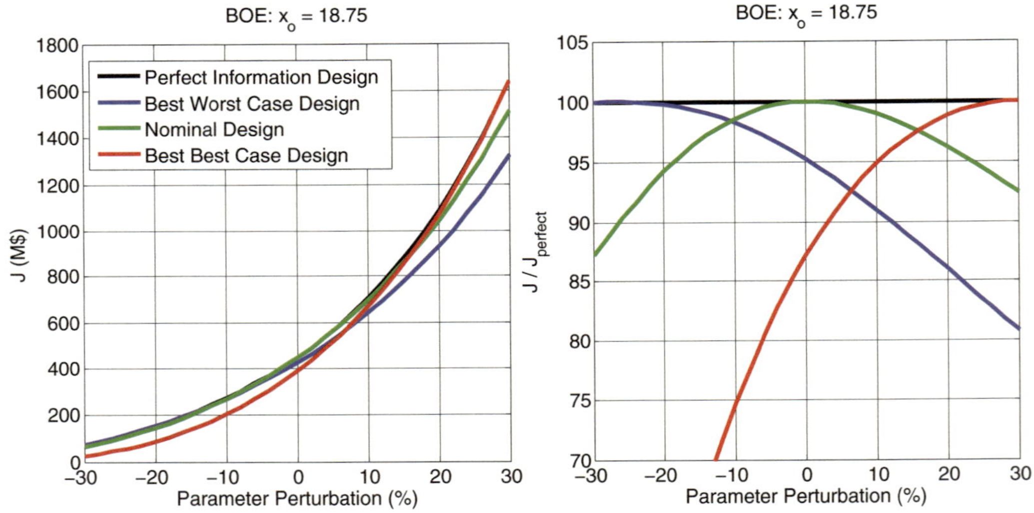

Fig. 4. Economic Inefficiencies for Various Optimal Control Laws: Extremal Perturbations (Small IC)

- $E[J] = 867.8$ for the Perfect Information Optimal Designs
- $E[J] = 772.0$ for the Best-Worst Case Design
- $E[J] = 838.0$ for the Nominal Design
- $E[J] = 800.4$ for the Best-Best Case Design

A uniform distribution has been assumed for the parameter uncertainty. The optimal perfect information control law is included for comparison purposes. Its performance can only be approximated over a range of parameter perturbations. This is because the (1) design plant parameters differ from those of the true plant and the desired target x_{ref} differs from the perfect information target x_e^*.

The following additional points are in order:

- Although the best-best case design appears worse in terms of percentages at off design conditions, it has a higher expected return across all cases versus the best-worst case design.
- A manager could readily design a policy that limited the worst case downside return to a certain percentage of the maximum possible. For example, if the manager wanted a worst case downside return no worse than 5% of the maximum possible, a policy should be designed around roughly a −2 % parameter perturbation.
- A manager may also be interested in implementing the following policy: $\max_\theta E[J(\theta)]$.

Sensitivity Analysis: Extremal Perturbation, Large Initial Condition. The expected value for each of the design cases considered are as follows:

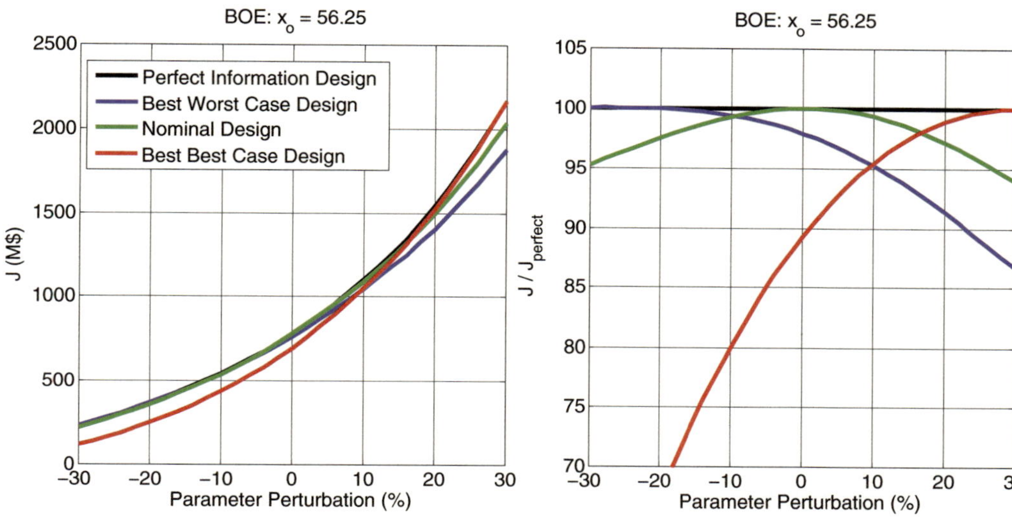

Fig. 5. Economic Inefficiencies for Various Optimal Control Laws: Extremal Perturbations (Large IC)

- $E[J] = 1368.2$ for the Perfect Information Optimal Designs
- $E[J] = 1284.5$ for the Best-Worst Case Design
- $E[J] = 1336.8$ for the Nominal Design
- $E[J] = 1259.3$ for the Best-Best Case Design

Finally, it should be noted that in contrast to the low initial condition study conducted, the Best-Best Case Design performs worse both in terms of the percentage possible and the expected return when compared to the Best-Worst Case Design.

Limit Cycles In the Presence of Uncertainty. Finally, consider figure 6. The optimal control law is based on the nominal BOE parameters. The initial condition is above the uncertain x_e^*. The simulation is conducted with a truth plant possessing a 10% reduction in k - hence the limit cycle behavior. The figure shows that: (1) The optimal control policy (in general) will exhibit limit cycle behavior when we have imperfect information; i.e. model parameters are not known exactly. (2) The optimal control policy (in general) will be very sensitive to finer sampling (ΔT smaller) under imperfect information; i.e. more oscillations (switching) will be exhibited as our x time samples are spaced closer together. The figure also shows that low pass filtering the optimal with a lag can be used to smooth oscillations a bit. To significantly reduce the oscillations, however, there is no easy fix. We either need a penalized $\dot{u}_p$ term within J to penalize switching or we need policies that are inherently more robust (like the ones we will describe subsequently). As such, this implies that, in practice, the optimal control policy is prohibitively expensive to implement in the presence of parametric uncertainty because of the inherent limit cycle behavior and the associated switching costs.

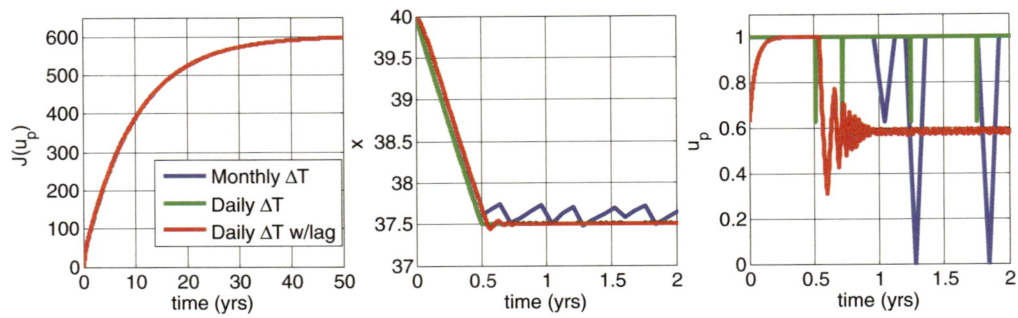

Fig. 6. Optimal Control Law Robustness: Limit Cycles In Presence of (-10% Capacity) Uncertainty

Motivation for Robust Control Laws. The above motivates the need for more robust control laws; i.e. control laws that (1) exhibit an acceptable return (i.e. return robustness) in the presence of anticipated (likely) parametric uncertainty; (2) do not exhibit limit cycle behavior in the presence of anticipated (likely) parametric uncertainty. As such, the above motivates the robust control laws to be considered in our presentation.

Control Law Implementation Issues. Unlike many control applications where controllers are implemented with great fidelity (within state-of-the-art digital computing units), controllers within a resource management system are implemented by an organization by setting rules for the fishery worker community (e.g. quotas, taxes (Clark, 1990, Chapter 8). As such, many types of uncertainties can be introduced by the organization. These could include any of the following: (1) parameter uncertainty, (2) additional uncertain actuation/sensing dynamics (e.g. lags, time delays, rate limiters, etc.), (3) nonlinearities (e.g. rate limiters, saturations, quantization, dead zones), (4) actuation/incentive errors (e.g. quota/tax miscalculations), (5) sensing, measurement, and estimation errors (e.g. sensor dynamics, biomass sampling/aliasing/quntization errors, noise, disturbances).

4. Robust control laws

The model under consideration is very simple. Many tools from the controls literature may be applied (e.g. classical control (Rodriguez, 2003), H-infinity (Rodriguez, 2004), feedback linearization, SDRE's, etc.). Given the introductory/tutorial nature of the paper, the simplicity of the model being used, as well as the fact that this text covers advanced control methodologies, we shall focus on simple control strategies from classical control theory. We will show that such control laws can be used to avoid limit cycles, increase robustness with respect to parametric uncertainty, and achieve returns that are close to those of the perfect information optimal control law.

Control System Structure. The structure of the control system may be visualized as shown in Figure 7.

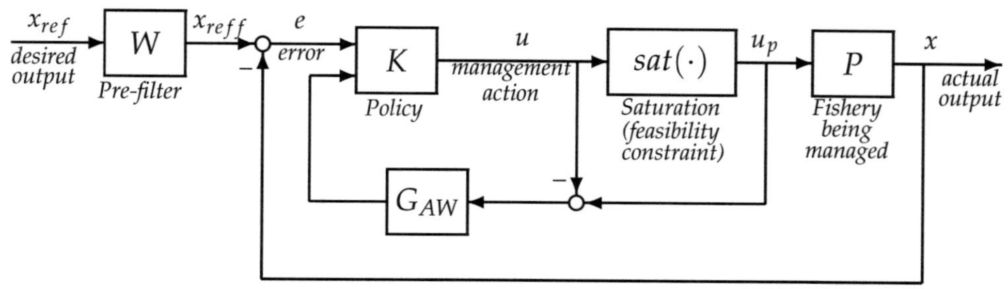

Fig. 7. Renewable Resource Management Problem Represented as a Standard Negative Feedback System with a Pre-Filter and Anti-Windup Logic

1. *Plant.* Here, P represents the plant under control. We shall use an LTI small signal model to approximate our nonlinear plant.
2. *Reference State or Command.* x_{ref} is the desired reference biomass state. Ideally, we would like to use $x_{ref} = x_e^*$. Parameter uncertainty prevents us from commanding the desired state. As such, we are forced to choose x_{ref} more judiciously. Given this, we will give special attention to maximizing our return under the worst case parameter uncertainty.
3. *PI Controller.* K is a proportional-plus-integral (PI) controller possessing the structure:

$$K(s) = \frac{g(s+z)}{s}\left[\frac{p_{ro}}{s+p_{ro}}\right] \tag{30}$$

   where $g > 0$, $z > 0$, and $p_{ro} > 0$. The integrator within the controller will ensure that step biomass commands are followed in the steady state while step input/output disturbances are rejected in the steady state. The $(s+z)$ term will ensure that the LTI plant-integrator pair will be stabilized. The term $\left[\frac{p_{ro}}{s+p_{ro}}\right]$ provides high frequency roll-off to ensure that high frequency sensor noise n is suitably attenuated.
4. *Command Pre-Filter.* W is a reference command pre-filter possessing the structure:

$$W(s) = \left[\frac{z}{s+z}\right] \tag{31}$$

   This pre-filter can be used to ensure that the overshoot to step reference commands is suitably bounded.
5. *Observer-Based Integrator Anti-Windup Logic.* Anti-windup logic is included so that the integrator in the PI controller does not windup. That is, the integrator is turned off so that it does not integrate constant errors which occur when the input to the plant is saturated (Aström, Hägglund). The structure of the anti-windup logic is as follows $\dot{x}_k = A_k x_k + B_k e + L(sat(u) - u)$ where L is an observer gain matrix. The PI controller with the anti-windup logic may be described by the following equations:

$$\dot{x}_1 = e + G_{AW}(sat(u) - u) \qquad \dot{x}_2 = gzx_1 - p_{ro}x_2 + ge \qquad u = p_{ro}x_2 \tag{32}$$

where G_{AW} is the anti-windup gain.

Nominal Design Methodology. The nominal design methodology can be described as follows (Rodriguez, 2003):

1. *Plant Approximant.* The following small signal LTI model $P \approx P_d \stackrel{\text{def}}{=} \frac{b}{s-a}$ will be used to approximate our nonlinear plant. Here, P_d is referred to as the *design plant*; i.e. the plant upon which we will base our control law design. While any design we obtain can be evaluated using plant approximants such as P_d, control designs must be evaluated with the actual nonlinear plant model.
2. *Controller Approximant.* Use the controller approximant $K \approx \frac{g(s+z)}{s}$ where $g > 0$ and $z > 0$.
3. *Nominal Open Loop Approximant.* Form the open loop transfer function approximant
$$L = P_d K \approx \frac{bg(s+z)}{s(s-a)} = \frac{n(s)}{d(s)}. \tag{33}$$
4. *Nominal Closed Loop Characteristic Equation.* Form the nominal closed loop characteristic equation
$$\Phi_{cl}(s) = d(s) + n(s) = s^2 + (bg - a)s + bgz = 0 \tag{34}$$
This polynomial has the "standard second order form"
$$\Phi_{cl}(s) = s^2 + 2\zeta\omega_n s + \omega_n^2 \tag{35}$$
where $\zeta = \frac{bg-a}{2\sqrt{bgz}}$ is the damping factor and $\omega_n = \sqrt{gz}$ is the undamped natural frequency. For stable nominal complex closed loop poles, we require $0 < \zeta < 1$.
5. *Closed Loop Poles.* Determine the nominal closed loop poles (assumed complex for rapid transient response):
$$s = -\zeta\omega_n \pm j\omega_n\sqrt{1-\zeta^2} \tag{36}$$
Given this, we will have nominal (local) closed loop exponential stability with an associated time constant $\tau = \frac{1}{\zeta\omega_n}$. The associated (approximate 1%) settling time is $t_s = 5\tau$.
6. *Standard Second Order Closed Loop Transfer Function and Percent Overshoot.* With the command pre-filter W, the associated closed loop transfer function takes the standard second order form:
$$T_{x_{ref}x} = \frac{WPK}{1+PK} \approx \frac{\omega_n^2}{s^2 + 2\zeta\omega_n s + \omega_n^2} \tag{37}$$
As such, the associated percent overshoot to a step reference command is given by
$$M_p = e^{-\zeta\omega_n t_p} = e^{-\left(\frac{\zeta\pi}{1-\zeta^2}\right)} \tag{38}$$
where $t_p = \frac{\pi}{\omega_n\sqrt{1-\zeta^2}}$ is the time at which the peak overshoot occurs.

7. *Damping Factor from Percent Overshoot Specification.* Determine ζ from the overshoot specification:

$$\zeta = \frac{|lnM_p|}{\sqrt{\pi^2 + |lnM_p|^2}} \tag{39}$$

8. *Undamped Natural Frequency from Settling Time Specification.* Determine ω_n from the settling time specification:

$$\omega_n = \frac{5}{\zeta t_s} \tag{40}$$

9. *PI Controller Parameters.* Determine the PI controller gain g and zero z from:

$$g = \frac{2\zeta\omega_n + a}{b} \qquad z = \frac{\omega_n^2}{bg} \tag{41}$$

10. *Controller Roll-Of Parameter.* Choose the roll-off parameter p_{ro} as follows:

$$p_{ro} = 10\omega_n \tag{42}$$

so that the added high frequency roll-off does not significantly degrade the nominal phase margin within the loop. It could also be selected in order to satisfy a specific sinusoidal steady state noise attenuation specification.

11. *Anti-Windup Gain.* Choose the anti-windup gain $G_{AW} > 0$ to be sufficiently large so that the integrator suitably shuts down in order to "maximally recapture" the dominant second order response characteristics described above. A family of gains is examined below.

5. Control law comparisons

In this section, we compare the properties of the nominal optimal control law with those for the robust policies based upon the nominal LTI plant model $P_d = \frac{b}{s-a}$.

5.1 Sample control law time responses

Within this section, sample time responses are provided for families of robust control laws (based upon the nominal LTI plant model) - families that approximate the performance of the nominal optimal control law. (Note: There will be an approximation gap when uncertainty is considered.)

Reference Biomass Tracking: Anti-Windup Gain Study. Figure 8 shows closed loop biomass tracking time responses for a family of robust control law designs where $\zeta = 1$, $t_s = 1$. The anti-windup gain G_{AW} is varied to control how well the responses approximate that of the optimal with no limit cycle behavior. As the anti-windup gain G_{AW} is increased, the responses come closer to the (nominal) optimal control law (with no limit cycle behavior). The limit cycle behavior of the (nominal) optimal has been cleaned up in order to improve the readability of the figure (see Figure 6).

A Note On Robustness with Respect to High Frequency Unmodeled Dynamics. It should be noted that as the speed of a policy is increased, the significance of unmodeled high frequency dynamics within the fishery or within the policy implementing organization/evironment (e.g. lags, time delays, rate limiters) becomes an issue to consider in final policy evaluation.

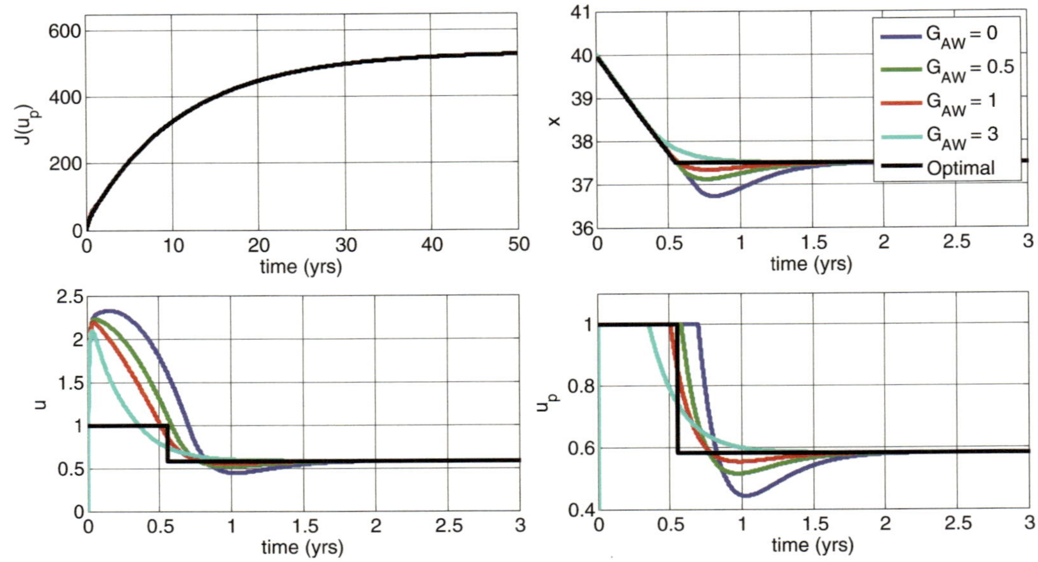

Fig. 8. Reference Biomass Tracking: Anti-Windup Gain Study ($\zeta = 1$, $t_s = 1$)

It is well known from fundamental robustness theory (Rodriguez, 2004; 2003) that fast control laws can result in closed loop oscillatory responses or instability when high frequency unmodeled dynamics are "significantly excited." This issue will be examined in future work.

Reference Biomass Tracking: Damping Factor Study. Figure 9 shows shows closed loop biomass tracking time responses for a family of robust control law designs where $t_s = 1$, $G_{AW} = 3$. The damping factor ζ is varied in order to control the speed of the response as

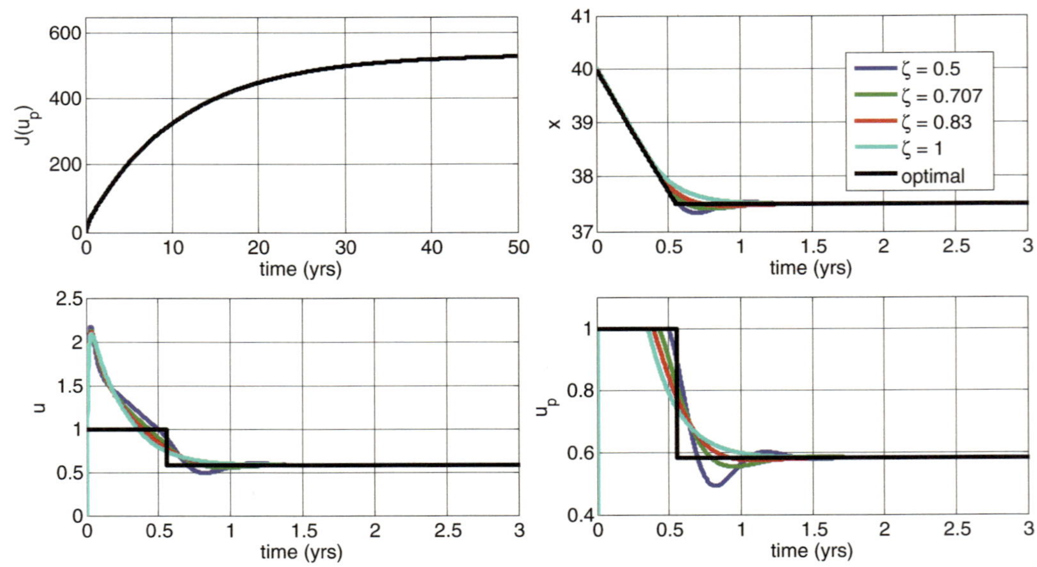

Fig. 9. Reference Biomass Tracking: Damping Factor Study ($G_{AW} = 3$, $t_s = 1$)

well as the undershoot. As the damping factor ζ is reduced, the response speeds up (getting closer to that of the (nominal) optimal with no limit cycle behavior), although the observed undershoot increases. The limit cycle behavior of the (nominal) optimal has been cleaned up in order to improve the readability of the figure (see Figure 6).

Reference Biomass Tracking: Settling Time Study. Figure 10 shows closed loop biomass tracking time responses for a family of robust control law designs where $\zeta = 1$ (critically damped, $M_p = 0$) and $G_{AW} = 3$. As the settling time t_s of the closed loop system is reduced, the responses come closer to the (nominal) optimal (with no limit cycle behavior). The limit cycle behavior of the (nominal) optimal has been cleaned up in order to improve the readability of the figure.

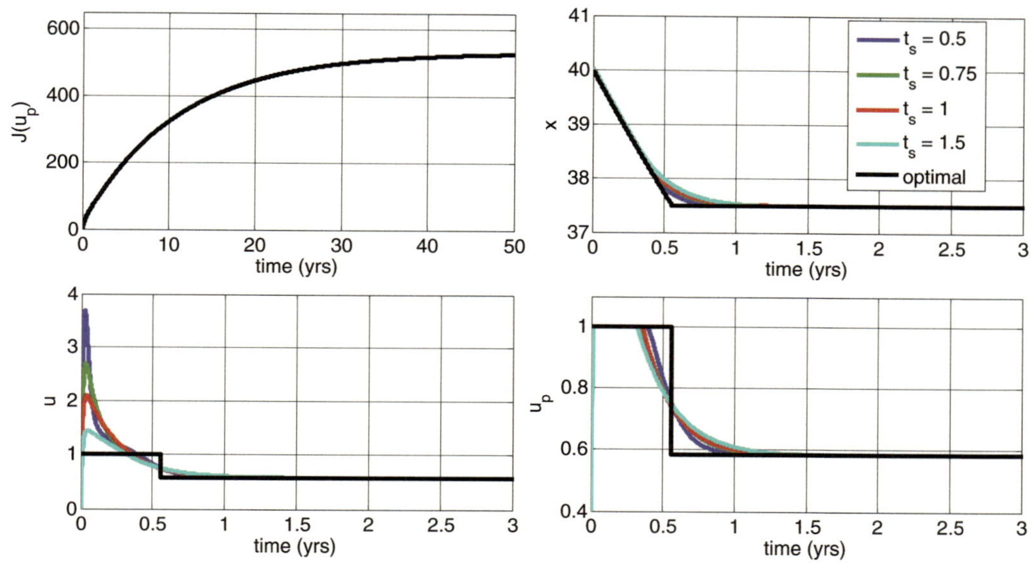

Fig. 10. Reference Biomass Tracking: Settling Time Study ($G_{AW} = 3, \zeta = 1$)

5.2 Utility of linear design methodology

In this section, we try to shed light on the utility of our linear time invariant (LTI) based robust control system design methodology and how linear simulation can be used to approximate/predict the behavior of the nonlinear simulations. All designs are based upon nominal parameter values.

Linear vs Nonlinear Biomass Tracking: x_o Near x_{ref}. Figure 11 compares linear and nonlinear closed loop biomass tracking simulations where the the initial condition (IC) is near the desired set point (target biomass). Four responses are shown for x and u_p: (1) purely linear; i.e. linear plant model, linear controller, and no saturation, (2) linear with plant saturation; i.e. linear plant model, linear controller, and plant saturation, (3) linear with anti-windup logic; i.e. linear plant model, linear controller, plant saturation, and anti-windup logic, (4) nonlinear; i.e. nonlinear plant model, linear controller, plant saturation, and anti-windup logic. Here, the reference command is very small ($x_{ref} = 0.375$), the control does not saturate, and all of the responses match one another. This shows that the "pure linear theory" suffices under small signal conditions (as expected).

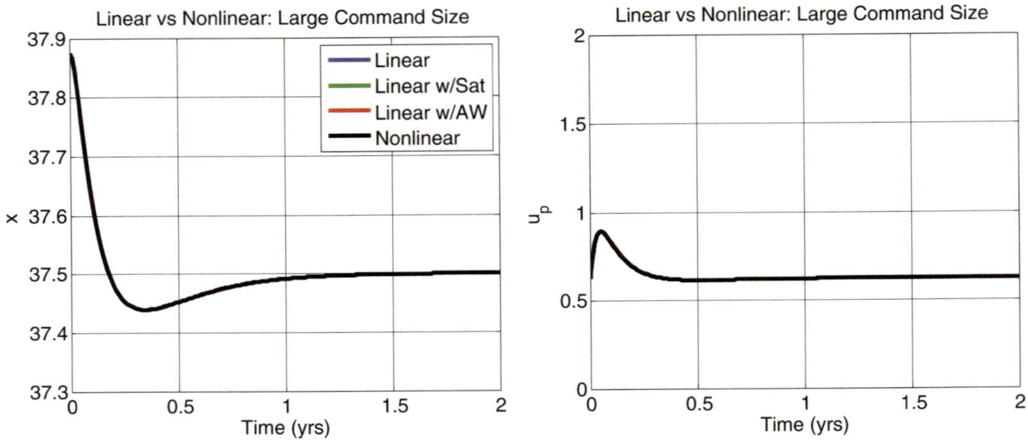

Fig. 11. Linear vs Nonlinear Biomass Tracking: x_o Near x_{ref}

Linear vs Nonlinear Biomass Tracking: x_o Far From x_{ref}. Figure 12 compares linear and nonlinear closed loop biomass tracking simulations where the the initial condition (IC) is

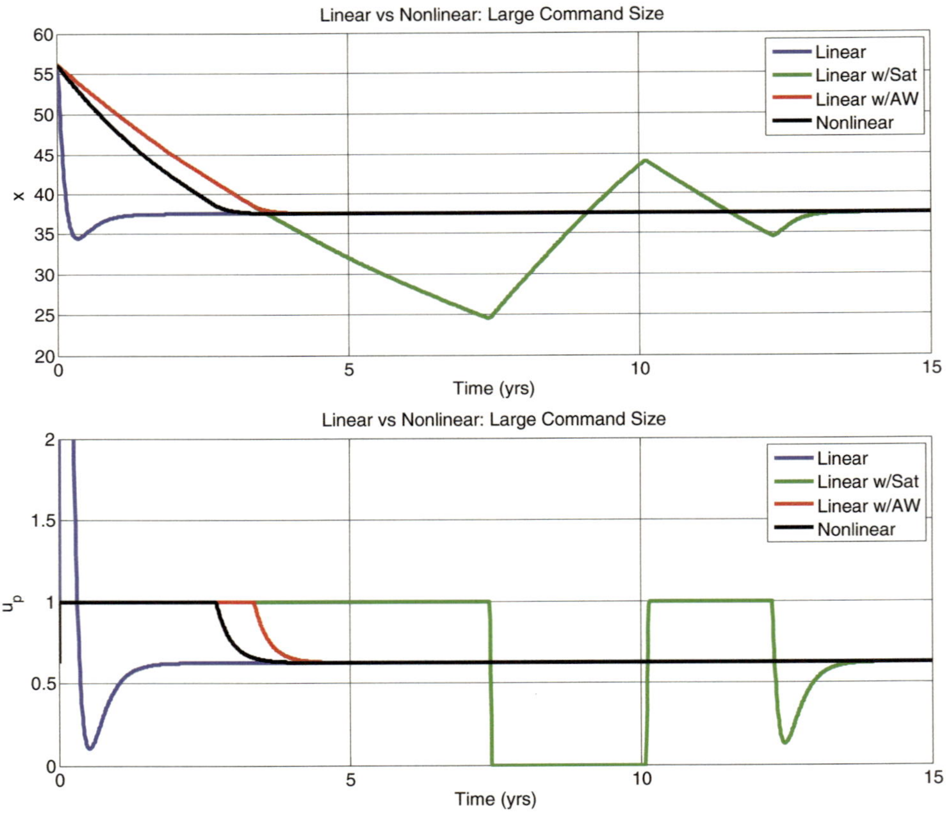

Fig. 12. Linear vs Nonlinear Biomass Tracking: x_o Far From x_{ref}

far from the desired set point (target biomass). Here, the reference command is large ($x_{ref} = 18.75$), the controls saturate, windup is exhibited in the linear w/Sat case, and we observe relatively good agreement between the linear (particularly linear w/AW) and nonlinear responses.

Biomass Tracking Robustness In Presence Of Capacity Uncertainty: Anti-Windup Gain Study. Figure 13 shows how our robust control laws can be adjusted to achieve the "flatter" economic inefficiency of the nominal optimal control law (see Figures 4-5). We observe the following:

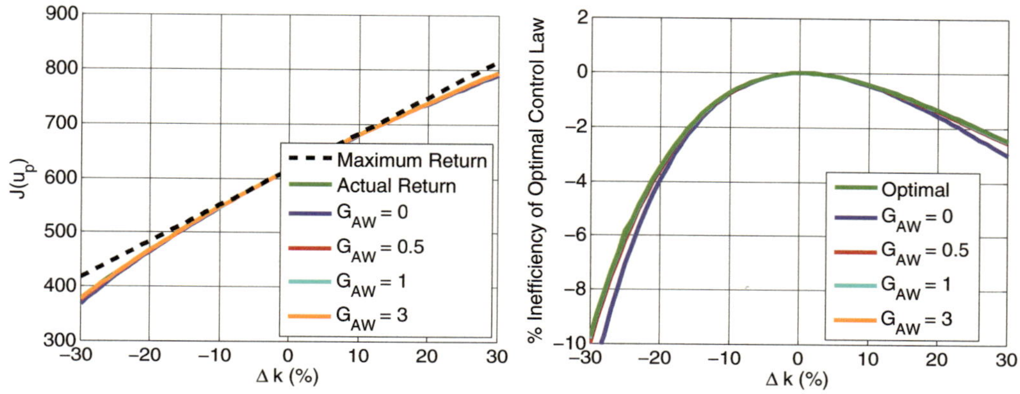

Fig. 13. PI Biomass Tracking Robustness (Capacity Uncertainty): Anti-Windup Gain Study ($\zeta = 1, t_s = 1$)

- With an anti-windup implementation, a PI control law can come arbitrarily close to matching the performance of the nominal optimal control law with imperfect information.
- Improving upon the nominal optimal control law with imperfect information requires some outer loop control logic as well as system identification to more appropriately select the reference/target biomass.

The observed performance gap (or inefficiency) is fundamentally because the target x_{ref} differs from the perfect information x_e^*; not because the nominal design plant differs from the truth plant. Closing the observed performance gap further requires an outer loop controller and/or parameter estimation techniques in order to get a more accurate target x_{ref} that is closer to the perfect information target x_e^*.

6. Summary and future directions

Summary. This chapter has shown how ideas from robust control may be applied to a fishery. It has been specifically shown how some small amount of income may be sacrificed for increased robustness with respect to uncertain fishery parameters.

Directions for Future Research. Future work will examine more complex models (e.g. decentralized, distributed), pros/cons associated with parameter estimation schemes, more complex robust control laws (e.g. use of receding horizon control for long-term management), robustness with respect to plant and controller uncertainty (parametric and dynamic).

7. References

Anderies, J.M., Rodriguez, A., Janssen, M. & Cifdaloz, O. (2007). Panaceas, uncertainty, and the robust control framework in sustainability science. *Proceedings of the National Academy of Sciences USA 104*, 15,194–15,199

Andersen, P. (1982). Commercial fisheries under price uncertainty. *Journal of Environmental Economics and Management 9*(1), 11–28

Aström, K. & Hägglund, T. (1995). PID Controllers: Theory, Design, and Tuning. *Int. Society for Measurement and Control*, (2nd Ed.), Research Triangle Park, N.C.

Babonneau, F., Vial, J.P. & Apparigliato, R.: Robust optimization for environmental and energy planning. *In: J.A. Filar, A. Haurie (eds.) Uncertainty and Environmental Decision Making, International Series in Operations Research & Management Science*, vol. 138, chap. 3, pp. 79–126. Springer (2010)

Belmiloudi, A. (2006). Minimax control problems of periodic competing parabolic systems with logistic growth terms. *International Journal of Control 79*(2), 150–161

Belmiloudi, A. (2005). Nonlinear optimal control problems of degenerate parabolic equations with logistic time-varying delays of convolution type. *Nonlinear Analysis-theory Methods & Applications 63*(8), 1126–1152

Ben-Tal, A., Nemirovski, A. (2008). Selected topics in robust convex optimization. *Mathematical Programming 112*(1), 125–158

Ben-Tal, A., Nemirovski, A. (2002). Robust optimization—methodology and applications. *Mathematical Programming 92*(3, Ser. B), 453–480. ISMP 2000, Part 2 (Atlanta, GA)

Ben-Tal, A., El-Ghaoui L. & Nemirovski, A. (2000) Robust semidefinite programming. *In: R. Saigal, L. Vandenberghe, H. Wolkowicz (eds.) Semidefinite Programming and Applications. Kluwer Dordrecht*

Ben-Tal, A., Nemirovski, A.(1998). Robust convex optimization. *Mathematics of Operations Research 23*(4), 769–805

Clark, C. (2007) The Worldwide Crisis in Fisheries: Economic Models and Human Behavior. *Cambridge University Press, Cambridge*, UK

Clark, C. (2006). Fisheries bioeconomics: why is it so misunderstood?, In: *Population Ecology 48*, 95-98

Clark, C., Munro, G. & U.R., S. (2006). Subsidies, buybacks, and sustainable fisheries. *Journal of Environmental Economics and Management 50*, 47–58

Clark, C.W. (1990). Mathematical Bioeconomics: The Optimal Management of Renewable Resources, 2 edn. *J. Wiley*, N.Y., U.S.A.

Clark, C. & Kirkwood, G.: On uncertain renewable resource stocks - optimal harvest policies and the value of stock surveys. *Journal of Environmental Economics and Management 13*(3), 235–244 (1986)

Dercole, F., Gragnani, A., Kuznetsov, Y.A. & Rinaldi, S. (2003) Numerical sliding bifurcation analysis: an application to a relay control system. *Ieee Transactions on Circuits and Systems I-Fundamental Theory and Applications 50*(8), 1058–1063

Gordon, H. (1954) The economic theory of a common property resource: the fishery. *Journal of Political Economy 62*, 124–142

Hansen, L.P. & Sargent, T.J. (2007) Robustness. *Princeton University Press*, Princeton, NJ

Holland, D. & Gudmundsson, E., J., G. (1999) Do fishing vessel buyback programs work? a survey of the evidence. *Marine Policy 23*, 47–69

Holling, C.S., Gunderson, L.H. (2002). Resilience and adaptive cycles. *In: L.H. Gunderson, C.S. Holling (eds.) Panarchy: Understanding Transformations in Systems of Humans and Nature*, chap. 2. Island Press

Holling, C.S. (1986) The resilience of terrestrial ecosystems, local surprise and global change. *In: C.W. C., R.E. Munn (eds.) Sustainable development of the biosphere*, pp. 292–317. Cambridge University Press, Cambridge

Holling, C.S.(1973). Resilience and stability of ecological systems. *Annual Review of Ecology and Systematics 4*, 1–23

ISRIC. (1990). Global assessment of the status of human induced soil degradation (glasdod). *dataset, International Soil Reference and Information Centre*, Wageningen, The Netherlands

Kendrick, D.A. (2005). Stochastic control for economic models: past, present and the paths ahead. *Journal of Economic Dynamics and Control 29*, 3–30

Koenig, E.F. (1984). Controlling stock externalities in a common property fishery subject to uncertainty. *Journal of Environmental Economics and Management 11*(2), 124–138

Lempert, R.J., Groves, D.G., Popper, S.W. & Bankes, S.C. (2006) A general, analytic method for generating robust strategies and narrative scenarios. *Management Science 52*(4), 514–528

Lempert, R.J., Schlesinger, M.E. (2000). Robust strategies for abating climate change - an editorial essay. *Climatic Change 45*(3-4), 387–401

Ludwig, D., Walker, B. & Holling, C.S. (1997). Sustainability, stability, and resilience. *Conservation Ecology 1*(1), (online) URL: http://www.consecol.org/vol1/iss1/art7/

Ludwig, D. & Walters, C. (1982). Optimal harvesting with imprecise parameter estimates. *Ecological Modelling 14*, 273–292

Moran, E.F., Ostrom, E. (eds.) (2006). Seeing the Forest and the Trees: Human-Environment Interactions in Forest Ecosystems. *MIT Press*, Cambridge, MA

Myers, R. & Worm, B. (2003). Rapid worldwide depletion of predatory fish communities. *Nature 423*, 280–283

Ogata, K. (1995). Discrete-Time Control Systems. *Prentice Hall*, (2nd Ed.), N.J.

Rodriguez, A.A., Cifdaloz, O., Anderies, J.M., Janssen, J. & Dickeson, J., (2010). Confronting Management Challenges in Highly Uncertain Natural REsource Systems: a Robustness-Vulnerability Trade-off Approach," *Environmental Modeling & Assessment*, Volume 16, Issue 1 (2011), Page 15.

Rodriguez, A.A. (2004). Analysis and Design of Multivariable Feedback Control Systems. *Control3D, LLC.*, Tempe, AZ

Rodriguez, A.A. (2003). Analysis and Design of Feedback Control Systems. *Control3D, LLC.*, Tempe, AZ

Schaefer, M. (1957). Some considerations of population dynamics and economics in relation to the management of the commercial marine fisheries. *Journal of the Fisheries Research Board of Canada 14*(5), 669–681

Sethi, G., Costello, C., Fisher, A., Hanemann, M. & Karp, L. (2005). Fishery management under multiple uncertainty. *Journal of Environmental Economics and Management 50*(2), 300–318

Shah, T., Molden, D., Sakthivadivel, R., Seckler. (2000). The global groundwater situation: Overview of opportunities and challenges. *Tech. rep., International Water Management Institute, Colombo, Sri Lanka*

Walters, C. (1986). Adaptive Management of Renewable Resources. *MacMillan Publishing Co.*
Weitzman, M. (2002). Landing fees vs harvest quotas with uncertain fish stocks. *Journal of environmental economics and management* 43(2), 325–338
Wilen, J., Homans, F. (1998). What do regulators do? Dynamic behavior of resource managers in the North Pacific Halibut Fishery 1935-1978. *Ecological Economics* 24(2-3), 289–298
Zhou, K., Doyle, J. (1998). Essentials of Robust Control. *Prentice Hall*

20

Robustness and Security of H_∞-Synchronizer in Chaotic Communication System

Takami Matsuo, Yusuke Totoki and Haruo Suemitsu
Oita University, Dannoharu, Oita
Japan

1. Introduction

In recent years, a large amount of work on chaos-based cryptosystems has been published (Kocarev (2001); Millérioux et al. (2008)). A general methodology for designing chaotic and hyperchaotic cryptosystems has been developed using the control systems theory (Grassi et al. (1999); Liao et al. (1999); Yang et al. (1997a;b)). The chaotic communication system is closely related to the concept of chaos synchronization. An overview of chaotic secure communication systems can be found in (Yang (2004)). He classified the continuous-time chaotic secure communication systems into four generations. In the third generation, the combination of the classical cryptographic technique and chaotic synchronization is used to enhance the degree of security. Specifically, Yang *et al.* proposed a new chaos-based secure communication scheme in an attempt to thwart the attacks (Yang et al. (1997a;b)). They have combined both conventional cryptographic method and synchronization of chaotic systems. Their cryptographic method consists of an encryption function (the multi-shift cipher), a decryption function (the inverse of the encryption function), a chaotic encrypter that generates the key signal for the encryption function, and a decrypter that estimates the key signal. The approach has a limitation since the cryptosystem design may fail if different chaotic circuits are utilized. So far, this generation has the highest security in all the chaotic communication systems had been proposed and has not yet broken. From the control theoretic perspective, the transmitter and the receiver in the chaotic communication system can be considered as the nonlinear plant and its observer, respectively. Grassi *et al.* proposed a nonlinear-observer-based decrypter to reconstruct the state of the encrypter (Grassi et al. (1999); Liao et al. (1999)). They extended the Chua's oscillator to the observer-based decrypter. The cryptosystem does not require initial conditions of the encrypter and the decrypter belonging to the same basin of attraction. If we can design a decrypter without the knowledge of the parameters of the encrypter, the chaos-based secure communication systems are not secure, because the parameters of the encrypter is selected as static secret keys in the cryptosystem. Parameter identification and adaptive synchronization methods may be effective for intruders in building reconstruction mechanisms, even when a synchronizing system is not available. Therefore, it is important for secure issues to investigate whether adaptive identifiers without the system information of encrypter can be constructed or not.
We have recently designed an observer-based chaotic communication system combining the cryptosystems proposed by Grassi *et al.* (Grassi et al. (1999)) and by Liao *et al.* (Liao et al. (1999)) that allows us to assign the relative degree and the zeros of its encrypter system (Matsuo et al. (2004)). Specifically, we constructed three cryptosystems based on a Chua's circuit by assigning its relative degree and zeros. The cryptosystem consists of

an encryption function (the multi-shift cipher), a decryption function (the inverse of the encryption function), a chaotic encrypter that generates the key signal for the encryption function, and a decrypter that estimates the key signal. The proposed cryptosystem allows us to assign the relative degree and the zeros of the encrypter dynamics by selecting an output vector that generates a transmitted signal as partial states of the encrypter. As in (Fradkov et al. (1997; 2000)), we can design an adaptive decrypter for minimum-phase systems with its relative degree 1. Therefore, the encrypter dynamics should be design such that its relative degree is more than two and its zeros are unstable so as to fail to synchronize the cryptosystem adaptively. At the same time, the designed cryptosystem should be robust with respect to uncertainties of the transmission lines such as a time delay, and noises. Suykens *et al.* (Suykens et al. (1997a;b)) presented a nonlinear H_∞ synchronization method for chaotic Lur'e systems based on the dissipativity of nonlinear systems to minimize the influence of the exogenous input such as the message signal and channel noises.

However, many proposed systems with robustness against parameter uncertainties and signal uncertainties are difficult to implement in practice with a reasonable degree of security. The basic difference between the conventional cryptography and the chaos cryptography is that the conventional encryption is defined discrete sets and the chaos encryption is defined on continuous sets. This makes the keyspace behavior of chaotic systems vary different that of conventional systems. Due to the continuous-value property, keys in chaotic cryptosystems form a key basin around the actual secret key.

When one key is very close to the real one, it could decrypt part or all of the ciphertext (Alvarez et al. (2006)). To avoid brute-force attacks, a secret parameter should be sensitive enough to guarantee the so-called avalanche property: even when the smallest change occurs in the parameter, the ciphertext will change dramatically (Alvarez et al. (2006)).

Various attacks such as the nonlinear forecasting, the return map, the adaptive parameter estimation, the error function attack (EFA), and inverse computation based on the chosen cipher attack, are proposed to recover messages from the chaotic ciphers (Zhou (2005)). Short (Parke et al. (2001); Short (1994; 1996)) and Guojie *et al.* (Guojie et al. (2003)) have proposed the attack strategies against chaotic communication systems. Short analyzed only the encrypter by using the nonlinear forecasting method that belongs to ciphertext-only attack when the attacker does not know the structure of the encryption system. They discussed the secure property of chaos communication based on chaotic parameter modulation from the chosen-ciphertext attack under the Kerckhoff principle (Guojie et al. (2003)). Guojie *et al.* discussed the secure property of chaos communication based on chaotic parameter modulation from the chosen-ciphertext attack under the Kerckhoff principle. We proposed chaotic communication systems using the adaptive control and robust control technologies (Matsuo et al. (2004; 2008)).

Wang *et al.* (Wang et al. (2004)) presented the error function attack to evaluate system security as an efficient cryptanalysis tool based on the public-structure and known-plaintext cryptanalysis. By defining the EFA function, an eavesdropper can scan the whole keyspace to find out the proper key that satisfies the EFA function with zero value. Since keys that are not identical with but are very close to the real one can be used to synchronize the two systems very well, a key basin around the actual secret key is formed. Once the eavesdropper knows the key basin, the correct key can be easily obtained through some optimization algorithms. To evaluate the security performance, Wang *et al.* also defined the key basin width by the distance between two trial keys located on the two sides of the key basin. The narrower than the whole keyspace the key basin width is, the higher the security of the cryptosysytem is. However, a systematic approach to get the key basin width is lacking. The brute-force-like calculations are needed to draw the shape of the EFA function. Thus, a considerable computing time is needed to get the key basin width. If the EFA function has numerous minima and a needle-like basin, the security level of the cryptosystem is high. In this case, the evolutionary optimization techniques such as the particle swarm optimization cannot find the secret key using the EFA function (Nomura et al. (2011)). Anstett *et al.* proposed a general framework based on identifiability for the cryptanalysis of chaotic cryptosystems (Anstett et al. (2006)). They also pointed out that cryptosystems involving polynomial nonlinearities are weak against a known plaintext attack.

In this chapter, we propose an H_∞ synchronizer in order to improve the robustness of chaotic communication systems with respect to delays in the transmission line based on the standard linear H_∞ control theory. To begin with, we derive an error system between the encrypter and the decrypter and reduce the design problem of the cryptosystem to the stabilization problem of a generalized plant in the robust control theory. Next, we give a synchronizer parameterization and an H_∞ synchronizer based on the robust control theory. Furthermore, the decrypter dynamics is designed via the linear controller parameterization to make the decrypter robust against disturbances in transmission line and/or sensitive to modeling errors of the decrypter. We present two design requirements on the robustness and the security. We need to design the free parameter such that both the requirements are satisfied. Since we cannot get this solution simultaneously, we design the dynamical compensator so as to satisfy the robustness requirement and then check the sensitivity to the key parameter mismatches whether the parameters in encrypter may play the role of the secret key or not, numerically. Finally, the proposed system is compared with that proposed by Grassi et al. using MATLAB simulations.

The following notation is used (Doyle et al. (1989)) :

$\mathcal{F}_l(G,Q)$: lower linear fractional transformation

$\left[\begin{array}{c|c} A & B \\ \hline C & D \end{array}\right] := C(sI - A)^{-1}B + D$

2. Observer-based chaotic communication system with free dynamics

Grassi *et al.* (Grassi et al. (1999)) proposed a nonlinear-observer-based cryptosystem that is an extension of the cryptosystem proposed by Yang *et al.* (Yang et al. (1997b)). The cryptographic method consists of an encryption function (the multi-shift cipher), a decryption function (the inverse of the encryption function), a chaotic encrypter that generates the key signal for the encryption function, and a decrypter that estimates the key signal. The transmitted signal through a public channel contains the nonlinear function that is equivalent to that of the encrypter. We add a dynamic compensator in the transmitted signal to the observer-based chaotic communication system proposed by Grassi *et al.* Figure 1 shows the relationship among the encrypter, the observer-based decrypter and the adaptive decrypter where we use the adaptive decrypter as a tool for ciphertext-only attacks.

The cryptosystem consists of an encryption function (the multi-shift cipher), a decryption function (the inverse of the encryption function), a chaotic encrypter that generates the key signal for the encryption function, and a decrypter that estimates the key signal.

- **Part 1 : dynamic encrypter**
  The chaotic encrypter is described by the following equations:

$$\dot{x} = Ax + b_2 f(x) + b_2 e_n \tag{1}$$

$$v = P(s)x \tag{2}$$

$$y = v + e_n + f(x) \tag{3}$$

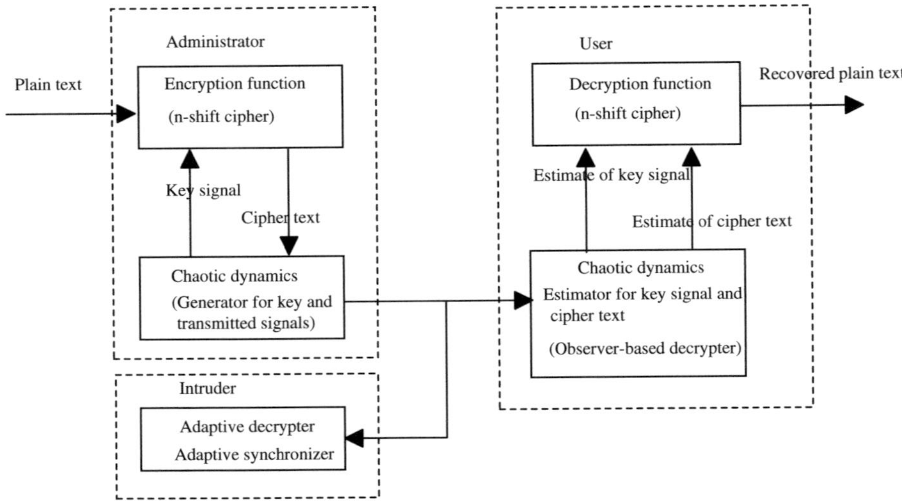

Fig. 1. Chaotic cryptosystem configuration.

where y is the transmitted signal that includes the nonlinear function, $P(s)$ is a transfer function that lets a decrypter synchronize the encrypter, and $s = \frac{d}{dt}$. We call this transfer function $P(s)$ a synchronizer.

- **Part 2 : encryption function**
  Given a plaintext signal $p(t)$, the ciphertext $e_n(t)$ is given by

$$e_n(t) = e_n(p(t), K(t)) \tag{4}$$

where $K(t)$ is a stream key signal that is generated by the encrypter dynamics and is given by the following equation:

$$K(t) = k^T x. \tag{5}$$

The signal e_n is a generic encryption function that makes use of the key signal and we choose a encryption function as the following n-shift cipher:

$$e_n(p(t), K(t)) = q(\cdots q(q(p(t), K(t)), K(t)), \cdots), K(t))$$

$$q(x,k) = \begin{cases} (x+k)+2h, & -2h \leq (x+k) \leq -h \\ (x+k), & -h < (x+k) < h \\ (x+k)-2h & h \leq (x+k) \leq 2h \end{cases}$$

- **Part 3 : dynamic decrypter with free dynamics**
  Given the encrypter, the decrypter used by an authorized user is the following observer:

$$\dot{\hat{x}} = A\hat{x} + b_2 e_y \tag{6}$$

$$\hat{v} = P(s)\hat{x} \tag{7}$$

$$e_y = y - \hat{v} = P(s)(x - \hat{x}) + e_n + f(x) \tag{8}$$

$$\hat{e}_n = y - (\hat{v} + f(\hat{x})) \tag{9}$$

where $\hat{e}_n$ is a recovered signal of the plain text.

- **Part 4 : decryption function**
  Using the estimated signals $\hat{K}(t)$ and $\hat{e}_n(t)$ by the decrypter, the estimate of the plaintext $\hat{p}(t)$ can be recovered by the following equations:

$$\hat{p}(t) = d(\hat{e}_n(t), \hat{K}(t)) \tag{10}$$
$$\hat{K}(t) = k^T\hat{x} \tag{11}$$

where $\hat{K}$ is an estimate of the stream key signal and d is the decryption function given by

$$\hat{p}(t) = q(\cdots q(q(\hat{e}_n(t), -\hat{K}(t)), -\hat{K}(t)), \cdots), -\hat{K}(t)).$$

3. Design of H_∞-synchronizer

3.1 Error equations and generalized system

If the transmitted signal is disturbed by an additional disturbance $w(t)$, the signal is rewritten by

$$\tilde{y}(t) = v(t) + e_n(t) + f(x(t)) + \omega(t) \tag{12}$$

When some of parameters of the dynamic encrypter are unknown, the dynamic decrypter constructed by a receiver based on the information of the encrypter has parametric uncertainties. The decrypter used by any receivers including intruders is given by

$$\dot{\hat{x}} = \tilde{A}\hat{x} + \tilde{b}_2 e_y \tag{13}$$
$$\hat{v} = \tilde{P}(s)\hat{x} \tag{14}$$
$$\hat{e}_n = \tilde{y} - (\hat{v} + \tilde{f}(\hat{x})) \tag{15}$$

Denoting the uncertainties of $\tilde{A}, \tilde{b}_2$ in the encrypter dynamics as Δ, the perturbed nonlinear function of $f(x)$ as $\tilde{f}(x)$, and the perturbation of the H_∞ synchronizer as $\tilde{P}(*)$, we assume that the decrypter with the uncertainties is given by

$$\dot{\hat{x}} = A\hat{x} + b_1\Delta + b_2 e_y \tag{16}$$
$$\hat{v} = \tilde{P}(s)\hat{x},\ e_y = \tilde{y} - \hat{v} \tag{17}$$
$$\hat{e}_n = e_y - \tilde{f}(\hat{x}) = \tilde{y} - (\hat{v} + \tilde{f}(\hat{x})). \tag{18}$$

A decrypter used by an authorized user satisfies $\Delta(t) = 0, f(\cdot) = \tilde{f}(\cdot), P(s) = \tilde{P}(s)$ since he knows all parameters of the encrypter. On the other hand, a decrypter used by an intruder has uncertainties in the encrypter dynamics, the nonlinear function, and the synchronizer.
In this chapter, we assume that the intruder knows the H_∞ synchronizer, $P(s) = \tilde{P}(s)$ but does not know the values of A, b_2, *i.e.* $\Delta(t) \neq 0$, and the nonlinear function, *i.e.* $f(\cdot) \neq \tilde{f}(\cdot)$.
Defining the estimation error of the decrypter as $e(t) = \hat{x}(t) - x(t)$, we have the following error system:

$$\dot{e}(t) = Ae(t) + b_1\Delta(t) + b_2\omega(t) - b_2\xi(t) \tag{19}$$
$$\xi(t) = P(s)e \tag{20}$$

We assign the estimation error of the key signal e_K or that of cipher text $\tilde{e}_n$ to the controlled output as follows:

$$\begin{aligned} e_K(t) &= \hat{K}(t) = k^T e(t) \\ \tilde{e}_n &= \hat{e}_n - e_n = \tilde{y} - \hat{v} - \tilde{f}(\hat{x}) - e_n \\ &= -\xi + (f(x) - \tilde{f}(\hat{x})) + \omega \end{aligned}$$

If $\lim_{t\to\infty} \omega(t) = 0$,$\lim_{t\to\infty}(f(x) - \tilde{f}(\hat{x})) = 0$ and $\lim_{t\to\infty} e(t) = 0$, then the plaintext can be recovered by the decrypter ,$\lim_{t\to\infty}(e_n(t) - \hat{e}_n(t)) = 0$.
Since $f(\cdot) = \tilde{f}(\cdot)$ for authorized users, we have

$$\begin{aligned} |\tilde{e}_n| &\le |\xi| + |f(x) - f(\hat{x})| + |w| \\ &\le |P(s)e| + \gamma\|e\| + |w|. \end{aligned}$$

Thus, if $\lim_{t\to\infty} w(t) = 0$ and $\lim_{t\to\infty} e(t) = 0$, then we attain the recover the plaintext, *i.e.* $\lim_{t\to\infty}(e_n(t) - \hat{e}_n(t)) = 0$.
For each controlled output, the generalized plant in Fig. 2 is defined as:

- When the controlled output is e_K, the generalize plant is

$$G_1(s) = \left[\begin{array}{c|cc} A & [b_1 \;\; b_2] & -b_2 \\ \hline k^T & 0 & 0 \\ I & 0 & 0 \end{array}\right] \tag{21}$$

- When the controlled output is the upper bound of $|\tilde{e}_n|$, the generalize plant is

$$G_2(s) = \left[\begin{array}{c|cc} A & [b_1 \;\; b_2] & -b_2 \\ \hline k^T & [0 \;\; 1] & -1 \\ I & 0 & 0 \end{array}\right]. \tag{22}$$

3.2 Synchronizer parameterization

To design the synchronizer based on the static output-feedback-based controller, we rewrite the generalized plant as in Fig.2. Since we can select the input of the synchronizer as arbitrary scalar signal, the signal in Eq.(2) is chosen as $v(t) = P(s)x(t) = P_o(s)c^T x(t)$, where c is an arbitrary vector.
We call a stabilizing compensator $P_o(s)$ for the generalized plant $G(s)$ the synchronizer of the chaotic cryptosystem. The design problem of the synchronizer is summarized as follows:

> Given a generalized plant $G(s)$ as in Fig. 2, parameterize all synchronizer $P(s)$ that internally stabilize $G(s)$.

We consider the n-th order generalized plant in Fig.3, where (A, B_2) is stabilizable and (A, C_2) is detectable;

$$G_o(s) = \begin{bmatrix} G_{11} & G_{12} \\ G_{21} & G_{22} \end{bmatrix} = \left[\begin{array}{c|cc} A & B_1 & B_2 \\ \hline C_1 & D_{11} & D_{12} \\ C_2 & D_{21} & 0 \end{array}\right] \tag{23}$$

and the p-th order dynamic stabilizing compensator,

$$P_o(s) = \left[\begin{array}{c|c} A_c & B_c \\ \hline C_c & D_c \end{array}\right]. \tag{24}$$

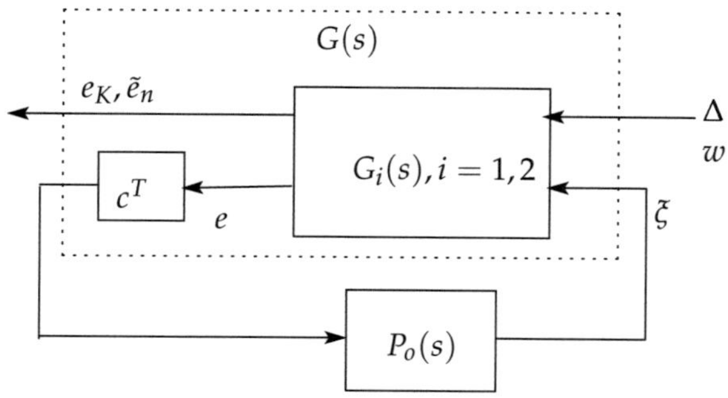

Fig. 2. Generalized plant and synchronizer in the chaotic communication system.

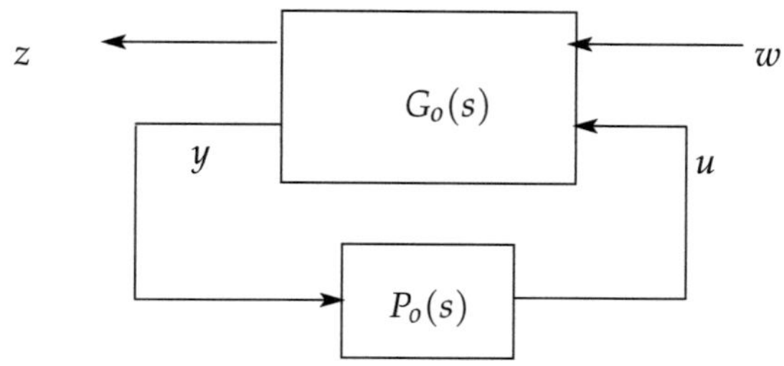

Fig. 3. Generalized plant and controller in the robust control theory.

For any choice of K_0, we can obtain the parameterization of $K(s)$ as follows (Matsuo et al. (1998)):

$$P_o(s) = \mathcal{F}_l(\tilde{P}_o(s), Q(s)) \tag{25}$$

$$\tilde{P}_o(s) = \left[\begin{array}{c|cc} A_K + H_0C_2 + B_2F_0 & -H_0 & B_2 \\ \hline F_0 & K_0 & I \\ -C_2 & I & 0 \end{array}\right] \tag{26}$$

$$Q(s) = \left[\begin{array}{c|c} A_{c22} & B_{c2} \\ \hline C_{c2} & D_{c2} \end{array}\right]$$

$$A_K = A + B_2K_0C_2$$

where

$$\begin{aligned} &A_{c22} : \text{stable} \\ &F_0 \text{ s.t. } A_K + B_2F_0 \text{ is stable} \\ &H_0 \text{ s.t. } A_K + H_0C_2 \text{ is stable.} \end{aligned}$$

Since $P_o(s)$ is a stabilizing compensator for each $Q(s) \in RH_\infty$, (25) is one of the parameterization of stabilizing compensators. This LFT form is equal to the Youla parameterization when the static output feedback gain, K_0, is selected as zero. When the generalized plant can be stabilized by a static output feedback gain, *i.e.* there exists an output feedback gain K_0 such that A_K is stable, we can set $H_0 = 0$, $F_0 = 0$. In this case, the parameterization of all stabilizing compensators is as follows (Matsuo et al. (1998)):

$$P_o(s) = \mathcal{F}_l(\tilde{P}_o(s), Q(s)) \tag{27}$$
$$= K_0 + Q(s)(I + C_2(sI - A_K)^{-1}B_2Q(s))^{-1} \tag{28}$$

where

$$\tilde{P}_o(s) = \left[\begin{array}{c|cc} A_K & 0 & B_2 \\ \hline 0 & K_0 & I \\ -C_2 & I & 0 \end{array}\right]. \tag{29}$$

In Fig. 2, since C_2 is replace to c^T, where c^T can be selected as an arbitrary vector, there exists a scalar k_0 such that $A_k = A - b_2k_0c^T$ is stable, as long as (A, b_2) is stabilizable. In this case, we can set $H_0 = 0$, $F_0 = 0$. The parameterization of all synchronizers in Fig. 2 is obtained as follows (Matsuo et al. (1998)):

$$P_o(s) = \mathcal{F}_l(\tilde{P}_o(s), Q(s)) \tag{30}$$

where $Q(s) \in RH_\infty$ and

$$\tilde{P}_o(s) = \left[\begin{array}{c|cc} A_k & 0 & -b_2 \\ \hline 0 & k_0 & 1 \\ -c^T & 1 & 0 \end{array}\right]. \tag{31}$$

We call this parameterization a synchronizer parameterization. By selecting $P_o(s)$ as constant gain k_0 *i.e.* $Q(s) = 0$, the proposed cryptosystem is equivalent to that proposed by Grassi *et al.*

3.3 Design problem of H_∞ synchronizer

The input-output relation of the generalized plant $G(s) = G_1(s)$ or $G_2(s)$ from the exogenous input $[\Delta \ w]$ to the controlled output z is given by

$$z = \mathcal{F}_l(G(s), P(s)) \left[\Delta(s) \ w(s) \right]^T \tag{32}$$
$$= \left[T_1(s) \ T_2(s) \right] \left[\Delta(s) \ w(s) \right]^T \tag{33}$$

The free dynamics $Q(s)$ is designed to make the decrypter robust against the disturbances in the transmission line of sensitive to the modeling errors of the decrypter by intruders. We present two design specifications:

1. **Robustness requirement:** The proposed decrypter can recover the plain text by the transmitted signals when the generalized plant with the synchronizer is internally stable. Moreover, the H_∞ synchronizer has an additional synchronization property with respect to plant uncertainties. To recover plain texts, the decrypter should be robust with respect to time delay uncertainties in the transmission line. Design the free parameter $Q(s)$ such that for a given γ_2,

$$\|T_2(s)\| < \gamma_2. \tag{34}$$

2. **Security requirement:** To attain the secure cryptosystem, the decrypter of the intruder should not synchronize the encrypter. Therefore, The free parameter $Q(s)$ is designed to

the error system sensitive to Δ. Design the free parameter $Q(s)$ such that for a given γ_1,

$$\underline{\sigma}\{T_1(j\omega)\} > \gamma_1, \text{ for } \omega \in [0, \infty). \tag{35}$$

However, since the generalized plant does not have a direct term from the uncertainty Δ to the transmitted signal $\tilde{y}$, (35) cannot be hold for all $\omega \in [0, \infty)$. Therefore, to satisfy the security requirement, we change the transmitted signal $\tilde{y}$ and the feedback term in the decrypter e_y as

$$\begin{aligned} \tilde{y}(t) &= v(t) + e_n(t) + f(x(t)) + \omega(t) + c^T A b_2 \\ e_y &= \tilde{y} - \hat{v} - c^T \tilde{A} \tilde{b}_2 \end{aligned}$$

In this case, the estimation error of the cipher text includes the direct term from the uncertainty to the transmitted signal as follows:

$$\tilde{e}_n' = \xi + (f(x) - \tilde{f}(\hat{x})) + w + \Delta'$$

In particular, when there is a perturbation in the nonlinear function, $f(x) \neq \tilde{f}(\hat{x})$ generates the direct term from the uncertainty to the transmitted signal.

We need to design the free parameter such that both the requirements are satisfied. Since we cannot get this solution, we design the dynamical compensator so as to satisfy the robustness requirement, and then check the security requirement whether the error system is sensitive to the modeling errors of the decrypter, *i.e.* the designed cryptosystem is secure against to attacks by intruders.

4. Simulations

We design a robust cryptosystem via Chua's circuits as in Yang *et al.* (Yang et al. (1997b)) and in Fradkov *et al.* (Fradkov et al. (2000)), and carry out simulations using MATLAB/Simulink.

4.1 Encrypter based on Chua's circuit

The chaotic encrypter based on the Chua's circuit is given by

$$\dot{x} = Ax + b_2 f(x_1) + b_2 e_n \tag{36}$$

$$\begin{aligned} y &= P(s)x + e_n + f(x) \qquad (37) \\ f(x_1) &= G_b x_1 + \frac{1}{2}(G_a - G_b)(|x_1 + 1| - |x_1 - 1|) \\ A &= \begin{bmatrix} -p_1 & p_1 & 0 \\ 1 & -1 & 1 \\ 0 & -p_2 & -p_3 \end{bmatrix} \\ b_2 &= \begin{bmatrix} -p_1 \\ 0 \\ 0 \end{bmatrix}, x = \begin{bmatrix} x_1 \\ x_2 \\ x_3 \end{bmatrix} \end{aligned}$$

We select the parameters in the Chua's circuit given by Liao *et al.* (Liao et al. (1999)) as $p_1 = 10$,$p_2 = 13.14$,$p_3 = 0.07727$,$G_a = -1.28$, and $G_b = -0.69$. The initial conditions are given by

$$x_1(0) = 1.1, x_2(0) = 0, x_3(0) = 0$$
$$\hat{x}_1(0) = 0, \hat{x}_2(0) = 0, \hat{x}_3(0) = 0$$

The encryption function is 30-shift cipher, the parameter h is equal to 1 and the key signal $K(t)$ is the second state variable x_2, *i.e.*

$$K(t) = \begin{bmatrix} 0 & 1 & 0 \end{bmatrix} x(t).$$

Moreover, we select $c^T = -\begin{bmatrix} 0 & 1 & 1 \end{bmatrix}$.
MATLAB has a built-in music file, `handel.mat`, with a short segment of Handel's *Messiah*. We use it as the plaintext signal.

4.2 Grassi-type system

In the encrypter presented by Grassi *et al.* (Grassi et al. (1999)), the dynamic synchronizer is simplified as $P_o(s) = k_0 = 0.8$.

4.3 Design of H_∞-synchronizer

The generalized plant $G_1(s)$ in designing the H_∞ synchronizer is shown in Fig. 4. The weighting function $W(s)$ in the exogenous signal is selected as $W(s) = 10 \times \frac{2.1Ls}{Ls+1}, L = 1 \times 10^{-3}$ and $\gamma = 0.75$ so as to stabilize the error system with time delay uncertainties. The H_∞ synchronizer is obtained by using MATLAB LMI toolbox as follows:

$$P(s) = \left[\begin{array}{c|c} a_k & b_k \\ \hline c_k & d_k \end{array}\right]$$

$$a_k = 10^5 \begin{bmatrix} -0.8762 & 1.9816 & -0.3890 & 0.0685 \\ 2.0890 & -4.7937 & 0.9380 & -0.1581 \\ -0.4425 & 1.0122 & -0.2074 & 0.0331 \\ -2.2132 & 5.3464 & -1.0531 & -0.0258 \end{bmatrix},$$

$$b_k = 10^5 \begin{bmatrix} 2.2024 & -0.0009 & 0.0114 \\ -5.3148 & 0.0046 & 0.0005 \\ 1.1245 & 0.0083 & 0.0008 \\ 5.8935 & 0.0186 & -0.1098 \end{bmatrix},$$

$$c_k = 10^3 \begin{bmatrix} -0.0549 & 0.2164 & -0.4198 & -5.4334 \end{bmatrix},$$
$$d_k = \begin{bmatrix} 0 & 0 & 0 \end{bmatrix}.$$

4.4 Nominal performance of H_∞-synchronizer

Figs. 5,6, and 7 show the responses of the Grassi-type decrypter and the H_∞-type decrypter of the nominal system. Fig 5 shows the plaintext and recovered signal for each decrypter. Fig 6 shows the transmitted signal and the estimation error of decrypter for each decrypter. Fig 7 shows the cipher text and the percentage error of the recovered signal for each decrypter. The nominal system means that the communication system has neither time delay nor parameter mismatches between the encrypter and the decrypter. The speed of response of the H_∞-type decrypter is faster than that of the Grassi-type.

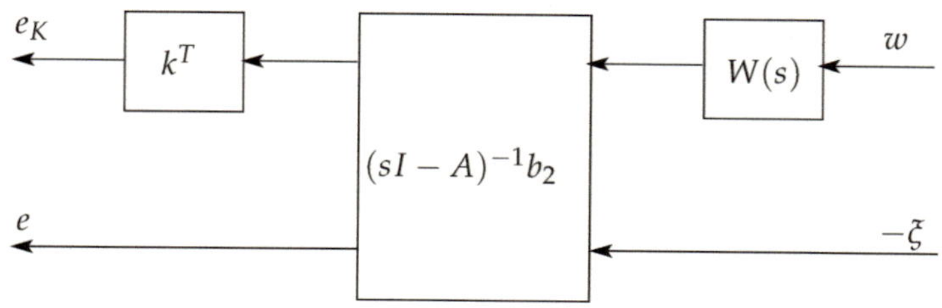

Fig. 4. Generalized plant $G_1(s)$.

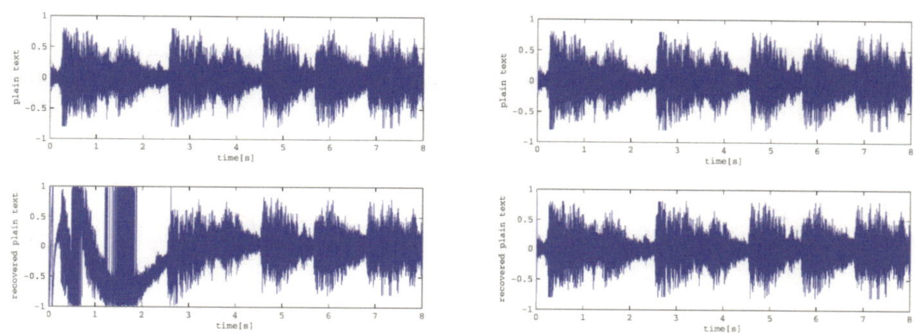

Fig. 5. The plaintext and the recovered plaintext in the nominal transmission line. Above left : the plaintext for the Grassi-type decrypter. Above right : the plaintext for the H_∞-synchronizer. Below left : the recovered plaintext by the Grassi-type decrypter. Below right : the recovered plaintext by the H_∞-synchronizer.

4.5 Robustness of H_∞-synchronizer against time delay in transmission line

Figs. 6 and 7 show the responses of the Grassi-type decrypter and the H_∞-type decrypter for the generalized plant $G_1(s)$in the presence of the time delay $L = 0.1$ in the transmission line, respectively. The responses of the H_∞-type decrypter for the generalized plant $G_2(s)$ in the presence of the time delay $L = 0.1$ in the transmission line is almost same as that for the generalized plant $G_1(s)$. The H_∞-type decrypter has a better robust performance to the time delay than the Grassi-type.

4.6 Security performance of H_∞-synchronizer

We assume that intruders have parameter mismatches in the decrypter. In this simulation, we consider the following parameter mismatches:

$$\hat{v} = P(s)\hat{x},\ \hat{e}_n = y - (\hat{v} + \tilde{f}(\hat{x}))$$

$$\tilde{A} = \begin{bmatrix} -p_1 & p_1 & 0 \\ 1 & -1 & 1 \\ 0 & -\tilde{p_2} & -p_3 \end{bmatrix}$$

$$\tilde{f}(x_1) = \tilde{G_b}x_1 + \frac{1}{2}(\tilde{G_a} - \tilde{G_b})(|x_1 + 1| - |x_1 - 1|)$$

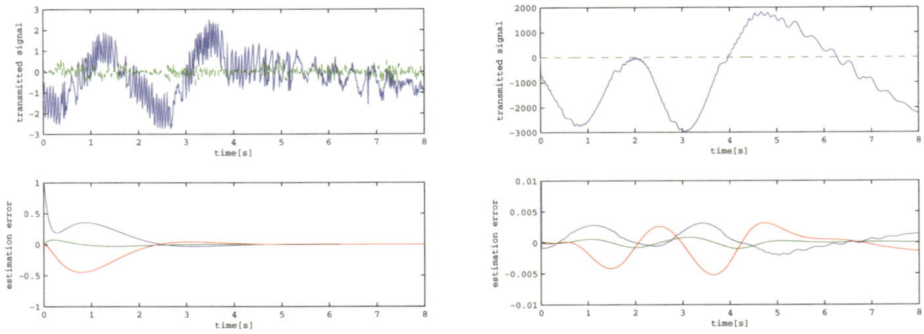

Fig. 6. The transmitted signal and the estimation error of decrypter in the nominal transmission line. Above left : the transmitted signal(solid line) and the plain text(dotted line) of the Grassi-type decrypter. Above right : the transmitted signal(solid line) and the plain text(dotted line) of the decrypter with the H_∞-synchronizer . Below left : the estimation errors of full states of the Grassi-type decrypter. Below right : the estimation errors of full states of the decrypter with the H_∞-synchronizer.

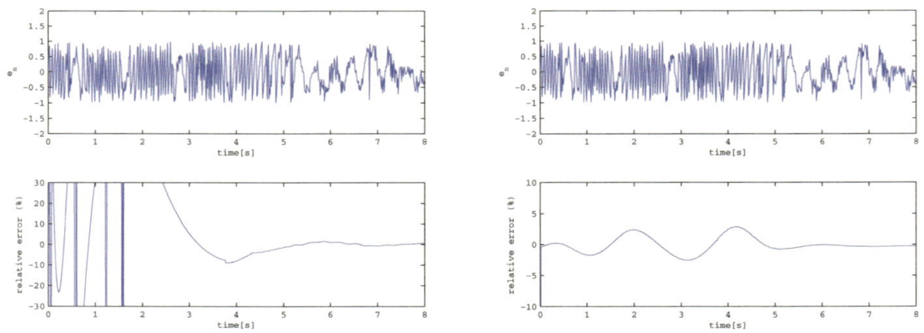

Fig. 7. The cipher text and the percentage error of the recovered signal in the nominal transmission line. Above left : the ciphertext in the Grassi-type decrypter. Above right : the ciphertext in the decrypter with the H_∞-synchronizer. Below left : the percentage error of the recovered signal of the Grassi-type decrypter. Below right : the percentage error of the recovered signal of the decrypter with the H_∞-synchronizer.

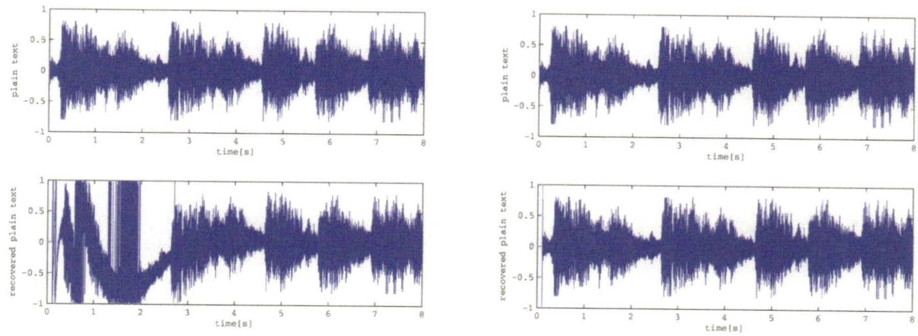

Fig. 8. The plaintext and the recovered plaintext in the transmission line with delay time. Above left : the plaintext for the Grassi-type decrypter. Above right : the plaintext for the H_∞-synchronizer. Below left : the recovered plaintext by the Grassi-type decrypter. Below right : the recovered plaintext by the H_∞-synchronizer.

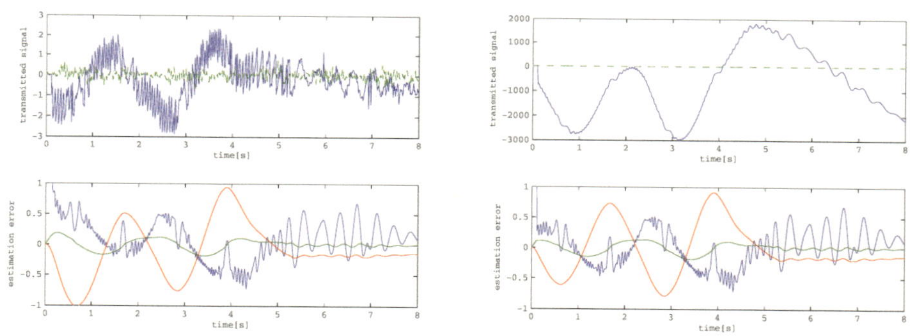

Fig. 9. The transmitted signal and the estimation error of the decrypter in the transmission line with delay time Above left : the transmitted signal(solid line) and the plain text(dotted line) of the Grassi-type decrypter. Above right : the transmitted signal(solid line) and the plain text(dotted line) of the decrypter with the H_∞-synchronizer . Below left : the estimation errors of full states of the Grassi-type decrypter. Below right : the estimation errors of full states of the decrypter with the H_∞-synchronizer.

In this simulation, we select the candidates of the static secret keys as the parameters p_2,G_a,G_b, and $P(s)$. Intruder A has the following parameter mismatch:

$$\tilde{p}_2 = 13.15, p_2 = 13.14$$
$$\tilde{G}_a = G_a = -1.28, \tilde{G}_b = G_b = -0.69$$
$$\tilde{P}(s) = P(s).$$

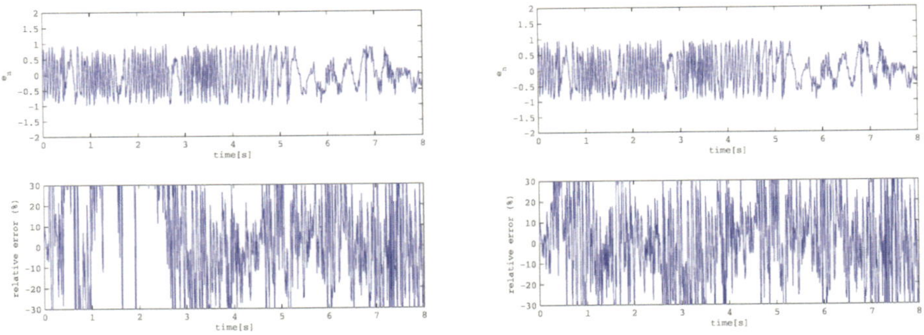

Fig. 10. The cipher text and the percentage error of the recovered signal in the transmission line with delay time. Above left : the ciphertext in the Grassi-type decrypter. Above right : the ciphertext in the decrypter with the H_∞-synchronizer. Below left : the percentage error of the recovered signal of the Grassi-type decrypter. Below right :the percentage error of the recovered signal of the decrypter with the H_∞-synchronizer.

Intruder B has the following parameter mismatches:

$$\tilde{p}_2 = p_2 = 13.14,$$
$$\tilde{G}_a = -1.3, \tilde{G}_b = -0.65$$
$$\tilde{P}(s) = P(s).$$

Figs. 11,12, and 13 show the responses of the H_∞-type decrypter used by the intruders A and B, respectively. The proposed synchronizer is sensitive to the parameter mismatches caused by Intruder A. The parameters in the dynamic encrypter may play the role of the secret key. However, Intruder A can identify the recovered wav file as the Handel's *Messiah* in spite of noisy sound. Fig. 14 shows the EFA function of the proposed H_∞-type decrypter. Since the width of the key basin in EFA function is not so narrow, the cryptosystem is not so secure.

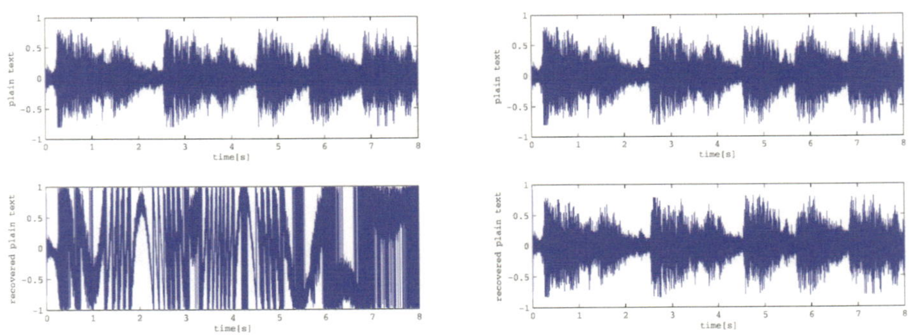

Fig. 11. The plaintext and the recovered plaintext by the intruders A and B. Above left : the plaintext. Above right : the plaintext. Below left : the recovered plaintext by Intruder A. Below right : the recovered plaintext by Intruder B.

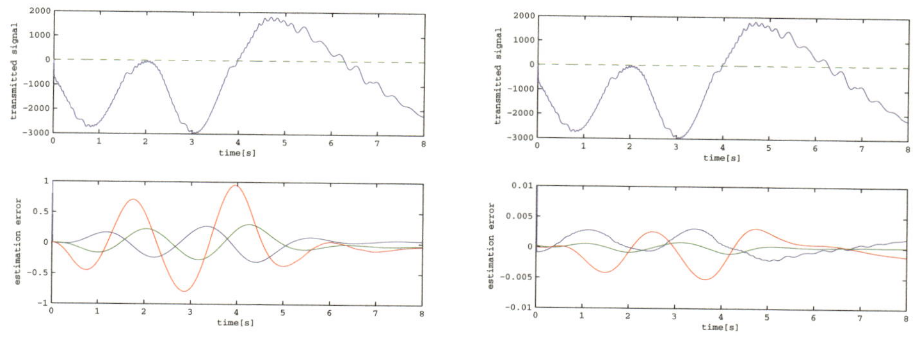

Fig. 12. The transmitted signals and the estimation errors of the intruders' decrypters. Above left : the transmitted signal(solid line) and the plain text(dotted line). Above right : the transmitted signal(solid line) and the plain text(dotted line). Below left : the estimation errors of full states by Intruder A. Below right : the estimation errors of full states by Intruder B.

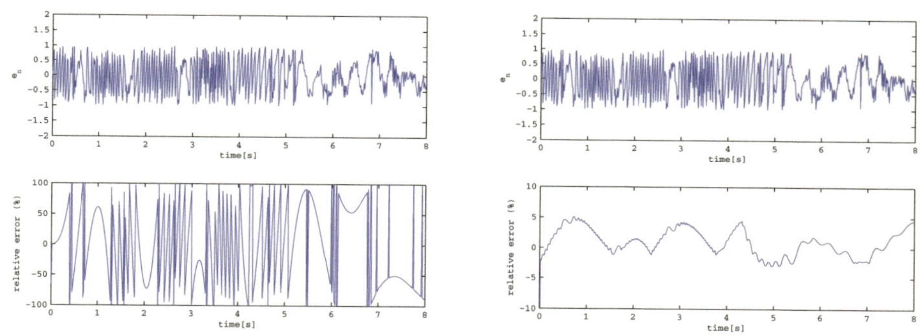

Fig. 13. The cipher text and the percentage error of the recovered signal in the transmission line with delay time. Above left : the ciphertext. Above right : the ciphertext. Below left : the percentage error of the recovered signal by Intruder A. Below right : the percentage error of the recovered signal by Intruder B.

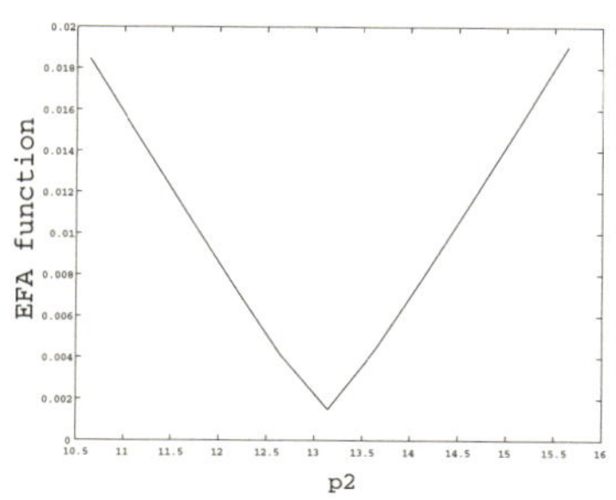

Fig. 14. The key basin of p_2 in EFA function for H_∞ synchronizer.

To improve the security of the H_∞ synchronizer, we select the secret key as a element of $P(s)$. Intruder C has the following parameter mismatch in the H_∞ synchronizer:

$$\begin{aligned}
&\tilde{p}_2 = p_2 = 13.14 \\
&\tilde{G}_a = G_a = -1.28, \tilde{G}_b = G_b = -0.69 \\
&\tilde{P}(s) = \left[\begin{array}{c|c} \tilde{a}_k & b_k \\ \hline c_k & d_k \end{array}\right] \\
&\tilde{a}_k(1,1) = a_k(1,1) + 450
\end{aligned}$$

The parameter mismatch of the element $a_k(1,1)$ is about 0.51%, because $a_k(1,1) = -0.8762 \times 10^5$.

Figs. 15,16, and 17 show the responses of the H_∞-type decrypter used by Intruder C. In this case, the decrypter with the parameter mismatch causes instability. The parameters in the H_∞ synchronizer $\tilde{P}(s)$ may play the role of the secret key.

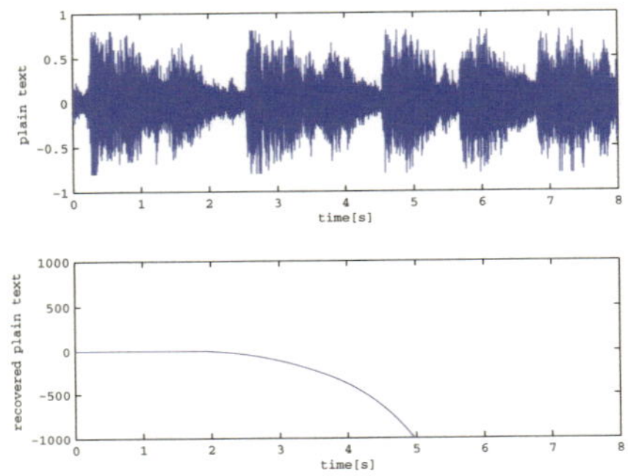

Fig. 15. The plaintext (top) and the recovered plaintext by Intruder C (bottom).

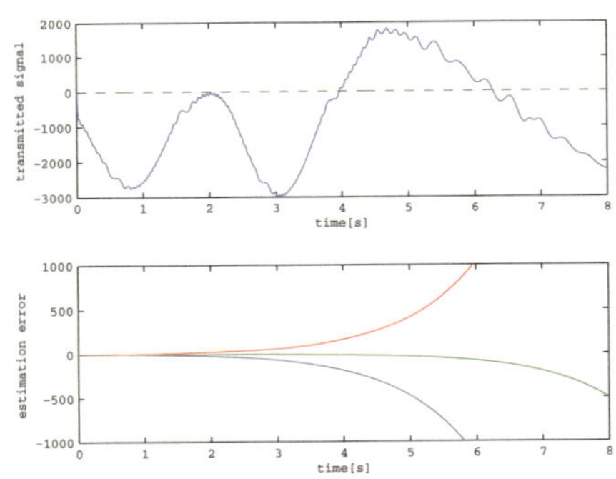

Fig. 16. The transmitted signal (top) and the estimation error of decrypter by Intruder C (bottom).

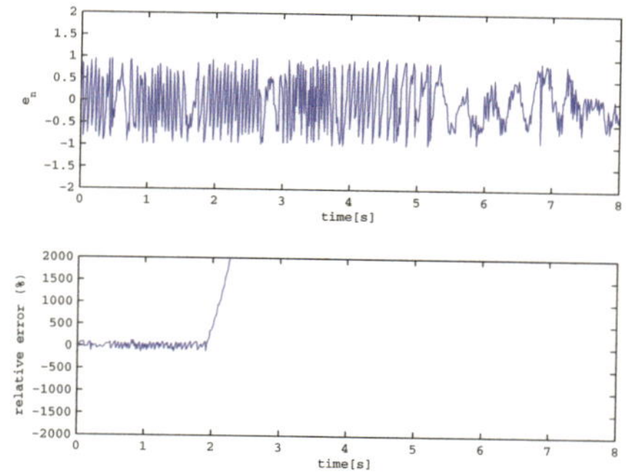

Fig. 17. The cipher text (top) and the percentage error of the recovered signal by Intruder C (bottom).

5. Conclusion

In this chapter, we added an observer-based chaotic communication system proposed by Grassi *et al.* to a dynamical compensator in its transmitted signal to improve the robustness of the cryptosystem with respect to delays in the transmission line. The proposed chaotic system has a good robust performance with respect to the time delay in the transmission line. Moreover, we checked the security in a point of parameters mismatch by an intruder.

6. References

Anstett,F.; Millerioux,G. & Bloch,G. (2006). Chaotic Cryptosystems : Cryptanalysis and Identifiability, *IEEE Trans. Circuits Syst. I*, Vol.53, No.12, pp.2673–2680

Alvarez,G. & Li, S. (2006). Some Basic Cryptographic Requirements for Chaos-Based Cryptosystems, *Int. J. of Bifurcation and Chaos*, Vol.16, No.8, pp.2129-2151

Cuomo,K.M. & Oppenheim,A.V. (1993). Synchronization of Lorenz-Based Chaotic Circuits with Applications to Communications, *IEEE Trans. Circuits Syst. I*, Vol.40, pp.626-633

Dedieu,H. & Ogorzalek,M.J. (1997). Identifiability and Identification of Chaotic Systems Based on Adaptive Synchronization, *IEEE Trans. Circuits Syst. I*, Vol.44, pp.948-962

Doyle,J.C.; Glover,K.; Khargonekar,P.P. & Francis,B. A. (1989). State-Space Solutions to Standard H_2 and H_∞ Control Problems, *IEEE Trans. Automat. Contr.*, Vol.34, No.8, pp.831-846

Fradkov,A.L. & Markov,A.Y. (1997). Adaptive Synchronization of Chaotic Systems Based on Speed Gradient Method and Passification, *IEEE Trans. Circuits Syst. I*, Vol.44, No.10, pp.905–912.

Fradkov,A.L., Nijmeijer,H., & Markov,A.Y. (2000). Adaptive Observer-Based Synchronization for Communication, *Int. J. of Bifurcation and Chaos*, Vol.10. No.12, pp.2807-2813

Grassi,G. & Mascolo,S. (1999). A System Theory Approach for Designing Cryptosystems Based on Hyperchaos, *IEEE Trans. Circuits Syst. I*, Vol.46, No.9, pp.1135-1138

Guojie,H. Zhengjin,F., & Ruiling,M. (2003). Chosen Cipher Attack on Chaos Communication Based on Chaotic Synchronization, *IEEE Trans. Circuits Syst. I*, Vol.50, No.2, pp.275-279

Kocarev,L. (2001). Chaos-Based Cryptography: a Brief Overview, *IEEE Circuits and Systems Magazine*, Vol.1, No.3, pp.6-21

Liao,T.L. & Huang,N.S. (1999). An Observer-Based Approach for Chaotic Synchronization with Applications to Secure Communications, *IEEE Trans. Circuits Syst. I*, Vol.46, No.9, pp.1144-1149

Matsuo,T. & Nakano,K. (1998). Robust Stabilization of Closed-Loop Systems by PID+Q Controller, *Int. J. of Control*, Vol.70, No.4, pp.631-650

Matsuo,T., Suemitsu,H., & Nakano,K. (2004). Zeros and Relative Degree Assignments of Adaptive Chaotic Communication Systems, *Int.J. of Bifurcation and Chaos*, Vol.14, No.12, pp.4233-4247

Matsuo,T.; Toshimitsu,Y.; & Suemitsu,H. (2008). H_∞-Synchronizer for Chaotic Communication Systems, *Int.J. of Bifurcation and Chaos* Vol.18, No.4, pp.1175-1187

Millérioux,G.; Amigó,J. M. & Daafouz,J. (2008). A Connection between Chaotic and Conventional Cryptography, *IEEE Trans. on CAS–I*, Vol.55, No.6, pp.1695-1703

Nomura,T.; Irie,T.; Suemitsu,H. & Matsuo,T. (2011). Stochastic Security Testing for Chaotic Communication Systems against Error Function Attack, *IEEJ Trans. on Electrical and Electronic Engineering*, Vol.6, No.5, in press

Parker,A.T. & Short,K.M. (2001). Reconstructing the keystream from a chaotic encryption scheme, *IEEE Trans. on CAS–I*, Vol.48, No.5 pp.624-630

Short,K.M. (1994). Steps toward unmasking secure communications, *Int. J. Bifurcation and Chaos* Vol.4, No.44, pp.959-977

Short,K.M. (1996). Unmasking a modulated chaotic communications scheme, *Int. J. Bifurcation and Chaos*, Vol.6-, No.2, pp.367-375

Suykens,J.A.K.; Vandewalle,J. & Chua, L.O. (1997a). Nonlinear H_∞ Synchronization of Chaotic Lur'e Systems, *Int.J. of Bifurcation and Chaos*, Vol.7, No.6, pp.1323-1335

Suykens, J.A.K.; Curran, P.F.; Vandewalle, J. & Chua, L.O. (1997b). Robust Nonlinear H_∞ Synchronization of Chaotic Lur'e Systems, *IEEE Trans. Circuits Syst. I*, Vol.44, No.10, pp.891-904

Wang,X; Zhan,M.; Lai,C.-H. & Gang,H. (2004). Error Function Attack of Chaos Synchronization Based on Encryption Schemes, *Chaos*, Vol.14, No.1, pp.128-137

Yang,T. & Chua,L.O. (1997a). Impulsive Control and Synchronization of Nonlinear Dynamical Systems and Application to Secure Communication, *Int. J. of Bifurcation and Chaos*, Vol.7, No.3, pp.645-664

Yang,T.; Wu,C.W. & Chua,L.O. (1997b). Cryptography Based on Chaotic Systems, *IEEE Trans. Circuits Syst. I*, Vol.44, pp.469-472

Yang,T. (2004). A Sruvey of Chaotic Secure Communication Systems, *Int.J. of Comput. Cogn.*, Vol.2, No,2, pp.81-130

Zhou,J.; Pei,W.; Huang,J.; Song,A. & He,Z. (2005). Differential-like Chosen Cipher Attack on A Spatiotemporally Chaotic Cryptosystem, nlin.CD/0506026